SOURCES OF GRAVITATIONAL RADIATION

Sources of Gravitational Radiation

PROCEEDINGS OF THE BATTELLE SEATTLE WORKSHOP

July 24 - August 4, 1978

edited by

LARRY L. SMARR

Center for Astrophysics and

Lyman Laboratory of Physics

Harvard University

Cambridge University Press

Cambridge

London . New York . Melbourne

Published by the Syndics of the Cambridge University Press
The Pitt Building, Trumpington Street, Cambridge CB2 1RP
Bentley House, 200 Euston Road, London NW1 2DB
32 East 57th Street, New York, NY 10022, USA
296 Beaconsfield Parade, Middle Park, Melbourne 3206, Australia

First published 1979

Printed in the United States of America
Printed and bound by Vail-Ballou Press, Inc.,
Binghamton, New York

ISBN 0 521 22778 X

CONTENTS

PREFACE

This book derives from lectures given at a two week summer workshop on Sources of Gravitational Radiation held July 24 - August 4, 1978 at the Battelle Seattle Research Center in Seattle, Washington. These lectures were carefully prepared by the authors to review their particular subject. I think they have done an exceptional job and I would like to thank them for the hard work they put in on each article. The conference and preparation of this book was jointly financed by the National Science Foundation, The National Aeronautics and Space Administration, and the Battelle Memorial Institute. Their generous support made the high quality of the workshop and this book possible. The preparation of the workshop was made easier for me by advice from the Board of Advisors: B.S. DeWitt, F.B. Estabrook, C.W. Misner, J.P. Ostriker, W.H. Press, D.N. Schramm, K.S. Thorne, R. Weiss, J.A. Wheeler, and J.R. Wilson. I am particularly grateful to D.M. Eardley, Associate Director of the workshop, for his vital help with organization throughout the year and a half required to come from planning to completion of this project. He was assisted by S.A. Teukolsky and J.W. York, Jr. The success of the workshop itself was in large measure due to the untiring efforts of both the Local Organizing Committee, J.M. Bardeen and P.C. Peters, and the Battelle staff, particularly Julia Swor, Jolene Kitzerow, and Jody Marshall. The University of Washington was quite helpful with many of the local arrangements. The workshop derived much stimulation from the participation of the Soviet delegation in relativistic astrophysics made possible by the cooperative program in physics between the National Academy of Sciences of the USA and the Academy of Sciences of the USSR. There was an enormous amount of post-workshop effort to produce this camera ready manuscript. Let me express my appreciation both to the authors and to their devoted secretaries who successfully worked so hard under the pressure of a deadline. Finally, I wish to thank the Harvard Society of Fellows who both supported me financially and gave me the research freedom necessary to complete this project.

Larry L. Smarr
Director of Workshop
Cambridge, Massachusetts
November, 1978

COPYRIGHT ACKNOWLEDGEMENTS

We wish to express our appreciation to the following sources for permission to reproduce copyrighted material.

From the Astrophysical Journal, published by the University of Chicago with copyright held by the American Astronomical Society we have reproduced: Figures 7a, 7b from J.P.A. Clark and D.M. Eardley, vol. 215, pp. 311-322 (1977); Figures 1,2,4, and 6 from R.A. Saenz and S.L. Shapiro, vol. 221, pp. 286-303 (1978); Figure 12 from K.A. Van Riper, vol. 221, pp. 304-319 (1978); Figures 10b, 13 from C. Cunningham, R.H. Price, and V. Moncrief, vol. 224, pp. 643-667 (1978); Figures 1,6,8,11 from S.L. Detweiler, vol. 225, pp. 687-693 (1978).

From Astronomy and Astrophysics, published by Springer-Verlag with copyright held by the European Southern Observatory we have reproduced: Figure 4 from P. Kafka and L. Schnupp, vol. 70, pp. 97-103 (1978); Figure 1 from J.P.A. Clark, E.P.J. van den Heuvel, and W. Sutantyo, vol. 72, 120-128 (1979).

LIST OF PARTICIPANTS

W.D. Arnett
University of Chicago

J.M. Bardeen
University of Washington

G. Barker
University of Texas

J.K. Beem
University of Missouri

V.A. Belinsky
Landau Institute for
Theoretical Physics - Moscow

B.K. Berger
Oakland University

R. Blandford
California Institute of
Technology

S.I. Blinnikov
Institute for Space Research
Moscow

S. Boughn
Stanford University

D.R. Brill
University of Maryland

W.L. Burke
University of California
Santa Cruz

C.M. Caves
California Institute of
Technology

J.M. Centrella
Cambridge University

J.P.A. Clark
Yale University

F.I. Cooperstock
University of Victoria

M.W. Cromar
University of Rochester

C.T. Cunningham
University of Utah

F. Dawson
Battelle Memorial Institute

F. de Felice
University of Alberta

P. D'Eath
Cambridge University

S.L. Detweiler
Yale University

D.H. Douglass
University of Rochester

R.W.P. Drever
Glasgow University

G.C. Duncan
Bowling Green State University

D.M. Eardley
Yale University

F.B. Estabrook
Jet Propulsion Laboratory

K.R. Eppley
University of Maryland

R. Epstein
Massachusetts Institute of
Technology

LIST OF PARTICIPANTS

F.J. Flaherty
Oregon State University

R.L. Forward
Hughes Research Laboratories

B.D. Gaiser
Stanford University

S. Gayer
University of California
Los Angeles

E.N. Glass
University of Windson

Y. Gursel
California Institute of
Technology

M. Haugan
Stanford University

R.W. Hellings
Jet Propulsion Laboratory

W.A. Hiscock
University of Maryland

D. Hobill
University of Victoria

J.N. Hollenhorst
Stanford University

J. Isenberg
University of Maryland

R.T. Jantzen
University of California
Berkeley

P. Kafka
Max-Planck-Institut für
Physik und Astrophysik

M. Karim
University of Ife, Nigeria

D. Kazanas
University of Chicago

L.S. Kegeles
University of Alberta

E.P. Liang
Stanford University

P. Lim
University of Victoria

L. Lindblom
University of Maryland

G. Lugo
University of California
Berkeley

C.W. Misner
University of Maryland

V. Moncrief
Yale University

D.K. Nadezhin
Institute of Applied Mathematics
Moscow

I.D. Novikov
Institute for Space Research
Moscow

J.P. Ostriker
Princeton University Observatory

D.G. Payne
Yale University

F. Pegoraro
Scuola Normale Superiore, Pisa

P.C. Peters
University of Washington

T. Piran
University of Texas

W.H. Press
Harvard University

J.R. Ray
Clemson University

LIST OF PARTICIPANTS

A. Rosenblum
Max-Planck-Institut für Physik und Astrophysik

V.D. Sandberg
California Institute of Technology

D.N. Schramm
University of Chicago

Z.F. Seidov
Shemakha Astrophysical Observatory - USSR

N.A. Sharp
Cambridge University

S.L. Shapiro
Cornell University

L.L. Smarr
Harvard University

A.A. Starobinsky
Institute for Space Research Moscow

R.A. Stokes
Batelle Pacific NW Laboratories

S. Stumbles
Cambridge University

A.H. Taub
University of California Berkeley

S.A. Teukolsky
Cornell University

K.S. Thorne
California Institute of Technology

M.S. Turner
Stanford University

K.A. Van Riper
University of Illinois

R.V. Wagoner
Stanford University

H.D. Wahlquist
Jet Propulsion Laboratory

R. Weiss
Massachusetts Institute of Technology

J.R. Wilson
Lawrence Livermore Laboratory

J.H. Winicour
University of Pittsburgh

R.S. Wolff
Jet Propulsion Laboratory

J.W. York, Jr.
University of North Carolina

N. Zamorano
University of Texas

M. Zimmermann
California Institute of Technology

INTRODUCTION

This book is an attempt to provide a comprehensive view of its subject: sources of gravitational radiation. This field has grown up as a number of fragmented specialities with little communication between them. In a rough manner one can group the major specialities into four areas: astrophysics, theoretical relativity, numerical relativity and experimental relativity. The two-week Seattle workshop was an attempt to bring the major workers in these areas together so that they could talk to and teach each other about their problems and points-of-view. The articles contained in this book are the written versions of the lectures given at the workshop. Each author has explained a particular part of the puzzle, but always with the finished picture in mind. The result is a unification of our knowledge and a new perspective for future research brought about by this holistic approach.

In a sense, this book can be considered a companion volume to that edited by Hawking and Israel and published simultaneously by Cambridge University Press. That book presents a much broader view of general relativity and provides a context in which this more detailed volume can be placed. In particular, the articles by Douglass and Braginsky on gravitational wave astronomy should be read as a companion to the articles in this volume on detectors. Actually, the first proof of the existence of gravitational radiation may come from observations of the binary pulsar system PSR1913+16. As Will describes in the Hawking-Israel volume, the decay of the orbit of this binary system by emission of gravitational radiation is predicted by Einstein's theory and observable within a few years time. However, most scientists will not be happy until an actual detection of gravitational radiation occurs. Therefore this book starts off with the subject of gravitational radiation detectors. It is followed by a section on the theoretical foundations required to study dynamic strong field gravity. This is applied first to black holes and then to neutron stars. Finally, a review of possible sources is made to compare with the possible detectors. There is so much material covered in these twenty-five articles, that I have prepared a more detailed guide to their contents and interconnections.

The first section gives a comprehensive discussion of the possibilities for experimental detection of gravitational radiation. There are two major classes of detectors: Acoustically and electromagnetically coupled antennas. The former includes the Weber-type aluminum bars and the high-Q monocrystals. The latter class contains two quite different detectors: laser interferometers and deep-space doppler tracking of spacecraft.

INTRODUCTION

Weiss describes the response to the three types of gravitational wave signals: impulsive, periodic, and stochastic. By treating the gravitational wave spectrum as a whole, he is able to produce three corresponding charts which compare the present and projected sensitivities of these detectors. The JPL group (Estabrook, et al.) describes the possibilities of using the microwave tracking capabilities of NASA's Deep Space Net to detect gravitational waves on the four interplanetary missions being flown in the next decade. This approach is particularly exciting because it will be exploring the low frequency part of the gravitational wave spectrum for the first time. At the other end of this spectrum, Thorne and coworkers give a beautiful exposition on the quantum nondemolition barrier awaiting experimentalists looking for stellar core collapse as a source of gravitational radiation. Finally, in Discussion Session I, an exploration of the experimental goals for gravitational wave astronomy is made by many of the participants in that quest. I think this comprehensive approach has shown that there are several very promising technologies for making gravitational wave astronomy a reality in the next two decades.

Presumably, the first sources of gravitational radiation seen will be strong field dynamic sources. In order to be able to calculate such sources one needs a unification of the abstract structrue of Einstein's equations with the practical techniques for numerical solution of partial differential equations. York gives an elegant review of the modern version of the 3+1 ADM split of spacetime into space plus time. This provides us with the equations to formulate the initial value problem, build a spacetime coordinate system, and to evolve the gravitational field. The key role in numerical relativity played by maximal time slices in avoiding singularities, discussed by York, is contrasted by Eardley to the new approaches for choosing slices which "wrap-up" around singularities. These techniques should bring the study of singularity structure and null infinity into union with numerical relativity. As Eardley shows, one is then in a position to investigate whether Cosmic Censorship can be violated by the formation of naked singularities in any realistic collapse or collision scenerio. The actual numerical techniques of finite differencing required to solve the partial differential equations of general relativity and hydrodynamics are reviewed by Smarr. This article is particularly directed towards those with no experience in finite differencing. The emphasis is on the concepts involved, such as stability and artificial viscosity, rather than on detailed recipes. In order to choose the best numerical methods, one would like to have a better feeling about how Einstein's equations relate to the more familiar equations of theoretical physics. After Smarr's discussion of some simple analogues, Belinsky and Zakharov discuss a powerful analytic technique, the method of inverse scattering, which directly relates the Einstein equations to the large family of nonlinear dispersive equations such as Korteweg-deVries and Sine-Gordon. This important new article begins the investigation of whether Einstein's equations can admit solitons. The common thread that runs through these articles is the constructive approach to general relativity. This unification of the spacelike, null, and numerical approaches occurred because of a desire to solve particular problems, whether about observable gravitational radiation or about the nature of this nonlinear field theory.

In the articles in the black hole section, these tools are put to work. Starting at the beginning, Novikov and Polnarev, investigate the formation of black holes in the early universe. The great advantage of numerical methods to answer the questions one really wants to ask, not just those that can be solved analytically, is beautifully illustrated here. The detailed interaction of equation of state, accretion flow, and shock waves can be studied to determine the final size of primordial black holes. Turning from calculation to observation, Blandford reviews the exciting new evidence for supermassive black holes in quasars and active galactic nuclei. Although not yet conclusive, an "establishment" picture is beginning to emerge of gas accretion onto supermassive black holes as the basic energizing mechanism. The question for gravitational wave astronomy is then: how often did gravitational wave bursts occur during the formation of these holes and how efficient was the process. Although Blandford can only sketch answers to these difficult questions, the basic radiating mechanism of black holes is becoming fairly clear. As Detweiler reviews, a disturbed black hole will radiate by free oscillations at certain resonant frequencies until it returns to a static Schwarzschild or stationary Kerr hole. By choosing particular ways of disturbing the hole, such as having a particle fall into it, one can study the gravitational radiation generated. Detweiler finds that spiraling into a black hole can be up to 200 times more efficient as a generator than head-on radial infall. A different perturbation calculation is to follow a slowly rotating deformed star collapsing into a black hole. As Moncrief and coworkers find, the resulting waveform is dominated by the ringing modes described by Detweiler. This picture is rounded out by Smarr's results of his and Eppley's calculation of the head-on collision of two black holes. Even though the calculation involves the full nonlinear theory with no well defined background, the resulting gravitational waveform is virtually indistinguishable from the perturbation results described by Detweiler. It would seem that the waveforms from black hole formation will tell us very little about the details of the formation, but they may precisely identify the mass and angular momentum of the resultant hole. Furthermore, this unexpected unity of the nonlinear and perturbation results, lead us to believe that the spiraling collision of two black holes may generate as much as $\sim$10% of Mc^2 in gravitational radiation. To investigate other spacetimes of interest requires a detailed study of the interrelation of gauge conditions and radiation formulae. This is discussed in detail by Smarr and by Eppley. Finally, D'Eath gives a very nice comparison of the analytic techniques for studying gravitational radiation produced by hyperbolic encounters.

Although black holes have been considered in the past to be the best sources of gravitational radiation, it is becoming increasingly clear that hydrodynamic effects can be quite efficient generating mechanisms. Here the study focuses on core collapse of stars to form neutron stars and supernovae. Arnett discusses the relevant physical processes which must be taken into account. There has been a rapid convergence on a unified picture of core collapse in the last year. The collapse proceeds roughly adiabatically because the neutrinos are trapped by the high opacities. As a result, negligible entropy increase occurs in the core which bounces at supernuclear densities. The shock which occurs at the outer boundary of the core is now believed to be the mechanism by which the envelope of the star is blown off as a supernova. A detailed model of such a collapse, using the new microphysics described by Arnett, is given by Wilson. As Arnett reviews, this process probably only occurs once in 10-30 years in our galaxy, necessitating gravitational wave detectors sensitive enough to see the Virgo cluster of galaxies to obtain an event rate of greater than one per year. It would be very desirable to observe core collapse jointly with neutrino and gravitational radiation detectors. This could happen if a core collapse occurred in our galaxy. However, whereas it does seem possible to build gravitational radiation detectors which can see the Virgo cluster, Kazanas and Schramm think it is unlikely that such neutrino detectors can be built. The competition between neutrinos and gravitational waves as dissipative mechanisms is reviewed by Kazanas and Schramm, particularly for small deviations from spherical symmetry. Shapiro extends this to the very non-spherical regime by constructing ellipsoidal "one-zone" models of collapsing cores with angular momentum or initial asymmetries. Although his models are Newtonian, they have detailed microphysics which probably provides a fairly accurate model of the gross behavior of the core. Since the equations are very simple, he is able to map out the dependence of the efficiency of both gravitational radiation and neutrino emission over a wide range of initial conditions. Most surprisingly he finds that if the collapse goes through several bounces, the gravitational wave efficiency can become quite high (0.01 Mc^2), even for slowly rotating cores. If verified by two dimensional general relativistic hydrodynamics codes such as described by Wilson, this could be the essential ingredient of success for ground based gravitational wave astronomy. Turner and Wagoner investigate the gravitational waves produced in a hydrodynamic bounce by studying gravitational perturbations off of slowly rotating collapses . This study of "the bounce" is extended to the full theory, but with the unphysical restriction to cylindrical symmetry, by Piran. He finds efficiencies as high as ∿65% are possible in extreme cases. The coupling of the full hydrodynamic equations to the full Einstein equations for two and three space dimensions are required to decide these issues of efficiencies and waveforms for realistic relativistic situations. Wilson gives a detailed account of how the equations are set up and finite differenced. His account of Eulerian hydrodynamics will be very useful even to those astrophysicists interested in purely Newtonian non-relativistic fluid flows and shock waves. The use of these codes to study both spherical and highly nonspherical collapse is demonstrated. Finally, even if core collapse were found to be too inefficient to observe gravitationally, Clark reminds us that binary systems can be sources of all three classes of gravitational radiation: periodic, burst, and stochastic. He shows that all of these sources, from classical binaries to colliding neutron stars, may be observable by gravitational radiation if sensitive enough detectors can be built. This section reveals the tight intermeshing of astrophysics, numerical relativity, and particle physics which is necessary to study sources of gravitational radiation. I think it is clear from the articles that a new sense of unity among these previously separate specialities is emerging. This will form a solid basis for future study.

In the final section, Ostriker attempts to summarize the vast array of possible astrophysical sources of gravitational radiation. His discussion of clusters of collapsed objects as sources covers an area not discussed elsewhere in this volume. It is clear that more work should be done on this interesting possibility. Ostriker emphasizes the importance of shocks in determining the final entropy of collapsed systems. While this is certainly true for planar collapse or collision, the results of Arnett and Wilson for spherical collapse, discussed above, seem to show that spherical cores are not shocked, but retain their initial entropy. Where the cross over occurs for non-spherical collapse is not certain yet. Presumably the techniques discussed by Wilson will answer this important question. The second Discussion Session attempts a quantitative approach to all possible sources of gravitational radiation. Using the three detector charts for periodic, impulsive, and stochastic sources drawn by Weiss, Epstein and Clark construct three source charts on the same scale, drawing from the discussion at the conference, but augmented by a great deal of post-conference research. This gives us a comprehensive view of the field of gravitational wave astronomy. By overlaying the three sets of detector and source charts, one can see where more work needs to be done in astrophysical, theoretical, numerical, and experimental relativity. These charts will no doubt be redrawn many times in the future. However, I think they symbolize the unification of many specialities into an overall physical approach to general relativity. This bodes very well for the future research in this field.

GRAVITATIONAL RADIATION - THE STATUS OF THE EXPERIMENTS AND PROSPECTS FOR THE FUTURE

Rainer Weiss
Massachusetts Institute of Technology

The equivalent of a gravitational Hertzian experiment using a terrestrial source and receiver is not feasible at present or in the forseeable future. The search for gravitational radiation for this reason has concentrated on detecting gravitational emission from astrophysical sources. These experiments could serve the dual purpose of establishing the existence and properties of the radiation as well as to open a new window on the universe which will probe relativistic astrophysics and cosmology.

The best estimates for the gravitational radiation incident on the earth from astrophysical sources are summarized in the notes of the second discussion section of this conference and in two current review articles by Tyson and Giffard (1978) and Douglass and Braginsky (1979).

The source estimates are divided into three categories. Impulsive sources such as isolated stellar collapses and collisions, periodic sources arbitrarily defined as those producing waves with a 1000 or more cycles such as binary stellar systems and stellar oscillations, and finally the radiation by aggregates of sources which could produce a stochastic background of gravitational waves such as might be generated by a turbulent or chaotic primeval universe.

This review of the status and prospects for detection is divided in the same manner. Acoustically coupled antennas are discussed first and then followed by a discussion of "free" mass electromagnetically coupled antennas both of the Doppler ranging and interferometrically sensed types.

ACOUSTICALLY COUPLED ANTENNAS

This type of antenna, pioneered by J. Weber (1969) and analyzed by Giffard (1976) has four components: the resonator, motion transducer, amplifier and output filter as schematized in Fig. 1. The resonator is characterized by its mass, m, operating temperature, T, oscillator quality factor, Q, resonant frequency ω_0 or equivalently by its length l and speed of sound, c_s, in the resonator material. The resonator is driven by both the gravitational radiation which is the signal, and stochastic forces due to thermal excitations as well as the amplifier noise transmitted through the transducer (back reaction forces). The transducer, if linear, can be specified by the electromechanical coupling matrix, (Hunt, 1954).

$$\begin{pmatrix} F \\ \phi \end{pmatrix} = \begin{vmatrix} Z_{11} & Z_{12} \\ Z_{21} & Z_{22} \end{vmatrix} \begin{pmatrix} u \\ I \end{pmatrix} \tag{1}$$

F is the force applied to the resonator by the transducer and u the velocity of the resonator, ϕ the voltage at the transducer output and I the current. Z_{11} is the mechanical impedance of the transducer, Z_{22} the electrical output impedance and Z_{12} the electromechanical coupling impedance. The matrix is symmetric in magnitude of the matrix elements $|Z_{12}| = |Z_{21}|$, which follows from reciprocity. Z_{12} is responsible for converting amplifier noise to a noise force on the resonator. A useful quantity, the electromechanical coupling efficiency, β, is defined as the ratio of the electrical energy stored in the transducer to the mechanical energy stored in the resonator, Gibbons and Hawking, (1971). In terms of the transducer matrix elements β is given by

$$\beta = \frac{|Z_{21}|^2}{|Z_{22}|\, m\omega} \tag{2}$$

Another quantity often used is the displacement sensitivity of the transducer

$$\alpha = |Z_{21}|\omega \quad \text{volts/cm} \tag{3}$$

The amplifier can be characterized by a series voltage noise source with spectral density $e_n^2(f)$ volts^2/Hz, a shunt current noise generator with spectral density $i_n^2(f)$ amp^2/Hz, followed by an ideal noise free amplifier. The input impedance of the amplifier is assumed large relative to the noise matching impedance defined as

$$|Z_{match}| = \left(\frac{e_n^2(f)}{i_n^2(f)}\right)^{1/2} \tag{4}$$

If the input of the amplifier is terminated in Z_{match} the voltage and current noise contributions become equal.

With the amplifier matched, the amplifier noise can be expressed in terms of the spectral power density of an equivalent thermal source at temperature T_{amp} given by

$$\frac{hf}{e^{hf/kT_{amp}} - 1} = \left(e_n^2(f) \; i_n^2(f) \right)^{1/2} \tag{5}$$

if the power gain of the amplifier is much larger than one (Weber, 1957, Heffner, 1962). In the Rayleigh-Jeans limit ($hf/kT \ll 1$), this expression reduces to the familiar Nyquist formula

$$T_{amp} = \frac{\left(e_n^2(f) \; i_n^2(f) \right)^{1/2}}{k} \tag{5a}$$

While if the amplifier is limited by spontaneous emission noise at its input - the quantum limit for a linear amplifier - the equivalent temperature becomes

$$T_{amp} = \frac{hf}{k \; \ell n 2} \tag{6}$$

which is 6.9×10^{-8} °K at 1 kHz.

The equivalent temperature and matching impedance completely characterize the amplifier in signal to noise calculations. In principle the electrical impedance transformation that matches the transducer to the amplifier can be noise free.

The amplifier does not have to operate at the audio frequency of the acoustic resonator, where low noise amplifiers are not readily available. The transducer can be a variable capacitor or inductor (or microwave cavity) which is driven by a high frequency carrier signal. Side bands are introduced in the output signal of the transducer with amplitude determined by the resonator motion and offset frequencies determined by the resonator frequency. The transducer becomes the first stage of a parametric amplifier and is then followed by a low noise amplifier at the carrier frequency. The utility of this scheme lies in the translation of the frequency to a region where a good amplifier exists.

The parametric amplifier in the ideal limit is governed by the Manley-Rowe (Manley and Rowe, 1956, Rowe, 1958) relations which state that the rate of signal photons at the output and input can at best be the same. The power gain is

$$\frac{P_{out}}{P_{in}} \leq \frac{\omega_{out}}{\omega_{in}} \tag{7}$$

and the equivalent noise temperature at ω_{in} can at best be

$$T_{in} = T_{out} \frac{\omega_{in}}{\omega_{out}} \tag{8}$$

where T_{out} is the noise temperature of the amplifier following the parametric converter, providing the converter is noise free. In principle then, if an amplifier can be made to operate at the quantum limit at some frequency by parametric conversion the quantum limit can be achieved at another frequency.

The final component of the system is a filter which can be optimized if the system noise spectrum, $G_{nn}(\omega)$, and the signal power spectrum $G_{ss}(\omega)$ are known. The optimal linear filter can be determined by using the Wiener-Hopf (Wiener, 1949) theorem which states that the transfer function of the output filter should be

$$T(\omega) \cong \frac{G_{ss}(\omega)}{G_{nn}(\omega) + G_{ss}(\omega)} \tag{9}$$

The optimization of the output filter becomes important when the signal to noise is small and requires knowledge of the signal power spectrum. This is one of the major experimental motivations for theoretical pulse shape predictions.

DETECTION CRITERIA FOR ACOUSTIC ANTENNA SYSTEMS

In this section the generalized formulations of the preceding section are applied to detection criteria for the three classes of gravitational wave sources. The system noise, expressed in the frequency domain, is used to derive an rms output noise which is then expressed in terms of the minimum detectable gravitational wave signal.

a) Impulsive Sources

The following calculation assumes that the gravitational wave pulse length is short compared to the relaxation time of the resonator. The output filter is not optimized but rather a simple integrator with integration time, t_{int}.

The rms noise in the antenna expressed as an equivalent energy in the resonator is given by

$$\Delta E = \frac{t_{int}}{m} \sum_{i=1}^{n} F_i^2(f) + \frac{m\,\omega_0^2}{4\,t_{int}}\, x^2(f) \tag{10}$$

where $F_n^2(f)$ is the spectral density of the stochastic forces on the resonator in $dynes^2/Hz$ and $x^2(f)$ is the spectral density of the equivalent displacement noise in cm^2/Hz due to the amplifier and transducer.

If the resonator is well isolated from external perturbations (ground noise, magnetic pulses, cosmic rays, etc.) the dominant stochastic forces are the thermal Nyquist forces given by

$$F_{th}^2(f) = \frac{4\,k\,T\,m\,\omega_0}{Q} \tag{11}$$

and the back reaction force due to the amplifier noise acting on the resonator through the transducer given by

$$F_{br}^2(f) = |Z_{12}|^2\, i_n^2(f) \tag{12}$$

The equivalent displacement noise spectral density is

$$x^2(f) = \frac{\left(e_n^2(f) + |Z_{22}|^2\, i_n^2(f)\right)}{|Z_{21}|^2\, \omega_0^2} \tag{13}$$

Combining these equations results in an expression for the rms noise in the system

$$\Delta E = \frac{t_{int}}{m}\left(\frac{4\,k\,T\,m\,\omega_0}{Q} + |Z_{12}|^2\, i_n^2\right) + \frac{m}{4t_{int}}\, \frac{\left(e_n^2(f)+|Z_{22}|^2 i_n^2(f)\right)}{|Z_{21}|^2} \tag{14}$$

The rms noise is minimized when t_{int} is chosen to make the two terms equal, this condition is given by

$$t_{int}^2 = \frac{m^2\left(e_n^2(f) + |Z_{22}|^2\, i_n^2(f)\right)}{\frac{16\,k\,T\,m\,\omega_0}{Q}\,|Z_{21}|^2 + 4\,|Z_{12}|^2\,|Z_{21}|^2\, i_n^2(f)} \tag{15}$$

The minimum noise becomes

$$\Delta E_{min} = \left[\left(\frac{e_n^2(f) + |Z_{22}|^2\ i_n^2(f)}{|Z_{12}|^2}\right)\left(\frac{4\,k\,T\,m\,\omega_0}{Q} + |Z_{12}|^2\ i_n^2(f)\right)\right]^{1/2} \tag{16}$$

where the reciprocity condition has been used.

Equation (16) can be simplified if re expressed in terms of β (Eq. (2)) and if the amplifier is matched to the transducer $|Z_{match}| = |Z_{22}|$, (Eq. (4)).

$$\Delta E_{min} = \left[\left(\frac{8\ kT_{amp}}{\beta}\right)\left(\frac{4\ kT}{Q} + 4\ kT_{amp}\ \beta\right)\right]^{1/2} \tag{17}$$

The minimum excitation energy in the resonator is related to the minimum detectable gravitational pulse flux by

$$F_{min}(\nu_0) = \frac{\Delta E_{min}}{\sigma_T} \tag{18}$$

where, σ_T, the total absorption cross section for a longitudinal resonator averaged over all polarization, is given by (Misner, Thorne and Wheeler, 1973)

$$\sigma_T = \frac{32}{15\pi}\frac{G}{c^3}\ m\ c_s^2 \tag{19}$$

The initial rms strain in the resonator is

$$\frac{\Delta \ell}{\ell} = \left(\frac{2\pi G}{c^3}\right)^{1/2} F^{1/2}(\nu_0) = \pi\left(\frac{15}{16}\right)^{1/2}\left(\frac{\Delta E_{min}}{Mc_s^2}\right)^{1/2} \tag{20}$$

Two limiting cases are important. The first is the case when the noise is dominated by the Nyquist forces so that the back reaction noise force can be neglected. This has been the case for all acoustic antennas constructed, including those now in operation and projected for the next several years. Assuming that the transducer is optimally matched to the amplifier, the optimum integration time becomes

$$t_{int}^2 = \frac{T_{amp}\ Q}{2\ T\ \beta\ \omega_0^2} \tag{21}$$

The minimum detectable flux and strain are

$$F(\nu) = \frac{15\pi\, c^3}{32\ G\ mc_s^2}\left(\frac{32\ kT_{amp}\, kT}{\beta Q}\right)^{1/2} \quad (22)$$

$$\frac{\Delta\ell}{\ell} = \pi\left(\frac{15}{16}\right)^{1/2} \frac{1}{(m\ c_s^2)^{1/2}}\left(\frac{32\ kT_{amp}\ kT}{\beta\ Q}\right)^{1/4}$$

Representative experimental parameters for this case are given in Table 2.

The other limiting case occurs when the back reaction noise dominates ($T \to 0$).

For this case the optimum integration time becomes

$$t_{int} = \frac{1}{\sqrt{2}\ \beta\omega_0} \quad (23)$$

and the minimum detectable energy change becomes

$$\Delta E = 4\ \sqrt{2}\ kT_{amp} \quad (24)$$

If the amplifier is quantum noise limited (Eq. 6) the minimum flux and strain are

$$F(\nu) \geq \frac{\sqrt{2}\ 15\pi}{(\ln 2)32}\ \frac{c^3}{G}\ \frac{\hbar\omega}{mc_s^2} \quad (25)$$

$$F_{GPU}(\nu) \geq 2.1 \times 10^{-7}\left(\frac{f}{kHz}\right)^{1/2} \frac{1}{\left(m/10^6\right)} \qquad \text{Aluminium}$$

$$\frac{\Delta\ell}{\ell} > \left(\frac{\sqrt{2}\ 15\ \pi^2}{\ln 2\ 16}\right)^{1/2}\left(\frac{\hbar\omega}{mc^2}\right)^{1/2}$$

$$\frac{\Delta\ell}{\ell} \geq 1.8 \times 10^{-20}\left(\frac{f}{kHz}\right)^{1/2} \frac{1}{\left(m/10^6\right)^{1/2}} \qquad \text{Aluminium}$$

Table 1. Useful Conversion Equations (cgs units).

$$\frac{c^3}{G} = 4 \times 10^{38} \text{ ergs/cm}^2\text{Hz} \qquad 1 \text{ GPU} = 1 \times 10^5 \text{ ergs/cm}^2\text{Hz}$$

IMPULSIVE SOURCES BANDWIDTH EQUAL TO FREQUENCY

$$F(\nu) = \frac{c^3}{2\pi G}\left(\frac{\Delta\ell}{\ell}\right)^2 = 6.4 \times 10^{37}\left(\frac{\Delta\ell}{\ell}\right)^2 \text{ ergs/cm}^2\text{Hz}$$

$$\frac{\Delta\ell}{\ell} = 4 \times 10^{-17}\left(\frac{F(\nu)}{\text{GPU}}\right)^{1/2}$$

IMPULSIVE MINIMUM RMS FLUX AND STRAIN IN TERMS OF SYSTEM EQUIVALENT TEMPERATURE (ALUMINIUM RESONATOR)

$$F_{\text{GPU}}(\nu) = 2.3 \frac{T_{\text{eff}}}{\left(\frac{m}{10^6}\right)} \qquad \frac{\Delta\ell}{\ell} = 6 \times 10^{-17}\left(\frac{T_{\text{eff}}}{\left(\frac{m}{10^6}\right)}\right)^{1/2}$$

RELATION BETWEEN FLUX AND STRAIN AMPLITUDE IN A MONOCHROMATIC WAVE

$$\frac{\Delta\ell}{\ell} = \left(\frac{G}{2\pi c^3}\right)^{1/2} \frac{I_g^{1/2}}{f} = 2 \times 10^{-20} \frac{I_g^{1/2}}{f}$$

SPECTRAL DENSITY OF STRAIN

$$\left(\frac{\Delta\ell}{\ell}(f)\right)^2 = \frac{4}{\pi}\frac{G}{c^3}\frac{I_g(f)}{f^2} = 3.2 \times 10^{-39} \frac{I_g(f)}{f^2} \text{ Hz}^{-1}$$

TYPE	f_o Hz	m gm $\times 10^6$	l cm	Q	T °K	t_{obs} sec	$x^2(f)$ cm^2/Hz	$\lvert Z_{12}\rvert^2$ cgs	β	T_{eq} °K	F(ν) min GPU	$\frac{\Delta \ell}{\ell}$
Weber type cylindrical bar PZT transducer Tyson (1973)	710	3.6	357	2.2×10^{5}	300	0.7	2.3×10^{-29}	1.6×10^{-2}	2×10^{-4}	20°	15	1.5×10^{-16}
Split bar PZT transducer Hough et al. (1975) Drever et al. (1973)	1050 ±400	0.3	155	6.6×10^{3}	300	6×10^{-4}	3×10^{-32}	8	.18	4.5	34	2.3×10^{-16}
Cryogenic aluminium bar squid variable inductable transducer Boughn et al. (1977)	1315	.68	200	6.6×10^{4}	4.4	0.3				0.39	1.3	4.5×10^{-17}
Quadrupole Antenna thermally tuned to pulsar NP0532 DC capacitive transducer Hirakawa et al. 1978)	60.2	.078 effec-tive	110 x 110	4.5×10^{3}	300		1.2×10^{-24}	3.1×10^{-2}	1.4×10^{-3}			$6\times10^{-17}/Hz^{1/2}$

Table 2. Representative Characteristics of Acoustic Antennae

b) Periodic Sources

The analysis for the performance of acoustic resonators driven by a periodic gravitational wave follows a comparable prescription. The spectral density of the displacement noise is

$$x_n^2(f) = |T(f)|^2 \sum_{i=1}^{m} F_{n_i}^2(f) + \frac{1}{|Z_{21}|^2 \omega^2}\left(e_n^2(f) + |Z_{21}|^2 i_n^2(f)\right) \qquad (26)$$

where the first term represents the sum of the stochastic forces on the resonator. $|T(f)|$ is the magnitude of the force to displacement transfer function of the resonator

$$|T(f)|^2 = \left|\frac{x(f)}{F(f)}\right|^2 = \left(m^2\left[(\omega_0^2 - \omega^2)^2 + \left(\frac{\omega\omega_0}{Q}\right)^2\right]\right)^{-1} \qquad (27)$$

The spectral density of the gravitational gradient force due to the wave is

$$F_g^2(f) = \left(ml\ \omega^2\left(\frac{\Delta\ell}{\ell}\right)\right)^2 \delta(f - f_0) \qquad (28)$$

The criterion for detection is that the signal power by greater than the noise

$$\int F_g^2(f)\ |T(f)|^2\, df > \int x_n^2(f)\, df \qquad (29)$$

Considering only the case where the resonator is tuned to the frequency of the source and the transducer impedance is matched to the amplifier, the minimum detectable wave strain becomes

$$\frac{\Delta\ell}{\ell} \geq \left[\frac{1}{2t_{int}\ m\ \omega_0^3\ l^2}\left(\frac{4kT}{Q} + 4kT_{amp}\beta + \frac{8kT_{amp}}{\beta Q^2}\right)\right]^{1/2} \qquad (30)$$

In this equation it is assumed that the Nyquist forces and back reaction forces are the only stochastic forces and that the bandwidth of the measurement is $\Delta f = 1/2\ t_{int}$ - where t_{int} is the total time of observation.

The demands on the transducer - amplifier combination are not as difficult to meet in detecting periodic sources as for impulsive ones. The interplay of the backreaction and broadband amplifier noise as seen in equation (30) is minimized when the transducer coupling is $\beta = \sqrt{2}/Q$, a lightly coupled transducer. Unlike the situation in detecting impulsive events, the strain limit continues to improve as $1/Q^{1/2}$ even when the amplifier noise dominates over the Nyquist noise.

At frequencies lower than 100 Hz the longitudinal resonators become impractically long - 50 meters at 60 Hz for aluminium, so that plate or mass loaded resonators have been used and are contemplated for future experiments. An example of present techniques is the room temperature experiment of Hirakawa et. al. (1978) (Table 2) which operates at a noise level of $\Delta\ell/\ell/\mathrm{Hz}^{1/2} \sim 6 \times 10^{-17}$ at 60.2 Hz, the Crab pulsar frequency.

One can contemplate quantum noise limited amplifiers for the detection of periodic sources. However, the optimal performance of the antenna will require that the physical temperature of the resonator be at the equivalent quantum temperature (Eq. 6); at 60.2 Hz this corresponds to 4×10^{-9} °K. Such low temperatures are not technically possible at present. Nevertheless, a single crystal resonator at 60 Hz with a Q of 10^9, an oscillating mass of 100 kg and length 1 meter could have a quantum noise limit $\Delta\ell/\ell/\mathrm{Hz}^{1/2} \sim 4 \times 10^{-25}$. A more realistic estimate for the next decade might be such an antenna operated at 0.1 °K which would have a limit of $\Delta\ell/\ell/\mathrm{Hz}^{1/2} \sim 2 \times 10^{-21}$.

c) Stochastic Gravitational Sources-Gravitational Radiometry

A detection scheme for a stochastic background of gravitational radiation requires the cross correlation of the output of at least two antennas, (Fig. 2). The premise is that the gravitational radiation noise signals are correlated in the antennas while the internal noise in each antenna due to other sources is uncorrelated and will average to zero in the cross correlation.

The cross correlation detection criterion is (Bendat, 1958)

$$\left(\frac{\Delta\ell}{\ell}\,(f)\right)^2_{grav} > \frac{\left(\Delta\ell/\ell\;(f)\right)^2_{internal}}{(\Delta f\; t_{int})^{1/2}} \tag{31}$$

where Δf is the bandwidth of the antenna system, t_{int} the post multiplication integration time, $(\Delta\ell/\ell(f))^2_{internal}$ is the spectral density of the internal antenna equivalent strain noise while $(\Delta\ell/\ell(f))^2_{grav}$ is the spectral density of the gravitational strain noise - the signal.

The signal to noise calculation is more complicated for stochastic sources than for the periodic and impulsive sources and therefore is carried out only approximately here.

With acoustic resonators the antenna will generally be followed by a filter to increase the detection bandwidth, (to suppress the resonance). The filter transfer function is

$$\left|\frac{V_{out}}{V_{in}}(f)\right|^2 = |S(f)|^2 .$$

A straightforward application of Eq. 31 gives the following detection criterion

$$\int_{\omega_1}^{\omega_1+\Delta\omega} \omega^4 \; |T(f)|^2 \; |S(f)|^2 \left(\frac{\Delta\ell}{\ell}(f)\right)^2_{grav} d\omega \geq$$

$$\frac{1}{m^2\ell^2\left(\Delta\, ft_{int}\right)^{1/2}} \left[\left(\frac{4\, kT\omega_0 m}{Q} + 4\, kT_{amp}\, \omega_0\, m\, \beta\right)\int |T(f)|^2\, |S(f)|^2\, d\omega \right.$$

$$\left. + \frac{8\, kT_{amp}\, m}{\beta\, \omega_0^2}\int \frac{|S(f)|^2}{\omega^2}\, d\omega\right] \qquad (32)$$

Some special cases are interesting. If the filter doesn't exist, the antennas measure only the spectral density in a bandwidth $\Delta f \sim \omega_0/\pi Q$ around the resonance frequency. The minimum is then

$$\left(\frac{\Delta\ell}{\ell}(f_0)\right)^2_{grav} > \frac{1}{m\ell^2\, \omega_0^3\, (\Delta f\, t_{int})^{1/2}} \left[\frac{4kT}{Q} + 4\, kT_{amp}\beta + \frac{8\, kT_{amp}}{\beta Q^2}\right] \qquad (33)$$

The optimization is as for a periodic source.

With a "whitening" filter which satisfies $|T(f)|^2\, |S(f)|^2 = 1/m^2$ and a detection bandwidth small compared to the center frequency, the minimum becomes

$$\left(\frac{\Delta \ell}{\ell}(f)\right)^2_{grav} > \frac{\omega_0}{m\ell^2 \langle\omega\rangle^4 (\Delta f t_{int})^{1/2}} \left[\frac{4kT}{Q} + 4\ kT\beta_{amp} + \frac{8\ kT_{amp}}{\beta} \left\{ \begin{array}{c} \left(\frac{\omega_0}{\langle\omega\rangle}\right)^2 \quad \omega_0 > \omega \\ \cdots\cdots \\ \left(\frac{\langle\omega\rangle}{(\omega_0)}\right)^2 \quad \omega_0 < \omega \end{array} \right\} \right] \tag{34}$$

which places strong demands on the transducer-amplifier combination much as in the detection of impulsive sources.

The only experiment that has attempted to measure the gravitational background noise in a manner similar to that described is Hough et. al. (1975) using a pair of split bar resonators (Table 2.). This experiment set a limit of $I_g(f) \sim 6 \times 10^5$ ergs/sec cm^2 Hz in a 160 Hz bandwidth near 1 kHz after an integration time of 90 hours. $((\Delta\ell/\ell(f))^2 \sim 2 \times 10^{-39}/\text{Hz})$.

A prospect for the future might be a longitudinal cryogenic resonator dominated by quantum amplifier noise. Taking a mass of 5×10^6 gms, $\beta \sim 1$, bandwidth $\sim 1/10\ f_0$, $f_0 \sim 1$ kHz and integration time of 10^6 seconds Eq. (34) and (6) would give

$$\left(\frac{\Delta \ell}{\ell}(f)\right)^2_{grav} \sim 10^{-49}/\text{Hz}$$

SUMMARY ON ACOUSTIC ANTENNAS

In the next few years it is expected that cryogenic single crystal and massive aluminium bars will yield impulsive source sensitivities in the range 10^{-2} to 10^{-3} GPU (rms) corresponding to impulsive strain sensitivities $\sim 10^{-18}$ (rms). Further advances will require the development of higher β transducers which have small losses, this is particularly important for the high Q single crystals. The state of amplifiers is better than that of transducers, there are several regions in the frequency spectrum where amplifiers approach the quantum limit, in particular maser amplifiers at 10 GHz and Josephson junction devices at 30 GHz. It is primarily for this reason that modulated microwave cavity transducer appear attractive.

An improvement of 10^4 in energy sensitivity over the performance of detectors available in the next few years is required before approaching the quantum limit using linear amplifiers. Nevertheless, it is a crucial question whether such a quantum limit exists (see K. Thorne et.al. article at this conference) as the prospective source intensities for reasonable event rates (1/month rather than 1/30 years) and confidence in the detection will require sensitivities smaller than the quantum limit.

In the search for periodic signals acoustic resonators, specifically cryogenic high Q single crystal, can approach interesting sensitivities, as the demands on the transducer amplifier combination are not as stringent as for impulsive sources. The obvious candidate is the Crab pulsar.

ELECTROMAGNETICALLY COUPLED ANTENNAS

The fundamental idea of electromagnetically coupled antennas is to measure the gravitational wave strain between a group of free masses using electromagnetic waves to determine their separations. The major attribute of such antennas is that the baselines can be made large, comparable to the gravitational wavelength. The large baselines increase the gravitational wave signal relative to many noise sources which do not scale with the size of the system, most importantly the uncorrelated stochastic forces on the antenna masses. Free mass antennas are broad band and can extend the observation of the gravitational radiation to low frequencies.

The discussion that follows is limited to antennas in which the electromagnetic wave travel time between antenna masses is less than the period of the gravitational waves. The inverse case is discussed in a companion article by the JPL group on Doppler ranging to spacecraft (this volume).

A schematic diagram of an element of an electromagnetically coupled antenna is shown in Fig. 3. The two arm element is essential for cancelling the effect of some noise sources in particular the phase fluctuations of the electromagnetic oscillator and also enhances the interaction with the polarization of tensor gravitational waves which produce a differential mode signal in this configuration.

For a monochromatic wave at frequency f, the measurable strain is given by

$$\frac{\Delta \ell}{\ell}(f)_{meas} = \frac{\Delta \ell}{\ell}(f)_{wave} \frac{\sin(2\pi\, t_{stor}/\tau_g)}{(2\pi\, t_{stor}/\tau_g)} \tag{35}$$

where $\Delta\ell/\ell(f)_{wave}$ is assumed much less than 1. ℓ is the physical antenna baseline while t_{stor} is the storage time of the electromagnetic waves in the antenna and τ_g is the period of the gravitational wave. It is assumed that the gravitational waves propagate at c so that for a single pass antenna $\tau_{stor}/\tau_g = \ell/\lambda$.

The criterion for detection of a gravitational wave strain spectral density is given by:

$$\left(\frac{\Delta \ell}{\ell}(f)\right)^2_{min} = \frac{1}{4\ell^2}\left[x_T^2(f) + 2\sum_{i=1}^{n}\frac{F_i^2(f)}{m^2\omega^4} + N_\lambda^2(f)\, c^2\, t_{stor}^2\right] \frac{(2\pi\, t_{stor}/\tau_g)^2}{\sin^2(2\pi\, t_{stor}/\tau_g)} \tag{36}$$

$x_T^2(f)$ is the spectral density of the differential displacement detector noise, $F_i^2(f)$ the spectral density of one of the stochastic noise forces on an antenna mass m, $N_\lambda^2(f)$ is the spectral density of uncorrelated index of refraction fluctuations in the antenna arms averaged over the length of the arms at the wavelength of the electromagnetic oscillator.

The transducer noise comes from amplitude and phase noise of the electromagnetic oscillator and amplitude noise in the electromagnetic receiver and amplifier.

$$x_T^2(f) = \frac{\lambda^2}{b^2}\left[\underbrace{\frac{4h\nu}{\pi^2\,\eta P_d}}_{\text{shot noise}} + \underbrace{\frac{4\,kT_{eff}}{\pi^2\,\eta P_d}}_{\text{receiver noise}} + \underbrace{8\,\tau^2\,\delta}_{\text{phase noise}}\right] \tag{37}$$

In this expression, P_d is the power at the detector, η, the quantum efficiency of the detector; T_{eff}, the noise temperature of the receiver; b, the number of beams in the antenna arms; τ, the difference in time for the light to travel from the source to the detector in the two arms; and δ, the frequency width of the oscillator. The same expression applies whether the antenna is an interferometer or incorporated into a heterodyning scheme as in

Doppler ranging.

For fixed oscillator power, the minimum transducer noise occurs for an optimum number of beams given by

$$b = \frac{2}{1 - R}$$

where R is the reflectivity of the mirrors. P_d becomes $1/e^2$ of the oscillator power. It is worth noting that increasing b (folding the arms) decreases $x_T^2(f)$; however, it also increases the storage time. The optimum storage time is 1/2 the gravitational wave period. At those frequencies where the stochastic forces contribute more than the transducer noise, the folding is of no value and the only way to reduce the minimum detectable signal is to increase the antenna length. The folding can be thought of as means of matching the transducer noise to the stochastic forces.

The detectability criteria for the three classes of sources follow from Eq. (36). In detecting transient events the proper filter would be determined from Eq. (9) if the pulse spectrum is known. In principle this kind of antenna could observe the pulse shape providing there is enough signal power. In the calculations that follow the filter assumed will be nothing fancier than an integrator with a time constant equal to the length of the pulse

$$\frac{\Delta\ell}{\ell}_{\text{impulsive}} \geq \left[\left(\frac{\Delta\ell}{\ell}(f)\right)^2_{\min} \frac{1}{2t_{\text{pulse}}}\right]^{1/2} \tag{38}$$

If the antenna length or storage time is optimized, the minimum detectable impulsive strain scales as $1/t^{3/2}_{\text{pulse}}$, up to the point where the stochastic forces dominate.

The detection criterion for periodic sources both known and unknown, as would be uncovered in a Fourier transformation of the antenna output, is given by

$$\frac{\Delta\ell}{\ell}_{\text{periodic}} \geq \left[\left(\frac{\Delta\ell}{\ell}(f)\right)^2_{\min} \frac{1}{2t_{\text{obs}}}\right]^{1/2} \tag{39}$$

where t_{obs} is the length of an unapodized record.

Finally, the minimum detectable gravitational noise spectral density in a multi antenna cross correlation experiment similar to the one discussed under acoustic antennas is given by

$$\left(\frac{\Delta\ell}{\ell}(f)\right)^2_{grav} = \frac{\left(\frac{\Delta\ell}{\ell}(f)\right)^2_{min}}{(\Delta f\ t_{int})^{1/2}} \qquad (40)$$

The detectability conditions are best discussed through examples of various antenna systems

Ground Based Antennas

Preliminary work on ground based interferometric antennas is described by Weiss (1972), Moss et al. (1971), and Forward (1978). At present research groups at the Max-Planck Institute in Munich, Glasgow University and at M.I.T. are engaged in constructing equal arm optical Michelson interferometers that have the mirrors suspended in high vacuum tanks. The interferometer arms are multi pass cavities of 1 to 10 meters in length. The interferometers are illuminated by argon ion lasers. The transducer noise (Eq. 37) for these antennas is entirely due to photon shot noise, $x^2(f) \sim 10^{-32}$ cm^2/Hz using $P_{osc} \sim 1$ watt, $\lambda \sim 5000$ °A, R = 99.5%, $\eta \sim 1/2$. The arm lengths have to be maintained equal to 10^{-3} cm in order that the laser phase instability does not contribute.

The dominant stochastic forces are seismic noise and Nyquist forces from the suspension (Weiss, 1972). The unattenuated seismic noise at reasonable locations on the earth has a displacement spectrum of

$$x^2_{seismic}(f) \sim \frac{3\times10^{-14}}{f^4}\ cm^2/Hz \qquad f > 10Hz$$

A one axis suspension could in principle provide an isolation above its resonance frequency, f_o, of

$$\frac{x^2_{ant}(f)}{x^2_{seismic}(f)} \sim \left(\frac{f_o}{f}\right)^4$$

if f_o is 1/2 Hz the attenuated ground noise spectrum becomes

$$x^2_{seismic}(f) \sim \frac{2\times10^{-15}}{f^8} \qquad cm^2/Hz$$

which crosses the transducer noise at 145 Hz.

The Nyquist noise of the suspension above the resonance frequency at 300 °K with $Q \sim 10^5$ and antenna masses of $m \sim 20$ kg is

$$x^2_{thermal}(f) \sim \left(\frac{1}{m^2 \omega^4}\right) \frac{4\ kTm\ \omega_0}{Q} \sim \frac{10^{-25}}{f^4} \quad cm^2/Hz$$

which crosses the transducer noise at about 60 Hz.

From this example it is clear that ground noise is the dominant problem in these antennas and cunning is needed in suspension design to provide seismic isolation without compromising the suspension Q. The near term prospects for these antennas operating above 300 Hz is a spectral strain sensitivity of $(\Delta\ell/\ell(f))^2 \sim 10^{-38}\ Hz^{-1}$.

A larger ground based version of such an antenna, although expensive, is a distinct possibility. To maintain the same detected power, the optics must scale in linear dimension as $\sqrt{\lambda\ell}$. The crossing point of the uncorrelated ground noise and Nyquist noise with the transducer noise is independent of the baseline. For example, a 10 km antenna with 15 passes per arm or a 1 km antenna with 150 passes would have an optimal storage time for 1 m sec pulses. The minimum detectable spectral density becomes $(\Delta\ell/\ell\ (f))^2 = 10^{-42}\ Hz^{-1}$ which corresponds to an rms impulsive strain sensitivity of $\Delta\ell/\ell \sim 2 \times 10^{-20}$ for millisecond pulses using a 1 watt laser. This limit could be reduced by another factor of 10 when 100 watt argon ion lasers become available in the near future.

If the ground noise isolation can be improved substantially with active isolation systems, such an antenna could measure the Crab pulsar with a strain sensitivity of $\Delta\ell/\ell \sim 2 \times 10^{-25}/\tau^{1/2}$ (days).

Space Antennas

Electromagnetically coupled antennas operated in space offer very large baselines and freedom from seismic noise; extending measurements of the gravitational radiation spectrum to frequencies lower than 10 Hz.

The microwave Doppler ranging experiments using the NASA deep space network described in the accompanying article by the JPL group, are a first step in space gravitational astronomy. In the near future the limiting noise in these experiments is due to the fluctuations in electromagnetic propagation through the solar plasma at microwave frequencies. The noise power of the propagation fluctuations varies as $1/\omega^4$ where ω is the electromagnetic frequency. The strong frequency dependence of the propagation noise as well as the smaller beam sizes possible with shorter wavelengths for fixed antenna dimensions argues for the eventual use of optical frequencies in space antenna systems even though the space qualified hardware is not now available.

Several optical space antennas have been suggested (Weiss et al., 1976). One design consists of masses loosely suspended to a large cubical frame. The masses mounted at the corners of the cube are the mirror mounts for 12 interferometers, a pair on each face. Peter Bender suggested the use of a frame to solve the problem of launching and maintaining the spacing of the antenna masses. The suspension of the masses relative to the frame is critical, they must isolate the masses from the thermal noise of the frame itself. At present cubical frames as large as 10 km on a side are being contemplated by the large space structures engineering group at NASA. For frames of this size the estimated stochastic forces of the solar radiation pressure, solar wind, interplanetary dust and cosmic ray background (Fig. 4) dominate over the thermal noise of the frame at $f < 1/100$ Hz.

Although the stochastic forces of the solar radiation pressure, solar wind and interplanetary dust are orders of magnitude larger than the cosmic ray background, they can in principle be shielded, while the penetrating cosmic ray proton flux cannot.

Assuming a 100 pass interferometer using a 1 watt laser and 10^6 gm antenna masses driven by the stochastic forces of the cosmic ray protons ($F^2(f) \sim 5 \times 10^{-25}$ dynes^2/Hz); the minimum detectable gravitational strain spectral density is

$$\left(\frac{\Delta\ell}{\ell}(f)\right)^2 \begin{cases} 10^{-44}\ \mathrm{Hz}^{-1} & 10^{-2} < f < 10^{2}\ \mathrm{Hz} \\ \dfrac{10^{-52}}{f^4}\ \mathrm{Hz}^{-1} & f < 10^{-2}\ \mathrm{Hz} \end{cases}$$

In this particular example a possible fundamental noise not considered in the general treatment of electromagnetically coupled antennas may become important, namely the recoil noise of the masses in the laser field. If the light beams in the two interferometer arms can be thought of as being uncorrelated then there is an additional stochastic force on the masses given by

$$F^2(f) = \frac{4\,b^2\,h\,P}{\lambda c} \tag{41}$$

which in this example is several 100 times larger than the proton noise.

There is a lively but unpublished controversy whether this

noise term actually exists in a balanced interferometer system.

A second design uses free flying masses in drag free shielded satellites. The control of the position of the interferometer arms against the gravitational gradient forces becomes a major concern.

The near equality of the interferometer arm path lengths must be maintained so that the phase noise of the laser remains smaller than shot noise. Using 1 watt as a typical laser power, the path length difference must be maintained to 10^{-2} cm if the laser linewidth is 100 kHz, typical of frequency unstabilized lasers. In any engineering planning of such an antenna the trade off between short term frequency control of the laser and precision station keeping of the masses will become a major factor in the design.

Baselines as large as 1000 km for a single pass interferometer may still be practical. Again assuming 10^6 gm masses and only the proton flux, the minimum detectable gravitational strain spectral density is

$$\left(\frac{\Delta \ell}{\ell}(f)\right)^2 \begin{cases} 10^{-44}\ \mathrm{Hz}^{-1} & 10^{-3} < f < 10^{2}\ \mathrm{Hz} \\ \dfrac{10^{-56}}{f^4}\ \mathrm{Hz}^{-1} & f < 10^{-3}\ \mathrm{Hz} \end{cases}$$

To further reduce the effect of the stochastic noise forces at low frequencies even larger baselines are necessary. Optical Doppler ranging (optical heterodyning) becomes an attractive option. Such a scheme could involve 3 drag free spacecraft placed at the earth-moon or earth-sun Lagrange points (Bender et al., 1979). The central spacecraft has a laser which transmits simultaneously to the other two spacecrafts equipped with laser transponders. The returned signals are beat against each other. The dominant noise is due to the finite power returned. Applying equation (37), the transducer noise is

$$\left(\frac{\Delta \ell}{\ell}(f)\right)^2 = \frac{\lambda^4}{d_1^2\, d_2^2}\left[\frac{h\nu}{\eta P_2} + \frac{h\nu}{\eta P_1}\right] \qquad (42)$$

where λ is the laser wavelength, d_1 and d_2 are the diameters of the telescope mirrors on the central spacecraft and the transponding spacecraft, P_1 and P_2 are the transmitted powers of the laser on the central spacecraft and the transponder. Taking:

$\lambda = 5 \times 10^{-5}$ cm, $d_1 = 2$ meters $d_2 = 1$ meter, P_1 and $P_2 \sim 1$ watt, $\eta = 1/2$, the transducer noise becomes

$$\left(\frac{\Delta \ell}{\ell}(f)\right)^2 \sim 10^{-44}\ \mathrm{Hz}^{-1}$$

Assuming the laser frequency width, δ, is 10^5 Hz the path lengths to the transponding spacecraft must be equal to 1/100 km so that the laser phase noise remains less than the amplitude noise. The estimated proton noise is of no consequence at frequencies larger than 1/day for 10^6 gm masses placed at separations as large as the earth-sun Lagrange points.

SUMMARY OF ELECTROMAGNETIC ANTENNAS

In the next few years several short baseline interferometers will come into operation. The estimated sensitivities of these antennas for detecting impulsive sources will be comparable with the acoustic resonators being developed concurrently, but the intereferometric antennas will have broader bandwidths. These antennas will be able to set new upper limits on the spectral density of gravitational radiation from continuous and periodic sources but not at a level which is expected to be astrophysically interesting (there may always be surprises). The development of these antennas is a means of testing the noise models and is an essential step for the planning of larger baseline systems both on the ground and in space.

A thorough analysis of large baseline systems on the ground and in space has not been carried out. Current thinking is that large baseline antennas on the ground using high power lasers hold the promise of astrophysically interesting sensitivities in the frequency range above 100 Hz with the application of straightforward engineering. The most uncertain factor is the low frequency performance, f smaller than 100 Hz, which depends on the success of ground noise isolation systems that can be more elaborate than in the example sited; for example multiple suspension systems and active isolation systems. The geophysical noise from atmospheric density fluctuations and both man made and naturally occurring gravitational gradients has not been studied sufficiently.

Electromagnetically coupled antennas in space are at present the best strategy for measuring the gravitational wave spectrum at periods longer than a hundred seconds. In the short term, multi frequency microwave Doppler ranging experiments are the only possible candidates with present day space technology. A detailed analysis of Doppler ranging systems applied to the detection of gravitational wave bursts is needed.

Optical systems in space are more promising, however, the necessary space technology is far from being developed. The noise calculations need considerably more work than is presented in this article, in particular the engineering questions associated with the shielding and the effect of gravitational gradients due to the planets as well as smaller space objects have to be studied.

Figures 5, 6 and 7 are a summary of the status and prospects of the performance of various antenna systems as applied to the search for impulsive, periodic and stochastic gravitational wave sources. The projections in the figures should be viewed in the spirit of educated guesses. The quantity h, the dimensionless gravitational wave amplitude, is twice the strain. All the figures are in terms of equivalent rms noise using the filtering techniques presented in the text. The projections of antenna performance for the detection of periodic and stochastic sources can be directly compared with the source predictions of the second round table of this conference. In estimating the detection probability of an impulsive source, using the results of Fig. 5, the effect of the pulse duty cycle must be included. If the noise in the antenna has a Gaussian distribution, the minimum impulsive h detectable becomes $h(\mathrm{rms}) \times \ell n\,(Rt_p)^{-1}$, where R is the event rate and t_p the pulse length.

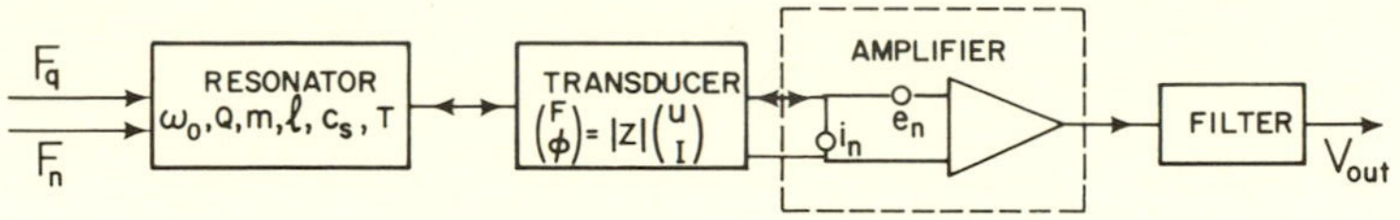

Figure 1. Basic components of an acoustically coupled antenna.

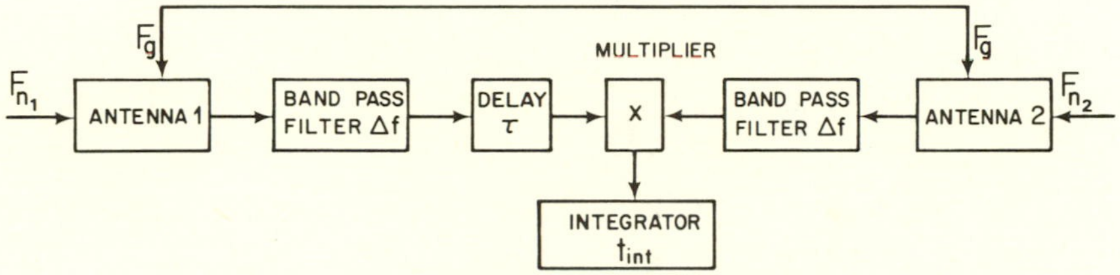

Figure 2. Schematic of a cross correlation experiment.

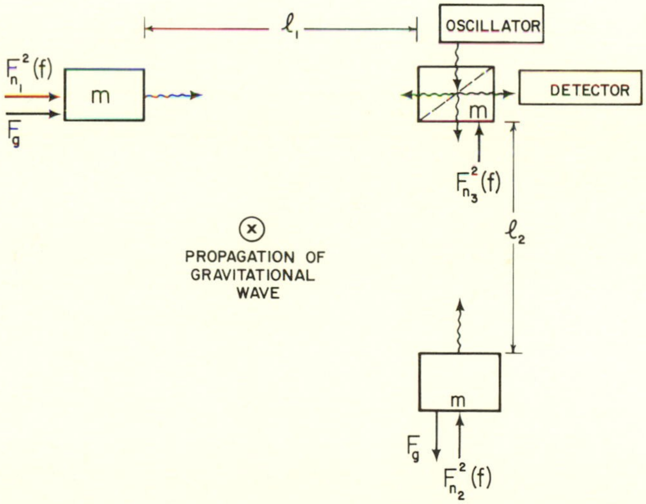

Figure 3. Schematic diagram of an element of an electromagnetically coupled antenna.

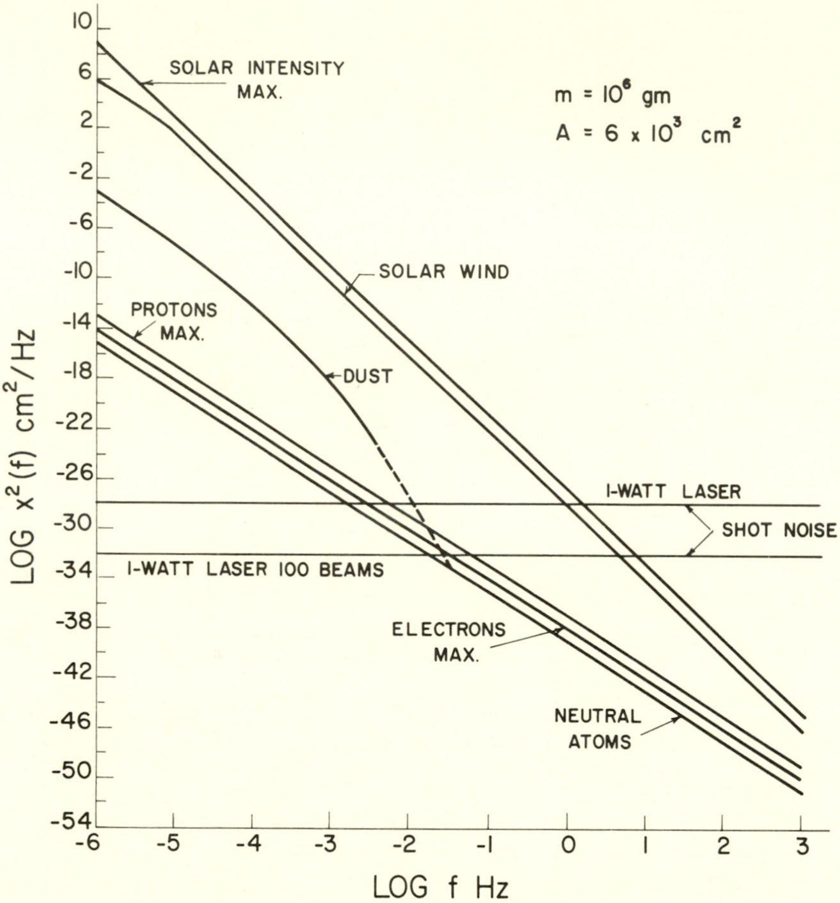

Figure 4. Estimates of displacement spectral density due to various stochastic forces in a space environment.

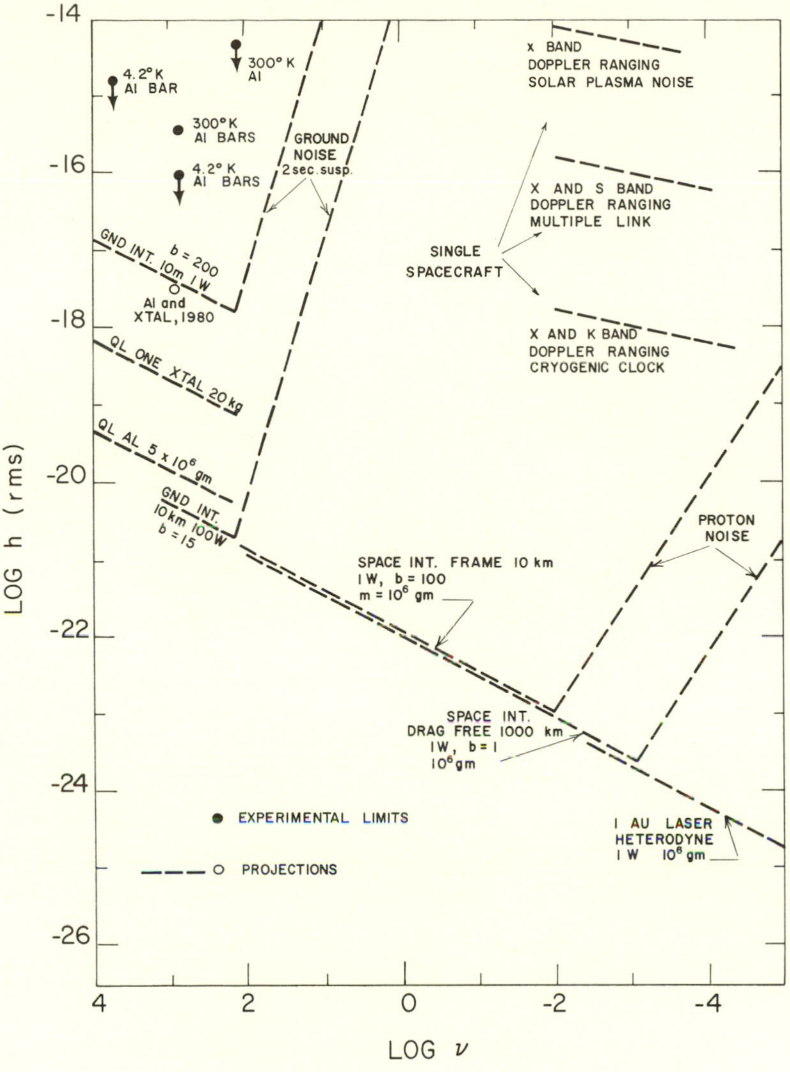

Figure 5. Experimental limits and projections for antenna performance in detecting impulsive sources. The figure presents rms values. To convert to detection probability rms values must be multiplied by the logarithmic duty cycle factor described in the text.

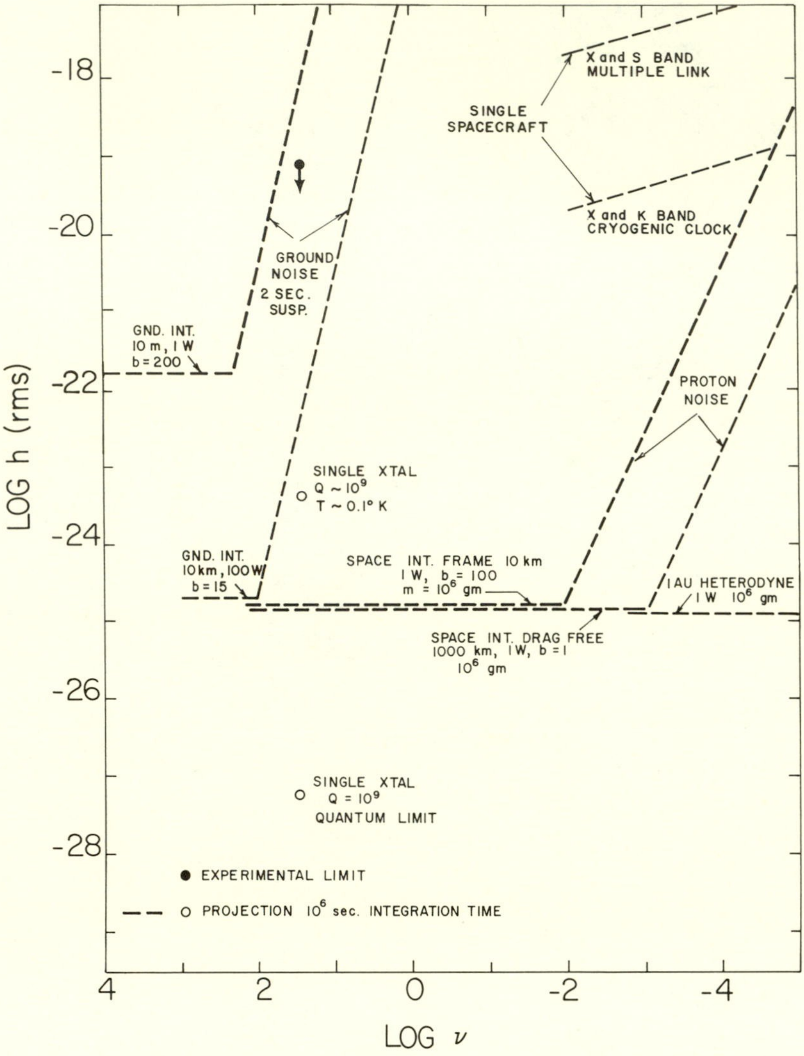

Figure 6. Experimental limits and projections for antenna performance in detecting periodic sources with an integration time of 10^6 seconds.

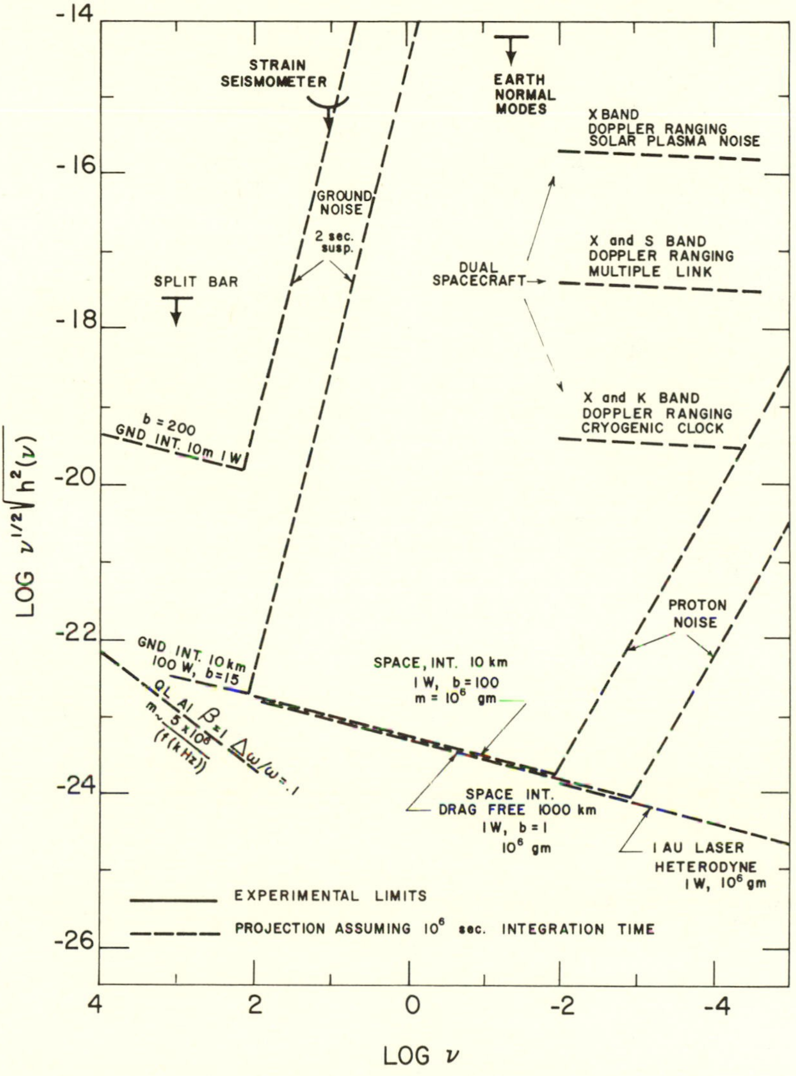

Figure 7. Experimental limits and projections for antenna performance in detecting a stochastic background using a bandwidth equal to the frequency and an integration time of 10^6 seconds.

REFERENCES

Bendat, J.S. (1958). Principles and Applications of Random Noise Theory. John Wiley, New York.

Bender, P.L., Drever, R.W.P., Faller, J.E. and Weiss, R. (1979). Detection Methods for Gravitational Waves from Rotating Binaries. To be published.

Boughn, S.P., Fairbank, W.M., Giffard, R.P., Hollenhorst, J.N., McAshan, M.S., Paik, H.J. and Taber, R.C., (1977). Observation of Mechanical Nyquist Noise in a Cryogenic Gravitational-Wave-Antenna. Phys. Rev. Let., 38, p. 454.

Douglass, D.H. and Braginsky, V.B., (1979). Gravitational Radiation Experiments. In Einstein Centenary Volume, ed. Hawking, S.W. and Israel, W., Cambridge University Press, Cambridge.

Drever, R.W.P., Hough, J., Bland, R. and Lessnoff, G.W. (1973). Search for Short Bursts of Gravitational Radiation. Nature, 246, p. 340.

Forward, R.L. (1978). Wideband Laser-Interferometer Gravitational-Radiation Experiment. Phys. Rev. D17, p. 379.

Gibbons, G.W. and Hawking, S.W. (1971). Theory of the Detection of Short Bursts of Gravitational Radiation. Phys. Rev. D4, p. 2191.

Giffard, R.P. (1976). Ultimate Sensitivity Limit of a Resonant Gravitational Wave Antenna Using a Linear Motion Detector. Phys. Rev. D14, P. 2478.

Heffner, H. (1962). The Fundamental Noise Limit of Linear Amplifiers. Proc. I.R.E. 50, p. 1604.

Hirakawa, H., Tsubono, K. and Fujimoto, M., (1978). Search for Gravitational Radiation from the Crab Pulsar. Phys. Rev. D17, p. 1919.

Hough, J., Pugh, J.R., Bland, R. and Drever, R.W.P., (1975). Search for Continuous Gravitational Radiation. Nature, 254, p. 498.

Hunt, F.V. (1954). Electroacoustics. Harvard Univ. Press, Cambridge, Mass.

Manley, J.M. and Rowe, H.E., (1956). Some General Properties of Non-Linear Elements I. General Energy Relations Proc. I.R.E. 44 p. 904.

Misner, C.W., Thorne, K.S. and Wheeler, J.A., (1973). Gravitation. W.H. Freeman, San Fransisco.

Moss, G.E., Miller, L.R. and Forward, R.L., (1971). Photon Noise Limited Laser Transducer for Gravitational Antenna. Appl. Optics 10, p. 2495.

Rowe, H.E., (1958). Some General Properties of Non-Linear Elements II, Small Signal Theory. Proc. I.R.E. 46, p. 850.

Tyson, J.A., (1973). Null Search for Bursts of Gravitational Radiation. Phys. Rev. Let. 31, p. 326.

Tyson, J.A. and Giffard, R.P. Gravitational Wave Astronomy. In Annual Review of Astronomy and Astrophysics, Vol. 16, 1978. Annual Reviews Inc., Palo Alto.

Weber, J., (1957). Maser Noise Considerations. Phys. Rev. 108, p. 537.

Weber, J., (1969). Evidence for Discovery of Gravitational Radiation. Phys. Rev. Let. 22, p. 1320.

Weiss, R., (1972). Electromagnetically Coupled Broadband Gravitational Antenna. Quart. Progr. Rep., Res. Lab. Elect., M.I.T., 105, p. 54.

Weiss, R. Bender, P.L., Misner, C.W. and Pound, R.V., (1976). Report of the Sub-Panel on Relativity and Gravitation, Management and Operations Working Group for Shuttle Astronomy, NASA.

Wiener, N., (1949). The Extrapolation, Interpolation and Smoothing of Stationary Time Series with Engineering Applications. John Wiley, New York.

GRAVITATIONAL RADIATION DETECTION
WITH SPACECRAFT DOPPLER TRACKING:
LIMITING SENSITIVITIES AND PROSPECTIVE MISSIONS

F.B. Estabrook, R.W. Hellings, H.D. Wahlquist & R.S. Wolff

Jet Propulsion Laboratory
California Institute of Technology
Pasadena, California

I. DOPPLER SHIFTS DUE TO PASSING GRAVITATIONAL RADIATION

At JPL we are investigating gravitational radiation detection schemes which use the superb microwave technology of the NASA Deep Space Net (DSN) (Estabrook & Wahlquist, 1978; Wahlquist, Anderson, Estabrook, & Thorne, 1977; Anderson & Estabrook, 1979). Precision Doppler tracking of interplanetary spacecraft carrying phase-coherent transponders is usually thought of as measurement of velocity, as contrasted with radar ranging, which directly measures distance. But more precisely, Doppler measurement is better understood as yielding the instantaneous phase of a returning signal, relative to that of the signal being sent. The rate of change of this is $\Delta\nu$, the Doppler shift of frequency. In addition to other physical causes, gravitational radiation passing through a Doppler link has a characteristic effect on the observed relative phase, or on $\Delta\nu$.

The rigorous expression for Doppler shift (on one leg) due to a passing plane polarized gravitational wave train h(t) has been shown in general (Estabrook & Wahlquist, 1975; Hellings, 1978) to be

$$\frac{\Delta\nu}{\nu_0} = \left\{ -\frac{1}{2}(1-\cos\theta)h(t) - \cos\theta h(t-[1+\cos\theta]\ell/c) + \frac{1}{2}(1+\cos\theta)h(t-2\ell/c) \right\} \sin 2\phi . \quad (1)$$

Here t is the instant of reception and θ is the angle between the vector from the earth to the spacecraft and the propagation direction of the gravitational wave. The factor $\sin 2\phi$ allows for arbitrary orientation of the leg with respect to the polarization direction of the gravitational wave. This expression applies to all electromagnetically tracked free mass antennas - spacecraft Doppler or laser interferometer (where two expressions like equation (1) would be differenced to give the interferometric response). If h(t) is a pulse of duration τ, then three pulses will be seen in the time series of $\Delta\nu/\nu_o$, each of size $\sim$h. Expression (1) gives their relative times and amplitudes, as functions essentially of one single parameter θ. If h(t) is periodic, the observed amplitude of the $\Delta\nu/\nu$ signal will vary as the spacecraft trajectory changes ℓ and varies the relative phase of the three contributions. If h(t) is stochastic, equation (1) provides for special autocorrelations of the signal. In any case, this very characteristic and unique "response function" is expected to allow the use of several sophisticated pattern recognition techniques in the data analysis.

In the limit that τ approaches $2\ell/c$ from below, the three contributions in (1) blend and cancel to first order. The round trip light time (which laser interferometer experimenters call the "storage time") thus divides these systems into two wavelength regimes. In the short wavelength case, with $\tau<2\ell/c$, the three contributions are separable (a computer study indicates $\tau\lesssim 2\ell/3c$ as a practical limit for resolving pulses), and the data analysis methods mentioned above may be applied. In this regime it is also clear from equation (1) that the system's sensitivity to a given h is not increased by increasing ℓ. A longer roundtrip light time only increases the wavelength for which the short wavelength regime applies. In the long wavelength case, with $\tau>2\ell/c$, the first order cancellation forces us to work with the second order residuum of equation (1). Assume h(t) is slowly varying and expand to give

$$h(t-1+\cos\theta\ \ell/c) \sim h(t)-(1+\cos\theta)(\ell/c)\dot{h}(t), \tag{2}$$

$$h(t-2\ell/c) \sim h(t)-2(\ell/c)\dot{h}(t). \tag{3}$$

Then we find

$$\Delta\nu/\nu_o \sim -(\ell/c)\dot{h}(t)\sin^2\theta\ \sin2\phi. \quad \tau>>2\ell/c. \tag{4}$$

With two legs, equation (4) would be differenced to give

$$\frac{\Delta\nu}{\nu_o} \sim \frac{\ell}{c}\ \dot{h}(t)\left[\sin^2\theta'\sin2\phi' - \sin^2\theta\sin2\phi\right]. \tag{5}$$

This is essentially the response function used by R. Weiss (Weiss, 1978) in his summary discussion at this conference. In this long wavelength regime an increase in ℓ does increase the sensitivity of the antenna, but the overall signal is always down by a factor $\dot{h}/h$ compared to (1) and the three pulse nature is lost. Both spacecraft Doppler detectors and laboratory laser interferometers may operate in both regimes, though, of course, the wavelength division is at laboratory distances for laser interferometers and at solar system distances for spacecraft.

One final note: the addition of an onboard clock and the use of both one-way and two-way data from both the earth and the spacecraft (Vessot & Levine, 1978) offers a means to perform time-correlations among ten manifestations of the single gravity wave pulse, rather than the three available from the single round trip link. This scheme therefore will further improve the prospects for pattern recognition as well as provide a measure of statistical independence via the second clock and the new paths through spacetime.

II. SPECTRAL CHARACTERIZATION OF SENSITIVITY LIMITS

The Doppler tracking system is probably most easily understood by reference to the accompanying figure (figure 1).

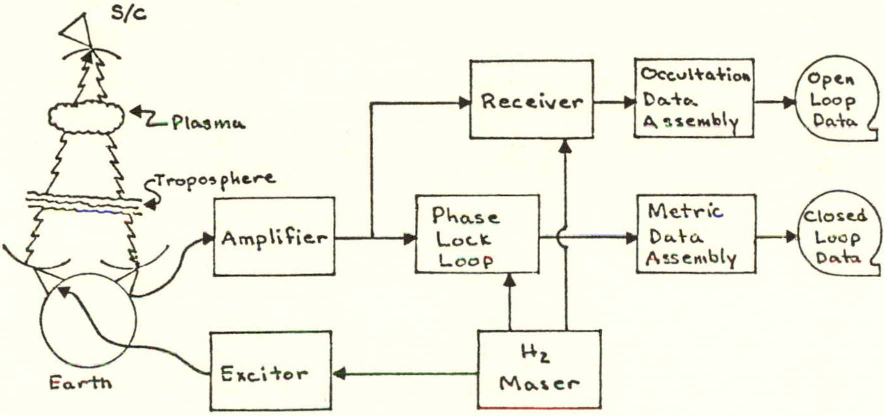

Figure 1.

The heart of the system is the hydrogen maser clock which both controls the frequency ν of the low noise excitor and provides the reference frequency for producing the Doppler tone $\Delta\nu = \nu' - \nu$, where ν' is the Doppler shifted frequency received by either the open loop or closed loop receivers. In the open loop data chain, the amplitude of this pure Doppler signal is sampled at regular times and quantified by the Occultation Data Assembly (ODA) to produce a tape record of voltage vs. time. This tape can then be postprocessed to reconstitute the Doppler signal and determine the frequency. It is a low noise process, but it is very time consuming and uses a lot of magnetic tape. In the closed loop data chain, a voltage controlled oscillator (VCO)in a phase locked loop tracks the incoming signal, delivering a Doppler tone to the Metric Data Assembly (MDA) which counts zero-crossings during some preset integration time, τ, and resolves the last partial cycle to a fraction of a radian. The accumulated phase measurements are recorded as a function of time on the closed loop data tapes. Much less costly than open loop data (this is, in fact, the normal spacecraft tracking mode), this closed loop process introduces white phase noise via the quantization associated with the resolver.

Any physical source of Doppler fluctuation, signal or noise, may be described as a time series $\Delta\nu(t_n)$, $n = 1,2,\ldots$. Especially when such fluctuations are Gaussian they can be sufficiently characterized by their spectral densities. Although our previous work has generally used the Allan variance in discussing instrumental limitations on sensitivity, we feel that plots of spectral density vs. f would also be useful, since similar spectral information is being developed by the laser experimenters. This approach is especially appropriate if gravitational radiation of known spectral content is expected. So, defining $y = \Delta\nu/\nu$ as the variable of interest, we measure the noise by giving the noise power spectral density, $S_y = S_y(f)$, where f is the Fourier spectral frequency and S_y is the power $\langle(\Delta\nu/\nu)^2\rangle$ per unit spectral bandwidth. The best current estimates of S_y from the various elements of the system are as follows.

1. Plasma Scintillation: Recent data from the Viking orbiter at solar opposition give (Armstrong, Woo & Estabrook, 1979):

$$S_y = 2 \times 10^{-26} f^{-0.6} \quad \text{(S-band. Measured)}$$

$$S_y = 2 \times 10^{-28} f^{-0.6} \quad \text{(X-band. Implied)}$$

Since a plasma is a dispersive medium, combined data at two frequencies allow for a calibration of the plasma effect. This does not appear to help very much when the dual frequencies are only used on downlink (though it does help to identify periods of low plasma noise), but a round trip dual band signal should reduce the plasma noise to $S_y \approx 10^{-30} f^{-0.6}$.

2. Troposphere Scintillation: The best available measurements of phase fluctuations in our fourier spectral region, using radio signals of 9 GHz through 33 GHz, have been reported by Thompson, et al, (Thompson, et al, 1975) along a 65 km path over the ocean and using a crystal controlled oscillator for timekeeping. Correcting to a 6.5 km effective path looking upward from Goldstone, their results would seem to imply

$$S_y = 1 \times 10^{-28} f^{-0.6}.$$

It is not clear how to further correct these results to apply to nighttime on the desert. Perhaps the best we can say at present is that troposphere scintillation was not observed in the Viking data and was therefore well below the S-band plasma scintillation given above.

3. Receiver Noise: Antenna, amplifiers, and receivers contribute phase noise given generally by

$$S_y = \frac{kT}{P\nu^2} f^2,$$

where T is the operating temperature, P is the received power, and ν is the radio frequency. At current typical values of $T = 20^{o}K$, $P = 2 \times 10^{-16}$ watts (corresponding to 40 watts transmitted at S-band and 10 at X-band),

$$S_y = 3 \times 10^{-25} f^2 \text{ at S-band}$$

$$S_y = 2 \times 10^{-26} f^2 \text{ at X-band}$$

4. MDA quantization error: The maximum size of the quantization error is $\Delta\nu/\nu = rF/\nu\tau$, where r is the fineness of the cycle resolution in seconds, F is the frequency of the signal delivered to the MDA, and τ is the integration time during which the fluctuation $\Delta\nu$ accrues. This error results in white phase noise, where spectral density has a high frequency cutoff, f_h, given by the reciprocal of the Doppler count time. Since the error is independent of f_h, the spectral density will depend inversely on it, giving

$$S_y = \frac{r^2F^2}{4\nu^2 f_h} f^2 .$$

In the current Doppler resolver, r is 2×10^{-9} sec and F is approximately 10^6Hz, a bias frequency inserted into the system to eliminate sign ambiguity at near-zero Doppler frequency. The resulting spectral density is

$$S_y \sim 3 \times 10^{-25} f^2/f_h \text{ at S-band}$$

$$S_y \sim 2 \times 10^{-26} f^2/f_h \text{ at X-band.}$$

This noise level has not yet been seen in actual system performance, however, and may in fact be dominated by other phase noise sources in the closed loop circuitry.

5. ODA quantization error: The maximum error here is given by

$$\frac{\Delta\nu}{\nu} = \left[\pi \sqrt{N} 2^{b-1} \nu\tau\right]^{-1} ,$$

where N is the number of samples in a Doppler count interval

($N = n/f_h$, where n is the sampling rate) and b is the number of bits of digitalization of the signal. The dependence on f_h now leads to a spectral density independent of bandwidth, and again we have white phase noise:

$$S_y = (\pi^2 n \nu^2 2^{2b-2})^{-1} f^2$$

The existing ODA, with 8-bit quantization and a sample rate of a few kiloherz, gives

$$S_y \sim 10^{-27} f^2 \text{ at S-band}$$

$$S_y \sim 10^{-28} f^2 \text{ at X-band}$$

6. Clock Jitter: The spectral density of clock jitter in the latest Smithsonian hydrogen maser clocks may easily be derived from R. Vessot's recent report (Vessot, Levine & Mattison, 1978). It is found to be

$$S_y = 10^{-27} + 10^{-26} f^2.$$

The report also indicates a slow drift which becomes prominent at frequencies below 3×10^{-4} hz, but points out that this appears to have been a result of temperature instability during the testing process. (See Hellings in Discussion Section 1 - this volume)

Two sources of noise we have not considered here are buffeting of earth and spacecraft. These do not appear to be stochastic processes and are difficult to characterize with a spectral density. It should be pointed out again here, however, that the Viking study mentioned above saw no significant noise due to buffeting above a level of 3×10^{-14} in $\Delta\nu/\nu$ and within a spectral frequency range of .03 hz to 2×10^{-3} hz.

The preceding noise sources have been collected into figure 2. The X-band value has been used throughout, and we take $f_h = 1$ hz.

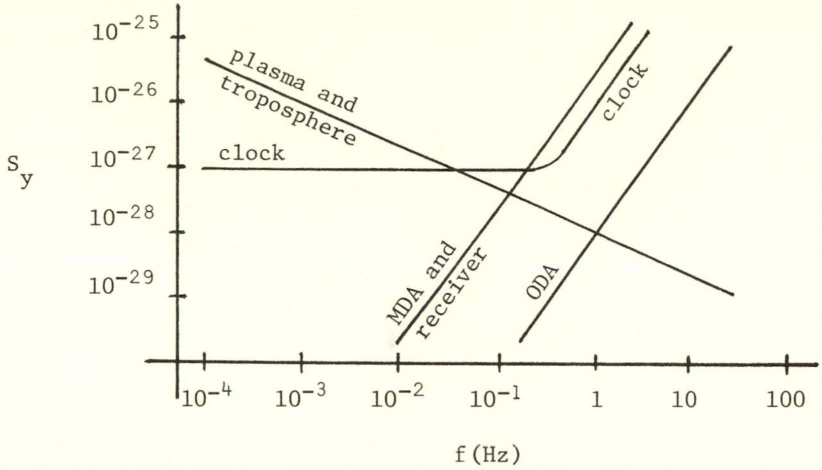

Figure 2

III. INTERPLANETARY MISSIONS AVAILABLE FOR GRAVITATIONAL RADIATION EXPERIMENTS

During the next decade NASA will be flying at least three, and possibly five, interplanetary missions which provide some capability of performing Doppler gravitational radiation experiments. The sensitivity and frequency ranges that each spacecraft can realize will depend on the telemetry characteristics of the mission and the earth-spacecraft distance at the time of measurement. In Tables I and II we briefly summarize each of the missions, its status and planned telemetry system, the most favorable times during each mission for the experiments (this time is defined to be when the sun-earth-spacecraft angle is $>150^{o}$, minimizing interference by solar wind plasma fluctuations), and the expected sensitivity and frequency ranges for each experiment period.

The missions outlined are either current flight projects or NASA study projects planned for launch within the next decade. We therefore do not discuss what might, solely for our purposes, seem to be the most obvious experiment: a multiple spacecraft mission consisting solely of several good X-band transponders, launched to large distances and having many overlapping tracking periods while in the antisolar direction, monitored simultaneously by independent ground stations, flying for years as a gravitational wave observatory.

TABLE I. MISSION CHARACTERISTICS

MISSION [1] (STATUS)	MISSION PROFILE	TELEMETRY [2]
VOYAGER (A)	Two 3-axis stabilized spacecraft (S/C) (V_1, V_2) on encounter trajectories to Jupiter, Saturn, Uranus (V_2) and eventually out of the Solar System. Launch dates: V_1 9/5/77; V_2 8/20/77 Jupiter Encounter: V_1 3/5/79; V_2 7/9/79 Primary mission objectives: to conduct exploratory investigations of planetary systems of Jupiter and Saturn, and interplanetary medium out to Saturn.	S ↑ ↓ X ↓
GALILEO (P)	Single (dual spinner type) S/C. Jupiter Orbiter with atmospheric probe. Primary mission objectives: 1. Determination of chemical composition and structure of Jovian atmosphere 2. Determination of the chemical composition and structure of the Galilean Satellites 3. Determination of the structure and dynamics of the Jovian Magnetosphere Launch: 1/4/82 Jupiter Orbit Insertion (JOI): 6/25/85	S ↑ ↓ X ↓ (↑ ?)
INTERNATIONAL SOLAR POLAR (ISP) (P)	Dual Launch, 1 NASA (dual spinner) S/C and 1 ESA (spinning) S/C to explore the 3-D space around the sun by flying two S/C (w. gravity assistance by Jupiter) out of the ecliptic and over the poles of the sun. Primary mission objectives are to get 3-D structure and time variation of solar wind, solar magnetic fields, coronal holes, photospheric, chromospheric and coronal features and sun spots. Launch: 2/3/83 (NASA S/C) First Polar Pass: 11/8/86 Jupiter Arrival: 5/25/84 Second Polar Pass: 7/29/87	S ↑ ↓ [3] X ↓ (↑ ?)
HALLEY-TEMPEL 2 COMET/ION DRIVE (CID) (S)	Single 3-axis stabilized S/C (Rendezvous S/C) with (spinning) Probe (for fly-through of Halley coma) Primary mission objectives: 1. Determine the chemical nature and physical structure and evolution of cometary nuclei 2. Characterize the evolving chemical and physical nature of the atmosphere, ionosphere, and dust envelope 3. Study the interaction of cometary atmospheres with the solar wind Launch: 8/1/85 Halley Encounter: 11/28/85 Tempel-2 Rendezvous: 7/18/88	S↑ ↓ X↑ ↓
SOLAR PROBE (S)	Single Drag-Free S/C. Primary mission to obtain data on solar wind, energetic particles, structure of solar magnetic field, and the composition, origin and evolution of interplanetary dust to within ∿3 solar radii of sun surface. Accurate orbit determination of the solar probe could yield a measurement of the quadrupole moment of the solar grav. field and perhaps higher accuracy tests of G.R. Launch: 1986 Arrive sun: 1990	S ↑ ↓ X ↑ ↓ onboard H-maser and independent downlink (?)

TABLE II. OBSERVING OPPORTUNITIES

MISSION	FAVORABLE EPOCHS(4)	SEP ANGLE(5) DURING EPOCH	RTLT(6) (SEC)	LOW FREQUENCY LIMIT (Hz)	EXPECTED(7) SENSITIVITY	MULTI-S/C OPPORTUNITIES
VOYAGER	V_1					
	1/2/79 - 2/20/79	150°-179°	3993-4300	3×10^{-4}	3×10^{-14}	none
	2/10/80 - 4/5/80	150°-178°	6287-6667	1.5×10^{-4}		
	V_2					
	1/5/79 - 2/23/79	150°-178°	3600-3840	3×10^{-4}		
	2/3/80 - 3/30/80	150°-179°	5287-5653	1.3×10^{-4}		
	2/28/81 - 4/23/81	150°-178°	7533-7880	1.2×10^{-4}		
GALILEO	3/5/82 - 4/24/82	150°-176°	1220-1287	7×10^{-4}	3×10^{-14} (without X↑)	with Solar Polar during 6/19-7/21/84
	6/1/83 - 7/21/83	150°-180°	3574-3680	5.5×10^{-4}	3×10^{-15} (with X↑)	
	6/19/84 - 7/11/84	150°-176°	3866-3986	5×10^{-4}		
SOLAR POLAR	4/1/83 - 5/24/83	150°-178°	333- 927	3×10^{-3}	3×10^{-14} (without X↑)	two S/C; also with Galileo during 6/19 - 7/21/84
	~6/2/84 - 7/21/84	150°-177°	4200-4307	2.4×10^{-4}	3×10^{-15} (with X↑)	
	6/19/85 - 7/5/85	150°-152°	3607-3660	2.8×10^{-4}		
HALLEY/TEMPEL-2	~1/22/87 - 3/3/87	150°-170°	2133-2233	4.7×10^{-4}	3×10^{-15}	possibly with Solar Probe
	~4/1/88 - 5/21/88	150°-160°	767-1067	1.3×10^{-3}		
SOLAR PROBE	TBD (depends on Launch date and propulsion system)	TBD	3333-4000	$\sim 3 \times 10^{-4}$	$\lesssim 10^{-16}$ (with on-board timekeeping and multiple links)	possibly with CID

NOTES

1. The five missions are listed according to status where: A ≡ Active (spacecraft currently flying); P ≡ Project (money has been funded and the spacecraft and science instruments are being designed and built); S ≡ Study phase (only study money has been funded; the mission is still being defined; none of the science instruments have been selected - in this category the Solar Probe mission is much less well defined than the CID mission, hence the entries TBD (to be determined).

2. S and X refer to the telemetry radio bands ($\sim$2.2 GHz and $\sim$ 8 GHz, respectively); ↑ ≡ uplink (from tracking station to spacecraft) and ↓ ≡ downlink (from spacecraft to tracking station). Thus, for example, Voyager has S- and X-band downlink, but only S-band uplink. A "?" beside an arrow indicates that that particular link has not yet been decided upon.

3. One ISP S/C is to be flown by the European Space Agency (ESA). It is presently planned to have only S↑X↓.

4. Defined by opposition (sun-earth-spacecraft angle at 180° $\pm$ 30°).

5. SEP ≡ sun-earth-spacecraft angle.

6. RTLT (round-trip-light-time).

7. Allan variance of $y = \Delta\nu/\nu_0$, the fractional Doppler shift. For flicker y noise, $S_y \sim f^{-1}$, Allan variance is independent of integration time. It is a conservative measure of variance appropriate when low frequency drifts are present and must be filtered. It corresponds to laboratory practice adopted by the precision timekeeping community for intercomparison of standards. Since tropospheric and solar plasma fluctuations have a similar spectral density of y, Allan variance is also a fair index of sensitivity limits imposed by propagation path effects. To precisely specify gravitational wave detection sensitivity one would of course need to know its spectral content. If, as is most obviously the case for c.w. sources, it differs significantly from flicker y, the detectable level of h may be considerably lower than the quoted Allan variance.

IV. CONCLUSIONS

The most exciting prospect in all of this is that a gravity wave experiment at wavelengths of 10^2 to 10^4 seconds now seems possible at a $\Delta\nu/\nu$ level of a few parts in 10^{15}. A cosmic closure density of stochastic background gravity waves in this bandwidth would produce a $\Delta\nu/\nu$ at just this level, as would the supermassive violent events envisioned by Thorne & Braginsky (Thorne & Braginsky,

1976). The current Voyager mission gives us an opportunity to test the system and learn how to analyze the data. The upcoming Galileo and Solar Polar missions, given the plasma noise improvements with X-band uplink, then bring us for the first time to the threshold of a reasonable possibility of detection. It must be emphasized, however, how critical the X-band uplink is on these missions. Without it, we spend the next ten or more years examining data from Voyager-like missions, an order of magnitude away from where we really hope to have a chance of detecting gravity waves. Finally, it is perhaps not too optimistic to suggest that the prospect of dual band round trip and a possible onboard clock on Solar Probe, along with third generation laboratory gravity wave detectors, give hope that the next decade may see the beginning of the age of gravity wave astronomy.

This presents the results of one phase of research carried out at the Jet Propulsion Laboratory, California Institute of Technology, under contract no. NAS7-100, sponsored by the National Aeronautics and Space Administration.

REFERENCES

Anderson, J.D. & Estabrook, F.B. (1979). Application of DSN Spacecraft Tracking Technology and Experimental Gravitation. J. Spacecraft and Rockets, in press.

Armstrong, J.W., Woo, R. & Estabrook, F.B. (1979). Interplanetary Phase Scintillation and the Search for Very Low Frequency Gravitational Radiation. Subm. to Ap. J.

Estabrook, F.B. & Wahlquist, H.D. (1975). Response of Doppler Spacecraft Tracking to Gravitational Radiation. GRG 6, pp. 439-447.

Estabrook, F.B. & Wahlquist, H.D. (1978). Prospects for Detection of Gravitational Radiation by Simultaneous Doppler Tracking of Several Spacecraft. Acta Astronautica 5, pp. 5-7.

Hellings, R.W. (1978). Testing Relativistic Theories of Gravity with Spacecraft Doppler Gravity Wave Detection. Phys. Rev. D. 17, p. 3158.

Thompson, et al (1975). Phase and Amplitude Scintillations in the 10 to 40 GHz Band. IEEE Transactions on Antennas and Propagation AP-23, p. 792.

Thorne, K.S. & Braginsky, V.B. (1976). Gravitational Wave Bursts from the Nucleii of Distant Galaxies and Quasars: Proposal for Detection Using Doppler Tracking of Interplanetary Spacecraft. Ap. J. 204, pp. L1-L6.

REFERENCES (Cont.)

Vessot, R.F.C., Levine, M.W. & Mattison, E.M. (1978a). Comparison of Theoretical and Observed Hydrogen Maser Stability Limitation Due to Thermal Noise and The Prospect for Improvement by Low Temperature Operation. Procedings of the Ninth Annual Precise Time and Time Interval Applications and Planning Meeting, NASA-Tech. Memo 78104, pp. 549-569.

Vessot, R.F.C. & Levine, M.W. (1978b). A Time-Correlated Four-Link Doppler Tracking System. A Close-Up of The Sun, pp. 457-997.

Wahlquist, H.D., Anderson, J.D., Estabrook, F.B. & Thorne, K.S. (1977). Recent JPL Work on Gravity Wave Detection and Solar System Relativity Experiments. Simposio internazionale sulla Gravitazione Sperimentale (Experimental Gravitation), Atti dei Convegni Lincei 34, pp. 335-350.

Weiss, R. (1978). Article in this volume.

THE QUANTUM LIMIT FOR GRAVITATIONAL-WAVE DETECTORS AND METHODS OF CIRCUMVENTING IT

Kip S. Thorne, Carlton M. Caves, Vernon D. Sandberg, and Mark Zimmermann

California Institute of Technology, Pasadena

and

Ronald W. P. Drever

Glasgow University, Glasgow, Scotland

1. HISTORICAL OVERVIEW

This lecture describes a serious obstacle, looming up about five years from now, in the arduous journey toward a 10^{-21} gravitational-wave-detection sensitivity. That obstacle, called the "quantum limit," was first recognized by Vladimir Braginsky (1970, 1977). The seriousness of the obstacle was demonstrated rigorously by Robin Giffard (1976). In brief, the obstacle is this: The Heisenberg uncertainty principle prevents one from monitoring the complex amplitude, $X = X_1 + iX_2$, of a mechanical oscillator more accurately than

$$|\Delta X|_{QL} = \left(\frac{\hbar}{m\omega}\right)^{\frac{1}{2}} \approx (3\times 10^{-19}\ \mathrm{cm}) \left(\frac{1\ \mathrm{ton}}{m}\right)^{\frac{1}{2}} \left(\frac{10^4\ \mathrm{s}^{-1}}{\omega}\right)^{\frac{1}{2}} . \tag{1}$$

Here m is the mass of the oscillator and ω is its angular frequency. This suggests that there is no hope to measure, with a Weber-type bar of length ℓ, gravitational waves stronger than

$$h_{QL} \approx 10\,\frac{|\Delta X|_{QL}}{\ell} \approx (3\times 10^{-20}) \left(\frac{1\ \mathrm{ton}}{m}\right)^{\frac{1}{2}} \left(\frac{10^4\ \mathrm{s}^{-1}}{\omega}\right)^{\frac{1}{2}} \left(\frac{100\ \mathrm{cm}}{\ell}\right) . \tag{2}$$

(Throughout this lecture $\hbar$ is Planck's constant divided by 2π; h is gravitational-wave strength.)

That in principle one can circumvent this quantum limit on h was shown by Braginsky & Vorontsov (1974). They proposed circumvention by probing the oscillator with "quantum-counting techniques." Although their original design for a quantum-counting sensor was flawed (Unruh 1977, 1978; Braginsky, Vorontsov & Khalili 1977), designs that are unflawed have been invented by

Unruh (1977, 1978) and by Braginsky et al. (1977)

An alternative method of circumventing the quantum limit (2) was discovered by the authors of this lecture (Thorne, Drever, Caves, Zimmermann & Sandberg 1978; cited henceforth as TDCZS). This method, which we call "back-action evasion," looks easier to implement in practice than quantum counting. A special variant of back-action evasion, called "the stroboscopic technique" was discovered independently by Braginsky, Vorontsov & Khalili (1978) in Moscow, and by our group (TDCZS).

Techniques such as quantum counting and back-action evasion, which circumvent the quantum limit (2), are sometimes called "quantum nondemolition techniques" because they attempt to monitor the oscillator without "demolishing" (i.e. perturbing) its quantum mechanical state. A general theory of quantum nondemolition techniques has been developed by Unruh (1979) and by Caves et al. (1979).

In section 2 of this lecture we elucidate the origin of the quantum limits (1) and (2) in a manner (due independently to Kafka 1977 and TDCZS) that suggests immediately how to circumvent them. In section 3 we show how, in real experiments, the quantum limits are replaced by "amplifier limits" which are even more serious. In section 4 we discuss the currently standard method of monitoring an oscillator ("amplitude-and-phase method"; i.e. method which measures X_1 and X_2 with equal precision); and we elucidate the way in which this method is constrained by the amplifier limit. In sections 5-7 we discuss specific quantum nondemolition techniques — quantum counting in section 5; and back-action evasion in sections 6 and 7. In section 8 we discuss the serious problems posed by Nyquist forces in the oscillator.

Elsewhere we shall publish a very long, detailed, and pedagogical discussion of the quantum limit, its generalization to "free-mass" gravitational-wave detectors, and its circumvention (Caves, Drever, Sandberg, Thorne & Zimmermann 1979; cited henceforth as CDSTZ).

2. THIRD-GENERATION BAR DETECTOR: A QUANTUM OSCILLATOR DRIVEN BY A CLASSICAL FORCE

2.1. The Oscillator

Consider the fundamental mode of a Weber-type bar detector for gravitational waves. The oscillations of the fundamental mode can be characterized by the displacement $x(t)$ of the end of the bar from equilibrium. Experimenters like to express x(t) in terms of a complex amplitude $X = X_1 + iX_2$:

$$x(t) = \text{Real}\left[(X_1 + iX_2)\, e^{-i\omega t}\right] = X_1 \cos\omega t + X_2 \sin\omega t \tag{3a}$$

where ω is the eigenfrequency of the fundamental mode. The complex amplitude changes very slowly (timescale $\tau_* \gg 1/\omega$) as a result of weak coupling to other modes of the bar ("Nyquist" forces and frictional forces). We shall ignore these changes until the last section of the lecture, because in principle they are irrelevant to the quantum limit. The complex amplitude also changes in response to a passing gravitational wave or other external force. To define X precisely at times when it is changing, we must specify a second relation in addition to (3a):

$$p(t)/m\omega = \text{Im}\left[(X_1 + iX_2)\, e^{-i\omega t}\right] = -X_1 \sin\omega t + X_2 \cos\omega t. \tag{3b}$$

Here p is the generalized momentum of the fundamental mode of the bar, and m is its mass. (p/m is equal to dx/dt, except when the bar is coupled to a momentum transducer.)

When one passes from classical theory to quantum theory, the position x and momentum p of the fundamental mode become canonically conjugate Hermitian operators $\hat{x}$ and $\hat{p}$; and the two parts of the complex amplitude X_1 and X_2 also become Hermitian operators $\hat{X}_1$ and $\hat{X}_2$. From the commutation relation $[\hat{x},\hat{p}] = i\hbar$ and the relation between $\hat{X}_1,\hat{X}_2$ and $\hat{x},\hat{p}$ (Eqs. 3 with hats inserted), one can easily derive the commutation relation

$$[\hat{X}_1,\hat{X}_2] = i\hbar/m\omega. \tag{4}$$

From this follows the Heisenberg uncertainty relation

$$\begin{aligned}\Delta X_1 \Delta X_2 \geq \tfrac{1}{2}\, |\langle[\hat{X}_1,\hat{X}_2]\rangle| &= \hbar/2m\omega \\ &= (2\times 10^{-19}\,\text{cm})^2 \left(\frac{1\ \text{ton}}{m}\right)\left(\frac{10^4\ \text{s}^{-1}}{\omega}\right) .\end{aligned} \tag{5}$$

Here ΔX_1 and ΔX_2 are the variances of X_1 and X_2 in whatever quantum state (pure or mixed) the fundamental mode may occupy. The uncertainty relation (5) in turn implies the quantum limit (1):

$$|\Delta X| \equiv \left[(\Delta X_1)^2 + (\Delta X_2)^2\right]^{\frac{1}{2}} \geq (\hbar/m\omega)^{\frac{1}{2}} . \tag{1'}$$

Prior to the work of Kafka (1977) and TDCZS, the quantum limit

(1') had been derived by analysis of various gedanken experiments (Braginsky 1970, Giffard 1976). Now that the more fundamental quantum limit (5) is known, one can more easily invent ways to evade its influence in gravitational-wave detection.

These quantum mechanical considerations are unimportant for the second-generation gravitational-wave detectors which will go into operation in the next several years. But third-generation detectors, five years or so hence, may well run up against the quantum limit. If so, we must regard such detectors as quantum mechanical devices — despite their enormous $\sim$ 1 ton sizes!

2.2. The Gravitational Waves

By contrast with future detectors, which may be quantum mechanical, the waves one seeks to detect are highly classical: A gravitational wave burst from a supernova in the Virgo cluster of galaxies (amplitude $h \sim 10^{-21}$, angular frequency $\omega \sim 10^4\ s^{-1}$, duration $\tau_{g\omega} \sim 2\pi/\omega$, solid angle of source at Earth $\Delta\Omega \sim 10^{-38}$ sterr) carries past Earth a graviton flux

$$\frac{dN}{dAdt} \sim \frac{1}{16\pi}\frac{c^3}{G}\frac{(\omega h)^2}{\hbar\omega} \sim 10^{26}\ \frac{\text{gravitons}}{\text{cm}^2\ \text{s}} \quad . \tag{6}$$

These gravitons can be regarded as occupying a handful of quantum states, with mean occupation number equal to the number density in phase space dN/d^3xd^3p multiplied by the quantum mechanical cell size of phase space $\frac{1}{2}(2\pi\hbar)^3$:

$$N_{occ} = \frac{(2\pi\hbar)^3}{2}\left[\frac{c^{-1}\,(dN/dAdt)}{(\hbar\omega/c)^3\Delta\Omega}\right] \sim 10^{75} \ . \tag{7}$$

Averaged over the beam width ($\sim$ 1 sterr) of the detector, the occupation number is

$$\langle N_{occ}\rangle_{B.W.} = N_{occ}\ (\Delta\Omega/\ 1\ \text{sterr}) \sim 10^{37} \ . \tag{8}$$

This is also the mean number of gravitons that interact with the detector during one cycle as the wave burst passes.

The occupation numbers (7) and (8) are enormous. From any point of view the gravitational wave is classical!

2.3. The Issue of the Quantum Limit

The issue of the quantum limit can now be stated succinctly: One wishes to measure a classical signal force (gravitational wave) by letting it drive a quantum mechanical oscillator, and by experimentally monitoring one or more oscillator observables. Does the oscillator's uncertainty relation (5) limit the precision with which the classical signal force can be measured?

The answer will be "yes, and the limit is potentially disastrous (Eq. 2)," if one uses standard amplitude-and-phase methods to monitor the oscillator; "yes, but the limit is not so serious as Eq. (2)" if one uses quantum counting; and "no, there is no limit in principle" if one uses back-action-evading techniques.

3. THE AMPLIFIER LIMIT

3.1. Example of Measuring System

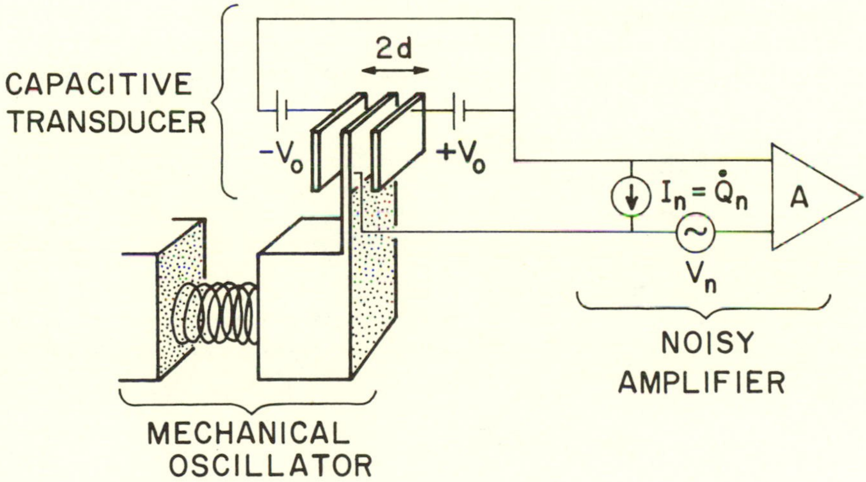

Figure 1. Schematic diagram of a system that makes "amplitude-and-phase" measurements of a mechanical oscillator.

Figure 1 is a schematic diagram of an idealized system that makes amplitude-and-phase measurements of a mechanical oscillator.

The transducer is a 3-plate capacitor. The outer plates, fixed in inertial space, are biased to potentials $\pm V_0$. The central plate, attached rigidly to the oscillator, acquires a voltage

$$V_{sig} = (V_0/d)x \tag{9}$$

when the oscillator is displaced a distance x from its central position. This signal voltage is fed into a voltage amplifier (idealized as having infinite input impedance), whose Thevinin equivalent circuit is shown in Fig. 1. The amplifier is necessarily noisy. Quantum mechanics forbids it to be free of noise (Weber 1959; Hefner 1962). The equivalent voltage noise source superimposes a voltage $V_n(t)$ on the output signal. The equivalent current noise source $I_n(t)$ drives a charge $Q_n(t) = \int I_n(t)dt$ onto the central plate of the capacitor. That charge superimposes on the signal a noise voltage Q_n/C, where C is the capacitance; it also interacts with the electric field in the capacitor to produce on the mechanical oscillator a "back-action force"

$$F = dp/dt = -(V_0/d)Q_n \ . \tag{10}$$

3.2. Quick Measurements

When this system is used to make quick measurements of x (measurement time $\tau \ll 1/\omega$) rather than amplitude-and-phase measurements, the rms error in the measured oscillator position is

$$\Delta x = \frac{1}{V_0/d}\left[(\Delta V_n)^2 + \frac{(\Delta Q_n)^2}{C^2}\right]^{\frac{1}{2}} = \frac{1}{V_0/d}\left[(S_V + \frac{1}{C^2}S_Q)\frac{1}{2\tau}\right]^{\frac{1}{2}} . \tag{11}$$

Here S_V and S_Q are the spectral densities of the amplifier's noise sources $V_n(t)$ and $Q_n(t)$; and $1/2\tau$ is the bandwidth of the measurement. A quick measurement kicks the oscillator (Eq. 10), producing an rms change of momentum

$$\Delta p \approx (V_0/d)(\Delta Q_n)\tau = (V_0/d)\left[S_Q\ (1/2\tau)\right]^{\frac{1}{2}}\tau \ . \tag{12}$$

No matter what the capacitance C may be, (11) and (12) imply the uncertainty relation

$$\Delta x \Delta p \gtrsim \tfrac{1}{2}(S_V S_Q)^{\frac{1}{2}} = \frac{\hbar}{e^{2\pi\hbar f/kT_n} - 1} \ . \tag{13}$$

Here we have expressed the product of the spectral densities in terms of the noise temperature T_n of the amplifier and the frequency $f \approx 1/2\tau$ of the output [e.g. Robinson 1974, Eq. (12.33)]. Quantum mechanics imposes the limit

$$T_n \geq 2\pi\hbar f/k \, \ell n \, 2 \tag{14}$$

on the noise temperature of any "linear amplifier" (Joseph Weber 1959; H. Hefner 1962) — a limit which, in (13), translates to

$$\Delta x \Delta p \gtrsim \tfrac{1}{2} \, (S_V S_Q)^{\frac{1}{2}} \approx kT_n/2\pi f \gtrsim \hbar \; . \tag{15}$$

Although this analysis applies to the special measuring system of Fig. 1, the final result (Eq. 15) is much more general: When one measures position x with any measuring system that has a transducer followed by an amplifier, the Heisenberg uncertainty relation gets replaced by (15). We shall call (15) the "amplifier limit" of the measuring system. Only when the amplifier has the minimum noise allowed by quantum mechanics ("ideal amplifier") does the amplifier limit reduce to the quantum limit $\Delta x \Delta p \gtrsim \hbar$. The best existing amplifiers at 30 GHz frequency (maser amplifiers) have noise temperatures ten times higher than the quantum limit (13). At lower frequencies the best amplifiers are even worse.

3.3. Slow Measurements

Quick measurements of the position of a Weber-type bar are not practical. For example, in the measurement system of Fig. 1 with $\tau << 1/\omega$, voltage breakdown prevents one from achieving a large enough electric field V_0/d to give reasonably small Δx. To achieve a reasonable precision one must integrate the transducer's output signal for a time $\tau \gtrsim 1/\omega$. Depending on how this integration is combined with external modulations and filtering, one achieves an "amplitude-and-phase" measurement, or a "back-action-evading" measurement, or perhaps something else. However, no matter how the signal processing is done, the final result will be information about the complex amplitude $X_1 + iX_2$ of the oscillator, information which is constrained by the "amplifier limit"

$$\begin{pmatrix}\text{area of}\\ \text{error box}\end{pmatrix} = \Delta X_1 \Delta X_2 \gtrsim \frac{\tfrac{1}{2}\,(S_V S_Q)^{\frac{1}{2}}}{m\omega} \approx \left(\frac{kT_n}{m\omega^2}\right) \gtrsim \frac{\hbar}{m\omega} \tag{16}$$

[generalization of (15); note that for short measurements, (3) guarantees the equivalence of (15) and (16)]. In (16) and below, if the amplifier output is at a frequency $\omega_e \neq \omega$, then T_n is the amplifier noise temperature referred back to frequency ω: T_n =

$(\omega/\omega_e) \cdot$ (actual noise temperature).

Actually, we do not know a fully general proof of the amplifier limit (16). However, we are rather certain of its generality, because of various specific examples and because of the way it fits into the logic of the rest of the topics discussed in this lecture.

4. AMPLITUDE-AND-PHASE MEASUREMENTS

All past and present measurements of Weber-type bars have been slow ($\tau \gtrsim 1/\omega$), and have had equal precisions $\Delta X_1 = \Delta X_2$ for the two parts of the complex amplitude. Such measurements we call "amplitude-and-phase measurements" because they seek information about both the amplitude $|X_1 + iX_2|$ and the phase $\varphi = \tan^{-1}(X_2/X_1)$ of the oscillator.

Amplitude-and-phase measurements are carried out in practice by feeding the output voltage of the transducer directly into a "linear" amplifier (Fig. 1) and then sending the amplifier output through a band-pass filter centered on the eigenfrequency ω of the bar's fundamental mode. From the output of the filter, one reads off the amplitude and phase of the oscillator, or equivalently one reads off $X_1 + iX_2$. (Sometimes the transducer output is upconverted to a higher frequency before being fed into the amplifier. For simplicity we shall ignore this possibility. The conclusions are basically the same with upconversion as without.)

For an amplitude-and-phase measurement the error box in the complex-amplitude plane is circular (Fig. 2a); no one direction is preferred over any other. The diameter of the error box is constrained by the amplifier limit

$$|\Delta X_1| = |\Delta X_2| \gtrsim |\Delta X|_{AL} \equiv \left[\frac{\frac{1}{2}(S_V S_Q)^{\frac{1}{2}}}{m\omega}\right]^{\frac{1}{2}} \approx \left(\frac{kT_n}{m\omega^2}\right)^{\frac{1}{2}} \gtrsim \left(\frac{\hbar}{m\omega}\right)^{\frac{1}{2}} \qquad (17)$$

(Eq. 16 with $\Delta X_1 = \Delta X_2$). Even with very gentle amplitude-and-phase measurements averaged over a long time, one cannot measure the complex amplitude more accurately than $|\Delta X|_{AL}$. The back-action of the amplifier perturbs the oscillator enough to keep the precision of measurement worse than $|\Delta X|_{AL}$.

A classical external driving force $F(t)$ ($\approx -\frac{1}{2}\ddot{h}\,\ell\, m$ if the force is due to a gravitational wave), acting for time τ, drives a change of the complex amplitude given by

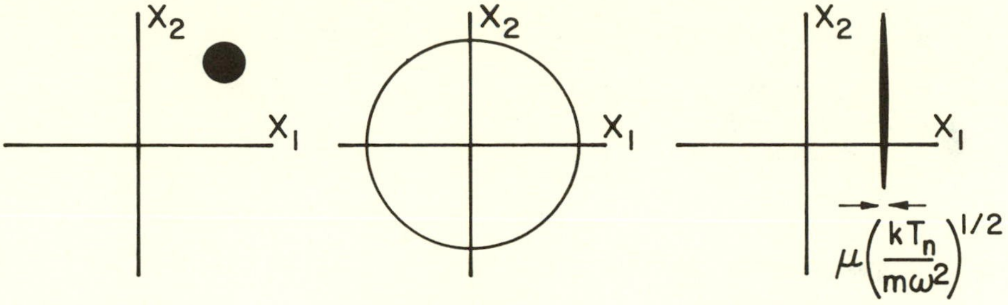

Figure 2. The error boxes, in the complex-amplitude plane, for three types of measurements of a mechanical oscillator. In all three cases the area of the error box exceeds the amplifier limit of $\sim (kT_n/m\omega^2)$ (Eq. 16).

$$\delta(X_1 + iX_2) = \int_0^t [F(t')/m\omega]\, ie^{i\omega t'}\, dt' \, . \tag{18a}$$

This change is detectable, with a specific amplifier plus optimal electronics and negligible other noise sources, if and only if

$$\frac{|\delta(X_1 + iX_2)|}{|\Delta X|_{AL}} \approx \frac{h/h_{QL}}{(kT_n/\hbar\omega)^{\frac{1}{2}}} \gtrsim 1 \tag{18b}$$

(cf. Eq. 2). Here the expression involving h refers to the case of a broad-band gravitational-wave burst; h_{QL} is the quantum limit of Eq. (2); and T_n is the noise temperature of the amplifier. Note that with real amplifiers ($kT_n/\hbar\omega \gtrsim 10$) the outlook for amplitude-and-phase measurements is even worse than we claimed in the introduction (Eq. 2)!

If the amplifier is "ideal" (i.e. if it has the minimum noise allowed by the Heisenberg uncertainty principle), then its back action drives the oscillator into a quantum mechanical "coherent state" (i.e. a state with a Gaussian probability distribution in X_1 and X_2, and with $\Delta X_1 = \Delta X_2 = (\hbar/2m\omega)^{1/2}$; cf. Secs 15.8-15.10 of Merzbacher, 1970). Hollenhorst (1979) has analyzed the optimal measurement strategy in this case, and has derived a precise probability distribution

$$\mathcal{P}(\alpha) = 1 - e^{-\alpha^2} \geq 0.5 \quad \text{if} \quad \alpha \geq 0.83 \tag{19a}$$

for successful detection of a classical force of strength

$$\alpha = \frac{|\delta(X_1 + iX_2)|}{2(\hbar/2m\omega)^{\frac{1}{2}}} \quad . \tag{19b}$$

5. QUANTUM-COUNTING MEASUREMENTS

One way, in principle, to circumvent the quantum limit on h is to measure the number of quanta (phonons) in the fundamental mode of the bar, instead of measuring the complex amplitude. The phonon number operator is

$$\hat{N} = \frac{\hat{H}_o}{\hbar\omega} - \tfrac{1}{2} = \frac{m\omega}{2\hbar}(\hat{X}_1{}^2 + \hat{X}_2{}^2) - \tfrac{1}{2} \tag{20}$$

where $\hat{H}_o$ is the Hamiltonian. The goal is to measure $\hat{N}$ without perturbing it significantly. To achieve this goal one must couple to the oscillator a transducer whose interaction Hamiltonian commutes with $\hat{N}$. The only oscillator observables that commute with $\hat{N}$ are $\hat{N}$ itself, and functions of $\hat{N}$; and they are all of quadratic order or higher in the complex amplitude.

Quadratic couplings are straightforward in principle; but they are fiendishly difficult in practice when the oscillator is mechanical, or even when it is electromagnetic with frequencies below the infrared. In fact, nobody has ever constructed a quantum counting device in those cases — though Unruh (1977, 1978) and Braginsky et al. (1977) have proposed designs that might someday do the job.

Figure 2b shows the uncertainty error box in the complex amplitude plane, for a quantum counting measurement which gives a value N_o with precision $\pm\frac{1}{2}$. The error box is an annulus. One can easily see from Eq. (20) that (i) the area of this error annulus is approximately the quantum limit $\hbar/m\omega$ (Eq. 16); (ii) the radial thickness of the annulus is smaller than the amplitude-and-phase error box (Fig. 2a) by a factor $1/\sqrt{N_o}$, which suggests that phonon counting might be more sensitive to gravitational waves by $1/\sqrt{N_o}$ than amplitude-and-phase measurements.

This higher sensitivity is indeed correct: Suppose that one had a phonon-counting device for a Weber-type bar, and suppose that a measurement at time $t = 0$ gave a value N_o for the number of phonons in the fundamental mode. Suppose further that a gravitational wave burst of strength α (Eq. 19b) were to act on the bar. Then one can show (Braginsky & Vorontsov 1974; CDSTZ; Hollenhorst 1979) that the probability for the wave to change the number of phonons is

$$\begin{pmatrix}\text{probability}\\ \text{of } \delta N \neq 0\end{pmatrix} = 1 - e^{-\alpha^2}\left[L_{N_o}(\alpha^2)\right]^2$$

$$\geq 0.5 \quad \text{if} \quad \alpha \geq 0.57/\sqrt{N_o} \quad \text{and} \quad N_o \gg 1. \tag{21a}$$

Here $L_{N_o}(\xi)$ is the Laguerre polynomial of order N_o. Comparison of (19a) and (21a) reveals the $1/\sqrt{N_o}$ better sensitivity of phonon counting.

Unfortunately, higher sensitivity does not guarantee higher precision: If the oscillator initially has occupation number N_o and a wave of strength α passes, then one cannot predict the new value of N. One can only predict a probability distribution:

$$\mathcal{P}(N_o \to N) = \frac{r!}{s!}\,\alpha^{2(s-r)}\,e^{-\alpha^2}\left[L_r^{(s-r)}(\alpha^2)\right]^2 . \tag{21b}$$

Here $s = \max(N_o, N)$, $r = \min(N_o, N)$, and $L_r^{(n)}$ is the generalized Laguerre polynomial. To map out this probability distribution and thereby determine α requires a huge number of mechanical oscillators all coupled to the same gravitational wave. If we have only one or a few oscillators, then (21b) implies that from a specific measured change $N_o \to N$ we can determine α only to within a factor of order 3, at the 90% confidence level. In semi-classical language: One cannot determine the wave strength α from a measured change in $\hat{N}$ because one does not know the relative phase of the gravitational wave and the oscillator; the phase of the oscillator is unknowable because it is canonically conjugate to the precisely measured quantity $\hat{N}$.

In summary: (i) Phonon counting is exceedingly difficult in practice because it requires quadratic coupling of the sensors to the antenna. (ii) In principle phonon counting can detect classical gravitational waves that are arbitrarily weak, if the detector initially is arbitrarily highly excited ($N_o \to \infty$). (iii) However, unless one has a huge number of detectors, by phonon counting one can measure the strength of the gravitational wave only to within a factor of ~ 3.

6. BACK-ACTION-EVADING MEASUREMENTS: ISSUES OF PRINCIPLE

6.1. The Basic Idea

The form of the uncertainty relation $\Delta X_1 \, \Delta X_2 \geq \hbar/2m\omega$ (Eq. 5) suggests a possible way to circumvent the quantum limit on h: Instead of monitoring X_1 and X_2 with equal precisions ("amplitude-and-phase measurement"), monitor X_1 with high precision $\Delta X_1 \ll \hbar/2m\omega$ and X_2 with low precision $\Delta X_2 \gg \hbar/2m\omega$, or vice versa. As we shall see, from high-precision measurements of X_1 alone it is possible in principle to infer with high precision the details of the gravitational radiation or other classical force that drives the oscillator.

Measurements with $\Delta X_1 \ll \Delta X_2$ we call "back-action evading" because they are carefully designed to permit X_1 to evade the back-action effects of the measuring system.

Figure 2 contrasts the error box for a back-action-evading measurement of an oscillator with the error boxes for quantum-counting measurements and for amplitude-and-phase measurements.

The basic idea behind a back-action-evading measurement is illustrated in Fig. 3. Suppose that at time t = 0 one has measured the oscillator's position x with high accuracy, producing by back action a large uncertainty in the momentum p (Fig. 3a).

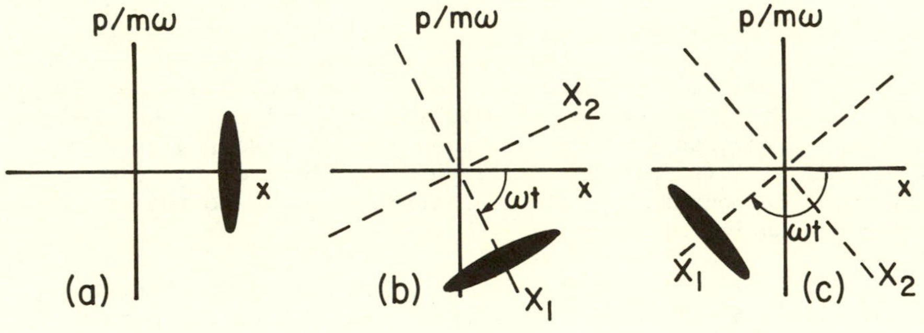

Figure 3. The error box for a back-action-evading measurement of X_1 as viewed in position-momentum phase space.

Then in the absence of further measurements or external forces, as time passes the error box rotates clockwise in x-p phase space with angular velocity ω ("harmonic-oscillator evolution"). If, at a later time (Fig. 3b), one attempts again to measure x with high precision ("amplitude-and-phase technique"), one will get a poorly predictable result. But if instead one tries to measure $X_1 \equiv$ [value of x in "rotating frame"]("back-action-evading technique"), one will get a highly predictable result, a result that takes full advantage of the precision of the original measurement.

The key mathematical point is that X_1 is a constant of the motion in the absence of external forces, while x is not. This means that if X_1 is measured accurately once, it remains precisely known thereafter, whereas if x is measured accurately, its precision thereafter gets mutilated by admixtures of the back-action-distorted momentum.

Generalizations of this idea are discussed and codified by Unruh (1979) and by CDSTZ.

6.2. The Precision of Back-Action-Evading Measurements

In principle, nonrelativistic quantum theory permits X_1 to be measured arbitrarily quickly and arbitrarily accurately. The easiest way to see this is (i) to recall that in principle the position of an oscillator can be measured arbitrarily quickly and accurately at any instant; and (ii) to notice that at any fixed moment of time when $\hat{X}_1$ is to be measured, one can perform a canonical transformation that converts $\hat{X}_1$ into the position of the oscillator.

For people who do not like this argument (and we have encountered many!), we give elsewhere a mathematical analysis of an explicit gedanken experiment which yields arbitrarily quick and accurate measurements of $\hat{X}_1$; see CDSTZ.

In principle, by a continuous sequence of arbitrarily quick and accurate measurements of $\hat{X}_1$, one can monitor with perfect accuracy a classical force F(t) acting on the oscillator. This fact is proved easily: (i) Let an initial measurement of $\hat{X}_1$ give a value ξ_0 and leave the oscillator in an eigenstate $|\xi_0\rangle$ of $\hat{X}_1$. (ii) Thereafter, the oscillator and measuring system evolve under the action of the Hamiltonian

$$\hat{H} = \tfrac{1}{2}m\omega^2(\hat{X}_1{}^2 + \hat{X}_2{}^2) - \hat{x}F(t) + K\hat{X}_1\hat{Q} + H_M(\hat{Q},\hat{\Pi}). \qquad (22)$$

Here the first term is the Hamiltonian of the free oscillator; the second term is the coupling to the classical force; the third is the coupling to the measuring system with K a coupling constant;

and the fourth is the Hamiltonian of the measuring system. (iii) In the Heisenberg Picture the state of the oscillator remains $|\xi_o\rangle$, but the operator $\hat{X}_1$ evolves in accordance with the Heisenberg equation

$$\frac{d\hat{X}_1}{dt} = \frac{\partial \hat{X}_1}{\partial t} - \frac{i}{\hbar}\left[\hat{X}_1, \hat{H}\right] = -\frac{F}{m\omega}\sin\omega t\,. \tag{23}$$

(iv) Consequently, at time t the state $|\xi_o\rangle$ of the oscillator is an eigenstate of

$$\hat{X}_1(t) = \hat{X}_1(0) - \int_0^t [F(t')/m\omega]\sin\omega t'\,dt' \tag{24}$$

with eigenvalue

$$\xi(t) = \xi_o - \int_0^t [F(t')/m\omega]\sin\omega t'\,dt'\,. \tag{25}$$

(v) This means that an accurate measurement of $\hat{X}_1$ at time t must yield the precise value $\xi(t)$ and must leave the state of the oscillator unchanged (except for an unpredictable and unimportant change of phase). (vi) Then from a continuous sequence of measurements of $\hat{X}_1$ one can monitor the evolution of $\xi(t)$; and from $\xi(t)$ one can compute, in principle with complete precision,

$$F(t) = -(\sin\omega t)^{-1}(m\omega d\xi/dt). \tag{26}$$

In practice the precision of one's measurements of $\xi(t)$ will be finite; and as a consequence the precision of one's inferred F(t) will oscillate in time between terrible precision when $\omega t \approx n\pi$ ($\sin\omega t \approx 0$) and good precision when $\omega t \approx (n+\frac{1}{2})\pi$. One can remedy this by coupling two different oscillators to the external force (gravitational wave) and by measuring $\hat{X}_1$ on one oscillator to give (25), and $\hat{X}_2$ on the other to give

$$\eta(t) = \eta_o + \int_0^t [F(t')/m\omega]\cos\omega t'\,dt'\,. \tag{27}$$

Then the precisions of the inferred F(t) will be complementary: when one oscillator is giving terrible precision the other will be good, and conversely. In the X_1-X_2 plane the error boxes for the two oscillators intersect in a very tiny region, and their intersection is driven by the force F(t) into exactly the same motion as would be an uncertainty-free, classical oscillator.

QUANTUM LIMIT FOR DETECTORS

In the absence of all noise except the amplifier, a perfect back-action-evading measurement will produce an error box that is elliptical, with

$$\Delta X_1 = \mu |\Delta X|_{AL} , \quad \Delta X_2 = \mu^{-1} |\Delta X|_{AL} , \quad \mu < 1. \tag{29}$$

Here $|\Delta X|_{AL} \approx (kT_n/m\omega^2)^{\frac{1}{2}}$ is the amplifier limit. In practice the value of μ will be determined by the details of the experimental setup. Of course, one will try to make μ as small as possible; and in principle (but not in practice) it can be arbitrarily small. A gravitational wave burst of strength α will be detectable only if

$$\alpha \gtrsim \mu \, (kT_n/\hbar\omega)^{\frac{1}{2}} ; \tag{30}$$

cf. Eq. (19). Hollenhorst (1979) has given a beautiful analysis of the "optimal strategy of measurement" for back-action-evading measurements with ideal amplifiers ($kT_n = \hbar\omega/\ln 2$). He shows that, if one uses the optimal strategy, and if one seeks to go far below the quantum limit $\mu \ll 1$, then the strength of a wave with random phase must be $\alpha > 1.3\ \mu$ if the probability of detection is to exceed 50%.

7. BACK-ACTION-EVADING MEASUREMENTS: PRACTICAL REALIZATIONS

Elsewhere (CDSTZ) we describe a wide variety of different experimental setups for back-action-evading measurements. Here we shall sketch only two: stroboscopic measurements using a microwave-cavity transducer; and slow, continuous measurements using a modulated capacitive transducer.

7.1. Stroboscopic Measurements

At times $t_n = (n\pi/\omega)$, X_1 is equal to $(-1)^n x$; see (3a). This suggests a simple, "stroboscopic" realization of back-action-evading measurements (Braginsky et al. 1978; TDCZS): Make a sequence of very quick ($\Delta t \ll 1/\omega$) measurements of position x at time $t = t_n$; compute $X_1 = (-1)^n x$ for each measurement; and, if necessary, average a large number of such measurements in order to get good accuracy.

Such stroboscopic measurements, in the idealized case of perfect quickness and perfect precision, can be described semiclassically as follows: One measures the oscillator's coordinate $x = X_1$ at $t = 0$, obtaining a precise value ξ_0, and in the process giving the momentum a huge, unknowable, uncertainty-principle kick. The kick causes x to evolve in an unknown way. However, because the oscillator's period is independent of its amplitude, after precisely a half cycle x must be precisely equal to $-\xi_0$. At $t = \pi/\omega$ a second pulsed measurement is made, giving precisely this value for $x = -X_1$, and again kicking the momentum by a huge unknowable amount. Subsequent pulsed measurements at $t_n = n\pi/\omega$ give values $(-1)^n \xi_0$ which are unaffected by the unknown kick of each measurement.

Braginsky (private communication) hopes to perform stroboscopic measurements, and thereby beat the quantum limit, using a sapphire-crystal gravitational wave antenna and a microwave-cavity transducer. One face of the microwave cavity is the niobium-coated end of the sapphire crystal. Consequently, the resonant frequency ω_e of the cavity depends on the position x of the end of the crystal (Braginsky, Panov, Petnikov & Popel'nyuk 1977). The cavity is driven by an electromagnetic generator, slightly off resonance, for a time $\Delta t \ll 1/\omega$, building up electromagnetic oscillations whose amplitude depends on $\omega_e(x)$. The electromagnetic energy is extracted through a hole in the cavity in a time $\sim \Delta t$; and from the amount of energy extracted one infers ω_e and thence x and thence $X_1(t_n)$. There are many complications and delicacies in such an experiment, but we cannot describe them here for lack of space and time.

In describing the ultimate precision of a practical stroboscopic measurement, it is useful to introduce the Gibbons & Hawking (1971) coupling constant

$$\beta \equiv \frac{\begin{pmatrix}\text{work done on transducer by antenna when} \\ \text{its face moves through distance } x\end{pmatrix}}{\begin{pmatrix}\text{energy stored elastically in antenna when} \\ \text{its face moves through distance } x\end{pmatrix}} \, . \tag{31}$$

(A more precise and more general definition of β is given in CDSTZ.) The ultimate precision $\mu = \Delta X_1/|\Delta X|_{AL}$ depends on β, and on the number of pulsed measurements $\omega\tau/\pi$ that one averages to get X_1 (τ is the averaging time). Specifically,

$$\mu \approx 1/(\beta\omega\tau)^{\frac{1}{4}} \tag{32}$$

is the best accuracy one can achieve for fixed β, ω, τ. To achieve this one must use, in addition to an optimal experimental setup, also an optimal pulse length

$$(\Delta t)_{\text{optimal}} \approx (1/\omega)(1/\beta\omega\tau)^{\frac{1}{2}} \, . \tag{33}$$

(These and other relations are derived in CDSTZ.) In practice it will be extremely difficult to achieve coupling constants β near unity. Thus, to circumvent the amplifier limit, i.e. to achieve $\mu < 1$, may require averaging over a large number of cycles of the mechanical oscillator, $\omega\tau \gg 1$.

7.2. Slow, Continuous Measurements

One can design a measuring system which, in principle, monitors X_1 with arbitrarily good time resolution $\tau \ll 1/\omega$ and precision $\mu \ll 1$; see CDSTZ. However, the experimental difficulty of achieving strong coupling ($\beta \gtrsim 1$) of transducer to oscillator makes such measuring systems look impractical. When $\beta \ll 1$ one can beat the amplifier limit only with slow measurements, $\tau \gg 1/\omega$.

Slow, continuous, back-action-evading measurements can be made with amplitude-and-phase type measuring systems (e.g. Fig. 1) that have been modified in a conceptually simply way: The output of the transducer is modulated by a suitable periodic signal, and then is passed through a suitable passive filter before entering the amplifier.

As an example, one can replace the batteries $\pm V_o$ of Fig. 1 by ideal voltage sources

$$\begin{aligned} \pm V(t) &= \pm V_o \cos \omega_e t \cos \omega t \\ &= \pm \tfrac{1}{2} V_o \left[\cos(\omega_e + \omega)t + \cos(\omega_e - \omega)t\right]. \end{aligned} \tag{34}$$

Here $\omega_e \gg \omega$ is an upconversion frequency at which the amplifier operates; and the phasing of this voltage must be regulated by an external clock. The output of the transducer will then be

$$\begin{aligned} V_{sig} &= [V(t)/d]x \\ &= (V_o/4d)X_1 \left[\cos(\omega_e + 2\omega)t + 2\cos\omega_e t + \cos(\omega_e - 2\omega)t\right] \\ &+ (V_o/4d)X_2 \left[\sin(\omega_e + 2\omega)t - \sin(\omega_e - 2\omega)t\right]; \end{aligned} \tag{35}$$

cf. (9), (34), and (3a). A passive bandpass filter, centered on frequency ω_e and with $\Delta\omega \ll \omega$, is placed between the transducer and the amplifier. As a result the amplifier receives a strong signal at frequency ω_e, with phase predictable in advance, and with amplitude proportional to X_1; it also receives weak signals at $\omega_e \pm 2\omega$, from which one can infer X_2. The amplifier's current noise feeds back through the filter and into the transducer, where (by the reciprocity theorem for the circuit) it kicks X_2 hard, and kicks X_1 only slightly. The upshot of all this is an excellent precision for the measurement of X_1 at frequency ω_e, and a poor precision for the measurement of X_2 at $\omega_e \pm 2\omega$.

If properly designed and constructed, such a measuring system can achieve a precision given by (29) with

$$\mu \approx 1/(\beta\omega\tau)^{\frac{1}{2}} . \tag{36}$$

Here β is the coupling constant and τ is the averaging time ($\Delta f = 1/2\tau$ is the bandwidth of the filter). Note that for $\beta\omega\tau \gg 1$ (i.e. $\mu \ll 1$), this precision is much better than that of a stroboscopic measurement (Eq. 32). The stroboscopic technique suffers because of its poor duty cycle.

8. NYQUIST NOISE IN THE OSCILLATOR

Any attempt to circumvent the quantum limit, by back-action-evading techniques, will face numerous experimental difficulties that were ignored in the above discussion. Some of these difficulties are discussed by CDSTZ. Here we mention only the worst one: Nyquist noise in the bar antenna, due to weak coupling of its fundamental mode to all the other modes.

The change of the fundamental-mode amplitude, produced by Nyquist forces during the averaging time τ, is

$$(\Delta X_1)_{\mathrm{Nyq}} \approx (kT/m\omega^2)^{\frac{1}{2}} (\omega\tau/Q)^{\frac{1}{2}} . \tag{37}$$

Here T is the temperature of the bar and Q is the quality factor of its fundamental mode. This change must be less than the desired precision, $\Delta X_1 = \mu_Q(\hbar/m\omega)^{1/2}$, of the back-action-evading measurement. (Here μ_Q is the amount below the <u>quantum</u> limit, not amplifier limit, that one wishes to go.) For a slow, continuous measurement this requires (Eqs. 17 and 36)

$$\begin{aligned} T/Q &\lesssim \beta\mu_Q^{\,4}(\hbar\omega/kT_n)(\hbar\omega/k) \\ &= \beta\mu_Q^{\,4}(\hbar\omega/kT_n)(5\times 10^{-8}\ {}^{\circ}\mathrm{K}) \qquad \text{if} \quad \omega/2\pi = 1\ \mathrm{kHz}. \end{aligned} \tag{38}$$

This same requirement is faced in any measurement (amplitude-and-phase, or back-action-evading) for which back-action forces are negligible compared to Nyquist forces. In particular, this is the requirement for all present amplitude-and-phase measurements of gravitational-wave detectors (regime $\mu_Q \gg 1$). Thus the constraints imposed by Nyquist noise will not suddenly become more stringent when one switches over from amplitude-and-phase to back-action-evading measurements, or when one uses back-action evasion to push into the sub-quantum-limit regime $\mu_Q \ll 1$.

Since one ultimately desires $\mu_Q \lesssim 0.1$, and since such difficulties as voltage breakdown probably force one to operate with $\beta \ll 1$, and since even with upconversion to gigahertz frequencies, the amplifier noise temperature T_n will likely exceed the quantum limit by $\gtrsim 10$, Eq. (38) is a very serious constraint! — serious, but perhaps not impossible with some years of effort.

In conclusion, back-action evasion offers a promising solution to the problem of the quantum limit; but it does not solve the other problems that currently plague gravitational-wave detection with Weber-type bars: weak coupling, $\beta \ll 1$; large amplifier noise, $kT_n/\hbar\omega \gg 1$; and large Nyquist noise, $T/Q \gtrsim 10^{-10}$ °K for the best crystals yet developed. These three problems will prevent one from reaching the quantum limit in the next few years; and once the quantum limit is reached, they will continue to plague "sub-quantum" back-action-evading measurements.

This research was supported in part by the National Aeronautics and Space Administration [NGR 05-002-256 and a grant from PACE] and by the National Science Foundation [AST76-80801 A01].

REFERENCES

Braginsky, V. B. (1970). Physical Experiments with Test Bodies. Nauka, Moscow. English translation published as NASA-TT F762, National Technical Information Service, Springfield, VA. See especially Eqs. (3.17) and (3.25).

Braginsky, V. B. (1977). The detection of gravitational waves and quantum nondisturbtive measurements. In Topics in Theoretical and Experimental Gravitation Physics, ed. V. De Sabbata and J. Weber, pp. 105-122. Plenum Press, London and New York.

Braginsky, V. B., Panov, V. I., Petnikov, V. G. & Popel'nyuk, V. D. (1977). The measurement of small mechanical oscillations using a capacitive pickoff device made from a superconducting resonator. Pribori i Tekhnika Eksperimenta, 1, 234.

Braginsky, V. B. & Vorontsov, Yu. I. (1974). Quantum mechanical limitations in macroscopic experiments and modern experimental technique. Usp. Fiz. Nauk, 114, pp. 41-53. English translation in Sov. Phys.—Uspekhi, 17, pp. 644-650.

Braginsky, V. B., Vorontsov, Yu. I. & Khalili, F. Ya. (1977). Quantum properties of a ponderomotive measuring device for electromagnetic energy. Zhur. Eksp. Teor. Fiz., 73, pp. 1340-1343. English translation in Sov. Phys.—JETP, 46, in press.

Braginsky, V. B., Vorontsov, Yu. I. & Khalili, F. Ya. (1978). Optimal quantum measurements in detectors of gravitational radiation. Pis'ma v. Zhur. Eksp. Teor. Fiz., 27, pp. 296-301. English translation in Sov. Phys.—JETP Letters, in press.

Caves, C. M., Drever, R. W. P., Sandberg, V. D., Thorne, K. S. & Zimmermann, M. (1979). On the measurement of a weak classical force coupled to a quantum mechanical oscillator. Rev. Mod. Phys., in preparation. Cited in text as CDSTZ.

Gibbons, G. W. & Hawking, S. W. (1971). Theory of the detection of short bursts of gravitational radiation. Phys. Rev. D, 4, pp. 2191-2197.

Giffard, R. P. (1976). Ultimate sensitivity of a resonant gravitational wave antenna using a linear motion detector. Phys. Rev. D, 14, pp. 2478-2486.

Hefner, H. (1962). The fundamental noise limit of linear amplifiers. Proc. I.R.E., 50, pp. 1604-1608.

Hollenhorst, J. N. (1979). Quantum limits on resonant mass gravitational radiation detectors. Phys. Rev. D, in press.

Kafka, P. (1977). Some remarks on gravitational wave experiments. In Proceedings of the International School of General Relativistic Effects in Physics and Astrophysics: Experiments and Theory (Third course), ed. R. Ruffini, J. Ehlers & C. W. F. Everitt. Max-Planck-Institut für Physik und Astrophysik, Munich.

Merzbacher, E. (1970). Quantum Mechanics, second edition. Wiley, New York.

Robinson, F. N. H. (1974). Noise and Fluctuations in Electronic Devices and Circuits. Clarendon Press, Oxford.

Thorne, K. S., Drever, R. W. P., Caves, C. M., Zimmermann, M. & Sandberg, V. D. (1978). Quantum nondemolition measurements of harmonic oscillators. Phys. Rev. Letters, 40, pp. 667-671.

Unruh, W. G. (1977). An analysis of quantum-nondemolition measurement. Unpublished manuscript.

Unruh, W. G. (1978). An analysis of quantum-nondemolition measurement. Phys. Rev. D, 18, in press.

Unruh, W. G. (1979). Quantum nondemolition and gravity wave detection. Phys. Rev. D, in press.

Weber, J. (1959). Masers. Rev. Mod. Phys., 31, pp. 681-710; esp. pp. 693-697.

DISCUSSION SESSION I: DETECTION OF GRAVITATIONAL RADIATION

Notes and Summary: Reuben Epstein

Massachusetts Institute of Technology

(Chairman: Rainer Weiss, Massachusetts Institute of Technology)

PREFACE

[J.P.A. Clark & R. Epstein: The format of this discussion session and the second session on sources is informal. The contributions are largely extemporaneous, except for introductory remarks by the chairmen and short reports by several individuals who had requested podium time in advance. These accounts are considerably condensed, especially where prompting and interruptions were frequent. Information from minor interruptions is included in the text attributed to the interruptee and is not cited to the interrupter. Statements within quotation marks are verbatim. Otherwise, we have taken great liberties with paraphrasing and reordering, where necessary, for brevity and clarity. It occasionally served these ends to condense a section of dialogue into a single statement cited to a number of individuals. We share responsibility for statements found in square brackets and the final form of the figures, and we apologize to our fellow participants for whatever of value has been lost in preparing these accounts.]

DETECTOR SENSITIVITIES

R. Weiss: To begin, I [refer you to the lectures by R. Weiss (this volume) for] the basic formulae for the limits of detection for transient sources, periodic sources, and a stochastic background.

The trouble with detecting transient sources is that the pulse duration, τ_p, is the effective limit on your integration time. Another point is that if one uses the optimum length, ℓ, of a broadband detection system, $\ell = \lambda/4$, we find that the minimum detectable amplitude, h_{min}, scales as $f^{3/2}$. Thus we need more complete low-frequency theoretical power spectra in order to determine if our chances for detecting gravitational pulses are better at longer wavelengths.

Tables 1a and 1b are based on sensitivity estimates mentioned previously [and, in some cases, altered during the course of this

Table 1a. RMS sensitivities for high-frequency ($10^4 - 10$ Hz) detectors. See Weiss (this volume) for symbol definitions. Detector specifications are listed in outline form, and sensitivities for impulsive, periodic, and radiometric (stochastic) sources are listed in tabular form. Given are detector parameters, frequencies, and anticipated dates of operation. Except where noted, times are in seconds and frequencies are in Hertz.

	Impulsive	Periodic	Radiometry ($\Delta B = f$)

I. Cryogenic ($\Delta B = f/Q$)
  A. Aluminum bars ($m \sim 5 \times 10^6$g, $Q \sim 2 \times 10^6$, $T \sim 1°K$), 10^3 Hz, 1980.

			(2 bar corr.)
GPU	10^{-3}	--	--
$\Delta \ell / \ell$	10^{-18}	--	--
$f^{1/2} \sqrt{\left[\frac{\Delta \ell}{\ell}(f)\right]^2}$	--	--	--

  B. Single crystal ($m \sim 2 \times 10^4$g, $Q \sim 10^9$, $T \sim 1°K$), 10^3 Hz, 1980.

			(2 bar corr.)
GPU	10^{-3}	--	--
$\Delta \ell / \ell$	10^{-18}	--	--
$f^{1/2} \sqrt{\left[\frac{\Delta \ell}{\ell}(f)\right]^2}$	--	--	--

  C. Aluminum acoustic, quantum limit ($m \sim 10^6$g, $Q \sim 10^6$, $\beta = 1$),
     19??.

	(10^3 Hz)	($\ell = 1$m, 60 Hz, $T_{bar} \sim 10^{-3}°K$)	(10^3 Hz, $\Delta B \sim f/10$, 2 bar corr.)
GPU	10^{-7}	--	--
$\Delta \ell / \ell$	10^{-20}	$10^{-22}/\tau_{int}^{1/2}$	--
$f^{1/2} \sqrt{\left[\frac{\Delta \ell}{\ell}(f)\right]^2}$	--	--	$10^{-22}/\tau_{int}^{1/4}$

II. Ground-based laser interferometer
  A. First-generation ($\ell \sim 10$m, $P \sim 1$W), $10^4 - 10^{1.5}$ Hz, 1980.

GPU	$10^{-5}f$	--	--
$\Delta \ell / \ell$	$10^{-19}/\tau_p^{1/2}$	$10^{-19}/\tau_{int}^{1/2}$	--
$f^{1/2} \sqrt{\left[\frac{\Delta \ell}{\ell}(f)\right]^2}$	--	--	$10^{-19}(f/\tau_{int})^{1/4}$

  B. Multipass, optimal length ($\ell \sim 10$ km, $P \sim 100$W), $10^4 - 10^{1.5}$ Hz, 1985-1990.

GPU	$10^{-11}f$	--	--
$\Delta \ell / \ell$	$10^{-22}/\tau_p^{1/2}$	$10^{-22}/\tau_{int}^{1/2}$	--
$f^{1/2} \sqrt{\left[\frac{\Delta \ell}{\ell}(f)\right]^2}$	--	--	$10^{-22}(f/\tau_{int})^{1/4}$

Table 1b. Sensitivities for low-frequency (10^1 Hz and below) detectors. See caption for Table 1a.

	Impulsive	Periodic	Radiometry ($\Delta B=f$)
III. Microwave Doppler tracking of spacecraft.			
A. Single spacecraft ($\ell<\lambda_g$, or $\ell>\lambda_g$ for "triple-pulse" signature), 10^{-2}-$10^{-3.5}$ Hz, 1980.			
$\Delta\ell/\ell$	$10^{-15}/\tau_p{}^{1/2}$	$10^{-15}/\tau_{int}{}^{1/2}$	--
	(τ_p and τ_{int} in hours, $\tau_p \leqslant \tau_{int} \lesssim 10^1$ hr)		
B. Multiple spacecraft (Receiver-noise limited, $\ell<\lambda_g$), $10^{-1.5}\rightarrow 10^{-3.5}$ Hz, 1990+.			
$\Delta\ell/\ell$	$10^{-16}/\tau_p{}^{1/2}$	$10^{-16}/\tau_{int}{}^{1/2}$	--
$f^{1/2}\sqrt{\left[\frac{\Delta\ell}{\ell}(f)\right]^2}$	--	--	$10^{-16}(f/\tau_{int})^{1/4}$
IV. Laser heterodyne tracking of multiple spacecraft (P~1W), 10^{-2} - 10^{-4} Hz, 1990+.			
$\Delta\ell/\ell$	$10^{-22}/\tau_p{}^{1/2}$	$10^{-22}/\tau_{int}{}^{1/2}$	--
$f^{1/2}\sqrt{\left[\frac{\Delta\ell}{\ell}(f)\right]^2}$	--	--	$10^{-22}(f/\tau_{int})^{1/4}$
V. Laser interferometer frame in space (P~1W, ℓ~10 km, multiple reflections), 10 - 10^{-3} Hz, 1990+.			
$\Delta\ell/\ell$	$10^{-22}/\tau_p{}^{1/2}$	$10^{-22}/\tau_{int}{}^{1/2}$	--
$f^{1/2}\sqrt{\left[\frac{\Delta\ell}{\ell}(f)\right]^2}$	--	--	$10^{-22}(f/\tau_{int})^{1/4}$

discussion] for various frequency ranges and styles of detectors. [The tables have been subsequently checked and filled in by R. Weiss.] These estimates summarize what is now happening and what you can imagine might happen. Note particularly that $[\Delta\ell/\ell(f)]^2$ is the strain spectral density and is not to be confused with the square of $\Delta\ell/\ell$.

L. Smarr: Now that Ray [Weiss] has provided a framework for comparing detection techniques, we will now read two letters from two esteemed colleagues who were unable to be present at our workshop. They will give their views on the goals of gravitational radiation astronomy.

LETTER TO THE WORKSHOP FROM JOHN A. WHEELER

L. Smarr: "Dear Larry [Smarr],

This is only to send warm good wishes for the conference you've worked so hard to organize.

I will regard it as truly very successful if you see to it, with the colleagues there assembled, that (1) beginning with this fall there is always someone, who at any given time has a gravitational wave detector in operation against the possibility that the next supernova goes off in his day or week; (2) a written statement of results from the conference, for wide publication, about advance planning for the next supernova.

...Best wishes...

John [Wheeler]"

LETTER TO THE WORKSHOP FROM V. I. BRAGINSKY

K. S. Thorne: V. I. Braginsky writes the following:

First, with regard to resonant aluminum and crystal bars, we must have more detailed waveforms in the kilohertz band from theorists. Experimenters need to know at least the following: How often do various kinds of catastrophic events occur (e.g. supernovae) in different kinds of galaxies? There is a need for some agreement among theorists on this point. They must know the most probable mean values of the signal frequencies in the 1-10 kHz band in order to know where to tune their antennas. Also, how are the frequencies of these bursts distributed? Is it best to concentrate near one frequency or to spread out? We must know the probable efficiencies of catastrophic sources, not just their potential efficiencies. For example, how often will supernovae be too symmetric to radiate with a detectable efficiency? Finally, what other observable events (e.g. electromagnetic and neutrino) accompany gravity wave events?

The same kinds of questions need to be answered for the low-frequency band where futuristic Doppler systems might operate. Theorists must provide the knowledge needed by the experimentalists who must ultimately apply optimal filters and the theory of pattern recognition to their data.

In my view, the key problem with resonant bar detectors is to develop and run quantum nondemolition devices (QND) to go beyond the quantum limit. Achieving the more immediate goal of the quantum limit is not enough.

If gravitational radiation is seen, how do we convince the world of this? We must be prepared with the following: coincidences among a large number of detectors worldwide, no significant correlations with terrestrial magnetic fluctuations or earthquakes, and good enough time resolution to get source directions based on differential signal delay. We would hope to have enough bandwidth to compare waveforms among different detectors and with theoretical predictions, and correlations with electromagnetic or neutrino events. How can we optimize the likelihood of satisfying these five criteria? "More theory is needed in the end so that we can take all this data, apply the theory of pattern recognition to it, and quantify the probability that gravitational radiation was actually discovered."

If we see nothing, then it is essential that we look for the known binary stars using detectors flown in space.

"In conclusion, Braginsky would like to emphasize his belief that the search for gravitational waves is a task that undoubtedly deserves the combined efforts of theorists and experimenters. Although this task is complicated, it will give much joy to those who try to solve it, both in the course of their work and when they finally succeed in solving it."

L. Smarr: In addition to searching for known periodic sources, one should do radiometry for noise from unknown sources. Isolating this background with a null measurement is important, even though the signal would be only noise.

R. Weiss: Gravitational and electromagnetic radiometry differ. In E&M, one can terminate an antenna at an almost arbitrarily low temperature; one can establish a zero. Here, a single antenna suffices. In gravitational radiometry, there is no way to shield such a low-temperature reference. The noise one sees in an antenna is a mixture of all sources of excitation. So the only way to do radiometry is by cross-correlating two antennas. The common mode signal is allegedly the gravitational wave signal.

THE ROLE OF QUANTUM NONDEMOLITION DEVICES

S. Boughn: I would like to comment from an experimentalist's viewpoint on the new kinds of detectors and quantum nondemolition (QND) devices for use at roughly kilohertz frequencies. I have summarized the past, present, and future on a temperature plot in Fig. 1. My point about QND detection is the following: The distinction between being linear-amplifier limited or QND limited is irrelevant until the quantum limit is reached. We still have seven orders of magnitude of sensitivity to go before we become amplifier limited. No measurement has ever been amplifier limited. We have a perfect amplifier, as far as we are concerned.

These new kinds of QND detectors will have the same transducer problems, etc., that our detector [at Stanford] has had for years, and their developers will have to fight just as hard as we will to jump these seven orders of magnitude to the quantum limit. It is very useful to think of new kinds of detectors as, for example, V. I. Braginsky has done, because they may not have the same technical difficulties we face, but their "QND-ness" has nothing to do with whether or not they will be able to jump to the quantum limit.

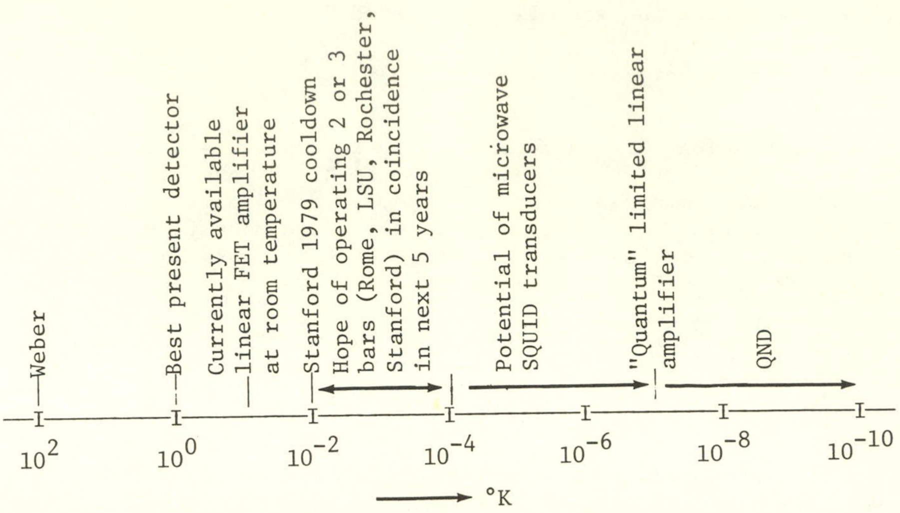

Fig. 1. A temperature/time line showing the progress of bars and other gravitational radiation detectors. The equivalent effective detector temperature indicated is proportional to the minimum detectable pulse energy. For the 5×10^3 kg bar at 10^{-2}°K contemplated for 1979, one can expect one accidental or "false alarm" per year at 10^{-1} GPU or $h \sim 10^{-17}$.

D. Long: Is bar temperature the limiting thing?

S. Boughn: Yes. The main problems faced by every such detector are thermal fluctuations in the bar and thermal noise in the amplifier. You can lower the temperature or raise the Q of the bar. When your transducer coupling time is very long, the limiting factor becomes the noise in the evolution of the bar. Right now, until the amplifier limit is reached, cooling by a factor would increase sensitivity by the same factor.

L. Smarr: Does all this carry over to crystal bars?

S. Boughn: Yes. Noise, effective temperature, etc., characterize crystals too. In comparing aluminum and crystal bars, you must also consider their respective cross-sections.

THE USE OF RMS QUANTITIES

R. Weiss: RMS quantities are system noise estimates. I would advocate their use. An h_{min} for a 90% certainty of an annual pulse not being an accidental is a 20-fold more conservative sensitivity level than an RMS h_{min} of the same value. RMS quantities are better.

D. H. Douglass: The RMS strain noise or noise temperature are the proper relevant quantities because they describe the antenna. When you quote an h or a burst intensity for a given event rate, you are assuming something about the waves, which we have not yet found.

COINCIDENCE DETECTION

Consensus: Must one have coincidences? What are the advantages of using several bars?

S. Boughn & R. Weiss: What you eliminate with coincidences is systematic, non-Gaussian noise such as from trucks and door slamming. For pure Gaussian noise, however, one gains relatively little with 2-bar coincidences, only a factor of 2 reduction in the value of the bar energy threshold that gives a particular accidentals rate, or, in other words, a factor of 2 gain in sensitivity.

R. Weiss: It is very hard to classify a systematic noise. That's the problem. You wouldn't believe a single bar anyway.

K. S. Thorne & D. H. Douglass: In principle, one could get that same factor of 2 [sensitivity improvement] by making the bar better, but a single bar cannot eliminate non-Gaussian spurious events.

CLOCK NOISE

R. Hellings: Some confusion has arisen in the analysis of the capabilities of spacecraft-Doppler gravitational radiation detection due to the various ways in which noise in the system may be characterized. I would like to discuss briefly the usefulness of three of these characterizations, applying my remarks to the newest H_2 maser clocks produced by R. F. C. Vessot, M. W. Levine, and E. M. Mattison (1978) at the Smithsonian Astrophysical Observatory.

A clock generates a signal of frequency ν, close to a perfectly constant frequency ν_0. The fluctuation, $\Delta\nu \equiv \nu-\nu_0$, is an instantaneous function of time, approximated by dividing the data record into contiguous narrow time segments and counting cycles during each segment. We can characterize the fluctuations in three ways:

1. The variance, $\langle\Delta\nu^2\rangle$, is an RMS time average of $\Delta\nu$ over all segments in the data record. This is the most intuitive measure of the noise, answering the experimenter's question, "What is the average size of the fluctuations?" The result, of course, may depend on the length of the data record (i.e. if there are long-term drifts) and on the number of segments into which the record has been divided.

2. The two-sample Allan variance (Allan, 1966), σ_ν, is the expectation value of the frequency difference between two successive time segments, rather than the expectation value of the fluctuations about the long-term average. This value will be essentially independent of the length of the record, though it will still depend on the size of the time segments.

3. The noise power spectral density, $S_\nu = S_\nu(f)$, is simply the power spectrum of the fluctuations observed in the record, given in units of (clock hz)^2/(spectral hz). This is a good physical measure of noise because various hardware noise sources tend to have characteristic and identifiable spectral distributions.

The relationships of $\langle\Delta\nu^2\rangle$ and σ_ν to S_ν can be written down by adapting a general expression derived by L. S. Cutler and C. L. Searle (1966) as follows:

$$\langle\Delta\nu^2\rangle = \frac{N}{N-1}\int_0^\infty df\; S_\nu(f)\,\frac{\sin^2(\pi f\tau)}{(\pi f\tau)^2}\left[1-\frac{\sin^2(N\pi f\tau)}{N^2\sin^2(\pi f\tau)}\right] \qquad (1)$$

$$\sigma_\nu^2 = 2\int_0^\infty df\; S_\nu(f)\,\frac{\sin^2(\pi f\tau)}{(\pi f\tau)^2}\left[1-\frac{\sin^2(2\pi f\tau)}{4\sin^2(\pi f\tau)}\right], \qquad (2)$$

where the record of length T is assumed to be divided into N segments, each of length τ. In both expressions, the term in brackets acts as a high-pass filter. However, as $T\to\infty$, $N\to\infty$, and the filter disappears in the formula for the total variance; so low-frequency contributions to the noise will make that integral diverge. The Allan variance avoids this divergence.

For those who are used to working only with the total variance, the two-sample variance may sometimes be deceiving as a measure of noise. If S_ν has large low-frequency contributions, then the two-sample variance will be smaller than the total variance. If S_ν has large high-frequency terms, then the two-sample variance will appear to be a pessimistic measure of the size of the noise, compared to the total variance. A more complete discussion of the behavior of the frequency standards and the methods to characterize their behavior is given by Vessot (1976).

Fig. 2 shows the two-sample Allan variance measured by Vessot et al. (1978) and Levine, Vessot, & Mattison (1978) for the latest SAO H_2 masers. The slopes indicate a combination of white phase noise ($S_y = h_2 f^2$, where $y \equiv \Delta\nu/\nu$) and white frequency noise ($S_y = h_0$), with $h_2 \sim 10^{-26}$ and $h_0 \sim 10^{-27}$. These results are very close to the predictions of stability made from the contribution of the

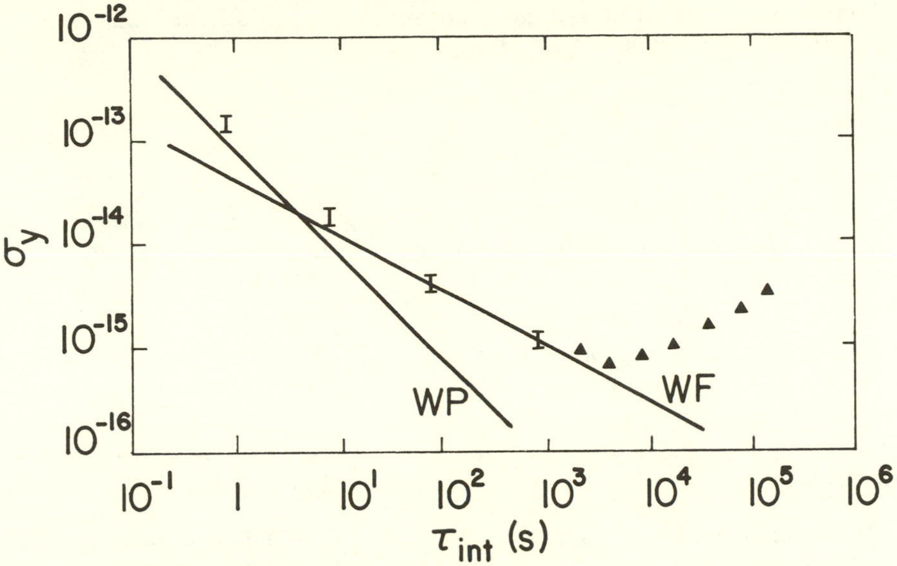

Fig. 2 Allan variance, σ_y, vs. integration time, τ_{int}, for R. Vessot's latest H_2 maser clock fitted to white phase (WP) noise, $S_y(f) = h_2 f^2$, $h_2 = 10^{-26}$, and white frequency (WF) noise, $S_y(f) = h_o = 10^{-27}$ (Vessot et al., 1978).

thermal noise to the maser oscillator system. Beyond $\tau \sim 1$ hr, Vessot reports that a residual systematic slow drift was apparent in the data that he attributes to a laboratory air conditioner failure 15 days before the data were taken. Based on the information in Fig. 2, I would suggest a number $[\Delta\ell/\ell]^2(f) \sim 10^{-27}$ Hz^{-1}, assuming integration times of $\tau_{int} \sim 10^3$ s, typical of Doppler tracking [which is roughly consistent with III.A. in Table 1b].

D. H. Douglass: I have three remarks.

The first remark has to do with the Doppler tracking of spacecraft. Consider the wave amplitude, h_p, of a pulse of duration τ_p given in terms of the phase, ϕ_p, measured by a Doppler tracking system operating at a tracking frequency, ν_o,

$$h_p \sim \frac{\phi_p}{2\pi\nu_o\tau_p} . \tag{3}$$

Phase, ϕ, is, in general, given by

$$\phi = 2\pi \int [\nu(t)-\nu_o]dt . \tag{4}$$

REUBEN EPSTEIN

I believe that the proper definition of burst noise, against which the pulse signal competes, is given by

$$\phi_{noise,p} = [\int_{1/\tau_p}^{\infty} S_\phi \, df]^{1/2} , \qquad (5)$$

where S_ϕ is the spectral density of phase fluctuations from all sources, e.g. solar wind, troposphere, clocks, amplifiers, etc. Another definition commonly used is the 2-sample Allan variance,

$$\phi_{noise,Allan} = [8 \int_0^{\infty} df \, S_\phi \, (\sin^2 \pi f\tau - \frac{\sin^2 2\pi f\tau}{4})]^{1/2} . \qquad (6)$$

This is used by the frequency and time community to compare clocks. It has nothing to do with gravitational radiation.

The Allan variance is always larger than the burst noise. It is too conservative. For solar wind and troposphere noise, it is known that $S_\phi \sim f^{-2.6}$, which gives $\phi_{noise,p}/\phi_{noise,Allan}$ = 0.13. For the $S_\phi \sim f^{-2}$ case considered earlier by R. Hellings this ratio is 1/4. "I would like Frank [Estabrook]...to never use the Allan variance again."

F. B. Estabrook (Frank): The Allan variance is for talking to clock people. This is important because clocks have been the limiting factor. Your message is that I've been conservative. I hope you're right!

WHITE DWARF OSCILLATIONS AND PROGRESS

D. H. Douglass: My second remark is that ordinary novae and unseen compact binaries should be given more attention as possible sources of gravitational radiation. [See "Discussion Session II" and J. P. A. Clark (this volume) for further discussion.]

For my third remark, I thought it might be interesting to consider the improvement of detector sensitivity vs. time, this is an order of magnitude every three years. Expressed in terms of energy sensitivity this works out to a rate of 6 orders of magnitude per ten years, which is not so bad. For comparison, in the field of radio astronomy, the average rate of improvement in energy sensitivity has been about one order of magnitude every ten years, spanning a 45-year period beginning with Jansky (John Kraus, 1978).

DISCUSSION SESSION I: DETECTION

MICROWAVE CAVITY DETECTORS

T. Piran: Could someone report on the proposal by Picasso and others for a microwave cavity detector?

F. Pegoraro: I am one of the authors of this proposal. The proposal deals with a method of detecting gravitational radiation with a microwave system. There is no specific system design, yet. The specifications are just orders of magnitude, and the noise analysis is not yet done. The system is similar to that proposed by Carlton Caves where a cavity has two modes whose nearly identical frequencies differ by an amount equal to the frequency of the gravitational radiation you wish to detect. The angular momenta of the modes differ by $\Delta\ell = 2$. Changes in the dielectric constant of the cavity due to the interaction of gravitational radiation changes the excitation level of each mode. One would detect gravitational radiation by measuring the occupation number of a previously unexcited mode.

At lower gravitational wave frequencies, such as 10^{-2} to 10^{2} Hz, the main problem is that the two modes are so close together that one sees the tail of the power distribution of the lower level when monitoring the upper level. This completely masks the effect of the gravitational radiation. What one must do, then, is to raise the Q of the cavity. To distinguish differences of 10^{2}-10^{3} Hz between modes of frequencies $\sim10^{9}$ Hz will require Q's of $\sim10^{13}$ or higher for $h \sim 10^{-23}$.

C. Caves: The microwave system is a transducer on a mechanical system which is the cavity. The biggest problem here, as with any mechanical detector, is Nyquist noise in the mechanical system.

CRYSTAL BARS

M. Karim: Can someone tell us about the status of the crystal antennas?

D. H. Douglass: We are studying the Q's of large silicon crystals which can be as large as 5 feet long, 3 inches in diameter, and weighing 20 kg, to see why Q's are what they are. We are simultaneously and independently developing a DC SQUID transducer. Speaking for V. I. Braginsky, he proposes a microwave cavity transducer and a sapphire (maybe now silicon) crystal antenna with one face of the crystal serving as a wall of the cavity. Crystal Q's have been observed up to $Q \sim 10^{10}$ at 1°K. Both our groups have yet to run their transducers with their crystals. We are not predicting when we will have a bar on the air. High Q will be even more important later because of the way it comes into the QND systems.

REUBEN EPSTEIN

SUMMARY, CONCLUSIONS, AND OUTLOOK

R. Weiss: Now Ron [Drever] will pull it all together for us.

R. Drever: I would like to start by saying that all the detection techniques discussed here have their place and that it is wrong and misleading to regard them as rivals. They cover different parts of the spectrum, etc., and they all should be developed.

Detection of gravitational radiation with spacecraft tracking would, at first, seem to be the most obvious and direct thing to try because it involves free masses. The $h \sim 10^{-16}$ that can be contemplated now by use of radio Doppler tracking induces a mere 10^{-2} mm motion of a spacecraft 1 A.U. away. Even sensing that would be a very impressive achievement! There could be important improvements and new experiments done by the cross-correlation of data from several spacecraft. Bursts have been predicted to be the most intense signals, but it is also practical to look for the stochastic background. One spacecraft could put an upper limit on this background, but two would be needed to discover it.

Greater sensitivity is needed for other anticipated sources. An amplitude of $h \sim 10^{-21}$ induces a relative motion of 1 angstrom in spacecraft 1 A.U. apart, which requires optical laser tracking to detect. This technique is good enough in principle for binary stars which must be there. These space techniques do not work as well for short-duration signals because of practical limitations due to the high power requirements of the lasers.

On the ground, the obvious first thing one would think of is to monitor the separation between two free masses, just as in space. But if you link the two masses with a piece of metal, you have what we call a "bar". Why on earth would you do this? First of all, a bar rings. The signal persists in the bar. This permits the use of a relatively poor transducer because you can look at the bar long enough to see a signal. That is the real reason why Weber used a bar and why others have followed. One useful thing which has first been attempted by H. Hirakawa, K. Tsubono, and M.-K. Fujimoto (1978) is the tuning of a resonant detector to a known continuous periodic source of known frequency [in this case, the Crab pulsar].

This test-mass "sandwich", or "bar", has snags. The link between the otherwise free masses puts thermal noise into the system. The range of detectable frequencies is limited by bars. Also, they cannot achieve the sensitivities we really need without overcoming their quantum limit.

The next technique is nearly-free masses whose separations are monitored by laser interferometers. Here there is no signal storage and thus the problem is one of not sensing fast enough. The interferometers do not store information. Fast sensing requires high-power lasers, so I think the final limitation here will involve the heating of the interferometer mirrors. Ground isolation is another serious problem. Thermal noise is not so serious a problem, except at low frequencies where it enters through the mass supports and residual gas. This explains why these devices can compete with cooled devices without actually cooling them. Their big advantage is that one can make the baseline very long, up to where the light-travel time matches the gravitational radiation period. This permits such a device to have as good a sensitivity as a QND bar detector without running into its own quantum limit. Also, one can observe a broad range of frequencies at once, which is especially important if you are not confident about source predictions. My hunch is that lasers will remain competitive with bars, neck and neck. Lasers will be best at lower frequencies, say 10 Hz - 10^3 Hz, because the interferometers will have more time to sense the signals.

In conclusion, I say, again, that all the different techniques should be pursued. There is a place for them all.

S. Boughn: I agree that we should pursue all possible ways of detecting gravitational radiation, but after having been at it for as long as I have, I have been impressed by the incredible difficulties of continuing in any one of these many directions. To take the experimental gravitational radiation program as it now stands and continue in all these directions toward workable detectors is going to be very expensive in time, labor, and money. Wouldn't it therefore be better to concentrate on fewer experiments?

R. Drever: I think I am a little more optimistic, perhaps, than most experimentalists in the field. I rather think, for example, that these interferometers will begin to work in the next year or two at an interesting sensitivity and will develop pretty rapidly, perhaps a little faster than the time scale you were suggesting. None of the problems look overwhelming [to me], so I am optimistic that the progress might not be that slow.

D. H. Douglass: "Steve [Boughn], I think you're wrong. Our field is not expanding in all different directions. The number of groups has remained constant in the past 3 or 4 years, and some groups are actually disappearing....I think you will be more optimistic after you get your first cooldown." [laughter]

L. Smarr: I think one very important thing to stress is exactly that which you so beautifully summarized, Ron [Drever]. In the past, there has been some disagreement on the best approach for detecting gravitational radiation. But now, I think, we have all stepped back and taken a long-range view of our field, [and we can see more clearly how we all fit in]. This was one major goal of this workshop. We have looked down the road a long way, and we can see how we're going to get there. Previously, there may have been an illusion of chaos, but now everything looks a lot more orderly. I think there is good reason to be optimistic.

R. L. Forward: "At least we are now able to draw the antenna sensitivity curves and the source [strength] curves on the same graph. Surely [laughter and applause] this means we have come a long way."

Unidentified: "That says it all."

REFERENCES

Allan, D. W. (1966). Statistics of atomic frequency standards. Proc. IEEE, 54, pp. 221-230.

Cutler, L. S. & Searle, C. L. (1966). Some aspects of the theory and measurement of frequency fluctuations in frequency standards. Proc. IEEE, 54, pp. 136-154.

Hirakawa, H., Tsubono, K. & Fujimoto, M.-K. (1978). Search for gravitational radiation from the Crab pulsar. Phys. Rev. D, 17, pp. 1919-1923.

Kraus, J. (1978). Private communication to D. H. Douglass.

Levine, M. W., Vessot, R. F. C. & Mattison, E. M. (1978). Performance evaluation of the SAO VLG-11 atomic hydrogen masers. In Proceedings of the 32nd Annual Frequency Control Symposium, CFA Preprint #999.

Vessot, R. F. C. (1976). Frequency and time standards. In Methods of Experimental Physics, ed. M. L. Meeks, 12 (Astrophysics), Part C, pp. 198-227. Academic Press, New York.

Vessot, R. F. C., Levine, M. W. & Mattison, E. M. (1978). Comparison of theoretical and observed hydrogen maser stability limitations due to thermal noise and the prospect for improvement by low-temperature operation. In Proceedings of the 9th Precise Time and Time Interval Conference (NASA Technical Memorandum 78104), CFA Preprint #895.

KINEMATICS AND DYNAMICS OF GENERAL RELATIVITY

James W. York, Jr.

Department of Physics and Astronomy
University of North Carolina at Chapel Hill

1 INTRODUCTION

In these lectures we shall look at spacetime from the point of view of the Cauchy problem. According to this approach, a classical gravitational field is the time history of the geometry of a spacelike hypersurface. To construct a gravitational field, one solves the initial-value problem, prescribes a reference system, and integrates the dynamical equations along the trajectories of the reference system. Initial-value and evolution equations or equations of state of any external sources must also be taken into account.

Carrying out this program requires the introduction of a space-plus-time (3 + 1) formalism. Because the problems are necessarily global, we shall not introduce explicit coordinates except when necessary. The emphasis will be on the geometry of the general problem as a guide to methods of calculation. No rigorous proofs are provided. These are found in the references cited. All notes and literature citations are given in Section 10.

We shall adopt the full Einstein equations from the beginning and introduce none of the largely successful approximation schemes that have been developed in the literature. I feel that it may be better in strong-field, high-velocity problems to leave sources and fields tightly interwoven and only after a solution is found to attempt to describe the results in terms of motion of sources, radiation reaction, radiated energy, etc. This approach is feasible only because of the present rapid development of numerical relativity using high-speed computers. I hope that these lectures will provide a suggestive guide to formulating such problems.

2 FOLIATIONS OF SPACETIME

A foliation $\{\Sigma\}$ is a family of three surfaces Σ that fills spacetime V such that locally the surfaces arise as the level surfaces of a scalar function. The foliation is spacelike if each of the slices is spacelike. The foliation represents a particular description of V that results from the evolution of a gravitational field and its sources. This description depends on the choice of initial data on a three-surface Σ , the construction of a family of slices $\{\Sigma\}$, and the choice of a congruence of curves passing through $\{\Sigma\}$ along which the data are propagated from one slice to the next by the equations of motion. I shall attempt to present this subject in a logical and general way that makes it useful for setting up calculations and perhaps also for exploring global structure numerically.

The foliation is described by a closed one-form $\Omega = \Omega_a E^a$ on V, with E^a denoting a general basis of forms on V and E_a denoting the reciprocal basis of vectors. Since the form is closed we have, with ∇ denoting the spacetime covariant derivative,

$$d\Omega = 0 \leftrightarrow \nabla_{[a}\Omega_{b]} = 0 . \tag{1}$$

Therefore, locally there is a scalar function τ such that $\Omega = d\tau$ or $\Omega_a = \nabla_a \tau$.

The spacetime metric enables us to define a norm $\|\Omega\|$ in terms of a strictly positive scalar function α, the lapse function, such that

$$\|\Omega\|^2 = g^{ab}\Omega_a\Omega_b = -\alpha^{-2} . \tag{2}$$

Hence, we have a normalized one-form ω associated with $\{\Sigma\}$ given by

$$\omega_a = \alpha\Omega_a , \quad \|\omega\|^2 = -1 , \tag{3}$$

where the minus sign denotes that the slices are spacelike. The unit one-form satisfies

$$\omega\wedge d\omega = 0 \leftrightarrow \omega_{[a}\nabla_b\omega_{c]} = 0 . \tag{4}$$

The unit normal vector field of the slices is $n = n^a E_a$. We define this by

$$n^a = -\, g^{ab}\omega_b \tag{5}$$

with the negative sign chosen so that locally n points in the direction of increasing τ and therefore satisfies

$$\langle \omega, n \rangle = 1 \tag{6}$$

where $\langle \ , \ \rangle$ denotes the interior product. Observe that

$$n_a = -\,\omega_a = -\,\alpha\Omega_a . \tag{7}$$

These conventions motivate the following notations for a vector W and a one-form μ :

$$W^{\hat{n}} = W^a \omega_a = - W^a n_a \ , \tag{8}$$

$$\mu_{\hat{n}} = \mu_a n^a \ . \tag{9}$$

The vector n can be interpreted as the four-velocity field of observers instantaneously at rest in the slices Σ. We shall call them the Eulerian observers because their motion follows the slices.

The metric $\gamma = \gamma_{ab} E^a \otimes E^b$ induced on the slices by g_{ab} is given by

$$\gamma_{ab} = g_{ab} + n_a n_b = g_{ab} + \omega_a \omega_b \ . \tag{10}$$

We refer to γ as a spatial tensor because $\gamma_{ab} n^b = 0$. It has a contravariant form

$$\gamma^{ab} = g^{ac} g^{bd} \gamma_{cd} = g^{ab} + n^a n^b \ . \tag{11}$$

Its mixed form is the unit operator of projection onto the slices:

$$\gamma^a_b \equiv \perp^a_b = \delta^a_b + n^a n_b = \delta^a_b - n^a \omega_b \ , \tag{12}$$

$$\perp^a_a \equiv \mathrm{tr} \perp \equiv \perp^a_b \perp^b_a = 3 \ . \tag{13}$$

The covariant derivative D induced by ∇ on the slices is defined by projection. For a scalar ψ we have $D_a \psi = \perp^b_a \nabla_b \psi$. For a vector W^a tangent to a slice ($W^a n_a = 0$), we define

$$D_a W^b \equiv \perp^b_d \perp^c_a \nabla_c W^d \equiv \perp \nabla_a W^b \tag{14}$$

with the extension to other spatial tensors defined similarly. In particular, one finds, as expected, that

$$D_a \gamma_{bc} = 0 \ . \tag{15}$$

It can be shown that D is independent of the behavior of n, and of the spatial tensors on which it operates, away from a given slice Σ.

It is now possible to define the curvature tensor Riem(γ) = $R^d{}_{cba}$ of a slice by requiring that for every spatial vector W, we have

$$D_{[a} D_{b]} W_c = \frac{1}{2} W_d R^d{}_{cba} \ , \tag{16}$$

$$n_d R^d{}_{cba} = 0 \ . \tag{17}$$

This amounts to the usual definition when worked out in a basis tangent to the slice. The Ricci tensor R_{ca} and the scalar curvature R follow from the standard definitions. (For the Ricci tensor, we contract the first and third indices.)

The imbedding of a slice and of the foliation also require for their description the extrinsic curvature tensor

$$K_{ab} = -\perp\nabla_{(a}n_{b)} = \perp\nabla_{(a}\omega_{b)} \tag{18}$$

From the equivalent expression

$$K_{ab} = -\frac{1}{2}\pounds_n\gamma_{ab} = -\frac{1}{2}\perp\pounds_n g_{ab}\ , \tag{19}$$

where $\pounds$ denotes the Lie derivative, we see that K is a "velocity" of the spatial metric. This velocity is defined with respect to the local proper time of the Eulerian observers of a slice. It is most important to recognize that K does not depend on the behavior of the vector field n away from a given slice. If another velocity of γ is defined by any other vector field passing through Σ , as is certainly permissible, then this new velocity will depend on the behavior of the vector on and near Σ. Hence, the tensors $\gamma(\Sigma)$, $K(\Sigma)$ characterize one slice Σ of the foliation. They will therefore be suitable for initial-value problems.

The complete geometry of a slice imbedded in spacetime is represented by the triple (Σ,γ,K). This triple will turn out to be an initial data set for V. The relation of initial data to the spacetime curvature is found by applying the projection of the commutator of ∇'s on an arbitrary spatial vector and on n . The result is the Gauss-Codazzi equations

$$\perp R_{abcd} = \mathcal{R}_{abcd} + K_{ac}K_{bd} - K_{ad}K_{bc}\ , \tag{20}$$

$$\perp R_{abc\hat{n}} = D_bK_{ac} - D_aK_{bc}\ . \tag{21}$$

These equations may also be derived as necessary and sufficient integrability conditions for the imbedding of (Σ,γ,K) in (V,g). They effectively supply 14 of the 20 algebraically independent components of Riem(g) and must hold for each and every member of the foliation $\{\Sigma\}$. The counting of independent components can be seen from the usual symmetries of the Riemann tensor plus the fact that in three dimensions the Riemann and Ricci tensors are equivalent:

$$\mathcal{R}_{abcd} = 2\,\gamma_{a[c}\mathcal{R}_{d]b} + 2\,\gamma_{b[d}\mathcal{R}_{c]a} + \mathcal{R}\gamma_{a[d}\gamma_{c]b}\ . \tag{22}$$

Four of the ten Einstein equations $G_{ab} = \kappa T_{ab}$ ($\kappa = 8\pi G$, $c = 1$) are completely expressible in terms of (20) and (21), which contain no second derivatives of the metric in a timelike direction. Hence, these will be initial-value equations or constraints. They are

$$2\, G_{\hat{n}\hat{n}} = R + (\mathrm{tr}K)^2 - K_{ab}K^{ab} = 2\,\kappa\rho, \tag{23}$$

$$\perp G^{a\hat{n}} = D_b(K^{ab} - \gamma^{ab}\mathrm{tr}K) = \kappa j^a\ , \tag{24}$$

where $\mathrm{tr}\, K = \gamma^{ab}K_{ab}$. Here $\rho = T_{\hat{n}\hat{n}}$ and $j^a = \perp T^{a\hat{n}} = -\perp T^a{}_{\hat{n}}$ are the energy and momentum densities determined by the Eulerian observers of Σ . The latter point is important to keep in mind in applications. For example, the pressure and density of a perfect fluid are ordinarily referred to observers at rest in the fluid rather than to the Eulerian observers of the foliation. Equation (23) is called the scalar or Hamiltonian constraint and (24) the vector or momentum constraint.

The remaining Einstein equations refer to the foliation itself rather than just to its individual slices. To form these equations we need the other six components of the spacetime Riemann tensor. For the computation it is helpful to note that the acceleration of the Eulerian observers is

$$a^a = n^b\nabla_b n^a\ , \quad a^a n_a = 0\ , \tag{25}$$

and that from using (4) we have

$$n_{[a}\nabla_b n_{c]} = 0 \leftrightarrow \perp \nabla_{[a}n_{b]} = 0\ . \tag{26}$$

Therefore we may write

$$\nabla_a n_b = -\, K_{ab} - n_a a_b\ . \tag{27}$$

Moreover, from $n_a = -\alpha\Omega_a$ and $\nabla_{[a}\Omega_{b]} = 0$ we find

$$a_b = D_b \ell n\, \alpha . \tag{28}$$

The computation of $\perp R_{a\hat{n}b\hat{n}}$ now follows from the above together with the definition of $\pounds_n K_{ab}$. This latter term contains second derivatives of the metric in the (timelike) unit normal direction.

However, $\pounds_n$ is not the natural orthogonal time derivative with which to propagate the tensors γ and K from one slice of the foliation to the next. The vector n is naturally paired with the normalized one-form $\omega_a = \alpha\Omega_a$ through (6) : $\langle\omega, n\rangle = 1$. But we may write this as

$$\langle\omega, n\rangle = \langle\alpha\Omega, n\rangle = \langle\Omega, \alpha n\rangle = 1\ . \tag{29}$$

Therefore $N^a = \alpha n^a$ is the natural orthogonal vector field that connects the slices of $\{\Sigma\}$. If locally $\Omega_a = \nabla_a\tau$, then the orthogonal proper time interval between level surfaces $\tau = \tau_1$ and $\tau = \tau_1 + \delta\tau$ is $\alpha\delta\tau$, where

$$\alpha = (-g_{ab}N^a N^b)^{\frac{1}{2}} . \tag{30}$$

An important property of N^a is that

$$\pounds_N \perp^a{}_b = 0 . \tag{31}$$

This implies that the time derivative operator $\pounds_N$ applied to any spatial tensor is a spatial tensor. Collecting the above results we compute $\pounds_N K_{ab}$ and find

$$\perp R_{a\hat{n}b\hat{n}} = \alpha^{-1}\pounds_N K_{ab} + K_{ac}K^c{}_b + \alpha^{-1}D_a D_b \alpha . \tag{32}$$

Before displaying the dynamical Einstein equations, we should note the important point that for a given foliation specified by Ω, the "dual" time vector N is by no means unique. We can satisfy $\langle\Omega, \text{vector}\rangle = 1$ by choosing any vector t of the form

$$t^a = N^a + \beta^a = \alpha n^a + \beta^a , \quad \beta^a n_a = 0 . \tag{33}$$

Hence there is an arbitrary spatial "shift vector" β at our disposal at each point. This vector represents the basic remaining kinematical freedom available in describing spacetime once a foliation is prescribed. For a definite choice of β, t becomes the tangent to a family of curves threading the slices of $\{\Sigma\}$. These curves are parameterized by τ if $\Omega_a = \nabla_a \tau$. Thus we write $t^a E_a = \partial/\partial\tau$ and

$$\langle d\tau, \partial/\partial\tau\rangle = 1 . \tag{34}$$

The curves are "tilted" relative to n if $\beta \neq 0$. See Figure 1.

The equation of motion of γ follows as an identity from the definition of K and t. Using the basic property $\pounds_t = \pounds_N + \pounds_\beta$, we find

$$\pounds_t \gamma_{ab} = -2\alpha K_{ab} + \pounds_\beta \gamma_{ab} . \tag{35}$$

In computing $\pounds_t K$, it is helpful to write the Einstein equations in the form

$$R_{ab} = \kappa(T_{ab} - \frac{1}{2} g_{ab} g^{cd} T_{cd}) . \tag{36}$$

We define the spatial stress tensor S by

$$S_{ab} = \perp T_{ab} . \tag{37}$$

Then we have

$$T_{ab} = S_{ab} + 2j_{(a}n_{b)} + \rho n_a n_b . \tag{38}$$

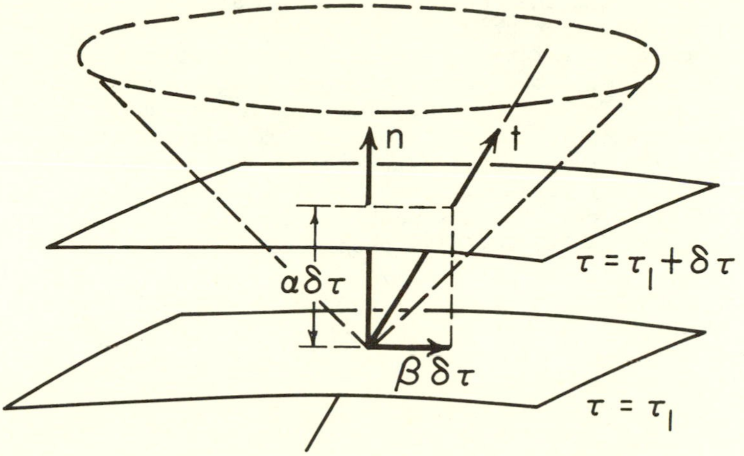

Fig. 1. Illustrated are parts of two nearby slices of the foliation $\{\Sigma\}$. The "time" vector is $t = \alpha n + \beta$, where α is the lapse function, β the shift vector, and n the unit normal. The dashed figure represents a local light cone. Not shown is the acceleration a corresponding to the four-velocity n of the Eulerian observers. We have $g(a,n) = g(\beta,n) = 0$; thus, both a and β are tangent to the first slice.

Using (38) and $g^{cd}T_{cd} = -\rho + \text{tr } S$ in $\perp R_{ab}$, we find

$$\pounds_t K_{ab} = -D_a D_b \alpha + \alpha[R_{ab} - 2K_{ac}K^c{}_b + K_{ab}\text{tr } K - \kappa(S_{ab} - \frac{1}{2}\gamma_{ab}\text{tr } S) - \frac{1}{2}\kappa\rho\gamma_{ab}] + \pounds_\beta K_{ab} \,. \tag{39}$$

Note that we have not inserted the constraints into (39). We can express the "conservation equations" $\nabla_b T^{ab} = 0$ in this language. A generalized "continuity equation" follows from $n_a \nabla_b T^{ab} = 0$:

$$\pounds_t \rho + \alpha D_b j^b = \alpha(S^{bc}K_{bc} + \rho \text{ tr } K) - 2j^b D_b \alpha + \pounds_\beta \rho \,. \tag{40}$$

A generalized "Euler equation" follows from $\perp \nabla_b T^{ab} = 0$:

$$\pounds_t j^a + \alpha D_b S^{ab} = \alpha(2K^{ab}j_b + j^a \text{ tr } K) - S^{ab}D_b\alpha - \rho D^a \alpha + \pounds_\beta j^a \,. \tag{41}$$

The equations for $\pounds_t S$ would follow from an equation of state or dynamical equations of the sources.

3 FRAMES AND FOLIATIONS

So far the spacetime bases of vectors and forms have been completely arbitrary and the constraints and equations of motion have been given in their most general form relative to a foliation described by a closed one-form Ω and a time derivative along t such that $\langle\Omega, t\rangle = 1$. The spacetime bases can be specialized in any desired way. For example, we could use a null basis or an orthonormal basis with n as the time-leg. The latter would define physical Eulerian frames adapted to $\{\Sigma\}$. Here, however, we shall introduce in the most general form "3+1" or <u>computational frames</u>. Later, in asymptotically flat spaces, we shall relate the computational frames to the "Minkowskian observers at spatial infinity".

We introduce on every slice Σ a basis of vectors e_i $(i = 1,2,3)$ that is tangent to Σ. Hence, $e_i = e^a{}_i E_a$ and

$$\langle\Omega, e_i\rangle = \Omega_a e^a{}_i = -\alpha^{-1} g(n, e_i) = 0 \ . \tag{42}$$

We require that each of these vectors satisfies

$$[t, e_i] = \pounds_t e_i = 0 \ , \tag{43}$$

i.e., the triad e_i is attached to t and is "dragged along" t. Since $\pounds_t \Omega = 0$, e_i remains tangent to the slices of the foliation. The triad is neither required to be orthonormal nor to be a coordinate basis $e_i = \partial/\partial x^i$, i.e., $[e_i, e_j](x) = e_k C^k{}_{ij}(x) \neq 0$ in general. However, we of course do not exclude such specializations.

The reciprocal basis of forms ρ^i on Σ is required to satisfy

$$\langle\rho^i, t\rangle = \rho^i{}_a t^a = 0 \ , \tag{44}$$

$$\langle\rho^i, e_j\rangle = \delta^i{}_j \ . \tag{45}$$

The one-forms ρ^i need not be closed. From the reciprocity of e_j and ρ^i we have $d\rho^i = -\tfrac{1}{2} C^i{}_{jk} \rho^j \wedge \rho^k \neq 0$ in general.

The relations $t = \alpha n + \beta$ and $\pounds_t e_i = 0$ give us an interpretation of the shift vector β. Consider a triad e_i with origin at an event p located in a slice Σ with local description $\tau = \tau_1 = \text{constant}$. If this triad is dragged along $t(p)$ to the nearby slice $\tau_2 = \tau_1 + \delta\tau$, its origin is now at an event p' in τ_2. On the other hand, if e_i were dragged along the normal connecting vector $N(p)$, its origin would be located at an event p'' in τ_2. The vector displacement $(p' - p'')$ evaluated on τ_1 is the shift vector $\beta(p)$. Thus, the shift vector measures the spatial velocity in τ-time of the triad e_i relative to the normal direction. This velocity measured in the local <u>proper</u> time $\alpha\delta\tau$ of the Eulerian observers is $v = \alpha^{-1}\beta$.

The basic equations (23), (24), (35), (39), (40), and (41) can now all be rewritten in the spatial basis by replacing a, b, c by i, j, k. The kinematic terms involving β become

$$\pounds_\beta \gamma_{ij} = D_i \beta_j + D_j \beta_i \,, \tag{46}$$

$$\pounds_\beta K_{ij} = \beta^k D_k K_{ij} + K_{ik} D_j \beta^k + K_{kj} D_i \beta^k \,, \tag{47}$$

$$\pounds_\beta \rho = \beta^i D_i \rho \,, \tag{48}$$

$$\pounds_\beta j^i = \beta^j D_j j^i - j^j D_j \beta^i \,. \tag{49}$$

As a technical point, we note that the preceding expressions (46)-(49) involving Lie derivatives use D, not ∂ . If we replace the D's by ∂'s, the expressions are not valid unless the $C^i{}_{jk}$'s all vanish, i.e., in a coordinate or holonomic spatial basis. In general, if we use ∂'s, then extra terms involving the $C^i{}_{jk}$'s appear. If we use D's, these extra terms have already been absorbed in the $\Gamma^i{}_{jk}$'s. This follows from the definition

$$D_i T^j{}_k = \partial_i T^j{}_k + T^\ell{}_k \Gamma^j{}_{\ell i} - T^j{}_\ell \Gamma^\ell{}_{ki} \,, \tag{50}$$

with

$$\Gamma^i{}_{jk} = \left\{ {i \atop jk} \right\} + \tfrac{1}{2}(\gamma^{i\ell}\gamma_{jm}C^m{}_{\ell k} + \gamma^{i\ell}\gamma_{km}C^m{}_{\ell j} - C^i{}_{jk}). \tag{51}$$

Here $\{\ \}$ is the usual Christoffel symbol defined in terms of γ . Of course, none of these remarks says that the Lie derivative depends on either a metric or a notion of parallel transport. The $C^i{}_{jk}$'s merely reflect the lack of commutivity of the spatial basis, which in turn is independent of metric and parallel transport.

As a final remark, we note that what has been done is to introduce the most general spacetime frames in which the standard 3+1 formulas are valid. This spacetime frame could be called "partially holonomic". Thus, suppose the vector frames we are discussing are denoted by λ_a with $\lambda_0 = t$ and $\lambda_i = e_i$. In general, $[\lambda_a, \lambda_b] = \lambda_c C^c{}_{ab}$. Our frames are defined in such a way that any of the C's with an index "0" in any position vanishes.

4 THE CAUCHY PROBLEM

We have described a spacetime V with metric g in terms of a foliation and have taken the point of view that (V,g) is known. We shall assume that g_{ab} is globally hyperbolic, which implies that

(A) There are no closed or "almost closed" causal paths.
(B) V possesses a Cauchy surface.
(C) The Cauchy surface can be described in terms of a

universal time function, i.e., one whose gradient is everywhere timelike. Thus, we may write $\Omega = d\tau$.

(D) The topology of V is $\Sigma \times \mathbb{R}$.

A Cauchy surface is defined as a submanifold of spacetime such that each causal path without endpoint intersects it once and only once.

It is known that suitable initial data are the quantities $\{\gamma, K, \rho, j\}$ on a Cauchy surface . There is always a development of these data for a finite time. Moreover, it has been established that there is, in a suitable sense, a maximal or "largest" development of the data.

Here we are looking at the Cauchy problem from the point of view that we do not yet know the development or maximal development. We would want to find a constructive procedure for finding the maximal development, or enough of it to answer questions of physical interest, such as what radiation would a distant observer record, do two black holes merge into one when they collide, etc. I do not know a fool-proof method of satisfying these requirements, but in these notes I shall suggest some fairly general points of view that may be helpful.

The form of the Einstein equations that have been used to establish the existence of Cauchy developments, Cauchy stability, etc. is not the form that will be used here. There one uses harmonic spacetime coordinates and the Einstein equations are written as a rigorously hyperbolic first or second order system. Here we shall speak of developments using the equations for $\pounds_t \gamma$ and $\pounds_t K$ with elliptic equations determining t in terms of α and β . Hence our system is not hyperbolic. Even the equations for $\pounds_t \gamma$ and $\pounds_t K$ may not always be hyperbolic in a strict sense, because our requirements on α and β will not always guarantee that t is timelike! All we "really" need is that $\alpha > 0$ so that the slices advance into the future. We know from using harmonic coordinates that a development exists and that any suitable coordinates, such as the ones we shall use, can be connected to the harmonic ones by a suitable transformation. Such a transformation would not alter the physics of the problem. However, the choice of variables and the description of spacetime is different. Our variables are known to be useful in finding numerical solutions of the Einstein equations and for other purposes, such as in the development of a canonical formalism for gravity.

The entire program of solving Einstein's equations constructively as a Cauchy problem hinges crucially on the construction of Ω and t . One part of the program, however, is in a more nearly definitive form. That is the initial-value problem, which is treated next. Even here there are still a number of important unanswered questions remaining for people to study.

Once the initial-value equations are solved, they automatically continue to hold throughout the development. This follows from the Einstein equations together with $\nabla_b T^{ab} = 0$. This fact is part of the Cauchy problem's being "well-posed".

5 INITIAL-VALUE PROBLEM

The algorithm for solving the constraints that I shall review here uses as its basic tool a conformal transformation of the metric γ of an initial slice Σ . We set

$$\gamma_{ij} = \phi^4 \hat{\gamma}_{ij} , \tag{52}$$

where the conformal factor ϕ is assumed to be a strictly positive scalar function. The "trial" metric $\hat{\gamma}$ is assumed to be given. The idea is to find a suitable ϕ , and hence the physical metric γ , from the constraints. From (52) we find that

$$\Gamma^i{}_{jk} = \hat{\Gamma}^i{}_{jk} + 2\phi^{-1}(\delta^i{}_j \hat{D}_k \phi + \delta^i{}_k \hat{D}_j \phi - \hat{\gamma}_{jk} \hat{\gamma}^{im} \hat{D}_m \phi) , \tag{53}$$

$$R = \hat{R}\phi^{-4} - 8\phi^{-5} \hat{\Delta}\phi , \tag{54}$$

where $\hat{R} = R(\hat{\gamma})$ and $\hat{\Delta} = \hat{\gamma}^{jk} \hat{D}_j \hat{D}_K$.

The definition (52) of a conformal transformation does not tell us how to treat K, ρ, and j. We have to invent rules for these quantities based on the ideas of getting the constraint equations into a simple form and on a reasonable physical interpretation of the results.

In treating K , it is convenient to work in terms of its trace-free and trace parts. The trace-free part is denoted by

$$A^{ij} = K^{ij} - \frac{1}{3} \gamma^{ij} \text{ tr K} \tag{55}$$

and the trace part is

$$\text{tr K} = \gamma^{ij} K_{ij} . \tag{56}$$

Our strategy will be to treat tr K as a given scalar function that is not subjected to a conformal transformation. However, A will be both conformally transformed and decomposed or split into a sum of a divergence-free, trace-free part and another trace-free part derived from differentiation of a vector. A question about the order of performing these two operations arises. Each method has certain advantages over the other and they do not lead to precisely equivalent initial data except in special cases. Mathematically, we may say that the diagram of the operations of conformal transformation and decomposition is not commutative. Of course, just how we obtain a suitable initial data set is not nearly as important as how we evolve it so as to obtain unambiguous physical results.

Here I shall outline one of these two methods that leads to the simpler equations. This means that we first perform a conformal transformation on γ and A and then decompose the transformed A in terms of the trial metric $\hat{\gamma}$. Thus, set

$$A^{ij} = \phi^{-10}\hat{A}^{ij} \tag{57}$$

along with (52). The reason for this is that the momentum constraints (24) can be written in terms of the covariant divergence of A and one finds identically for all $\phi > 0$ that

$$D_j A^{ij} = \phi^{-10}\hat{D}_j\hat{A}^{ij} , \tag{58}$$

and of course $\mathrm{tr}\, A = \hat{\mathrm{tr}}\, \hat{A} = 0$.

Any trace-free symmetric tensor such as $\hat{A}$ can be split into two pieces by the following prescription. Set

$$\hat{A}^{ij} = \hat{A}_*^{ij} + (\hat{\ell}W)^{ij} \tag{59}$$

where

$$(\hat{\ell}W)^{ij} = \hat{D}^i W^j + \hat{D}^j W^i - \frac{2}{3}\,\hat{\gamma}^{ij}\hat{D}_k W^k , \tag{60}$$

$$\hat{D}_j\hat{A}_*^{ij} = \hat{\mathrm{tr}}\, \hat{A}_* = 0 , \tag{61}$$

$$\hat{D}_j(\hat{\ell}W)^{ij} = \hat{D}_j\hat{A}^{ij} , \tag{62}$$

$$\hat{D}_j(\hat{\ell}W)^{ij} \equiv (\hat{\Delta}_\ell W)^i = (\Delta W)^i + \frac{1}{3}\,\hat{D}^i(\hat{D}_j W^j) + \hat{R}^i{}_j W^j . \tag{63}$$

The "Laplacian" $\hat{\Delta}_\ell$ acting on vectors is elliptic and formally self-adjoint, just as is the ordinary covariant Laplacian $\hat{\Delta} = \hat{D}^k\hat{D}_k$ acting on scalars. Note that $\hat{A}_*$ and $(\hat{\ell}W)^{ij}$ are formally orthogonal in the Riemannian measure defined by $\hat{\gamma}$ on Σ:

$$\int_\Sigma \hat{A}_*^{ij}\,(\hat{\ell}W)^{k\ell}\hat{\gamma}_{ik}\hat{\gamma}_{j\ell}(\det\hat{\gamma})^{\frac{1}{2}}d^3x = 0 . \tag{64}$$

This follows from integration by parts and neglect of boundary terms. This turns out to be rigorous L^2-orthogonality when carefully analyzed.

The above results show how γ and K are handled. We regard $\hat{\gamma}$, $\mathrm{tr}\, K$, and $\hat{A}_*$ as given. The latter could be obtained, for example, by splitting an arbitrary symmetric trace-free tensor T according to the above method and setting $T_* = A_*$. The constraints are supposed to determine ϕ and W. However, we must still address the question of how ρ and j are to be treated.

Just as for K, we must invent the rules for ρ and j. If these quantities arise from another field coupled to gravity, the conformal properties of this field and the nature of the coupling must be taken into account. These will give us formulas for ρ and j. Examples of this type are treated in the literature and will not be repeated here. Instead, we shall just consider the simple but practical case in which the source is treated phenomenologically as a continuum of "matter" (e.g., a fluid) with constitutive relations among ρ, j and the stresses S (e.g., an equation of state $p = p(\rho)$, p = pressure). It is assumed that such relations do not restrict the initial values of ρ and j. (The stresses do not enter the gravity initial-value equations.)

There are several "more fundamental" arguments available, but here we shall reason pragmatically. The momentum constraints relate $D_j A^{ij}$ and j^i. Because of (52), therefore we set

$$j^i = \phi^{-10}\hat{j}^i \; . \tag{65}$$

To guarantee that the local "dominance of energy" condition on the sources, i.e.,

$$\rho^2 - \gamma_{ij} j^i j^j \geq 0 \; , \tag{66}$$

will hold for all $\phi > 0$, we therefore choose

$$\rho = \phi^{-8}\hat{\rho} \tag{67}$$

because then

$$\rho^2 - \gamma(j,j) = \phi^{-16}[\hat{\rho}^2 - \hat{\gamma}(\hat{j},\hat{j})] \; . \tag{68}$$

Actually, we shall see later that we need for technical reasons $\rho = \phi^{-s}\hat{\rho}$ with $s > 5$ even in the simplest cases. We shall also assume the local non-negativity of energy, $\rho \geq 0$, which is equivalent to $\hat{\rho} \geq 0$ for $\phi > 0$.

Here we should like to point out that in certain simple problems, the algorithm $\rho = \phi^{-6}\hat{\rho}$ has an advantage over (67) because one has "mass conservation" in the sense that

$$\int_\Sigma \rho (\det \gamma)^{\frac{1}{2}} d^3x = \int_\Sigma \hat{\rho} (\det \hat{\gamma})^{\frac{1}{2}} d^3x \; . \tag{69}$$

This follows from $(\det \gamma)^{\frac{1}{2}} = \phi^6 (\det \hat{\gamma})^{\frac{1}{2}}$.

Collecting the above formulas and inserting them into the constraints (23) and (24) yields

$$8\hat{\Delta}\phi - \hat{R}\phi + (\hat{A}_{ij}\hat{A}^{ij})\phi^{-7} - \tfrac{2}{3}(\mathrm{tr}\, K)^2\phi^5 + 2\kappa\hat{\rho}\phi^{-3} = 0 \; , \tag{70}$$

$$(\hat{\Delta}_\ell W)^i - \tfrac{2}{3}\phi^6 \hat{D}^i \,\mathrm{tr}\, K - \kappa\hat{j}^i = 0 \; . \tag{71}$$

Assuming that we can solve this quasi-linear elliptic system for the "potentials" ϕ and W, the initial-data set is obtained by setting

$$\gamma_{ij} = \phi^4 \hat{\gamma}_{ij} , \tag{72}$$

$$K_{ij} = \phi^{-2}[\hat{A}_{*ij} + (\hat{\ell}W)_{ij}] + \frac{1}{3}\phi^4\hat{\gamma}_{ij} \text{ tr } K , \tag{73}$$

$$\rho = \hat{\rho}\phi^{-8}, \; j^i = \hat{j}^i\phi^{-10} . \tag{74}$$

The analysis and solution of (70) and (71) are simplified by the assumption that tr K satisfies ∂_i tr K = 0 on the initial slice Σ. In particular, this includes the well-known maximal slicing condition tr K = 0 , which will be discussed further. The condition that tr K is position-independent on Σ decouples (70) and (71) in the sense that one may then solve (71) independently of ϕ and substitute the solution into (70). One describes this initial restriction on tr K by saying that the mean extrinsic curvature is constant. It need not, of course, be held constant in the development of the initial data.

Existence, uniqueness, and linearization stability of initial data that satisfy the constraints have been reviewed comprehensively in several recent articles and will not be dealt with in detail here. Clearly, if ∂_i tr K = 0 , then (71) is a linear system independent of (70) and can be analyzed successfully, and even solved analytically in some cases, in terms of a combination of standard mathematical physics and modern analysis. Equation (70), on the other hand, is more difficult, and is a subject of ongoing research, because of its essential nonlinearity. A variety of "good" results have been established.

To close this section, I wish to illustrate one aspect of the nonlinear equation (70) that shows that even in the simplest of cases, the source energy density ρ cannot in general be freely specified on the initial slice. We have to use a "trial" density $\hat{\rho}$ instead.

Suppose that the initial data are to be momentarily static ($K = j = 0$) and conformally flat. Then we have $\gamma_{ij} = \phi^4 f_{ij}$, f = flat three-metric, and we denote the scalar Laplacian formed from f by $\bar{\Delta}$. It is assumed that the global topology of Σ is Euclidean. Then (70) becomes

$$\bar{\Delta}\phi + \frac{1}{4}\kappa\hat{\rho}\phi^{-3} = 0 . \tag{75}$$

We assume that $\hat{\rho}$ is smooth, non-negative, and has compact support. We assume as an asymptotic condition that $\phi \to 1$ as $r \to \infty$ and also that a positive solution ϕ exists. Now perturb the trial energy density by setting $\delta\hat{\rho} = \varepsilon\lambda$ for small ε. The flat metric f is fixed. We have that $\delta\phi = \varepsilon u$ with $u \to 0$ as $r \to \infty$.

Linearization of (75) shows that u satisfies

$$(\bar{\Delta} - \tfrac{3}{4}\kappa\hat{\rho}\phi^{-4})u = -\tfrac{1}{4}\kappa\phi^{-3}\lambda \ . \qquad (76)$$

Standard elliptic theory shows us, from the fact that $(-\hat{\rho}\phi^{-4}) \le 0$, that a unique u exists. Moreover, if $\lambda = 0$, we must have $u = 0$ so that the original solution of ϕ of (75) was unique. Thus, the linearization of (75) is "good". This conclusion can be seen to follow also whenever the trial density $\hat{\rho}$ and the physical density ρ are related by $\rho = \phi^{-s}\hat{\rho}$ for $s > 5$.

On the other hand, if we had freely specified ρ instead of $\hat{\rho}$, we would have had

$$\bar{\Delta}\phi + \tfrac{1}{4}\kappa\rho\phi^{5} = 0 \ . \qquad (77)$$

Similar assumptions and linearization yields

$$(\bar{\Delta} + \tfrac{5}{4}\kappa\rho\phi^{4})u = -\tfrac{1}{4}\kappa\phi^{5}\lambda' \ , \qquad (78)$$

where now $\delta\rho = \varepsilon\lambda'$. However, by hypothesis, $(\rho\phi^{4}) \ge 0$ and no $u \to 0$ as $r \to \infty$ will exist in general. This is because the homogeneous equation $(\lambda' = 0)$ does not satisfy a maximum principle and in fact its "solutions" u would tend to be oscillatory. Hence, the linearization of (77) is in general unsatisfactory. This says that in general the original equation (77) possesses no admissible solution ϕ with ρ a specified function. This kind of reasoning, in more sophisticated form, is used to analyze (70) in more general cases. The general lesson seems to be that initial data should be specified in "conformal equivalence classes" and not given *a priori* The latter idea carries over also to cases where the initial slice is compact instead of topologically Euclidean.

6 ASYMPTOTIC CONDITIONS

If the foliation $\{\Sigma\}$ has slices that are not "compact without boundary", then questions of appropriate boundary conditions on the data $\{\gamma, K, \rho, j\}$ and the kinematical variables $\{\alpha,\beta\}$ arise. The two kinds of boundaries that are of interest here are the "boundary at infinity in spacelike directions" (spatial infinity) and an "inner" boundary or boundaries ∂B with two-sphere topology that can arise when there are "throats", horizons, or vacuum-matter interfaces. It is possible that the ∂B's could be moving boundaries. We have not at present settled all the issues about these inner boundaries. However, many of the problems in which such questions arise involve partial differential equations that can be posed in terms of variational principles whose boundary terms suggest physically and mathematically natural choices. These ideas are illustrated in the next section where we discuss the choices of α and β .

In this section I shall discuss boundary conditions at spatial infinity. Spatial infinity will be thought of as a standard metric two-sphere with an arbitrarily large radius on each slice times the real line: $S^2 \times \mathbb{R}$. The subject will be dealt with in the traditional language of the asymptotic fall-off of various quantities at large spacelike distances from an "origin". Experience in other branches of mathematical physics seems to indicate that the structure of spatial infinity as some kind of manifold (or even as an ideal point) attached to spacetime can be prescribed with confidence only when the general asymptotic behavior of solutions of the relevant differential equations has been determined. Therefore we shall talk about the limits of certain quantities as the distances in a slice become arbitrarily large. We shall also have to consider the conservation of these asymptotic limits in the evolution of the data on a slice.

Suppose we have a spacetime V with metric g that is asymptotically flat. This spacetime has topology $\Sigma \times \mathbb{R}$, where $\{\Sigma\}$ is the foliation. There are two cases I shall consider: (1) The topology of each slice is $\mathbb{R}^3$ (Euclidean). In this case we assume that there is a diffeomorphism F (with inverse F^{-1}) from V to a region M of Minkowski spacetime with flat metric f (Riem (f) = 0) . (2) The topology of a slice Σ is not Euclidean. In this case I assume that there is for each Σ a compact set B, and an "exterior region $\Sigma_{ext} = \Sigma - B$, such that $\Sigma_{ext} \times \mathbb{R}$ is related by a diffeomorphism to a region of Minkowski spacetime that is the exterior of a timelike world tube.

The important question for us is how F relates the slices Σ (or Σ_{ext}) to spacelike slices of Minkowski spacetime (M, f). The most interesting spacelike slices of (M, f) for our purposes are the (future and past) "mass hyperboloids" (with tr K = constant) and the standard x^0 = constant hyperplanes (with tr K = 0). The former are asymptotic to (future and past) null cones. Here we shall deal only with the x^0 = constant hyperplanes. We shall assume that F puts the slices Σ and the x^0 = constant hyperplanes into correspondence. The metric of these hyperplanes is f_{ij} (flat) in an arbitrary three-basis and we assume, moreover, that $\partial_0 f_{ij} = f_{0i} = 0$, $f_{00} = -1$, so that the standard Minkowski foliation <u>and</u> time vector field (time-translation Killing vector of f_{ab}) are being used.

With these preliminaries given, we shall define "asymptotically Minkowskian" initial data by requirements on both the intrinsic geometry of Σ <u>and</u> its imbedding (extrinsic geometry). The metric of Σ is compared to f_{ij} (via F) and its extrinsic curvature is compared to that of the x^0 = constant hyperplanes, which vanishes. We define r as the Euclidean distance from some origin, as determined by f_{ij}. Then for r sufficiently large, we can write

$$\gamma_{ij} = f_{ij} + h_{ij} = f_{ik}(\delta^k{}_j + h^k{}_j) \ , \tag{79}$$

$$h^k{}_j = O(r^{-1}) \ . \tag{80}$$

The mixed form of h is used in order that the fall-off can be stated in basis-independent form. For the same reason, we shall introduce the three unit orthogonal translational Killing vectors $k^i{}_{(j)}$ of f_{ij} and denote by $\pounds_{(i)}$ the Lie derivative along the ith Killing vector. Thus $\pounds_{(i)} f_{jk} = 0$. A further requirement on may then be stated as

$$\pounds_{(i)} h^j{}_k = O(r^{-2}) \ . \tag{81}$$

The extrinsic curvature is required to satisfy, in parallel with (81),

$$K^i{}_j = O(r^{-2}) \ . \tag{82}$$

Equations (81) and (82) deal in essence with "asymptotic forces" in the sense that they involve first derivatives of the metric. Sometimes, one also finds it necessary to restrict explicitly the "tidal forces" by "curvature conditions"

$$\pounds_{(i)} \pounds_{(j)} h^k{}_\ell = O(r^{-3}) \ , \tag{83}$$

$$\pounds_{(i)} K^j{}_k = O(r^{-3}) \ . \tag{84}$$

The asymptotic requirements (79)-(82) are motivated by the fact that they guarantee that the total energy and total linear momentum associated with $\{\gamma, K\}$ are finite and well defined by the standard two-sphere integrals.

$$E = \lim (2\kappa)^{-1} \oint D^j (h^i{}_j - \delta^i{}_j \operatorname{tr} h) d^2 S_i \tag{85}$$

$$P_j k^j{}_{(i)} = \lim (\kappa^{-1}) \oint (K^m{}_j - \delta^m{}_j \operatorname{tr} K) k^j{}_{(i)} d^2 S_m \tag{86}$$

Here "lim" denotes the limit as the radius of the two-sphere of integration becomes arbitrarily large. In these expressions, D and tr could be replaced by their flat-space counterparts. These expressions are "gauge-invariant" in the sense that if the correspondence F between physical spacetime slices and Minkowski spacetime hyperplanes is changed smoothly in such a way as to preserve (79)-(82), while introducing no "asymptotic Lorentz boost", then the values of (85) and (86) are unaffected. In the present context, "no asymptotic Lorentz boost" means that the difference between the Jacobian of F (J) and the Jacobian (I) of the identity transformation is $O(r^{-1})$ with the derivatives in Σ of (J-I) of $O(r^{-2})$, etc. Of course, this restriction still permits an infinite set of transformations representing asymptotic gauge degrees of freedom, e.g., in Minkowskian coordinate language

$$x^{a'} = x^a + \varepsilon\xi^a \ , \quad \partial_b\xi^a = 0(r^{-1}) \ , \quad \partial_c\partial_b\xi^a = 0(r^{-2}) \ , \qquad (87)$$

where ε is an infinitesimal parameter. Unfortunately, the gauge invariance of E and P does not carry over to the angular momentum surface integral, which is just the right side of (86) with the translational Killing vector k of f replaced by a rotational Killing vector of f . The gauge freedom in defining the rotations permits not only the usual constant translations of the spacetime origin, but also the infinite set of "supertranslations", which are essentially a class of functions on the two-sphere. (This problem originally arose in defining the angular momentum at null infinity as is well known.) To define the angular momentum, in the absence of an exact rotational spacetime Killing vector symmetry, one must impose further "asymptotic gauge conditions" beyond (79)-(82). Such conditions will be mentioned below. They forbid gauge changes such as (87).

Appropriate fall-off conditions on ρ and j that are compatible with (79)-(82) and the constraints are easily stated. These are motivated by the same kind of finiteness requirements that arise in Newtonian physics. Roughly, we must have that ρ and j fall off faster than r^{-3} . It usually suffices to assume the slightly stronger conditions

$$\rho = 0(r^{-4}) \ , \quad j_i k^i{}_{(\ell)} = 0(r^{-4}) \ . \qquad (88)$$

It is possible to state all of the above conditions rigorously in terms of the "weighted Sobolev spaces" that were introduced recently. It is in the context of such spaces that precise existence and uniqueness theorems concerning solutions of the constraints on asymptotically flat slices have been proved. This work has received considerable attention in the recent literature and will not be described here.

Physical interpretation of the conformal transformations, splittings, and of the kinematical conditions of the next section are simplified by the introduction of asymptotic gauge or coordinate conditions. This means that among the allowed correspondences F between the events of physical spacetime and those of Minkowski spacetime, we select particular ones that satisfy (79)-(82) plus four further conditions. We try to choose the simplest and most natural ones in the present context.

Three of these can be viewed as spatial coordinate conditions. In (79) we write

$$h_{ij} = (h_{ij} - \tfrac{1}{3}f_{ij}f^{mn}h_{mn}) + \tfrac{1}{3}f_{ij}f^{mn}h_{mn}$$

$$\equiv \psi_{ij} + \frac{1}{3} f_{ij} \, \overline{\mathrm{tr}}\, h \ , \quad \overline{\mathrm{tr}}\, \psi = 0 \ , \tag{89}$$

where an over-bar denotes the flat-space operator using f_{ij}. The "quasi-isotropic" gauge condition is a condition on the divergence of ψ :

$$k^{i}{}_{(\ell)} \bar{D}^{j} \psi_{ij} = O(r^{-3}) \ . \tag{90}$$

If (90) holds, then the total energy integral is

$$E = -\lim (4\kappa^{-1}) \oint (D^{i}\phi) d^{2}S_{i} \ , \tag{91}$$

which shows that the scalar potential ϕ can be viewed as a generalization of the ordinary Newtonian potential that satisfies Poisson's equation. In (91), either flat space (f_{ij}) or curved space (γ_{ij}) operators can be used. Using the latter form and assuming that the topology of the slices is $\mathbb{R}^3$, the constraint equation (70) and (91) are related using Gauss's theorem. Thus one can show that the total energy E is the "monopole" of ϕ and the $O(r^{-1})$ part of the metric is

$$\gamma_{ij} = f_{ij}\left(1 + \frac{\kappa E}{4\pi r}\right) + \psi_{ij} \ , \tag{92}$$

which is similar to the standard isotropic Schwarzschild coordinates, where $\psi_{ij} = 0$ and $(1 + \kappa E/16\pi r)^{4} = 1 + \kappa E/4\pi r + \ldots$.

Our fourth asymptotic gauge condition can be viewed as a time coordinate condition in the sense that it involves the relation of the labels of the physical slices τ = constant and the corresponding Minkowski hyperplanes x^0 = constant. This relationship can always be set up in such a way that

$$\mathrm{tr}\, K = O(r^{-3}) \ , \tag{93}$$

which can be referred to as "asymptotically maximal". Using (90), (93), and analyzing the constraints and the surface integral (86) for the constant total linear momentum P , one finds that the $O(r^{-2})$ part of K_{ij} as a symmetric covariant tensor or quadratic form is

$$K = \frac{3\kappa}{16\pi r^{2}} [\underset{\sim}{P} \otimes \underset{\sim}{\nu} + \underset{\sim}{\nu} \otimes \underset{\sim}{P} - \langle \underset{\sim}{\nu}, \vec{P} \rangle (\underset{\sim}{f} - \underset{\sim}{\nu} \otimes \underset{\sim}{\nu})] \tag{94}$$

where $\vec{\nu}$ is the unit normal vector of the two-sphere of integration, $\underset{\sim}{\nu}$ is the corresponding one-form $(\underset{\sim}{\nu} = d\underset{\sim}{r})$, and $\underset{\sim}{P}$ is the total momentum one-form. Note that $(\underset{\sim}{f} - \underset{\sim}{\nu} \otimes \underset{\sim}{\nu})_{ij}$ is the metric of the two-sphere.

The $O(r^{-(2+\varepsilon)})$ parts of K $(\varepsilon > 0)$ can be regarded as giving various "momentum multipoles". A particular $O(r^{-3})$ term is of interest, the part carrying angular momentum. Let $\underset{\sim}{J}$ be the spatial angular momentum two-form $(J_{ij} = -J_{ij})$ with the angular momentum vector J being its spatial dual. Define the "interior product" of J and $\vec{v}$ by $\langle J,\vec{v}\rangle_i = v^k J_{ki}$. Then the angular momentum-carrying piece of K_{ij} is of $O(r^{-3})$ and has the form

$$K = \frac{3\kappa}{8\pi r^3}\left(\langle \underset{\sim}{J},\vec{v}\rangle \otimes \underset{\sim}{v} + \underset{\sim}{v} \otimes \langle \underset{\sim}{J},\vec{v}\rangle\right) . \tag{95}$$

This form of the part of K that carries angular momentum depends on using the asymptotic gauges (90) and (93). As previously remarked, the surface integral for J requires the use of some asymptotic gauge-fixing conditions in order to make it meaningful, unlike the integrals for E and P . If the spacetime happens to admit an exact rotational Killing symmetry, then no questions of gauges, location of origin, etc., arise. For example, in the Kerr metric in Boyer-Lindquist coordinates, the leading term of K is $O(r^{-3})$ and can be written in precisely the form (95).

7 DEGREES OF FREEDOM

When general relativity is viewed as a dynamical theory of gravity, one sees that the spacetime metric g naturally separates into the ten functions γ, α, and β relative to a given foliation and choice of time vector. In the Einstein equations, the only second time derivatives that appear are those of γ ; thus, we have the dynamics of spacelike three-geometry, "geometrodynamics". The lapse and shift correspond to kinematical degrees of freedom. The dynamical degrees of freedom are found among the γ_{ij}'s. However, the initial-value problem shows that γ_{ij} need only be given up to an overall variable conformal factor ϕ^4 , with ϕ determined by the constraints. Therefore, it is convenient to regard the "conformal metric"

$$\tilde{\gamma}_{ij} \equiv (\det\gamma)^{-1/3}\gamma_{ij} \ ; \quad \det\tilde{\gamma} = 1 \ , \tag{96}$$

as containing the dynamical degrees of freedom. We may regard (96) as a separation of the "configuration coordinates" γ_{ij} of the gravitational field into $(\det\gamma)^{\frac{1}{2}}$ and $\tilde{\gamma}_{ij}$.

In the Hamiltonian form of general relativity, one easily shows that $(\det\gamma)^{\frac{1}{2}}$ and $\operatorname{tr} K$ are canonically conjugate variables (modulo a constant). Hence by a canonical transformation, it follows that the variables $(\operatorname{tr} K, \tilde{\gamma}_{ij})$ may equally well be regarded as the independent configuration variables. We also note in this connection that all of the six quantities $(\operatorname{tr} K, \tilde{\gamma}_{ij})$ commute as required for configuration variables. This set gives six numbers at each event on a slice Σ , which is four too many, because there are really just two dynamical degrees of freedom of the field at each point.

Suppose we have a system with configuration variables q^A . The number of true degrees of freedom of such a system is given by definition by the number of linearly independent velocity variables $\dot{q}^A$ that are compatible with all constraints. To obtain the real degrees of freedom, we fix appropriately the values of all the superfluous velocities. This does not necessarily mean explicitly eliminating certain q^A's, because the constraining relations may be non-integrable. Similarly, in a field theory there may be gauge degrees of freedom (equal to the number of initial-value constraints). We may eliminate these by fixing the values of the correct number of velocity variables.

In the next section, we shall assume the constraints have been satisfied. By the above argument, we shall then fix the gauge degrees of freedom by imposing four conditions on the velocities ∂_τ tr K and $\partial_\tau \tilde{\gamma}_{ij}$. These will lead to equations for α and β, but they will not restrict our initial choice of spacetime coordinate labeling schemes. In this way only the correct number of dynamical degrees of freedom will actually be evolved and useful coordinate freedom will remain.

8 LAPSE FUNCTION

8.1 Construction of Foliation

The first order of business is to prescribe a method of generating a foliation from a given initial data slice. The simplest method is to use the timelike geodesics normal to the slices and to label the slices with the local proper time of the Eulerian observers ($\alpha = 1, \beta = 0$). However, these so-called "synchronous" reference systems have long been known to develop coordinate singularities by focusing of the normal geodesics. Their use has been carefully criticized in a recent study. These frames do not cover much of the development of an initial data set and the idea of piecing such frames together successively toward the future (or past) does not seem to be helpful, either.

Long ago Lichnerowicz realized that an effective anti-focusing condition must be used. He suggested maximal slicings: tr $K = 0$. These are appropriate in asymptotically flat spacetimes but not in closed universes (compact slices without boundary), where at most one such slice exists. In closed universes we may use a condition that is nearly as simple: tr K = constant (a different constant for each slice). In neither of these cases do we know for certain in the general case that enough of the maximal development will be covered in order to calculate what we want to know (nature of formation of singularities or horizons, gravitational radiation, etc.). However, these slicings have worked well in examples so far. The development of a constructive prescription for generating a foliation covering the maximal Cauchy development is the number one problem in the dynamical approach to the solution of Einstein equations. It is a problem that has only recently begun to receive the attention it deserves.

Here I shall suggest a general way of looking at this problem that may be useful. We focus attention on the velocity variable $\partial_\tau \operatorname{tr} K = \pounds_t \operatorname{tr} K$. The function $\operatorname{tr} K$ itself is here regarded as an initial datum (just as in the initial-value problem). The reason for this approach is that the equation for $\partial_\tau \operatorname{tr} K$ controls the lapse function through a sequence of linear elliptic equations that have the same form on each slice. We turn what was traditionally regarded as a hyperbolic dynamical equation for $(\det \gamma)^{1/2}$ (because $\partial_\tau \operatorname{tr} K \propto \partial_\tau \partial_\tau (\det \gamma)^{1/2} + \ldots$) into elliptic equations of condition on the lapse function α. Thus, suppose we require $\partial_\tau \operatorname{tr} K = u(\tau,x)$ where u is a given function. Then we have from the Einstein equations for $\pounds_t \gamma$ and $\pounds_t K$ that

$$\partial_\tau \operatorname{tr} K = u = -\Delta\alpha + \alpha[R + (\operatorname{tr} K)^2 + \tfrac{1}{2}\kappa(\operatorname{tr} S - 3\rho)] + \beta^i D_i \operatorname{tr} K. \tag{97}$$

If we insert the Hamiltonian constraint into (97) we find

$$\partial_\tau \operatorname{tr} K = u = -\Delta\alpha + \alpha[K_{ij}K^{ij} + \tfrac{1}{2}\kappa(\rho + \operatorname{tr} S)] + \beta^i D_i \operatorname{tr} K . \tag{98}$$

Note that if the "strong energy condition" $(\rho + \operatorname{tr} S) \geq 0$ holds, then the linear operator on α has the form $(\Delta - P(x))$, with $P(x) \geq 0$, so that the standard maximum-minimum principles apply.

8.2 Deformation and Variational Principles

We can view the equations for $\partial_\tau \gamma$ and $\partial_\tau K$ as defining small deformations of Σ imbedded in spacetime. This is analogous to the deformation of a thin curved "shell" in three-dimensional space. In this case α defines a normal deflection of the shell and β defines a tangential strain. Thus, $\partial_\tau \gamma$ defines a "stretching strain tensor" consisting of two parts: (1) $-2\alpha K$. A normal deflection α will produce stretching if the shell is bent ($K \neq 0$) in its initial state. (2) $D_i\beta_j + D_j\beta_i$. This describes tangential stretching. The extrinsic curvature K defines the bending of the shell before deformation and $\partial_\tau K$ defines the "bending strain tensor". We shall identify $\partial_\tau \operatorname{tr} K$ with a "volume bending strain".

We now propose to determine α by minimizing what may be called a "free energy of volume bending strain"

$$\mathcal{H}[\alpha] = \tfrac{1}{2}\int_\Sigma (\partial_\tau \operatorname{tr} K)^2 \, dv . \tag{99}$$

Clearly, varying $\mathcal{H}$ with respect to α will lead to $\partial_\tau \operatorname{tr} K = 0$. However, the interesting point is that the equation resulting is of fourth order, as expected from the analogy to bending a shell and finding the equilibrium solution.

We write (98) in the form

$$-\partial_\tau \operatorname{tr} K = (\Delta - P)\alpha - \beta^i D_i \operatorname{tr} K$$
$$= \mathcal{m}\alpha + Q \; , \tag{100}$$

where $\delta Q/\delta\alpha = 0$ and $\mathcal{m}$ obeys a maximum principle provided that the strong energy condition holds.

Consider the case when Σ is asymptotically flat. Then

$$\delta\mathcal{H} = \int dv\delta\alpha[\mathcal{m}(\mathcal{m}\alpha + Q)] + \oint(\mathcal{m}\alpha + Q)\delta(D^i\alpha)d^2S_i$$
$$- \oint\delta\alpha[D^i(\mathcal{m}\alpha + Q)]d^2S_i \; . \tag{101}$$

The Euler-Lagrange equation for $\mathcal{H}$ is therefore the fourth order "biharmonic" type equation

$$\mathcal{m}(\mathcal{m}\alpha + Q) = 0 \; . \tag{102}$$

For the "boundary at infinity" we have $\alpha \to 1$ as $r \to \infty$. (This is equivalent to the existence of a force that deflects the thin shell in our analogy.) Using the previously stated asymptotic conditions on the other variables, we see that $(\mathcal{m}\alpha + Q) = O(r^{-3})$. "Inner" boundary conditions are found by requiring that the surface integral terms vanish. One set that is compatible with our later work is (1)α is free at ∂B and either $(\mathcal{m}\alpha + Q) = 0$ on ∂B or $\partial/\partial\nu(\mathcal{m}\alpha + Q) = 0$ on ∂B, where $\vec{\nu}$ is the unit normal of ∂B . (2) $\partial\alpha/\partial\nu$ is fixed on $\partial B[\delta(\partial\alpha/\partial\nu) = 0]$, e.g., $\partial\alpha/\partial\nu = 0$ on ∂B .

In any case, with the vanishing of all the surface integrals and $(\mathcal{m}\alpha + Q) = O(r^{-3})$, we see that (102) is satisfied if and only if the second order equation $\mathcal{m}\alpha + Q = 0$ holds. Thus, we find $\partial_\tau \operatorname{tr} K = 0$, which is just (98) with $u = 0$.

If Σ is compact without boundary, then in general there will be no positive solutions of $(\mathcal{m}\alpha + Q) = 0$ and we shall have to add an extra term, i.e., $\partial_\tau \operatorname{tr} K \neq 0$ for closed slices, as expected.

We shall now deal with the second order equation (98) in various cases. We shall discuss the "preferred cases" when tr K = constant. As previously mentioned, when $D_i \operatorname{tr} K = 0$ on a slice, the constraints are effectively decoupled. Moreover, we see in (98) that the term containing the shift vector β disappears, so that the lapse and shift conditions are also effectively decoupled.

8.3 Closed Slices

Here we have tr K = const. (non-vanishing in general) and we may take ∂_τ tr K = u = a positive constant in (98) in order to obtain $\alpha > 0$ and maintain the constancy of tr K. The constant value of tr K is different for different slices. The constant value of u may also be different for successive slices if desired. However, if we pick u = 1 for all slices, then we have the possibility of setting tr K = τ , so that tr K serves as a natural time coordinate.

8.4 Asymptotically Null Slices

It is possible in at least some asymptotically flat spacetimes (e.g., Schwarzschild spacetime) to set tr K = a negative constant and u = 0 . Such slices are similar to the "mass hyperboloids" of Minkowski spacetime and are asymptotically null. The lapse grows "like r" at large distances.

8.5 Maximal Slices

The most widely studied slicing criterion for asymptotically flat spacetimes is the maximal foliation defined by

$$\mathrm{tr}\, K = 0 = \partial_\tau\, \mathrm{tr}\, K \,. \tag{103}$$

This yields a linear elliptic equation for α to be solved on each slice:

$$\Delta\alpha = \alpha[K_{ij}K^{ij} + \tfrac{1}{2}\kappa(\rho + \mathrm{tr}\, S)] \tag{104}$$

with $\alpha \to 1$ as $r \to \infty$. From the fact that tr K = $-\nabla_a n^a$, this states that the Eulerian world lines are like those of an incompressible (and irrotational) "test fluid".

There are two variational principles of interest for maximal slices. The first gives the reason for its name. Consider the functional defining the volume of any bounded portion S of the slice Σ :

$$\mathrm{vol}(S) = \int_S (\det \gamma)^{\frac{1}{2}} d^3x \,. \tag{105}$$

Consider any small deformation of Σ along a vector $\lambda^a = \ell n^a + \sigma^a$, $n_a\sigma^a = 0$. Assume that $\ell = \sigma^a = 0$ on the boundary ∂S of S. Then

$$\mathrm{vol}(S) = \int_S d^3x (\det \gamma)^{\frac{1}{2}}[-\ell\, \mathrm{tr}\, K] \tag{106}$$

since $\sigma^a = 0$ on ∂S. Since ℓ is arbitrary, we have tr K = 0 . This is a maximum condition as one can see geometrically or by examining the second variation. (If the signature of g_{ab} were

Euclidean, we would have a minimal volume condition, i.e., a classical "Plateau problem".) In general, proving the existence of maximal spacelike slices is a mathematically difficult problem. However, they are known to exist in a (functional) neighborhood of Minkowski data $(\gamma = f,\ K = 0)$ and in many other cases of interest (Schwarzschild spacetime, Kerr spacetime, etc.).

The other variational principle concerns (104), which is used to construct a maximal slicing step-by-step starting from a given one. This equation is derived by varying α in

$$F[\alpha] = \frac{1}{2}\int_\Sigma [(D_i\alpha)(D^i\alpha) + P\alpha^2]dv \tag{107}$$

with $dv = (\det\gamma)^{\frac{1}{2}}d^3x$ and $P = K_{ij}K^{ij} + \frac{1}{2}(\rho + \mathrm{tr}\ S)$. With $\delta\alpha = 0$ at infinity and an "inner boundary" ∂B, we find

$$\delta F = -\int_\Sigma dv\delta\alpha[(\Delta - P)\alpha] + \oint_{\partial B}\delta\alpha(D^i\alpha)d^2S_i\ . \tag{108}$$

We find that $\delta F = 0$ gives (104) if either (a) Dirichlet condition: α is fixed on ∂B ($\delta\alpha = 0$ on ∂B), or (b) Neumann condition ("natural boundary condition"): $\nu^i D_i\alpha = 0$ on ∂B, ν^i = unit normal of ∂B. Clearly, with $P \geq 0$, the Euler-Lagrange equation (104) associated with (107) defines a minimum of $F[\alpha]$. The actual value of the minimum depends on whether the Dirichlet or the Neumann boundary condition is used. Note that these boundary conditions are compatible with the earlier ones associated with (99).

In interpreting this procedure, it is helpful again to think of deformations of the slice in spacetime as being analogous to deformations of a curved thin shell in ordinary space. If $G = c = 1$, we see that the integrand of $F[\alpha]$ is a kind of kinematical curvature (dimension = $(\text{length})^{-2}$) or "bending energy" density associated with a normal deflection of the surface from one maximal volume configuration to another, under prescribed boundary conditions. Observe that $F[\alpha]$ has the dimensions of length or energy. We think of it as a volume bending energy or "work".

Both kinds of boundary conditions can be illustrated in the Schwarzschild-Kruskal spacetime. Let the initial maximal slice be the time-symmetry slice in the Kruskal diagram ($v = t_{\mathrm{Schw}} = 0$). In case (a), take $\alpha = 0$ at the throat $r = 2m$ (r = standard Schwarzschild radial coordinate). Then the solution of (104) is $\alpha = (1 - 2m/r)^{\frac{1}{2}}$ for $r \geq 2m$. Continuation of this procedure yields the standard t_{Schw} = constant slices that cover only the exterior of the black hole. The value of $F[\alpha]$ on the initial slice is computed to be $2\pi m$.

Again starting with the time-symmetry slice, take the Neumann condition $\partial_r\alpha = 0$ at the throat. In this case the solution of (104) on the initial slice is $\alpha = 1$. Of course, α does not remain constant on the later slices. The "extended maximal foliation" of Schwarzschild spacetime that results here is well known. These maximal slices reach from spatial infinity and on into the interior of the black hole. The "last" such maximal slice that reaches infinity is $r = 3m/2$, thus stopping short of the singularity at $r = 0$. Notice that the value of $F[\alpha]$ on the initial slice is zero when we use the natural (Neumann) boundary condition. Thus the minimum achieved here is less than when the Dirichlet condition is used. At the same time, this foliation covers much more of the maximal development of the Cauchy data of the time-symmetric slice.

The "volume bending energy" idea is useful in understanding these results. In the first case $\alpha = 0$ at the throat and therefore there must (in our shell analogy) be a "force" tangential to ∂B to hold the shell fixed there while it is displaced a unit amount a great distance away. Naturally, then there must be some bending ($D_i\alpha \neq 0$) and bending energy ($2\pi m$) present in order to maintain an extreme value of the volume (surface area for the shell). In the second case, there is no tangential force at ∂B, so the surface is allowed to slip freely "upward" as it is displaced upward elsewhere. It is easier then to keep the volume (surface area) fixed, thus resulting in less bending and less bending energy.

I would like to finish this discussion with a conjecture based on the above examples and analogies. Conjecture: On an asymptotically flat Cauchy data slice with a well defined inner boundary, ∂B, we will find: (a) $F[\alpha]$ is least when the Neumann condition is used. (b) The resulting maximal foliation will cover more of the maximal development with the Neumann condition on ∂B than with the Dirichlet condition. (Obviously, part (b) is the more important aspect of this conjecture from the point of view of spacetime constructions.)

8.6 Collapse of the Lapse

In the extended maximal slicing of Schwarzschild and in other examples that have been studied, the lapse function tends to go to zero rapidly as a function of time in regions where a spacelike singularity is developing. However, it is known that in general maximal slicing is not really "avoiding" spacelike singularities but is rather avoiding the vanishing of the volume elements of the associated Eulerian observers, as implied by its definition. In cases such as the Schwarzschild spacetime, the latter behavior and the former occur together.

There is a simple model that involves the maximum principle for the lapse equation, plus some guesswork, that is instructive. We consider (104) in vacuum with $\mathrm{tr}\, K = 0$ initially:

$$\Delta\alpha - R\alpha = 0 . \tag{109}$$

Here we have $R \geq 0$. We assume a flat metric $\gamma = f$ in spherical coordinates and that R is a positive constant $= R_0$ for $r \leq a$ and zero for $r > a$. We decide not to worry about the fact that now R and γ do not maintain their usual relationship. We can solve (109) under these hypotheses exactly. (We can also solve exactly when the edges of the ball of scalar curvature are rounded by a correct r^{-4} fall-off, but this does not improve the simple picture substantially.)

The solution of (109) is now

$$\alpha = (\cosh x_0)^{-1} x^{-1} \sinh x \ , \quad 0 \leq x \leq x_0 \ , \tag{110}$$

$$\alpha = 1 + x^{-1}(\tanh x_0 - x_0) \ , \quad x \geq x_0 \ , \tag{111}$$

$$x = rR_0^{\frac{1}{2}} \ , \quad x_0 = aR_0^{\frac{1}{2}} . \tag{112}$$

Note that a more general definition of x_0 can be given as

$$x_0 = \int_0^\infty R^{\frac{1}{2}} (\gamma_{rr})^{\frac{1}{2}} dr = aR_0^{\frac{1}{2}} \tag{113}$$

and that x_0 is a natural dimensionless measure of the strength of the scalar curvature. In accordance with the maximum principle, α has a positive minimum at $x = r = 0$ ($\partial_r \alpha = 0$ here):

$$\alpha_{min} = (\cosh x_0)^{-1} . \tag{114}$$

As x_0 increases, we have $\alpha_{min} \sim e^{-x_0}$. Thus, α goes to zero exponentially as x_0 increases.

The simplest guess about time-dependence in such a model is that x_0 increases linearly with the time τ that labels the "maximal" slices; this is also the proper time of observers at spatial infinity. Thus we guess that

$$x_0 = \tau/\tau_e + \text{const.} \ , \quad \tau_e = \text{const.} \tag{115}$$

so that at late times,

$$\alpha_{min}(\tau) \sim \exp(-\tau/\tau_e) . \tag{116}$$

One would also guess that τ_e (e-folding time) is $\approx m$ (mass of source).

Comparing this to the actual <u>extended</u> maximal slicing of Schwarzschild spacetime involves identifying $r = 0$ in the model with the throat in the real spacetime. We find for late times $\alpha_{min} \approx \exp(-0.97\ x_0)$ for x_0 defined by the integral in (113), very close to the value $\exp(-x_0)$ of the model. In the actual spacetime one also finds in a separate "test" that the linear relation $x_0 \approx \tau(1.77\ m)^{-1} + (0.34)$ fits to within a few percent so that the guess of simple time dependence (115) works well. Inserting this relation into the actual data for α_{min} versus x_0, we find $\tau_e \approx 1.82m$, so again the ideas of the model are not far off the mark.

Finally, look at the proper free-fall time from $t_{Schw} = 0$ for the collapse of the throat to the singularity. This is computed analytically to be πm. On the other hand, an exact calculation shows that the proper time from the initial slice to the last maximal slice at $r = 3m/2$ is $(\pi/3 + \sqrt{3}/2)m \approx 1.91m$, which is, of course, well short of the free-fall time to the singularity = πm. The model calculation also yields a time

$$\int_0^\infty \alpha_{min}(\tau)d\tau \approx 2.87m \tag{117}$$

which is less than πm. Thus, the singularity is "avoided".

This model is of course overly simplified and should not be taken too seriously. However it does suggest several things:

(A) Maximal slicing may halt the evolution in some region, depending on how fast the "line integrated curvature" (113) grows as a function of the time that labels the maximal slices, which is the time at infinity since $\alpha \to 1$ as $r \to \infty$.

(B) The lapse equation obeys a maximum-minimum principle. Thus in general on each maximal slice there are positive minima where $D_i\alpha = 0$. The maximum of α is at infinity where $\alpha = 1$ and there are no points where $\alpha \le 0$. At each minimum, the local Eulerian observer is in free fall but his Eulerian neighbors are necessarily accelerated outward to prevent local convergence of their world lines. We may compute

$$\int_0^\infty \alpha_{min} d\tau = T \tag{118}$$

along each world line of a minimum of α. If in any of these integrals α_{min} goes to zero as $\tau^{-(1+\varepsilon)}$, $\varepsilon > 0$, (or faster) at late times, then $T < \infty$ and the evolution will halt in a neighborhood of the minimum point. Thus, a suitable collapse of the lapse halts the evolution via maximal slices locally. In our example, we had the rapid collapse $\alpha_{min} \sim e^{-\tau/\tau_e}$. One expects such "haltings" to occur in maximal slicings. I know of no precise general criterion as to when this leads to "singularity avoidance".

(C) The model suggests that the functional relations $\alpha_{min} = \alpha_{min}(\tau)$, $x_0 = x_0(\tau)$, and $\alpha_{min} = \alpha_{min}(x_0)$ may be closely approximated by rather simple formulas. In general cases, a suitable generalization of x_0 in (113) presumably can be found. Perhaps this is already known mathematically, in terms of "a priori bounds" or "majorizations" of solutions of elliptic equations. One can see that this entire subject is ripe for further study.

9 SHIFT VECTORS

9.1 Methods of Choosing Shift Vectors

Regardless of how a foliation is constructed, one must still choose a shift vector β to complete the description of spacetime. People have often simply chosen $\beta = 0$ to simplify the equations. However, this choice can make one miss a chance to simplify the interpretation of the results. The histories $\gamma(\tau)$ and $K(\tau)$ depend not only on the choice of a slicing and on the choice of an initial spatial basis or coordinate system, but also on how the basis or coordinates are moved from one slice to the next, i.e., on the full specification of $t = \alpha n + \beta = \partial/\partial\tau$.

Besides $\beta = 0$, one can identify three ways of choosing a shift vector. One is to correlate t and the four-velocity of matter sources. Another is to keep the three-metric functions γ_{ij} in a preferred form, e.g., to use β to preserve a manifest symmetry of γ_{ij} . Finally, one may choose β to maintain certain differential "gauge conditions" on the velocity of γ_{ij}. We shall outline the first and third of these three methods below.

9.2 Comoving Shift Vectors

If there is a continuum of matter, one can arrange that each spatial differential volume element of the matter retains fixed spatial coordinate values, which is a "comoving" condition. Thus, let u^a be the unit four-velocity field of the matter. The spatial three-velocity of the matter determined by the Eulerian observers is $w^a (w^a n_a = 0)$, where

$$u^a = (1 - w^2)^{-\frac{1}{2}}(n^a + w^a) \ . \tag{119}$$

The velocity w^a is computed by the Eulerian observers using their local proper time. This velocity is more conveniently described in terms of τ-time , which gives $\tilde{w}^a = \alpha w^a$. Clearly, spatial coordinate frames will be comoving with the matter if we choose $\beta^i = \tilde{w}^i$. Then the spatial velocity of matter relative to the computational frames $\{t, e_i\}$ is zero. For other choices of the shift, the spatial velocity of matter is $\tilde{w}^i - \beta^i$. The comoving choice is useful in many cases, but if the matter is (spatially) rotating, then $\tilde{w}^i = \beta^i$ would cause the spatial coordinate grid eventually to wind up into a mess. In vacuum, of course, the comoving condition gives us no guidance.

9.3 Minimal Distortion Shift Vectors

This criterion for β is based directly on the gravity field configuration variable $\tilde{\gamma}$ (conformal metric), just as the choice of the lapse was based on $\mathrm{tr}\, K$. Again we must work with the velocity variable $\pounds_t\tilde{\gamma}$. This is a trace-free tensor density (of weight $-\frac{2}{3}$). For convenience, we shall use instead

$$\Sigma_{ij} = \pounds_t\gamma_{ij} - \frac{1}{3}\gamma_{ij}\ \mathrm{tr}\ \pounds_t\gamma$$
$$= (\det\gamma)^{1/3}\pounds_t\tilde{\gamma}_{ij}\ . \tag{120}$$

From (35) and (46), we have

$$\Sigma_{ij} = -2\alpha(K_{ij} - \frac{1}{3}\gamma_{ij}\ \mathrm{tr}\ K) + (\ell\beta)_{ij}\ . \tag{121}$$

Note that $[-(K - \frac{1}{3}\gamma\ \mathrm{tr}\ K)]$ is the usual shear of the unit normal field n.

We call Σ_{ij} the "distortion tensor" because it measures the change of shape of a small "spheroid" dragged along t from τ = constant to a nearby slice τ = constant + $\delta\tau$. The only control we have over Σ_{ij} is through the choice of β because γ and K are assumed to be given at this stage, as well as α.

Recall our previous "thin shell" analogy. Then Σ_{ij} is the shear strain intrinsic to the shell as it is deformed. It refers to the relative stretching or change of shape (not size) as it undergoes deformation. The first term on the right hand side of (121) is the shear stretching caused by the normal displacement α. (In the theory of deformation of flat plates, such a term would be of second order, i.e., $[(D_i\alpha)(D_j\alpha) - \frac{1}{3}\gamma_{ij}(D^k\alpha)(D_k\alpha)]$, because there $K = 0$ for a flat plate in Euclidean space. Here it is of first order since the shell is already extrinsically curved before the deformation takes place, $K \neq 0$.) The second term $(\ell\beta)$ is the shear stretching or distortion caused by tangential displacement of the shell.

We pose the choice of β as an equilibrium problem. The shift is chosen to minimize the distortion in a global sense. We define what is for a shell quite literally (apart from a constant shear modulus) the (Helmholtz) free energy density ($\sim(\mathrm{length})^{-2}$) of stretching, i.e., $\Sigma_{ij}\Sigma^{ij}$. The integral of this quantity is the shear stretching energy:

$$\mathcal{S}[\beta] = \frac{1}{2}\int_\Sigma \Sigma_{ij}\Sigma^{ij}dv = \frac{1}{8}\int_\Sigma [4\alpha^2(K_{ij} - \frac{1}{3}\gamma_{ij}\ \mathrm{tr}\ K)(K^{ij} - \frac{1}{3}\gamma^{ij}\ \mathrm{tr}\ K)$$
$$-4\alpha(\ell\beta)^{ij}(K_{ij} - \frac{1}{3}\gamma_{ij}\ \mathrm{tr}\ K) + (\ell\beta)_{ij}(\ell\beta)^{ij}]dv\ . \tag{122}$$

Varying β^i with $\delta\beta^i = 0$ at infinity gives

$$\delta\mathcal{S}[\beta] = -\int_\Sigma dv\,\delta\beta^i D^j\Sigma_{ij} + \oint_{\partial B}\Sigma^i{}_j\delta\beta^j d^2S_i\ , \tag{123}$$

where Σ_{ij} is a shorthand for (121). If the boundary term vanishes, the Euler-Lagrange equations for $\mathcal{S}[\beta]$ are the "minimal distortion equations"

$$D^j \Sigma_{ij} = 0 \tag{124}$$

or, equivalently,

$$(\Delta_\ell \beta)^i = 2D_j[\alpha(K^{ij} - \tfrac{1}{3}\gamma^{ij} \text{ tr } K)] \ , \tag{125}$$

where $(\Delta_\ell \beta)^i \equiv D_j(\ell\beta)^{ij}$ is the "vector Laplacian" introduced earlier. The velocity of the conformal metric is seen to be covariantly divergence-free and trace-free with this β . Inserting the momentum constraints into this purely geometrical criterion gives

$$(\Delta_\ell \beta)^i = 2\alpha\kappa j^i + \tfrac{4}{3}\alpha D^i \text{ tr } K + (D_j\alpha)(K^{ij} - \tfrac{1}{3}\gamma^{ij} \text{ tr } K) \ . \tag{126}$$

Note the similarity of (126) and the momentum constraint (71) for the vector potential W . See Figure 2.

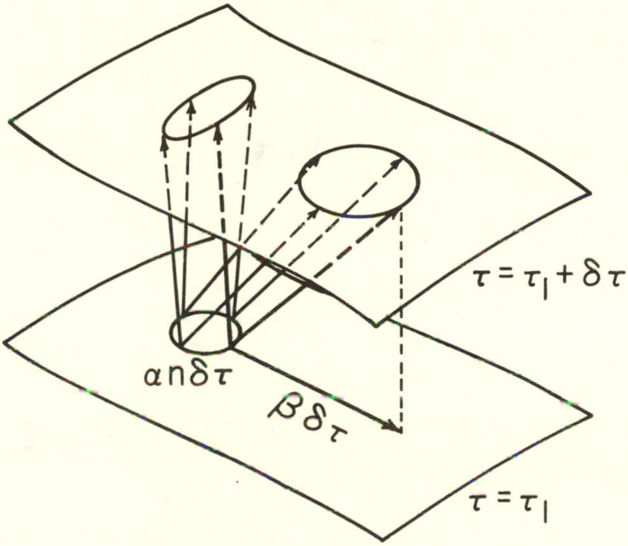

Fig. 2. This schematic diagram illustrates the use of the minimal distortion shift vector to reduce coordinate shear. If a small sphere is transported along the normal direction from $\tau = \tau_1$ to $\tau = \tau_1 + \delta\tau$, it will in general be sheared into an ellipse. On the other hand, if the minimal distortion shift vector is used, then some of the shear can be removed. The remaining shear depends only on the spacetime geometry and choice of slicing, but not on the spatial coordinates.

It is easy to see that this choice of shift yields a minimum for $S[\beta]$, which is non-negative. The value of the minimum depends on the choice of inner boundary conditions. We have (a) β

is fixed on ∂B so $\delta\beta = 0$ there. This is a Dirichlet condition. (b) The "natural" or Neumann-like condition $\Sigma^i{}_j \nu^j = 0$, ν^j = unit normal of ∂B. This can be written in terms of $\pounds_\nu \beta$ on ∂B.

Again, I believe that in many cases the Neumann condition will be preferable. As an example, consider the standard maximal slicing of the Kerr metric outside the outer horizon, using Boyer-Lindquist coordinates on the initial slice. Here, the unique solution of the minimal distortion equation is the standard Boyer-Lindquist shift vector if we impose the Neumann condition at the outer horizon $r = r_+$. It is this shift vector whose value on r_+ defines the angular velocity of the rotating black hole as seen from infinity. In this case the shear stretching "energy" of the slice is zero. This is because the rotating "shell" is allowed to "slip freely" around at $r = r_+$. Therefore, if one obtained a solution with a Dirichlet condition (in which β was not fortuitously chosen to be the Boyer-Lindquist shift), the value of S would be greater than zero and the resulting spacetime metric would be more complicated than is necessary. There is more "shear stretching" because the shell cannot slip freely at $r = r_+$ if β is fixed there.

A criterion similar to minimal distortion would come from minimizing the <u>total</u> "stretching energy"

$$\int_\Sigma \gamma^{ij}\gamma^{k\ell}(\pounds_t \gamma_{ik})(\pounds_t \gamma_{j\ell})dv \ .$$

Here the solution has the form $D^j \pounds_t \gamma_{ij} = 0$. Below I give reasons why the minimal distortion shift may be preferred in general.

9.4 Quasi-Isotropic Coordinates

In Section 6 we introduced as asymptotic gauge conditions

$$k^i_{(\ell)} \bar{D}^j \psi_{ij} = O(r^{-3}) \ , \quad \mathrm{tr}\, K = O(r^{-3}) \ . \tag{127}$$

It is easy to see that the first of these conditions deals with the conformal three-metric $\tilde{\gamma}$. If we use asymptotically Cartesian coordinates, then the first condition is equivalent to

$$\tilde{\gamma}_{ij,j} = O(r^{-3}) \ . \tag{128}$$

Writing out the minimal distortion shift condition and assuming that for large distances $\alpha = 1 + O(r^{-1})$ and $\beta = O(r^{-1})$, we see that

$$(\partial \tilde{\gamma}_{ij}/\partial \tau)_{,j} = O(r^{-3}) \tag{129}$$

so that (128) is preserved in the evolution.

Moreover, if $\text{tr } K = O(r^{-3})$ on the initial slice and we choose $u = \partial_\tau \text{ tr } K = O(r^{-3})$, e.g., $u = 0$, then the trace condition is also preserved in evolution. Therefore, $\partial_\tau \text{ tr } K = D^j \pounds_t \tilde{\gamma}_{ij} = 0$ yields a diffeomorphism, with generating vector field $t = \alpha n + \beta$, that preserves the quasi-isotropic coordinate conditions. It is of interest to note that this type of frame is customarily used in approximating the metric in PPN approximation schemes. Here we have seen how such conditions are imposed by working always with the exact theory rather than with various approximations to the exact theory. Our approach seems much better suited to solving exact strong-field high-velocity problems.

9.5 Radiation Gauge

One of the most difficult problems of general relativity is identifying dynamical degrees of freedom in a reliable way. I believe the methods outlined here may be quite helpful in this respect, because they are founded on the exact dynamical theory (using tr K and $\tilde{\gamma}$ as configuration variables) and on a rigorous analysis of the initial-value constraints (where tr K and $\tilde{\gamma}$ are freely specifiable data). Of course, I make no claims that this is the only way. Another method is to work at null infinity, which is quite natural in radiation problems. However, the present analysis is suited to the standard (as opposed to the "characteristic") Cauchy problem.

Here I shall sketch the fact that our conditions $\partial_\tau \text{ tr } K = D^j \pounds_t \tilde{\gamma}_{ij} = 0$, when linearized, produce the ADM and Dirac radiation gauges in a simple way.

Write the spacetime metric as

$$g_{ab} = f_{ab} + h_{ab} , \tag{130}$$

with the standard time variable of Minkowski spacetime: $f_{00} = -1$, $f_{0i} = 0$, $\partial_\tau f_{ij} = 0$. We allow any curvilinear spatial coordinates. An over-bar denotes operations with respect to f_{ij}. The perturbations of the spatial metric are

$$h_{ij} = \psi_{ij} + \frac{1}{3} f_{ij} \, \overline{\text{tr}} \, h , \quad \overline{\text{tr}} \, \psi = 0 . \tag{131}$$

The conformal three-metric is, in the linear approximation,

$$\tilde{\gamma}_{ij} = (\det f)^{-1/3} (f_{ij} + \psi_{ij}) . \tag{132}$$

The linearization of our α and β gauges read

$$\partial_\tau \, \overline{\text{tr}} \, K = 0 , \tag{133}$$

$$\bar{D}^j \partial_\tau \tilde{\gamma}_{ij} = \partial_\tau \bar{D}^j \tilde{\gamma}_{ij} = (\det f)^{-1/3} \partial_\tau \bar{D}^j \psi_{ij} = 0 . \tag{134}$$

The ADM "isotropic" radiation gauge can be written as

$$\overline{\mathrm{tr}}\, K = 0 \tag{135}$$

$$(\delta^{k}{}_{i}\bar{\Delta} - \tfrac{1}{4}\bar{D}^{k}\bar{D}_{i})(\bar{D}^{j}\psi_{jk}) = 0 \ . \tag{136}$$

The latter is a form simpler than, but equivalent to, that originally postulated by ADM, as it is equivalent to

$$\bar{D}^{j}\psi_{ij} = 0 \ . \tag{137}$$

The ADM conditions (135) and (136) are clearly special cases of our linearized conditions (133) and (134). The latter imply $h^{00} = 0$ and $h^{0i} = 0$ in vacuum. Thus,

$$h^{0a} = 0 \ , \tag{138}$$

$$h^{ij} = \tfrac{1}{3}f^{ij}\,\overline{\mathrm{tr}}\, h + h^{ij}{}_{tt} \tag{139}$$

where we have written $\psi^{ij} = h^{ij}{}_{tt}$ to emphasize its spatially trace-free and divergence-free character. In vacuum, the linearized constraints give $\overline{\mathrm{tr}}\, h = 0$ and we obtain the usual "TT" wave gauge of the linearized vacuum theory.

Dirac's coordinate conditions were given as

$$\mathrm{tr}\, K = 0 \ , \tag{140}$$

$$\tilde{\gamma}^{ij}{}_{,j} = 0 \ . \tag{141}$$

The latter is awkward and not spatially covariant. Its covariant form is clearly

$$\bar{D}_{j}\tilde{\gamma}^{ij} = 0 \tag{142}$$

which agrees with (137) of ADM. We see that (140) and (142) are special cases of our gauge.

As a final remark in connection with a "radiation gauge" interpretation of the minimal distortion condition, note that the integrand of the associated variational integral $\mathcal{S}[\beta]$ is in the "wave zone" essentially just the familiar expression

$$\dot{h}_{+}{}^{2} + \dot{h}_{\times}{}^{2}$$

that is interpreted as gravitational wave energy, where $+$ and $\times$ denote the two independent states of polarization of a gravity wave of amplitude h .

A conclusion that we can draw from this discussion is that it is reasonable to identify the radiative (or, more generally, dynamical) degrees of freedom with the time dependence of $\tilde{\gamma}$, when we evolve along a congruence defined by $\partial_\tau \operatorname{tr} K = D^j \pounds_t \tilde{\gamma}_{ij} = 0$, especially when the slices are also maximal (tr K = 0).

We may go slightly further and conjecture that a good generalized definition of a "global rest frame" is when

(A) The foliation is maximal with $\alpha \to 1$ at infinity.
(B) β satisfies the minimal distortion criterion with $\beta \to 0$ at infinity.
(C) The gauge-invariant total linear momentum surface integral vanishes.
(D) The asymptotic spatial gauge condition $k^i{}_{(\ell)}\bar{D}^j\psi_{ij} = O(r^{-3})$ holds on an initial slice.

These requirements appear also to eliminate asymptotic gauge transformations and thus allow the spatial angular momentum to be defined by a surface integral of the Cauchy data.

9.6 Harmonic Coordinates

I have made no mention in these notes of the well-known space-time harmonic coordinate conditions

$$\partial_b[(-\det g)^{\frac{1}{2}} g^{ab}] = 0 \; . \tag{143}$$

These can also be written in a "background metric" formalism that makes them look covariant. They result in mathematically nice simplifications of the Einstein equations. All the "hard" theorems about existence of maximal Cauchy developments, Cauchy stability, etc., (that I am aware of) have been proved using harmonic coordinates. If they hold on an initial data slice, they give a "reduced" form of the Einstein equations that is hyperbolic and that, remarkably, preserves both the constraints and the harmonic conditions in the evolution. In effect, they impose hyperbolic (rather than elliptic) equations for α and β . In this form (143) becomes

$$(\partial_\tau - \beta^j\partial_j)\alpha + \alpha^2 \operatorname{tr} K = 0 \; , \tag{144}$$

$$(\partial_\tau - \beta^j\partial_j)\beta^i + \alpha^2[\gamma^i{}_j \partial_j \ln\alpha + \gamma^{jk}\Gamma^i{}_{jk}(\gamma)] = 0 \; . \tag{145}$$

These equations are hyperbolic, coupled, and non-linear. Because I do not know their geometrical meaning in terms of foliations and time vectors (Ω and t), I do not know if they possess computational and interpretive (as opposed to purely mathematical) advantages. This is something that has still to be discovered. They may have a deeper meaning than has been evident up to now.

10 NOTES

These notes follow the text in order by sections. They are intended to give representative references and to provide helpful further commentary on the material discussed in the preceding sections of this chapter.

10.1 Introduction

The 3+1-dimensional approach to Einstein's equations is treated by Arnowitt, Deser & Misner (1962), Wheeler (1963) and Misner, Thorne & Wheeler (1973). I use the "spacelike" conventions of the latter reference. Units are chosen so that $c = 1$, $\kappa = 8\pi G$.

Numerical relativity is treated extensively in the chapters of this book by Eppley, Smarr, and Wilson.

10.2 Foliations of Spacetime

An excellent example is given by Reinhart (1973). However, note in this article that the one-form that defines the extended maximal foliation of Schwarzschild-Kruskal spacetime does not enter the null cone at $r = 2m$.

The projection operator approach and corresponding use of Lie derivatives is related to material found in Schouten (1954), Stachel (1962; 1969), and Ehlers *et al*. (1961).

The Gauss-Codazzi equations are treated in many references, for example, Eisenhart (1926), Schouten (1954), and Hicks (1965).

10.3 Frames and Foliations

This material is standard differential geometry adapted to the 3+1 formalism.

10.4 The Cauchy Problem

I have mentioned only a few key points in this very extensive subject. For details, see Choquet-Bruhat (1962), Hawking & Ellis (1973), Fischer & Marsden (1979), and Choquet-Bruhat & York (1979).

10.5 Initial-Value Problem

The pioneering work on the conformal approach was given by Lichnerowicz (1944). See also Choquet-Bruhat (1962).

The idea of separating out $\mathrm{tr}\, K$ when it is not zero and using it as a "time variable" is found in York (1972).

The two methods of combining conformal transformations and covariant splittings were discussed by York (1973b). The method not used here is found in Ō Murchadha & York (1974a). The method used in this article is treated rigorously by Choquet-Bruhat & York (1979). See also Ō Murchadha & York (1973).

The conformal scaling of sources is dealt with in Ō Murchadha & York (1973; 1974a), Isenberg, Ō Murchadha & York (1976), and Isenberg & Nestor (1977).

For the momentum constraint when $\partial_i \operatorname{tr} K = 0$, see Choquet-Bruhat & York (1979).

Existence and uniqueness of solutions ϕ of the generalized Lichnerowicz or scale equation on compact slices are demonstrated in Ō Murchadha (1973) and Ō Murchadha & York (1973). They used the methods of Choquet-Bruhat & Leray (1972). A recent summary in the context of Sobolev spaces is given in Choquet-Bruhat & York (1979). Recent results on asymptotically flat slices are found in the latter reference, in Cantor (1977), and in Chaljub-Simon & Choquet-Bruhat (1978). An important example of the ϕ equation (with $K = \rho = 0$) was analyzed carefully by Brill (1959). A fundamental example using non-Euclidean topology was given by Misner (1960).

Useful general results on elliptic equations and maximum principles are found in Miranda (1970), Ladyzhenskaya & Ural'tseva (1968), and Protter & Weinberger (1967).

Linearization stability of solutions of the constraints is demonstrated in Moncrief (1975a; 1976), in Fischer, Marsden & Moncrief (1978), and in Ō Murchadha & York (1974b).

10.6 Asymptotic Conditions

Much of this material is an updated version of Arnowitt *et al*. (1962). See also Arnowitt (1964), Misner (1964), and Smarr & York (1978a). Geometrical treatments of spatial infinity were given by Geroch (1972), Sommers (1978), and Ashtekar & Hansen (1978).

For the interpretation of ϕ as a generalized Newtonian potential, see Brill (1959) and Ō Murchadha & York (1974a, 1974c; 1976). The interpretation of W^i as a vector potential is found in Ō Murchadha & York (1974a; 1976) and has antecedents in Arnowitt *et al*. (1962).

10.7 Degrees of Freedom

This method of enumerating and separating dynamical and kinematical degrees of freedom is due to the author. Important antecedents are in Dirac (1959) (using tr K = 0), Brill (1959) (using $K_{ij} = 0$), and Arnowitt et al. (1962). Properties of the spatial conformal curvature tensor were helpful guides; see York (1971) and Arnowitt et al. (1962). See also Regge & Teitelboim (1974), Hanson, Regge & Teitelboim (1976), and Smarr & York (1978a; 1978b).

A gauge invariant approach to "true degrees of freedom" in a perturbative sense is found in Moncrief (1975a; 1975b; 1976) and Fischer & Marsden (1978; 1979).

10.8 Lapse Function

An extensive dicussion and many examples are given in Smarr & York (1978b).

A careful criticism of synchronous coordinates is given in Barrow & Tipler (1978).

The maximal slicing condition was first proposed by Lichnerowicz (1944) and later by Dirac (1959) and Deser (1967), among others. The tr K = constant condition was proposed by York (1972) for compact slices and subsequently by a number of people for noncompact slices. See, for example, Brill, Cavelle & Isenberg (1976), Goddard (1977), and Eardley & Smarr (1978). The general use of functions of K_{ij} to define time slices goes back to Arnowitt et al. (1962) and Kuchař (1970; 1971).

Existence of tr K = 0 and tr K = constant slices is demonstrated in Choquet-Bruhat (1975a; 1975b), Cantor et al. (1976), and Choquet-Bruhat, Fischer & Marsden (1978). For an important remark about the proofs for tr K = 0 , see Choquet-Bruhat & York (1979). For extremal properties of tr K = constant slices, see Brill et al. (1976), Flaherty & Brill (1976), and Goddard (1977).

The variational principle and boundary conditions for the maximal equation (involving $\Delta\alpha$) are discussed in Smarr & York (1978a). The fourth-order equation and the "volume bending energy" analogies are new to the present work. The general relationship to normal and tangential deformations of plates and thin shells and their equilibrium conditions was suggested by the material in Landau & Lifshitz (1959).

The extended maximal slicing of Schwarzschild spacetime was found independently by Estabrook et al. (1973) and Reinhart (1973).

The "collapse of the lapse" is discussed in detail in Smarr & York (1978b). The flat-space model is based on a similar model for demonstrating positive energy of pure gravity waves by Brill (1959) and Wheeler (1963). Graphs comparing the model slicing to the actual maximal slicing of Schwarzschild spacetime are given in Smarr & York (1978b).

10.9 Shift Vectors

Comparisons of the kinematics of the unit timelike vectors n^a, u^a, and $(-t_b t^b)^{\frac{1}{2}} t^a$ are given in Smarr & York (1978b).

The use of the shift to preserve a symmetry or a "preferred" form of the spatial metric is demonstrated in the chapter by Wilson in this book.

Minimal distortion shift vectors and similar criteria were proposed by York & Ō Murchadha (1974). An interpretation related to "factoring out" the action of infinitesimal diffeomorphisms in the space of conformal metrics is found in York (1974) and Fischer & Marsden (1977). The variational approach is found in York & Smarr (1975) and Smarr & York (1978a). The latter includes a discussion of boundary conditions. By using group-theoretical methods, Jantzen (1978) has found independently that minimal distortion shift vectors are natural in identifying dynamical degrees of freedom in homogeneous cosmologies.

The "quasi-isotropic" asymptotic coordinate conditions were proposed by the author in a series of lectures given in 1976 at the College de France, Paris. See also Regge & Teitelboim (1974) and Smarr & York (1978a).

The material related to the "radiation gauge" interpretation is based on Smarr & York (1978a), as are the remarks on harmonic spacetime coordinates.

10.10 Acknowledgments

I am indebted to numerous colleagues for helpful discussions of the material described in this article, in particular, M. Cantor, Y. Choquet-Bruhat, N. Ō Murchadha, L. Smarr and P. Sommers. The work reported here has been supported by the National Science Foundation, the Mattern Theoretical Physics Fund of the University of North Carolina, and a grant from the Université de Paris VI.

REFERENCES

Arnowitt, R. (1964). Asymptotic coordinate conditions, the wave front theorem, and properties of energy and momentum. In Conference Internationale sur les Théories Relativistes de la Gravitation, ed. L. Infeld, pp. 356-364. Gauthier-Villers, Paris.

Arnowitt, R., Deser, S. & Misner, C. W. (1962). The dynamics of general relativity. In Gravitation, ed. L. Witten, pp. 227-265. Wiley, New York.

Ashetekar, A. & Hansen, R. (1978). A unified treatment of null and spatial infinity in general relativity. I. Universal structure, asymptotic symmetries, and conserved quantities at spatial infinity. J. Math. Phys., 19, pp. 1542-1566.

Barrow, J. D. & Tipler, F. J. (1978). Demythologizing the singularity studies of Belinskii, Lifshitz & Khalatnikov. Preprint.

Berger, M. & Ebin, D. (1969). Some decompositions of the space of symmetric tensors. J. Diff. Geom., 3, pp. 379-302.

Brill, D. (1959). On the positive definite mass of the Bondi-Weber-Wheeler time-symmetric gravitational waves. Ann. of Phys., 7, pp. 466-483.

Brill, D., Cavelle, J. & Isenberg, J. (1976). Extremal properties of tr K = constant hypersurfaces. Unpublished report.

Cantor, M. (1977). The existence of non-trivial asymptotically flat initial data for vacuum spacetimes. Commun. Math. Phys., 57, pp. 83-96.

Cantor, M., Fischer, A., Marsden, J., Ō Murchadha, N. & York, J. (1976). The existence of maximal slicings in asymptotically flat spacetimes. Commun. Math. Phys., 49, pp. 897-902.

Chaljub-Simon, A. & Choquet-Bruhat, Y. (1978). Solutions asymptotiquement euclidiennes de l' équation de Lichnerowicz. C. R. Acad. Sci.(Paris), 286, pp. 917-920.

Choquet-Bruhat, Y. (1962). The Cauchy problem. In Gravitation, ed. L. Witten, pp. 130-168. Wiley, New York.

Choquet-Bruhat, Y. (1975a). Sous variétés maximales, ou à courbure constante, de variétés lorentziennes. C. R. Acad. Sci. (Paris), 280, pp. 169-171.

Choquet-Bruhat, Y. (1975b). Maximal submanifolds and manifolds with constant mean extrinsic curvature of a Lorentzian manifold. C. R. Acad. Sci. (Paris), 281, pp. 577-580.

Choquet-Bruhat, Y., Fischer , A. & Marsden, J. (1978). Maximal hypersurfaces and positivity of mass. In Proceedings of the 1976 Summer School "Enrico Fermi", ed. J. Ehlers, in press.

Choquet-Bruhat, Y. & Leray, J. (1972). Sur le problème de Dirichlet quasi-linéaire d'ordre 2. C. R. Acad. Sci. (Paris), 274, pp. 81-85.

Choquet-Bruhat, Y. & York, J. W. (1979). The Cauchy problem. In Einstein Centenary Volume, eds. P. Bergman, J. Goldberg, & A. Held, in press.

Deser, S. (1967). Covariant decompositions of symmetric tensors and the gravitational Cauchy problem. Ann. Inst. Henri Poincaré, 7, pp. 149-188.

Dirac, P. A. M. (1959). Fixation of coordinates in the Hamiltonian theory of gravitation. Phys. Rev., 114, pp. 924-930.

Eardley, D. & Smarr, L. (1978). Time functions in numerical relativity. I. Marginally bound dust collapse. Preprint.

Ehlers, J., Jordan, P., Kundt, W. & Sachs, R. (1961). Beiträge zur relativischen mechanik. Akad. Wiss. Lit. Mainz abh Math.-Nat. Kl., 11, pp. 1-47.

Eisenhart, L. P. (1926). Riemannian Geometry. Princeton University Press, Princeton.

Estabrook, F., Wahlquist, H., Christensen, S., DeWitt, B., Smarr, L. & Tsiang, E. (1973). Maximally slicing a black hole. Phys. Rev. D., 7, pp. 2814-2817.

Fischer, A. & Marsden, J. (1975). Deformations of the scalar curvature. Duke Math. J., 42, pp. 519-547.

Fischer, A. & Marsden, J. (1977). The manifold of conformally equivalent metrics. Can. J. Math., 29, pp. 193-209.

Fischer, A. & Marsden, J. (1978). Topics in the dynamics of general relativity. In Proceedings of the 1976 Summer School "Enrico Fermi", ed. J. Ehlers, in press.

Fischer, A. & Marsden, J. (1979). The initial value problem and the dynamical formulation of general relativity. In General Relativity: An Einstein Centenary Survey, eds. S. Hawking and W. Israel, in press. Cambridge University Press, Cambridge.

Fischer, A., Marsden, J. & Moncrief, V. (1978). Linearization stability of the Einstein equations. Preprint.

Flaherty, F. J. & Brill, D. (1976). Isolated maximal surfaces in spacetime. Commun. Math. Phys., 50, pp. 157-163.

Geroch, R. (1972). Structure of the gravitational field at spatial infinity. J. Math. Phys., 13, pp. 956-968.

Goddard, A. J. (1977). A generalization of the concept of constant mean curvature and canonical time. Gen. Rel. and Grav., 8, pp. 525-532.

Hanson, A., Regge, T. & Teitelboim, C. (1976). Constrained Hamiltonian Systems. Academia Nazionale dei Lincei, Rome.

Hawking, S. W. & Ellis, G. F. R. (1973). The Large Scale Structure of Space-time. Cambridge University Press, Cambridge.

Hicks, N. (1965). Notes on Differential Geometry. Van Nostrand, Princeton.

Isenberg, J. & Nestor, J. (1977). Extension of the York field decomposition to general gravitationally coupled fields. Ann. of Phys., 108, pp. 368-386.

Isenberg, J., Ō Murchadha, N. & York, J. W. (1976). Initial-value problem of general relativity III. Phys. Rev. D., 13, pp. 1532-1537.

Jantzen, R. (1978). The dynamical degrees of freedom in spatially homogeneous cosmology. Preprint.

Kuchař, K. (1970). Ground state functional of the linearized gravitational field. J. Math. Phys., 11, pp. 3322-3334.

Kuchař, K. (1971). Canonical quantization of cylindrical gravitational waves. Phys. Rev. D., 4, pp. 955-986.

Ladyzhenskaya, O. A. & Ural'tseva, N. N. (1968). Linear and Quasi-Linear Elliptic Equations. Academic Press, New York.

Landau, L. & Lifshitz, E. M. (1959). Theory of Elasticity. Addison-Wesley, Reading, Massachusetts.

Lichnerowicz, A. (1944). L'integration des équations de la gravitation relativiste et le problème des n corps. J. Math. Pures et Appl., 23, pp. 37-63.

Miranda, C. (1970). Partial Differential Equations of Elliptic Type. Springer-Verlag, Berlin.

Misner, C. W. (1960). Wormhole initial conditions. Phys. Rev., 118, pp. 1110-1112.

Misner, C. W. (1964). Waves, Newtonian fields, and coordinate functions. In Conference Internationale sur les Théories Relativistes de la Gravitation, ed. L. Infeld, pp. 189-205.

Misner, C. W., Thorne, K. S. & Wheeler, J. A. (1973). Gravitation. Freeman, San Francisco.

Moncrief, V. (1975a). Spacetime symmetries and linearization stability of the Einstein equations, I. J. Math. Phys., 16, pp. 493-498.

Moncrief, V. (1975b). Decompositions of gravitational perturbations. J. Math. Phys., 16, pp. 1556-1560.

Moncrief, V. (1976). Spacetime symmetries and linearization stability of the Einstein equations, II. J. Math. Phys., 17, pp. 1893-1902.

Monroe, D. (1978). Local transverse-traceless tensor operators in general relativity. Preprint.

Ō Murchadha, N. (1973). Existence and uniqueness of solutions of the Hamiltonian constraint of general relativity. Thesis, Princeton University.

Ō Murchadha, N. & York, J. W. (1973). Existence and uniqueness of solutions of the Hamiltonian constraint of general relativity on compact manifolds. J. Math. Phys., 14, pp. 1551-1557.

Ō Murchadha, N. & York, J. W. (1974a). Initial-value problem of general relativity I. Phys. Rev. D., 10, pp. 428-436.

Ō Murchadha, N. & York, J. W. (1974b). Initial-value problem of general relativity II. Phys. Rev. D., 10, pp. 437-446.

Ō Murchadha, N. & York, J. W. (1974c). Gravitational energy. Phys. Rev. D., 10, pp. 2345-2357.

Ō Murchadha, N. & York, J. W. (1976). Gravitational potentials: a constructive approach to general relativity. Gen. Rel. and Grav., 7, pp. 257-261.

Protter, M. & Weinberger, H. (1967). Maximum Principles in Differential Equations. Prentice-Hall, Englewood Cliffs, New Jersey.

Regge, T. & Teitelboim, C. (1974). Role of surface integrals in the Hamiltonian formulation of general relativity. Ann. of Phys., 88, pp. 286-318.

Reinhart, B. L. (1973). Maximal foliations of extended Schwarzschild space. J. Math. Phys., 14, p. 719.

Schouten, J. A. (1954). Ricci-Calculus. Springer-Verlag, Berlin.

Smarr, L. & York, J. W. (1978a). Radiation gauge in general relativity. Phys. Rev. D., 17, pp. 1945-1956.

Smarr, L. & York, J. W. (1978b). Kinematical conditions in the construction of spacetime. Phys. Rev. D., 17, pp. 2529-2551.

Sommers, P. (1978). The geometry of the gravitational field at spacelike infinity. J. Math. Phys., 19, pp. 549-554.

Stachel, J. (1962). Lie derivatives and the Cauchy problem in the general theory of relativity. Thesis, Stevens Institute of Technology.

Stachel, J. (1969). Covariant formulation of the Cauchy problem in generalized electrodynamics and general relativity. Acta Phys. Polon., 35, pp. 689-709.

Wheeler, J. A. (1963). Geometrodynamics and the issue of the final state. In Relativity, Groups, and Topology, eds. C. DeWitt and B. DeWitt, pp. 316-520. Gordon and Breach, New York.

York, J. W. (1971). Gravitational degrees of freedom and the initial-value problem. Phys. Rev. Letters, 26, pp. 1656-1658.

York, J. W. (1972). Role of conformal three-geometry in the dynamics of gravitation. Phys. Rev. Letters, 28, pp. 1082-1085.

York, J. W. (1973a). Conformally invariant orthogonal decomposition of symmetric tensors on Riemannian manifolds and the initial-value problem of general relativity. J. Math. Phys., 14, pp. 456-464.

York, J. W. (1973b). Initial-value problem and dynamics of general relativity. Lecture notes, Princeton University.

York, J. W. (1974). Covariant decompositions of symmetric tensors in the theory of gravitation. Ann. Inst. Henri Poincaré, 21, pp. 319-332.

York, J. W. & Ō Murchadha, N. (1974). Physical interpretation of initial data. Bull. Am. Phys. Soc., 19, p. 509.

York, J. W. & Smarr, L. (1975). Maximal slices and minimal shear. Bull. Am. Phys. Soc., 20, p. 544.

GLOBAL PROBLEMS IN NUMERICAL RELATIVITY

Douglas M. Eardley

Departments of Physics and Astronomy
Yale University

1. INTRODUCTION

The main outlines of the non-quantum theory of general relativity are by now fairly well understood. However, there remains the difficulty of making detailed predictions due to the subtleties and complications of the theory. There also remains a number of unanswered questions, mostly having to do with spacetime singularities and positivity of mass-energy. The conventional conclusions are that a black hole is the only possible final state of a gravitationally collapsing body which is unable to support itself by pressure forces, that spacetime singularities occur during gravitational collapse only inside black holes, and that total mass-energy is positive for all bodies.

However none of these conclusions have been proven in general. It is still theoretically possible, although perhaps unlikely, that "naked singularities" can form outside of, or instead of, black holes; and that such a singularity, or some other exotic kind of body, would have negative total mass-energy. These unlikely possibilities are worth bearing in mind as exotic gravitational-wave sources until they are disposed of theoretically although most relativists, including myself, do not believe they exist. Except for the Schoen & Yau (1978) Theorem discussed below, there has been rather little progress on these questions recently, and I can refer to the articles of Penrose (1973) and Geroch (1973) and to the book of Hawking and Ellis (1973), for a thorough review.

Numerical relativity has already shown its ability to predict gravitational radiation and to determine the final state for astrophysically interesting systems;see the articles of Wilson, Smarr, Eppley, and Piran in this volume. Numerical relativity also promises to be a very useful "experimental" technique to study global questions in general relativity; again, see Smarr's articles. This article will mention three connections between numerical relativity and the global method in relativity which specifically involve singularities, null infinity, and mass-energy.

2. SINGULARITIES

Numerical relativity may be defined as the numerical solution of Einstein's equations as an initial-value problem. Therefore one is immediately restricted to the construction of "globally hyperbolic" spacetimes; for discussion see York (this volume), Eardley & Smarr (1978), and Geroch (1970). Conversely, one can in principle construct any globally hyperbolic spacetime from suitable initial data through numerical relativity.

This restriction may be a strong one for the construction of spacetime singularities; one of the early victories of global relativity was to eliminate global hyperbolicity from the hypotheses of the singularity theorems, but one gives up this victory in turning to numerical methods. In consequence, the largest spacetime (M, ds^2) one can construct numerically from some given initial data, called the maximal development, may fail to reach some or even all of the singularities. Specifically, the future boundary or "roof" of (M, ds^2) will consist in general of a union of singularities G^+, Cauchy horizons H^+, and infinities $\mathcal{I}^+$ or I^+. Spacetime will be extendible across any H^+ that occur into a larger spacetime (M', ds^2) which is not globally hyperbolic; but (M', ds^2) cannot be constructed numerically because it cannot be reached by continuous Cauchy evolution starting from the initial data for (M, ds^2). Numerical relativity can study the singularities G^+ of (M, ds^2), but cannot study any further singularities lying in (M', ds^2). Well-known examples of such Cauchy horizons H^+ are the inner horizons at $r = r_-$ inside charged or rotating stationary black holes (Hawking & Ellis 1973).

However, there is some evidence for the conjecture that the Cauchy horizons H^+ are always unstable under small variations in the initial data, and do not occur in the generic case (Simpson & Penrose 1973). Then in the generic case spacetime would always obey the "Strong Cosmic Censorship Hypothesis", i.e. be globally hyperbolic. This is a stronger form of the standard "Cosmic Censorship Hypothesis" that no singularities occur outside of black holes ("naked singularities"), i.e. that all of spacetime outside of black holes is globally hyperbolic. Numerical relativity will construct a generic spacetime within the class of assumed Killing symmetries, because of roundoff and discretization error. If the Strong Cosmic Censorship Hypothesis is true, then numerical relativity will construct singularities G^+ but no Cauchy horizons H^+, and will in fact allow the study of all generic singularities. Conversely, numerical methods may shed a good deal of light on this question by construction of examples (see Simpson & Penrose 1973).

A key issue in the construction of singularities is the choice of a slicing condition that will in fact allow the construction of the maximal development (M, ds^2) of the initial data. Up to now, numerical relativists have generally used the maximal slicing condition (see York, this volume; note that "maximal" is being used in two different senses). This condition seems to have the property, desirable for studies of gravitational radiation but undesirable for the present purpose, that it generates slices which avoid the singularity, i.e. which stay outside of some neighborhood N(0) of the singularity. On the other hand, one's slices must not actually intersect the singularity, because the intrinsic and extrinsic geometry of the slice will then become singular and will halt the evolution before all of (M, ds^2) is constructed.

The tr K = constant slicing condition (also known as the constant-extrinsic-curvature slicing condition) seems at present to be the best bet for the investigation of singularities, although we are very far from a general proof that it will always work. The property that is necessary but unproven, and which I will henceforth assume to be true, is that for any value of constant k, however large, there is a neighborhood N(k) of the singularity G^+ that is intersected by no complete slice S of constant extrinsic curvature with $\mathrm{tr}\, K \leq k$. To be specific, the slices S here must be Cauchy slices of the interior of the black hole; or if no black hole exists, Cauchy slices of spacetime (M, ds^2). Conditions which are equivalent to this one have been discussed by Eardley & Smarr (1978; "crushing singularities"), Marsden & Tipler (1978; "strong curvature singularities"); the latter prove some existence theorems assuming such a condition. Conversely , some such condition seems at present necessary to prove existence.

The foliation then is to consist of slices S(k) with tr K = k, and variable k between slices, with $k \to \infty$ to the future. The very great advantage of such slicing, if it exists, is that one is guaranteed to approach the true singularity as $k \to \infty$, thanks to a special case of one of the singularity theorems (Hawking & Ellis, p. 274). Applied here, the theorem states that no future-directed causal curve in the globally hyperbolic spacetime (M, ds^2) has length greater than 3/k to the future of S(k), assuming the strong energy condition. Therefore as $k \to \infty$, the S(k) sweep to the future in (M, ds^2), necessarily exhaust the maximal development, and hence approach the singularity G^+ (or Cauchy horizon H^+) in the limit. (The loose idea of "limit of S(k)" is made precise in Appendix A.) On the other hand, the existence of the "zone of avoidance" N(k) for each k guarantees that no S(k) intersects G^+ or H^+.

Therefore, if the foliation S(k) for k arbitrarily large can be constructed, one has constructed a true singularity (or Cauchy horizon) and not a mere coordinate singularity. The next requirement is for diagnostic, invariant quantities to evaluate at the singularity in order to characterize it. The slices S(k) are invariantly defined so quantities constructed out of their γ_{ij} and K_{ij} are also invariant. There has been very little work done on this requirement, and I shall make just one likely suggestion: that one evaluate the exponents p_i (i = 1, 2, 3) that occur in the Lifshitz-Khalatnikov (1963) and Belinsky-Khalatnikov-Lifshitz (1970) approximations for singularities. Here these may be defined as the eigenvalues of the spatial matrix

$$\gamma^{ik} K_{kj}/\mathrm{tr}\, K \ . \tag{1}$$

They obey

$$\Sigma_i \, p_i = 1; \tag{2a}$$

and <u>if</u> matter and 3-curvature terms in Einstein's equations vanish sufficiently fast near the singularity,

$$\Sigma_i \, p_i^{\,2} \to 1. \tag{2b}$$

Values for the lim p_i can be interpreted in various ways (these remarks are intended to be suggestive; they are not rigorous):

1. In spherical symmetry either lim p_i = (-1/3, 2/3, 2/3) or lim p_i = (1, 0, 0). The latter case may be a Cauchy horizon, the former not. The Schwarzschild singularity has (-1/3, 2/3, 2/3), the Cauchy horizon inside a Reissner-Nordström black hole (1,0,0).

2. In axial symmetry without rotation (ϕ an ignorable coordinate, spacetime invariant under $\phi \to -\phi$) any values of lim p_i consistent with Eqs. (2ab) may arise. One expects the limit to exist because one eigenvector of (1) will point in the ϕ-direction by symmetry, and will be surface-forming, so that the singularity may be expected to be "velocity-dominated" or "Kasner-like". Again, the appearance of (1, 0, 0) should arouse suspicion of a Cauchy horizon.

3. In axial symmetry with rotation, or in the absence of symmetry, none of the eigenvectors of (1) will in general be surface-forming, and one might expect a "general oscilliatory" or "mixmasterlike" singularity, in which the lim p_i do not exist. In this case the spatial correlation length of S(k) may tend to zero as $k \to \infty$, so that numerical treatment may become very difficult.

3. NULL INFINITY

In an asymptotically flat spacetime, a complete maximal slice automatically reaches out to spatial infinity I^0. A complete slice S(k) with tr K = k $\neq$ 0 automatically intersects future null infinity $\mathcal{I}^+$ (if k < 0) or past null infinity $\mathcal{I}^-$ (if k > 0). The intersection of S(k) and $\mathcal{I}^+$ (specializing to the case k < 0) will be a 2-dimensional surface C at infinity, called a <u>cut</u>. A promising idea is to use tr K = k slicing in a computer code. As the slice S(k) evolves in spacetime, the cut C moves to the future along $\mathcal{I}^+$. Then one would automatically have an outgoing-wave boundary condition at $\mathcal{I}^+$, and one could use the standard theory of null infinity (Bondi, van der Burg & Metzner 1962, Sachs 1962, Penrose 1974) to define and to measure numerically the radiation. In particular, there is a complex asymptotic shear σ^o defined on any cut C, the time derivative of which is gravitational-wave amplitude. The germ of this idea goes back to Hernandez & Misner (1966).

However, Goddard (1977) suggests that there may be a roadblock to this program: He conjectures that it may be a very restrictive condition on a cut C that it bound a tr K = k slice S(k) in spacetime, so that a generic asymptotically simple spacetime might admit no such S(k) at all. In Appendix B I give a result which indeed points in this direction, although is much weaker than Goddard's conjecture: If a cut C bounds a slice S(k) which is smooth of at least differentiability class C^4 at $\mathcal{I}^+$, then the asymptotic shear of C must obey the restriction

$$0 = \mathrm{Re}[(\eth\,\eth - \dot{\sigma}^o)\,\bar{\sigma}^o] \qquad (3)$$

(For notation see Penrose 1974.) This result is weaker than the conjecture because it does not prove that such slices S(k) do not exist if (3) fails, only that they cannot be smooth; also existence is not proven if (3) holds. However the result shows that some modifications must be made in the standard discussion of null infinity, which does assume that the metric can be expanded in powers of $1/r$ near $\mathcal{I}^+$; if (3) fails then the metric cannot be expanded past $O(1/r^2)$ because of terms in $(\ln r)/r^3$, when using the S(k). There are at least three ways around this problem:

1. Keep the slicing condition and generalize the standard expansion at null infinity to an asymptotic expansion in terms $(\ln r)^m/r^n$. This generalization would complicate the discussion somewhat, and may or may not be valid.

2. Modify the slicing condition to

$$\mathrm{tr}\,K = k - 3b/(kr^3) \qquad (4)$$

where b is any smooth function on S which tends to the R.H.S. of Eq. (3) as $r \to \infty$. Then the standard expansion in factors $1/r^n$ is possible.

3. In axial symmetry without rotation, there exists a 2-parameter family of "good cuts" C with $\sigma^o = 0$. For these (3) holds, so that the standard expansion is possible for tr K = k slicing in this special symmetry. All currently running codes for asymptotically flat spacetime assume this symmetry.

The main advantages of S(k) slicing are the clean boundary condition for radiation and the economical zoning that one then can use in the wave zone. The method should therefore be considered for problems in which the limitation on total number of mesh points is severe and for which radiation is the primary output, e.g., the 3-dimensional black-hole-collision problem. However, there is a good deal of theoretical work to be done before the method is ready for the computer.

4. MASS-ENERGY

Mass-energy cannot be defined locally in general relativity. The best one can do is to associate a mass-energy M(A) with a 2-surface A; we will take A to be spacelike, topologically spherical, and embedded in an asymptotically flat spacetime. If S is any spacelike hypersurface with A as its outer boundary, then M(A) can be thought of as the total mass-energy residing in S. The fact that M(A) is independent of the choice of S is a consequence of the law of conservation of mass-energy. Now no completely general and satisfactory definition of M(A) for all A in a spacetime is known (to me anyway); but various definitions are in use and seem to give good results numerically, when used carefully. Here I shall list a number of properties that any such definition should obey and review a particular definition due to Hawking (1968) with some nice properties.

1. A point p in spacetime must have zero mass, in that lim M(A) = 0 if A shrinks to p, for instance if A is a shrinking geodesic sphere in a given slice S.

2. A metric 2-sphere A in Minkowski spacetime should have M(A) = 0.

3. In any spherically symmetric spacetime there is an invariant "mass function" M(A) (Cahill & McVittie 1970). Any general definition should reduce to this special case in spherical symmetry.

3'. In particular, in Schwarzschild spacetime of mass m, we should have M(A) = m for any (centered) 2-sphere.

4. If S is an asymptotically flat slice, r any standard radial coordinate is S, and A(r) any large coordinate sphere in S, then we should have lim M(A(r)) = M_{ADM} as $r \to \infty$.

Here M_{ADM} is the ADM mass of the slice (see York, this volume), which should be independent of slice S in a given spacetime.

5. If S is an asymptotically null slice, that is, a slice that crosses $\mathcal{I}^+$ (or $\mathcal{I}^-$) in some cut C, r any standard radial coordinate, and A(r) any large coordinate sphere in S, then we should have lim M(A(r)) = M_B(C) as $r \to \infty$.

Here M_B (C) is the Bondi mass (Penrose 1974; see also Tamburino & Winicour 1966) evaluated on the cut C. Bondi et al. (1962) and Sachs (1962) proved that M_B decreases to the future: most generally if C_1 is any cut lying to the future of C in $\mathcal{I}^+$, then $M_B(C_1) \leq M_B(C)$. This result signifies that the radiation crossing $\mathcal{I}^+$ between C and C_1 always carries positive mass-energy. One would like to believe that $M_{ADM} \geq M_B(C)$, and that as C tends to spatial infinity I^0 to the past along $\mathcal{I}^+$, $\lim M_B(C) = M_{ADM}$. These beliefs have not been proven.

6. If A is an apparent horizon (for present purposes, the outermost outer-marginally-trapped surface in a slice S) then we should have $M(A) = [\text{Area}(A)/16\pi]^{1/2} \equiv M_{irr}$, the irreducible mass of the horizon. Some motivation for this definition of M_{irr} comes from Hawking's Area Theorem (Hawking & Ellis 1973). If a slice S contains an apparent horizon A, and if further any 2-surface A' lying outside A in S obeys Area (A') $\geq$ Area (A) then the mass M_{hole} of the final stationary black hole to which the system evolves must obey $M_{hole} \geq M_{irr}$.

The Positive Mass Conjecture in general relativity asserts that $M_{ADM} \geq 0$ for any asymptotically flat slice S, and that

$M_B(C) \geq 0$ for any asymptotically null slice S which intersects $\mathcal{I}^+$ in a cut C; and that $M_{ADM} = 0$ or $M_B(C) = 0$ if and only if S is an

initial data set for flat spacetime. Schoen & Yau (1978) recently proved this conjecture for an asymptotically flat slice S under the restriction that S be a maximal slice. Other cases still remain open. A more general form of the Positive Mass Conjecture is that $M_{ADM} \geq M_{irr}$ or $M_B(C) \geq M_{irr}$, where M_{irr} is the irreducible mass of an apparent horizon A in S. Penrose (1973) pointed out that if one could construct an initial data set on an S which violated these inequalities, then the conventional viewpoint on black holes would be in serious trouble. Such an hypothetical S has come to be called "a counterexample to Cosmic Censorship"; no such counterexamples are known and a number of special cases have in fact been ruled out by Gibbons, and Jang & Wald (1977). However, the proof that no such S exist would not prove the Cosmic Censorship Hypothesis.

The final and key requirement on any definition of M(A) is that of "local positivity":

7. If A' is "bigger" than A in the sense that $A \subset S'$ for some achronal hypersurface S' with outer boundary A', $M(A') \geq M(A)$.

In fact, 7. is too strong; I suspect no M(A) at all exists that satisfies 1.-7. In order to progress it seems to be necessary to restrict the relation of A' to A further. As an example, consider the particular definition $M_H(A)$ due to Hawking (1968):

$$M_H(A) = [\text{Area }(A)/16\pi]^{1/2}\,(1 - \int \rho\mu\, dA/2\pi) . \tag{5}$$

Here $\int dA$ is the integral over A, ρ is the convergence of the outer null normal $\vec{\ell}$, and $-\mu$ is the convergence of the inner null normal $\vec{n}$, in the notation of Newman & Penrose (1962). This definition satisfies 1. - 6. above, and it also satisfies a restricted form of local positivity:

7'. Let A obey the restrictions

$$\rho < 0,\ \mu < 0 \tag{6}$$

at each point on A. Then define as a constant on A, $r \equiv [\text{Area }(A)/4\pi]^{1/2}$, and rescale $\vec{\ell}$ so that $\rho = -1/r$. The quantity r then extends to a luminosity distance along the outgoing future-directed null hypersurface S normal to A, and r defines a one-parameter family of level surfaces A(r) in S, of which A is one. Under change of A(r) along this family

$$dM_H(A(r))/dr = (4\pi)^{-1} \int [\Phi_{11} + 3\Lambda + |\alpha + \bar{\beta}|^2 - r\mu(|\sigma|^2 + \Phi_{00})]$$

$$\geq 0 \tag{7}$$

where we assume the Dominant Energy Condition (Hawking & Ellis 1973) to ensure that the matter terms are nonnegative.

Therefore M_H is nondecreasing along the family A(r) in S. Similarly, M_H is nondecreasing along the one-parameter family A(r) in S', the outgoing past-directed null hypersurface normal to A, where r is a similarly defined luminosity distance along S'. (In fact, more general spacelike families A(r) can be defined generated by the vector field $\vec{q} = P\vec{\ell} - Q\vec{n}$, where P(r) and Q(r) are nonnegative and constant on each A(r), and must satisfy $P\rho + Q\mu = -1/r$, where r is luminosity distance. Then M_H is nondecreasing along A(r).)

The restrictions (6) are very serious, and the existence of suitable families A(r) can be established only in very special spacetimes. When such a family exists, some interesting results can be proven.

Proposition: Let spacetime contain a point p, whose future light-cone $J^+(p)$ admits a luminosity distance r with level surfaces A(r) which obey (6). (In particular, $J^+(p)$ has no caustics.) Then $M_B(C) \geq 0$ on the cut C at which $J^+(p)$ intersects $\mathcal{I}^+$.

Proof: Use 1., 5., and 7'., above. This is a proof of positivity of Bondi mass, under a serious restriction. In particular r must tend to a standard coordinate at $\mathcal{I}^+$ as in 5. (but it turns out that r need not tend to a standard radial coordinate at p).

Corollary: In a spherically symmetric spacetime, if I^+ is in the domain of dependence of I^-, then $M_B \geq 0$ on any cut of I^+.

Geroch (1973) gave a similar argument on a maximal slice S. Later, Jang & Wald (1977) used Geroch's argument against certain counterexamples to Cosmic Censorship. Hawking's mass can be used in a similar way. For instance, Penrose (1973) proposed a possible class of counterexamples to Cosmic Censorship involving an impulsive shell of null fluid imploding along a past null hypersurface S' in Minkowski spacetime. Because of the gravitational field set up by the fluid, spacetime is not flat to the future of S' and an apparent horizon A can form in S'.

Now restrict to the subclass in which S' admits a family A(r) from A to I^-, such that $\rho \leq 0$ everywhere, and such that r tends to a standard radial coordinate at I^-. From 5., 6., and 7'. it follows that $M_B \geq M_{irr}$, where M_B is the total Bondi mass of null fluid. Therefore this subclass of counterexamples is ruled out.

Two cautions must be recognized if M_H is to be used in a computer code. First, A should be carefully chosen so as to be as nearly spherical as possible. Second, the form (5) is ill conditioned at large radius due to near-cancellation of large terms. At large radius the equivalent form

$$M_H(A) = (2\pi)^{-1} \, [\text{Area }(A)/16\pi]^{\frac{1}{2}} \int (-\Psi_2 + \Lambda + \Phi_{11} - \sigma\lambda)\, dA$$

should be used.

5. CONCLUSION AND ACKNOWLEDGEMENT

Global questions are already being studied by numerical means. Every time a collapse calculation gives the expected result, namely a neutron star or black hole rather than something more exotic, the conventional viewpoint is further validated. Almost no numerical work has been done on singularities; there seems to be a great opportunity here to help answer many of the remaining questions about the detailed nature of singularities, as reviewed recently by Ellis (1978). In some preliminary numerical experiments, borrowing the 1-dimensional collapse code described by Wilson (this volume), I succeeded in using the S(k) slicing discussed in Section 2 to push close to the singularity, but this work is barely begun.

I am very grateful to many colleagues for discussions and correspondence, to the National Science Foundation for research support, and to Lawrence Livermore Laboratory for hospitality.

APPENDIX A: LIMITS OF SLICES

Here I mention a definition of the loose idea "the sequence S(k) of slices approaches the singularity". Consider a globally hyperbolic spacetime (M, ds^2), taken spatially compact for simplicity. Form as usual the topological space $\mathcal{S}$ of all Cauchy slices S. There is a natural spacetime distance $d(S_1, S_2)$ between any two Cauchy slices S_1 and S_2, namely the maximum length of any causal curve from any point in S_1 to any point in S_2. In fact, $(\mathcal{S}, d)$ is a metric space, and can be completed to form a complete metric space $(\bar{\mathcal{S}}, d)$. Assuming (M, ds^2) is future-incomplete, $(\bar{\mathcal{S}}, ds^2)$ contains a unique element S^+ which can be interpreted as the future boundary or "roof" of (M, ds^2). The statement "$S(k) \to S^+$" now has precise meaning in the topology of $(\bar{\mathcal{S}}, d)$.

APPENDIX B: TROUBLE AT NULL INFINITY

I shall show that a slice S(k) with tr K = constant = k can be formally expanded in powers of 1/r around $\mathcal{I}^+$ in vacuum, if and only if the asymptotic shear of the bounding cut C of S(k), denoted c or σ^o, obeys

$$0 = \mathrm{Re}[(\eth\eth - \dot{c})\bar{c}] \tag{8}$$

Here $\dot{c} = \partial_u c$ and $\eth\eth\bar{c} = \tfrac{1}{2}(\csc\theta\,\partial_\theta + i\csc^2\theta\,\partial_\phi)^2(\bar{c}\sin^2\theta)$.
Proof: Following the notation of Sachs (1962), define S(k) as the level surface $\tau = 0$ of the time function $\tau(u, r, \theta, \phi)$, such that the cut C lies at $u = 0$. The equation for tr K is

$$k = \mathrm{tr}\,K = \quad [g^{ab}\tau_{,b}(-g^{cd}\tau_{,c}\,\tau_{,d})^{-1/2}]_{,a} \tag{9}$$

Expand g_{ab} in powers of 1/r as usual, impose the vacuum Einstein equations, expand τ as

$$\tau = u + \sum_{n=1}^{\infty} r^{-n} f_n(\theta\phi) \tag{10}$$

and thereby write (9) as a hierarchy of equations

$$(r^{4-n} f_n)_{,r} = r^{3-n} J_n(\theta\phi);\ n = 1, 2, 3, \ldots \tag{11}$$

where each $J_n(\theta\phi)$ ("junk factor") is known after the equations for 1,..., n-1 are solved. Solving (9) successively we obtain

$$f_1 = -9/(2k^2);\ f_2 = 0;\ f_3 = 81/(8k^4) - 15|c|^2/(4k^2);$$

but then at n = 4

$$f_4 = 81\,\ln r/(4k^4)\ \mathrm{Re}[(\eth\eth - \dot{c})\bar{c}] + D(\theta,\phi). \tag{12}$$

The solution cannot proceed unless the term in ($\ln$ r) vanishes, so that (8) is necessary. No further (ℓn r)'s appear for other n, so (8) is also sufficient QED.

This method of formal power series cannot show either existence or nonexistence. Nevertheless this result clearly suggests the conjecture that slices S(k) will exist for any C, but will be smooth (no ℓn r) only if (8) holds. In the general case they will be C^3 but no better at $\mathcal{I}^+$. The boundary conditions on S(k) are the "Dirichlet-like" cut C, and the "Neumann-like" function D which appears as a function of integration in (12). One expects to be able to specify one, but not both of these conditions.

REFERENCES

Belinsky, V. A. Khalatnikov, I. M. & Lifshitz, E. M. (1970). Oscilliatory approach to a singular point in relativistic cosmology. Advan. Phys., 19, pp. 525-573.

Bondi, H., van der Burg, M. G. J. & Metzner, A. W. K. (1962). Gravitational waves in general relativity VII. Waves from axi-symmetric isolated systems. Proc. Roy. Soc. A, 269, pp. 21-52.

Cahill, M. E. & McVittie, G. C. (1970). Spherical symmetry and mass-energy in general relativity. I. General theory. J. Math. Phys., 11, pp. 1382-1391.

Eardley, D. M. & Smarr, L. (1978). Time functions in numerical relativity I. Marginally bound dust collapse. Harvard Center for Astrophysics preprint, #1036.

Ellis, G. F. R. (1978). Singularities in general relativity theory. Comments Astrophys., VIII, pp. 1-7.

Geroch, R. (1970). Domain of dependence. J. Math. Phys., 11, pp. 437-449.

Geroch, R. (1973). Energy extraction. Ann. N.Y. Acad. Sci., 224, pp. 108-117.

Goddard, A. J. (1977). Some remarks on the existence of spacelike hypersurfaces of constant mean curvature. Math. Proc. Camb. Phil. Soc., 82, pp. 489-495.

Hawking, S. W. (1968). Gravitational radiation in an expanding universe. J. Math. Phys., 9, pp. 598-604.

Hawking, S. W. & Ellis, G. F. R. (1973). The Large Scale Structure of Space-time. Cambridge Univeristy Press, Cambridge.

Hernandez, W. C., Jr. & Misner, C. W. (1966). Observer time as a coordinate in relativistic spherical hydrodynamics. Astrophys. J., 143, pp. 452-464.

Jang, P. S. & Wald, R. M. (1977). The positive energy conjecture and the cosmic censor hypothesis. J. Math. Phys.,18, pp. 41-44.

Lifshitz, E. M. & Khalatnikov, I. M. (1963). Investigations in relativistic cosmology. Advan. Phys., 12, pp. 185-249.

Marsden, J. E. & Tipler, F. J. (1978). Maximal hypersurfaces and foliations of constant mean curvature in general relativity. University of California at Berkeley preprint.

Newman, E. & Penrose, R. (1962). An approach to gravitational radiation by a method of spin coefficients. J. Math. Phys., 3, pp. 566-578.

Penrose, R. (1973). Naked singularities. Ann. N. Y. Acad. Sci., 224, pp. 125-134.

Penrose, R. (1974). Relativistic symmetry groups. In Group Theory In Non-Linear Problems, ed. A. O. Barut, pp. 1-58. D. Reidel, Dordrecht.

Sachs, R. K. (1962). Gravitational waves in general relativity VIII. Waves in asymptotically flat spacetime. Proc. Roy. Soc. A, 270, pp. 103-126.

Schoen, R. & Yau, S.-T. (1978). On the proof of the positive mass conjecture in general relativity. University of California at Berkeley preprint.

Simpson, M. & Penrose, R. (1973). Internal instability of a Reissner-Nordström black hole. Inter. J. Theor. Phys., 7, pp. 183-197.

Tamburino, L. A. & Winicour, J. H. (1966). Gravitational fields in finite and conformal Bondi frames. Phys. Rev.,150, pp. 1039-1053.

BASIC CONCEPTS IN FINITE DIFFERENCING OF PARTIAL DIFFERENTIAL EQUATIONS

Larry Smarr

Harvard-Smithsonian Center for Astrophysics and
Lyman Laboratory of Physics
Harvard University

1. INTRODUCTION

This survey article is meant as a simple introduction to the basic concepts and problems which arise in numerical solution of partial differential equations. The material is by and large not original (except for my treatment of Einstein's equations), but I felt that an introduction to this subject would be useful. In section 2 I show how one can think of the Einstein equations and the hydrodynamic equations in terms of simpler model equations. In section 3 I describe the method of finite differences and the new problems which arise with finite difference equations. The notion of artificial viscosity is treated at length in section 4 with emphasis on techniques used by other authors in the volume (Eppley, Piran, Wilson). Finally, in section 5 a brief description of the methods for solving elliptic equations in given.

I will not give extensive bibliographic references but rather refer the reader to reviews I have found very useful: Whitham (1974), Richtmyer (1963), Richtmyer and Morton (1967), Roache (1972), Kreiss and Oliger (1973), Crowley (1975), and White (1976).

2. PARTIAL DIFFERENTIAL EQUATIONS

2.1 Hyperbolic, Parabolic, Elliptic

There are three broad classes of second order partial differential equations (PDE's): hyperbolic, parabolic, and elliptic. These classes have completely different mathematical properties and represent quite different physical processes. It is therefore not surprising that the numerical methods for solving them are quite different.

Hyperbolic equations represent propagation. A simple example is:

$$\partial_t \partial_t \phi - c^2 \partial_x \partial_x \phi = 0 \qquad (1)$$

where $\phi = \phi(t,x)$. This becomes a well-posed problem when initial data $\phi(0,x)$ and $\partial_t\phi(0,x)$ are given. The speed of propagation is given by c.

Parabolic equations represent <u>diffusion</u>. Here the simplest example is:

$$\partial_t\phi - \nu\partial_x\partial_x\phi = 0 \tag{2}$$

which is well-posed if supplemented by the initial condition $\phi(0,x)$. The rate of diffusion is governed by the diffusion coefficient ν.

Finally, elliptic equations represent <u>equilibrium</u>. If $\phi = \phi(x,y)$, then Laplace's elliptic equation becomes:

$$\partial_x\partial_x\phi + \partial_y\partial_y\phi = 0 \tag{3}$$

Here there is no coefficient since both derivatives have the same sign. This is not an initial value problem, but rather a boundary value problem in which one must specify complete boundary information for ϕ.

I will mainly concentrate on hyperbolic and elliptic PDE's which arise naturally in general relativity and hydrodynamics. It will turn out that the addition of parabolic terms to both these types of PDE's are necessary for some numerical solution techniques.

2.2 The Wave Equation

It is surprising how much of the structure of the Einstein equations can be explained by studying equation (1). I will sketch a few elementary properties of this simple equation and then parallel the discussion with the full Einstein equations. To formulate a Hamiltonian approach to an equation of motion, one singles out time to appear only in first order. We can accomplish this by introducing a new variable k(t,x) and rewriting (1) as:

$$\partial_t\phi = -k \tag{4a}$$

$$\partial_t k = -c^2\partial_x\partial_x\phi \tag{4b}$$

Note that in this form both first and second derivatives are present and that space and time are on different footings. By introducing another auxiliary variable $d(t,x) \equiv \partial_x\phi$, one can rewrite (4) as a first order system:

$$\partial_t\phi = -k \tag{5a}$$

$$\partial_t d = -\partial_x k \tag{5b}$$

$$\partial_t k = c^2 \partial_x d \tag{5c}$$

where I have assumed c is a constant. Now only first derivatives appear. If one writes this in vector form, one finds

$$\partial_t \begin{pmatrix} \phi \\ d \\ k \end{pmatrix} + \begin{pmatrix} 0 & 0 & 0 \\ 0 & 0 & -1 \\ 0 & -c^2 & 0 \end{pmatrix} \partial_x \begin{pmatrix} \phi \\ d \\ k \end{pmatrix} + \begin{pmatrix} -k \\ 0 \\ 0 \end{pmatrix} = 0 \tag{6}$$

The dynamics is contained in the pair (d,k). Note that the only nonzero elements of the velocity matrix multiplying ∂_x are the two off diagonal elements. This second order equation written as a first order system (see Courant and Hilbert (1962) for this general procedure) suggests we consider an even simpler model equation:

$$\partial_t \phi + v \partial_x \phi = 0 \tag{7}$$

This first order hyperbolic equation in fact will provide most of our insight into much more complicated systems.

If one wishes to prove rigorous theorems about the existence and uniqueness of solutions to the second order wave equation, then the methods one uses will be entirely different if one uses form (1), form (4), or form (5). Similarly the numerical techniques (discussed in later sections) to solve the wave equation will depend on which form we consider.

2.3 Einstein's Equations

It was clear even to Einstein that in certain limits his field equations reduce to wave equations for gravitational radiation. Rather than retrace this familiar result of the linearized theory, I want to show how the full Einstein equations can still be thought of as a wave equation. Furthermore, I will do so in the 3+1 language reviewed by York in this volume. This discussion is meant as an introduction to the more detailed articles by Eppley, Piran, Wilson and myself herein. I will not discuss how the shift vector $\beta_i \equiv g_{oi}$ enters into the equations (see Wilson or York this volume), but I will leave in the time slicing freedom of the lapse function α.

Traditionally, the 3+1 form of Einstein's equations are written in Hamiltonian form (York equations (35), (39)). This is analogous to the wave equation in form (4). Here the role of k is played by the extrinsic curvature tensor K_{ij}. Note that the 3-dimensional Ricci tensor R_{ij} can be written as

$$R_{ij} = -\tfrac{1}{2}\gamma^{k\ell}\,(\partial_i\partial_j\gamma_{k\ell} + \partial_k\partial_\ell\gamma_{ij} - \partial_i\partial_\ell\gamma_{kj} - \partial_k\partial_j\gamma_{i\ell})$$
$$-\gamma^{k\ell}\gamma_{mn}\,(\Gamma^m_{k\ell}\Gamma^n_{ij} - \Gamma^m_{kj}\,\Gamma^n_{i\ell}) \tag{8}$$

so that it serves the role of $-c\partial_x\partial_x\phi$ in equation (4). This form of the Einstein equations was used by Smarr, et al.(1976), Eppley and Smarr (1978), Wilson (this volume), and Piran (this volume).

If we eliminate K_{ij} from the Einstein equations, we can write them as a full second order system (Smarr (1975)) in analogy to equation (1):

$$\partial_t\partial_t\gamma_{ij} + 2\alpha^2 R_{ij} = \gamma^{\ell m}(\partial_t\gamma_{\ell i}\partial_t\gamma_{mj} - \tfrac{1}{2}\partial_t\gamma_{\ell m}\partial_t\gamma_{ij}) \tag{9}$$
$$+ \partial_t\gamma_{ij}\partial_t\,\ell n\alpha + 2\alpha D_i D_j\alpha$$

Here D_i is the spatial covariant derivative. If one substitutes R_{ij} from equation (8) into this, one sees that a wave equation results with velocity tensor

$$(c^2)^{k\ell} = \alpha^2\gamma^{k\ell} \tag{10}$$

This is just what one would intuitively expect for the coordinate velocity tensor $dx^k dx^\ell/dt^2$ since if the speed of light is unity then

$$1 \simeq \frac{\text{proper distance}}{\text{proper time}} = \frac{\gamma_{k\ell}dx^k dx^\ell}{\alpha^2 dt^2} = \frac{\gamma_{k\ell}}{\alpha^2}\,(c^2)^{k\ell} \tag{11}$$

This was the form of the Einstein equations used numerically by Estabrook, et al. (1973) and Smarr (1975). The major disadvantage are 1) that the natural initial value variables are (γ_{ij}, K_{ij}) not $(\gamma_{ij}, \partial_t\gamma_{ij})$ and 2) that the term $\partial_t\,\ell n\alpha$ appears. From the kinematic viewpoint time derivatives of α should not appear in the evolution equations (see York this volume). We add in passing that most rigorous studies of the Einstein equations as a second order system (e.g. Bruhat (1962)) use the 4-dimensional form and assume harmonic coordinates. For numerical relativity, one uses instead elliptic gauge conditions such as maximal time slicing (York-equation (104)). The main difference is that for harmonic coordinates the second spatial derivatives on γ_{ij} (equation 8) decouple whereas for elliptic gauges they do not.

As shown by Hahn and Lindquist (1963), one can rewrite the Einstein equations (9) as a first order quasilinear system. They used the 3-dimensional Christoffel symbols as the auxiliary variable d in equation (5). As remarked by Smarr (1977), one can just as well use $d_{ijk} \equiv \partial_i\gamma_{jk}$, in which case the Einstein equations become:

$$\partial_t \gamma_{ij} = -2\alpha K_{ij} \tag{12a}$$

$$\partial_t d_{ijm} = -2\partial_m(\alpha K_{ij}) \tag{12b}$$

$$\partial_t K_{ij} = \alpha R_{ij} + \alpha(KK_{ij} - 2K_{i\ell}K^{\ell}{}_j) - D_i D_j \alpha \tag{12c}$$

with R_{ij} given by (8) written in terms of d_{ijm}.

On comparing (12a,b,c) with the scalar wave equation (5a,b,c), one sees that the latter has all the terms that the former has except for the quadratic nonlinearities $d \cdot d$ and $K \cdot K$. One sees that γ_{ij} plays the role of ϕ, with d_{ijm} and K_{ij} as d and k. The kinematic terms in α in (12) are present because we have allowed a time slicing freedom in (12) we have not allowed in (5). If one sets $\alpha=1$ (geodesic slicing) then the parallel is particularly vivid. This form of the Einstein equations was used numerically by Hahn and Lindquist (1963), Eppley (this volume), Smarr (1977), and Centrella (1977).

Again, the first order form of Einstein's equations has been rigorously studied only in 4-dimensional form and with harmonic coordinates (Fischer and Marsden (1972)). From the 3+1 approach above it is suggestive that a model equation of the form

$$\partial_t \phi + V\partial_x \phi + \phi^2 = 0 \tag{13}$$

might represent many features of the Einstein equations. One could argue that since V depends on γ_{ij} through York's equation (39) that $V\partial_x \phi$ should be replaced by $\phi\partial_x \phi$ - a nonlinearity which causes steepening of waves. However, the dynamic content of the Einstein equations are carried in (d,k), so the nonlinearity caused by V will probably be weak. This is an area that requires more study.

2.4 Hydrodynamic Equations

These equations will be discussed in detail by Wilson, so I will confine myself to showing how they can be written as a first order system. For one dimension, the Eulerian compressible adiabatic equations governing a inviscid fluid are

$$\partial_t \rho + u\partial_x \rho + \rho\partial_x u = 0 \tag{14a}$$

$$\partial_t u + u\partial_x u + \rho^{-1}\partial_x P = 0, \tag{14b}$$

where ρ is the matter density and u is the velocity of the matter flow. If the pressure is a function of ρ only, $P = P(\rho)$, then as a matrix equation these become:

$$\partial_t \begin{pmatrix} \rho \\ u \end{pmatrix} + \begin{pmatrix} u & \rho \\ a^2\rho^{-1} & u \end{pmatrix} \partial_x \begin{pmatrix} \rho \\ u \end{pmatrix} = 0 \tag{15}$$

where $a^2 = \partial P/\partial\rho$ is the speed of sound squared. The velocity matrix clearly depends on both (ρ,u) so a scalar equation which models this is

$$\partial_t\phi + \phi\partial_x\phi = 0 \tag{16}$$

The nonlinearity in the velocity term is what causes steepening of wave profiles, breaking of waves, and shocks. Note this is not like the major nonlinearity of Einsteins' equations (12) which occur in the quadratic undifferentiated terms.

In any real fluid there is viscosity which introduces a diffusion term into the equations. In terms of our model equation, we now have Burgers equation:

$$\partial_t\phi + \phi\partial_x\phi = \nu\partial_x\partial_x\phi \tag{17}$$

The present of the diffusion term (a parabolic equation (2) if we neglect the transport term $\phi\partial_x\phi$) causes a damping of wave profiles and thus counters the tendency for shocks to form. Burger's equation is one of a whole class of nonlinear dispersive equations including the Korteweg-de Vries, Sine-Gordon, and nonlinear Schrodinger equations (see e.g. Whitham (1974)). In special cases, the nonlinear terms can just balance the dispersive terms and cause solitary waves or solitons. In what sense one can think of Einstein's equations as nonlinear dispersive equations is not completely clear. In a major new article, Belinsky and Zakharov (this volume) explore the possibility of solitons in general relativity (see also Harrison (1978)).

In summary, we have seen how the structure of the Einstein equations can be understood in terms of more familiar equations such as wave motion or hydrodynamics. We have derived a series of model scalar equations (7), (13), (16), and (17) which capture the essence of the more complicated equations. In the next few sections we will study how to apply finite difference techniques to these classes of equations.

3. FINITE DIFFERENCE EQUATIONS

3.1 Discretization

Except in very simple cases, it is impossible to obtain analytic or closed-form solutions to the PDE's discussed above. An alternative approach is to use the method of finite differences to find solutions to a finite differenced equation (FDE) analogue to the given PDE. This procedure has two separate steps. First, the spacetime continuum is replaced by a discrete spacetime lattice of grid points. Second, the derivatives in the PDE are replaced by a particular finite difference approximation (FDA). These two steps should be thought of as quite separate. The exact solution to the FDE depends on both the specification of the grid and on the finite difference approximation used. Note that for a given PDE, there are an infinite number of possible finite difference analogues each with its own exact solution. A very large number of these solutions of FDE's will bear little resemblence to the solution of the PDE. This is because of instabilities which plague FDE's. We will see that one must choose certain regimes of discretation $(\Delta x,\Delta t)$ relative to dimensional quantities of the physical solution (wavelength, speed of propagation, etc.) if we wish the FDE solution to be a good approximation to PDE solution. Furthermore, we shall see that even if we make Δx and Δt very small, there are some FDE's whose solutions are always unstable and far from the PDE solution. This relative richness of solutions of FDE's compared to PDE's is what makes the subject of finite differencing seem so complex. In a sense, it is the introduction into a physical problem of new dimensionless ratios $(\Delta t/\Delta x, \Delta x/\lambda$, etc.) which makes the discrete problem have so many more degrees of freedom.

3.2 Truncation Errors

A typical spacetime lattice (one space dimension plus time) is shown in Figure 1.

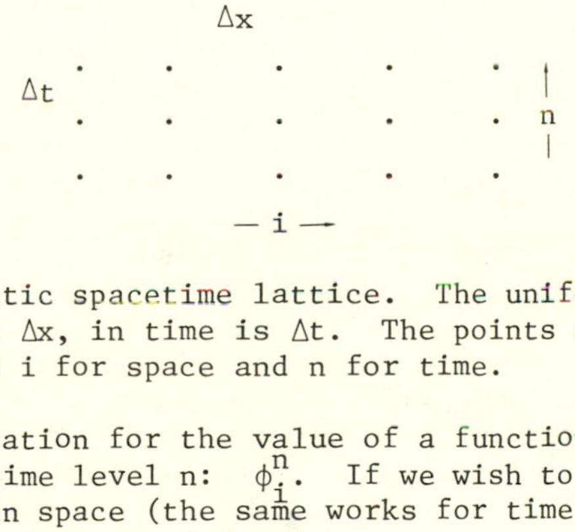

Figure 1. A schematic spacetime lattice. The uniform spacing in space is Δx, in time is Δt. The points are labeled by integers i for space and n for time.

We use standard notation for the value of a function ϕ defined at space index i and time level n: ϕ_i^n. If we wish to approximate ϕ at a nearby point in space (the same works for time), we can expand in a Taylor series around (n,i):

$$\phi_{i+1}^n = \phi_i^n + \left(\frac{\partial\phi}{\partial x}\right)_i^n \cdot \Delta x + \tfrac{1}{2}\left(\frac{\partial^2\phi}{\partial x^2}\right)_i^n \cdot (\Delta x)^2 + O(\Delta x)^3 \tag{18a}$$

$$\phi^n_{i-1} = \phi^n_i - \left(\frac{\partial\phi}{\partial x}\right)^n_i \Delta x + \tfrac{1}{2}\left(\frac{\partial^2\phi}{\partial x^2}\right)^n_i (\Delta x)^2 + O(\Delta x)^3 \tag{18b}$$

To obtain approximations for the derivatives of ϕ we can add and subtract these equations. If we can allow errors as large as $O(\Delta x)$, where we assume $\Delta x << 1$, then we can use forward (or backward) FDA's:

$$\left(\frac{\partial\phi}{\partial x}\right)^n_i = \frac{\phi^n_{i+1} - \phi^n_i}{\Delta x} + O(\Delta x) \tag{19}$$

If we wish higher accuracy, we can use centered differences:

$$\left(\frac{\partial\phi}{\partial x}\right)^n_i = \frac{\phi^n_{i+1} - \phi^n_{i-1}}{2\Delta x} + O(\Delta x)^2 \tag{20}$$

Similarly, we can solve for the second derivative:

$$\left(\frac{\partial^2\phi}{\partial x^2}\right)^n_i = \frac{\phi^n_{i+1} - 2\phi^n_i + \phi^n_{i-1}}{(\Delta x)^2} + O(\Delta x)^2 \tag{21}$$

The FDA, equation (19), is said to be <u>first order accurate</u> and to have a <u>truncation error of order Δx</u>. The FDA, equations (20-21), is said to be <u>second order accurate</u> and to have a <u>truncation error of order $(\Delta x)^2$</u>. Higher order accurate FDA's can be calculated in an analogous manner. Although much investigation is currently underway on fourth order methods (see e.g. Turkel, *et al.* (1976)), most numerical work still uses either first or second order accurate schemes.

The truncation error is the major source of inaccuracy in finite difference calculations (assuming one is using a stable scheme-see below). Notice that the truncation error does *not* give the absolute error in a calculation with fixed $(\Delta x, \Delta t)$, but instead tells how that error is reduced as $(\Delta x, \Delta t)$ approach zero. One must be aware that in practice this limit $(\Delta x, \Delta t) \to 0$ can not be taken because of computer limitations in speed and memory. This is why one wishes to use a higher order FDA. Unfortunately, experience has shown that one usually trades off accuracy for stability. Thus, near a steep gradient (i.e. a shock) one may be forced to use only first order accurate methods in order that the scheme stay stable (see below).

3.3 Propagation Errors

For this discussion let us assume we are trying to choose a FDE to represent the simplest hyperbolic equation (7). Then this FDE can be symbolically written as

$$\phi^{n+1} = G\phi^{n} \tag{22}$$

where G is an operator which connects the values of the function ϕ on one time level with the value on the next. This will be quantified in section 4. Depending on the properties of G (which are determined by the nature of the FDA used) a variety of propagation errors can occur. By propagation error I mean that the solution to the FDE exhibits a wave behavior not shared by the solution to the PDE. It should be noted in passing that by assuming ϕ^{n+1} is determined only by the values of ϕ^{n} on a previous time step, I have limited my discussion to the explicit schemes. Also in use are implicit schemes which solve simultaneously for the values of ϕ^{n+1} (see Roche pp. 83-87).

First, consider the amplitude of G. If $|G| = 1$, then the wave amplitude will be propagated without change, just as in the physical solution. However, if $|G| > 1$ the solution's amplitude will grow without bound. This is termed an unstable solution. If $|G| < 1$ the solution will decrease in amplitude as it propagates, a dissipative solution. Note that in general the value of $|G|$ will depend on the dimensionless ratios $\Delta t/\Delta x$ and $\lambda/\Delta x$, so that the character of amplitude propagation depends on 1) the FDA used which gives the formula for G, 2) the grid used which gives $(\Delta x, \Delta t)$, and 3) the solution desired specified by say the wavelength λ in units of Δx.

The phase of the operator G also is important. In general, one will have $\text{Im}(G) = f(\lambda/\Delta x, \Delta x/\Delta t)$ so that dispersion errors or phase errors will occur. This can change both the effective velocity of propagation of the signal as well as its shape. Thus, even if $|G| = 1$, but $\text{Im}(G) \neq 0$, changes in amplitude can occur in propagation because of dispersion erros (see e.g. Roche pp. 56-57).

The lesson is that any FDA has such inherent errors, Even if one had a scheme in which G=1 for the simple equation (7), as soon as the velocity v becomes a function of position or time, or as soon as new terms are added to the PDE, G becomes complex. Thus, one learns to live with a certain level and type of error. In some problems phase error may be irrelevant, in others all important. The FDE chosen depends very much on which aspect of the problem one is modeling is most important. For a detailed comparison of phase errors and stability for most major differencing schemes see Turkel (1974).

3.4 Nonlinear problems

The stability problems become even more difficult for nonlinear equations (such as either hydrodynamics or Einstein's equations). This is because nonlinear terms cause an exchange between short and long wavelength components of ϕ. For instance, in the model equation (16), the nonlinear term $\phi\partial_x\phi$ causes a transfer in Fourier space from low frequencies to high frequencies. This results in a wave profile steepening until a vertical profile or shock occurs. Since the various propagation errors discussed in the previous section depend on $(\lambda/\Delta x)$, the character of the FDA can change as the solution advances in time. A FDA which was stable for profiles with shallow slopes can become unstable as the slope steepens.

Another varient of this problem is the nonlinear instability or aliasing error (see e.g. Roche p. 81). Assume one is concerned only with the long wavelength part of the solution to a nonlinear equation. As the evolution continues, the cascading to smaller wavelengths occurs. When this hits $\lambda \sim 2\Delta x$, there is nowhere for the components to go, so they pile up at $\lambda \sim 2\Delta x$. This pile up continues until the amplitude in Fourier space gets larger than the long wavelength components. Thus, a solution can be stable for a time, then all of a sudden it goes unstable at short wavelengths. This is dramatically shown in Figure 8.1 of Richtmyer and Morton (1967).

For these reasons, nonlinear problems usually require artificial viscosity which damps out short wavelengths. This is the way Nature usually deals with nonlinearity. However, even in linear problems the notion of artificial viscosity arises. To see why we must make a more detailed analysis.

4. STABILITY AND ARTIFICIAL VISCOSITY

4.1 Von Neumann Stability Analysis

A very powerful method for analyzing the propagation errors described in section 3.3 was introduced by von Neumann in 1944 (see e.g. Roche pp. 42-45). Let us perform a Fourier decomposition in the spatial direction:

$$\phi_i^n \equiv \xi^n e^{I\thei i} \qquad I^2 = -1 \tag{23}$$

$$\theta = k_x \Delta x = 2\pi(\Delta x/\lambda)$$

The phase angle θ is the dimensionless variable which occurs because of the discretation of space. We can now rigorously define the operator G mentioned above:

$$\phi_i^{n+1} = \xi^{n+1} e^{I\theta i}$$
$$\xi^{n+1} = G\xi^n \tag{24}$$

The <u>von Neumann stability criterion</u> then requires

$$|G| \leq 1 \tag{25}$$

to assure boundedness in time. I will apply this test to a variety of the standard FDA to illustrate some of the major concepts of finite differencing.

4.2 Implicit Artificial Viscosity

Recall the simple model equation (16). I will not be concentrating on the nonlinear aspect to start with, but instead simply regard the coefficient ϕ in $\phi\partial_x\phi$ as a "velocity" v of transport (equation (7)). One of the simplest FDA of this equation is to forward time difference and center space difference (FTCS):

$$\frac{\phi_i^{n+1} - \phi_i^n}{\Delta t} = -\frac{\phi_i^n(\phi_{i+1}^n - \phi_{i-1}^n)}{2\Delta x} \tag{26}$$

Linearize by assuming $\phi_i^n \simeq v$ in the velocity term. Plug in the Fourier decomposition (23) and find (using $e^{I\theta i} = \cos(i\theta) + I\sin(i\theta)$):

$$\xi^{n+1} = \xi^n \left[1 - \frac{v\Delta t}{\Delta x} I \sin\theta\right] \equiv G\xi^n \tag{27}$$

Calculating $|G|$ we find:

$$|G|^2 = 1 + \left(\frac{v\Delta t}{\Delta x}\right)^2 \sin^2\theta \geq 1 \tag{28}$$

Thus, we have the surprising result that the straightforward FTCS method is <u>unstable for any Δt</u> regardless how small.

The way in which the instability manifests itself is shown in Figure 3-6 of Roache. If one has a small amplitude error:

$$\varepsilon_i = \begin{cases} +1 & i \text{ even} \\ -1 & i \text{ odd} \end{cases} \tag{29}$$

then the amplitude of ε_i grows monotonically in time. This simple instability was one of the first encountered by Eppley and myself in the finite differencing of Einstein's equations (see Eppley 1975).

The instability must be cured by altering the FDA used. First, let us explore what happens when we change the time derivative FDA. Lax and Friedrichs showed in the early 1950's (Roache pp. 242-244) that the simple replacement of the time difference in equation (26) by:

$$\partial_t \phi \rightarrow \frac{\phi_i^{n+1} - \frac{1}{2}(\phi_{i+1}^n + \phi_{i-1}^n)}{\Delta t} \tag{30}$$

will stabilize the FTCS method. To see why let us rewrite, by adding and subtracting $\phi_i^n/\Delta t$, the Lax-Friedrichs differencing of $\partial_t\phi + \phi\partial_x\phi = 0$, so that it regains the form of the FTCS, equation (26), plus remainder terms:

$$\frac{\phi_i^{n+1} - \phi_i^n}{\Delta t} + \phi_i^n \left[\frac{\phi_{i+1}^n - \phi_{i-1}^n}{2\Delta x}\right] = \frac{(\Delta x)^2}{2\Delta t}\left[\frac{\phi_{i+1}^n - 2\phi_i^n + \phi_{i-1}^n}{(\Delta x)^2}\right] \tag{31}$$

The remainder terms are precisely the differencing of a Laplacian multiplied by a diffusion coefficient

$$\nu_{eff} = (\Delta x)^2/2\Delta t. \tag{32}$$

That is, the Lax-Friedrichs method of differencing the inviscid transport equation (16) is equivalent to the FTCS differencing of the viscous Burgers equation (17) with a viscosity coefficient that goes to zero in the continuum limit. Thus, a prescription for stabilizing is to add a parabolic diffusion term to the time derivative. This <u>artificial viscosity</u> is an <u>implicit</u> one since the differential equations were left unchanged and a particular FDA was chosen for the time or space derivatives. Note that there is nothing sacred about the FTCS form which we compared the Lax-Friedrichs form to. All one can do is note that there is a set of differenced terms in one form which does not occur in the other form.

Let us now see that the implicit artificial viscosity did indeed stabilize the FTCS scheme. An easy calculation finds that G for the FDE (31) is:

$$G = \cos\theta - I\frac{v\Delta t}{\Delta x}\sin\theta \tag{33}$$

$$|G|^2 = \cos^2\theta + \left(\frac{v\Delta t}{\Delta x}\right)^2 \sin^2\theta \tag{34}$$

Thus, the Lax-Friedrichs scheme is stable if

$$\frac{v\Delta t}{\Delta x} < 1 \quad \text{or} \quad \Delta t < \frac{\Delta x}{v}\ . \tag{35}$$

This is the famous Courant-Friedrichs-Lewy (1928) stability criterion which holds for all explicit hyperbolic FDE's. The generality of this condition can be seen by considering the propagation characteristics for the system. In Figure 2, one has a spacetime diagram on which the characteristic x=vt of the differential equation is drawn. For a fixed Δt, the CFL time limit says that the finite difference cone must lie outside the physical propagation cone.

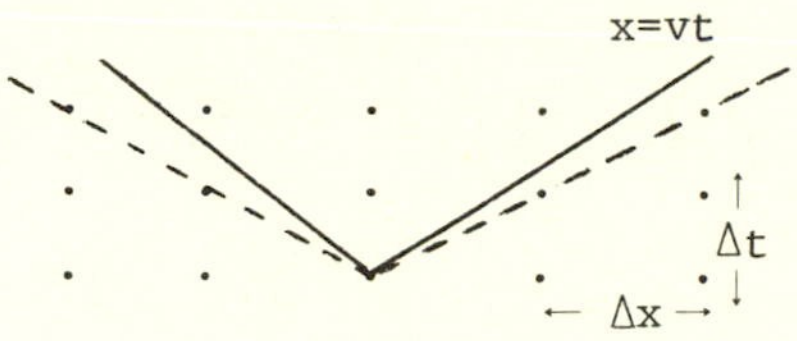

Figure 2. Physical propagation cone and stable finite difference cone.

If this were not the case, signals could not freely propagate and would "pile up" causing an instability. The manner in which this instability manifests itself is quite different than the "static" FTCS instability. For the CFL instability the errors ε_i alternate in sign and grow very rapidly (Figure 3-6 of Roache).

The Lax-Friedrichs method was used for the two black hole collision problem by Hahn and Lindquist (1964). Eppley and I found that it caused far too much damping or diffusion of the metric coefficients. After much experimentation we used a form similar to equation (31), but with a smaller effective viscosity coefficient:

$$\nu_{eff} = K(\Delta x)^2/2\Delta t \qquad K \simeq .01 - .02 \tag{36}$$

This kept our code stable and produced very little damping (see Eppley and Smarr 1978).

It should be mentioned that there are much more sophisticated variants of the old Lax-Friedrichs scheme, such as two-step Lax-Wendroff, Rusanov method, Burstein's method, MacCormack's method, etc. (see Roache pp. 242-256). The above was meant as an introduction to the concept of implicit artificial viscosity rather than as a recipe for computing. For a comparison of methods see Sod (1878).

4.3 Transport Errors

Our model equation (7) can be thought of as a scalar version of a hyperbolic wave equation or as a transport operator: $\partial_t + v \cdot \partial_x$. If it is a wave equation, then considerations of the last section are most important, i.e. one concentrates on the propagation cone. On the other hand, if there is a local velocity v of matter flow which occurs in an Eulerian hydrodynamics equation, then it is more appropriate to concentrate on transport properties. These remarks extend to Einstein's equations which have both wave parts and transport (shift vector) parts (see York's and Wilson's articles in this volume).

The problem with the Lax-Friedrichs method is that both the time and space differences are centered in space. Thus, a disturbance ε_i^n will effect equally ϕ_{i+1}^{n+1} and ϕ_{i-1}^{n+1}. If the velocity term in $v\partial_x\phi$ is positive, then only ϕ_{i+1}^{n+1} should be effected as the matter flow moves forward.

This suggests that we can both better model the transport properties and perhaps stabilize the FTCS scheme if we use backward or upwind differencing:

$$\frac{\phi_i^{n+1} - \phi_i^n}{\Delta t} = -v_i^n \begin{cases} \dfrac{\phi_i^n - \phi_{i-1}^n}{\Delta x} & \text{if } v_i^n > 0 \\[2ex] \dfrac{\phi_{i+1}^n - \phi_i^n}{\Delta x} & \text{if } v_i^n < 0 \end{cases} \tag{37}$$

This is manifestly only first order accurate in both time and space. Therefore it is less accurate in truncation error than Lax-Friedrichs (second order in space), but more accurate in transport properties.

That it stabilizes FTCS is suggested by rewriting it as FTCS plus remainder, in which case one finds it is of the same form as equation (31), but with an effective viscosity (see e.g. Crowley (1975) p.28):

$$\nu_{eff} = \left|\frac{v\Delta x}{2}\right| \tag{38}$$

The von Neumann G factor is:

$$G = 1 - \left|\frac{v\Delta t}{\Delta x}\right|(1-\cos\theta) - I\left|\frac{v\Delta t}{\Delta x}\right| \sin\theta \tag{39}$$

with amplitude

$$|G|^2 = 1 - 2\left|\frac{v\Delta t}{\Delta x}\right|\left(1 - \left|\frac{v\Delta t}{\Delta x}\right|\right)(1-\cos\theta) \tag{40}$$

We have stability if the CFL criterion, equation (35), is fulfilled. Besides being less accurate in truncation error than Lax-Friedrichs, the upwind scheme damps high frequencies more severely. The shortest allowed wavelength is $\lambda \sim 2\Delta x$ or $\theta \sim \pi$.

We find that for Lax-Friedrichs (for any value of $v\Delta t/\Delta x$):

$$|G(\theta=\pi)|^2 = 1 \tag{41}$$

while for upwind one has (for $|v\Delta t/\Delta x| < 1$):

$$|G(\theta=\pi)|^2 = 1 - 4\left|\frac{v\Delta t}{\Delta x}\right|\left(1 - \left|\frac{v\Delta t}{\Delta x}\right|\right) < 1 \tag{42}$$

There are several methods to make upwind differencing more accurate. The basic idea is to note that the fluid carries to ϕ_i^{n+1} the quantity located $v\Delta t < \Delta x$ upwind from ϕ_i^n whereas the upwind scheme always carries the quantity from a distance Δx upwind. If one is running with $v\Delta t/\Delta x < 1$ this can be a substantial error. Therefore, one can improve accuracy by interpolating ϕ between i-1 and i before transport. This method is described in detail by Wilson in this volume. An unfortunate problem is that in the vicinity of a steep gradient or shock in ϕ, this second order accurate transport method becomes unstable and one has to revert to upwind first order accurate differencing.

4.4 Explicit Artificial Viscosity

We have seen by now that one stabilizes a FDE by adding a diffusion term to the difference equation. For the implicit schemes described above, this term arises "automatically" by the choice of FDA for the time or space derivatives. There is an alternative approach which was introduced by von Neumann in 1944 and described in von Neumann and Richtmyer (1950). This widely used method is to add an extra term Q to the pressure P in the momentum equation (14a):

$$\rho \frac{\partial u}{\partial t} + \rho u \frac{\partial u}{\partial x} = - \frac{\partial (P+Q)}{\partial x} \tag{43}$$

This Q has the form

$$Q = \nu_{eff} \frac{\partial u}{\partial x} \tag{44}$$

and therefore is a diffusion term in u.

If ν_{eff} is independent of u, then this is called a linear Q. However, the reason it was introduced is because of the tendency of the nonlinear equation (43) to form shocks or discontinuities in u. In order to combat this tendency only in the immediate neighborhood of the gradient $\partial u/\partial x$, a quadratic Q can be used by taking ν_{eff} as:

$$\nu_{eff} = K\rho(\Delta x)^2|\partial u/\partial x| \tag{45}$$

where K is an adjustable parameter. Comparing this effective viscosity with that of upwind ($\nu_{eff} \sim u$) or Lax-Friedrichs (ν_{eff} independent of u), one sees that this artificial viscosity is much more sensitive to the solution than the others. This is why it is often favored.

A phenomenon which occurs when discontinuous (step function) solutions are numerically propagated is ringing. Even in the linear PDE (7), a step function will not be propagated intact. What happens is that a numerical overshoot and damped ringing appears near the discontinuity. This profile can be shown to be the integral of an Airy function (Hedstrom (1975); Chin (1975); Chin and Hedstrom (1976)). The use of Q factor damps out this ringing and spreads the shock front over several grid zones.

Of course, in a linear PDE, if the initial condition is not discontinuous, this problem will not arise. However, nonlinear equations such as hydrodynamics convert smooth initial conditions into discontinuous ones, so the Q is always needed if compression is occuring. For the use of a Q see Wilson's lectures this volume. In two spatial dimensions it is sometimes advantageous to use a tensor Q_{ij} which allows for shear viscosity (see e.g. Maenchen and Sack (1964)).

4.5 Staggered Leapfrog

One may very well ask why any artificial viscosity is required in Einstein's equations, especially in vacuum. The answer is that one does not require it if the spatial gradients and nonlinear term remain sufficiently small (Smarr 1977). An example is the evolution of pure gravitational waves described by Eppley and by Piran in this volume. The classic example of such a scheme is the staggered leapfrog (Roache pp. 53-61).

For our model equation (7), this difference scheme is:

$$\frac{\phi_i^{n+1} - \phi_i^{n-1}}{2\Delta t} = - v \cdot \frac{\phi_{i+1}^n - \phi_{i-1}^n}{2\Delta x} \tag{46}$$

This is second order accurate in both space and time. Notice that the time levels in the time derivatives "leapfrog" over the time level on which the spatial derivatives are taken. The von Neumann analysis is more complicated because G is now a matrix:

$$\begin{pmatrix} \xi^{n+1} \\ \xi^{n} \end{pmatrix} = G \begin{pmatrix} \xi^{n} \\ \xi^{n-1} \end{pmatrix}$$

$$G = \begin{pmatrix} -2I\frac{v\Delta t}{\Delta x}\sin\theta & 1 \\ 1 & 0 \end{pmatrix} \tag{47}$$

The criterion is that G have eigenvalues Γ such that $|\Gamma| \leq 1$. One finds (see Roache p.55) that

$$\Gamma = -I\left(\frac{v\Delta t}{\Delta x}\right)\sin\theta \pm \sqrt{1 - \left(\frac{v\Delta t}{\Delta x}\right)^2 \sin^2\theta} \tag{48}$$

It can be seen that we again require the CFL criterion (35). In this case

$$|\Gamma|^2 = 1 \qquad \text{for } v\Delta t/\Delta x < 1 \; . \tag{49}$$

Thus, we see the great advantage of leapfrog is that no amplitude dissipation occurs. It is usually said that the leapfrog method has no artificial viscosity. However, as discussed in section 3.3 above, this method has dispersion errors since $\mathrm{Im}(\lambda)$ depends on $(\lambda/\Delta x)$ and $(\Delta x/\Delta t)$. These can mimic dissipation particularly if $v\Delta t/\Delta x < 1$ (see Roache pp. 56-57).

Thus, leapfrog is very useful if the solutions are smooth and no shocks develop. For this reason, Eppley and I used leapfrog extensively for pure gravitational waves (Smarr (1977) and Eppley-this volume). It was found that in our multidimensional codes (2 space + 1 time dimension), a small amount of artificial viscosity was needed as curvatures get stronger. For instance, we used this method on the two black hole collision problems, but always found that a nonlinear instability occured (see section 3.4 above). This leapfrog scheme is also subject to a "mesh drifting" instability since the time levels are decoupled by (46). It would be particularly nice if fourth order methods could be used for gravitational waves since higher accuracies on coarser grids would result.

5. ELLIPTIC EQUATIONS

Finally, I will briefly discuss numerical solution of elliptic equations such as the maximal slicing or other gauge equations (see e.g. Wilson - this volume). Again, the key idea is to add a diffusion term to the elliptic equation. However, whereas one added $\partial_x \partial_x \phi$ to the hyperbolic equations discussed above, here one adds $\partial_t \phi$ to the elliptic equation. The "time" here is "relaxation time." One starts with a good guess of the solution $\phi(x,y)$ and then diffuses to the point that $\partial_t \phi \to 0$. This steady state is then a solution of the desired elliptic equation (3). There are other methods for solving elliptic equations (such as ADI), but I have not used them. I refer the interested reader to Roache and references therein.

Our model equation is then:

$$\partial_t \phi = \partial_{xx} \phi + \partial_{yy} \phi \tag{50}$$

It turns out that for parabolic equations the FTCS differencing method is stable (for two space dimensions) if $\Delta t \leq (\Delta x)^2/4$. Thus, using the time difference for $\partial_t \phi$ from (26) and space differences from (21), one obtains for $\Delta x = \Delta y$:

$$\phi^{n+1}_{i,j} = \phi^n_{i,j} + \tfrac{1}{4}\left[\phi_{i+1,j} + \phi_{i-1,j} + \phi_{i,j+1} + \phi_{i,j-1} - 4\phi_{i,j}\right] \tag{51}$$

where the two subscripts (i,j) refer to the spatial (x,y) grid. Notice that I have not specified the time levels of ϕ on the right hand side. If they are all $\phi^n_{i,j}$, then the method is referred to as Jacobi's method. If, however, as one sweeps through the (x,y) grid one uses the newest values of ϕ available (i.e. some will be $\phi^{n+1}_{i,j}$), then the method is referred to as Gauss-Seidel and it converges twice as fast as Jacobi.

The convergence can be speeded up even more by introducing an overrelaxation factor ω $(1 \leq \omega \leq 2)$ in front of the square brackets in (51). In fact, the rate can be increased by orders of magnitude if ω is chosen close to an optimal value ω_0 which depends on the coefficients of the differential equation and on the grid spacing. (see Roache figure 3-16). This resonance-like behavior is, of course, very useful for numerical solution of elliptic equations when one has to solve a new equation on each time slice of the spacetime. The old solution is a good guess for the new one and the old ω_0 will be near the new one. For a discussion of "tactics and strategy" on solving elliptic equations see Roache pp. 119.

This method of simultaneous overrelaxation (SOR) is the one used by all codes for the Einstein equations to date. There the elliptic equations are far more complicated than our model (3), because all possible first and second spatial derivative terms are present and because each derivative has a coefficient which varies spatially over the grid. Furthermore, the boundary conditions, which determine the solution to the elliptic equation are updated on each time level by the hyperbolic equations. Nevertheless, one is not plagued by the instabilities that haunt the hyperbolic equations.

This review has only scratched the surface of a immensely complicated subject. I hope it will lead more people to think about these problems and give nonparticipants some flavor of why progress sometimes seems so slow.

REFERENCES

Bruhat, Y. (1962). The Cauchy problem. In Gravitation: An Introduction to Current Research, ed. L. Witten. John Wiley, New York.

Centrella, J. (1977). Inhomogeneous cosmologies in the ADM formalism. In Abstracts of Contributed Papers: GR8. University of Waterloo, Ontario.

Chin, R.C.Y. (1975). Dispersion and Gibbs phenomenon associated with difference approximations to initial boundary-value problems for hyperbolic equatons. J. Comp. Phys., 18, pp. 233-247.

Chin, R.C.Y. & Hedstrom, G.W. (1976). A dispersion analysis for difference schemes: Tables of generalized Airy functions. Preprint UCRL-78668, Lawrence Livermore Laboratory.

Courant, R., Friedrichs, K., & Lewy, H. (1928). Uber die Partiellen Differenzengleichurgen der Mathematischen Physik. Math. Annalen, 100, pp. 34-74. English translation in IBM Journal, March, 1967, pp.215-234.

Courant, R. & Hibert, D. (1962) Methods of Mathematical Physics, vol. II. John Wiley, New York.

Crowley, W.P. (1975) Numerical methods in fluid dynamics. Preprint UCRL-51824. Lawrence Livermore Laboratory.

Eppley, K.R. (1975). The numerical evolution of the collision of two black holes. Ph.D. Dissertation, Princeton University.

Eppley, K.R. & Smarr, L. (1978a). The collision of two black holes II. Evolution of metric functions. Preprint.

Eppley, K.R. & Smarr, L. (1978b). The collision of two black holes III. Gravitational radiation. Preprint.

Estabrook, F., Wahlquist, H., Christensen, S., DeWitt, B.,

Smarr, L., & Tsiang, E. (1973). Maximally slicing a black hole. Phys. Rev., D7, pp. 2814-2817.

Fischer, A.E. & Marsden, J.E. (1972). The Einstein evolution equations as a first-order quasi-linear symmetric hyperbolic system I. Comm. Math. Phys., 28, pp. 1-38.

Hahn, S.G. & Lindquist, R.W. (1964). The two-body problem in geometrodynamics. Ann. Phys., 29, pp. 304-331.

Harrison, B.K. (1978). Backlund transformations for the Ernst equation of general relativity. Phys. Rev. Lett., 41, pp. 1197-1199.

Hedstrom, G.W. (1975). Models of difference schemes for $u_t + u_x = 0$ by partial differential equations. Math. Comp., 29, pp. 969-977.

Kreiss, H. & Oliger, J. (1973). Methods for the Approximate Solution of Time Dependent Problems. Global Atmospheric Research Programme Publications Series No. 10.

Maenchen, G. & Sack, S. (1964). The tensor code. In Methods in Computational Physics, vol. 3. ed. by Alder, B., Fernback, S., & Rotenberg, M. Academic Press, New York.

Richtmyer, R.D. (1963). A survey of difference methods for non-steady fluid dynamics. Nat. Center for Atms. Res. Technical Note 63-2.

Richtmyer, R.D. & Morton, K.W. (1967). Difference Methods for Initial-Value Problems. John Wiley, New York.

Roache, P.J. (1972). Computational Fluid Dynamics. Hermosa Publishers, Albuquerque

Smarr, L.L. (1975). The structure of general relativity with a numerical illustration: the collision of two black holes. Ph.D. Dissertation, University of Texas at Austin.

Smarr, L. (1977). Spacetimes generated by computers: Black holes with gravitational radiation. Ann. N.Y. Acad. of Sci., 302, pp. 569-604.

Smarr, L., Cadez, A., DeWitt, B., Eppley, K. (1976) Collision of two black holes: Theoretical framework. Phys. Rev. D14, pp. 2443-2452.

Sod, G.A. (1978). A survey of several finite difference methods for systems of nonlinear hyperbolic conservation laws. J. Comp. Phys., 27, pp. 1-31.

Turkel, E. (1974). Phase error and stability of second order methods for hyperbolic problems I. J. Comp. Phys., 15, pp. 226-250.

Turkel, E., Abarbanel, S., Gottlieb, D. Multidimensional difference schemes with fourth-order accuracy. J. Comp. Phys., 21, pp. 85-113.

von Neumann, J. & Richtmyer, R.D. (1950). A method for the numerical calculation of hydrodynamic shocks. J. Appl. Phys., 21, pp. 232-257.

White, J.W. (1976) An elementary introduction to finite difference equations. Preprint UCRL-52067, Lawrence Livermore Laboratory.

Whitham, G.B. (1974). Linear and Nonlinear Waves. John Wiley, New York.

THE INTEGRATION OF EINSTEIN'S EQUATIONS, BY THE METHOD OF INVERSE SCATTERING, AND EXACT SOLITON SOLUTIONS.

V. Belinsky and V. Zakharov

Landau Institute for Theoretical Physics, Moscow

A full paper with this title has been submitted to the JETP (July 1978). At the Battelle/Seattle Workshop, Prof. Belinsky presented the following summary of this paper.

1. INTRODUCTION

Einstein's equations are considered in the case when only two independent variables are present. This case includes colliding plane waves, stationary axially symmetric metrics, cylindrical waves, etc. We can set

$$g_{ik} = \begin{pmatrix} -f & 0 & 0 & 0 \\ 0 & g_{11} & g_{12} & 0 \\ 0 & g_{12} & g_{22} & 0 \\ 0 & 0 & 0 & f \end{pmatrix}; \qquad (a,b,c,d = 1,2) \quad x^0,x^1,x^2,x^3 = t,x,y,z \tag{1.1}$$

$$-ds^2 = f(-dt^2 + dz^2) + g_{ab}dx^a dx^b$$

Introduce the conventions

lightlike coords.:

$$t = \xi - \eta \qquad g_{ab} \Rightarrow \text{matrix } g$$
$$z = \xi + \eta \qquad g^{ab} \Rightarrow \text{matrix } g^{-1} \tag{1.2}$$

and notation:

$$\det g \equiv \alpha^2 ; \qquad A \equiv -\alpha g_{,\xi} g^{-1} ; \qquad B \equiv \alpha g_{,\eta} g^{-1} \tag{1.3}$$

The vacuum equations are then

$$A,_{\eta} - B,_{\xi} = 0 \tag{1.4}$$

and

$$\begin{aligned}(\ell nf),_{\xi} &= (\ell n\alpha),_{\xi\xi}/(\ell n\alpha),_{\xi} + \mathrm{Tr}A^2/4\alpha\alpha,_{\xi} \\ (\ell nf),_{\eta} &= (\ell n\alpha),_{\eta\eta}/(\ell n\alpha),_{\eta} + \mathrm{Tr}B^2/4\alpha\alpha,_{\eta}\end{aligned} \tag{1.5}$$

Equations (1.5) are consistent by virtue of (1.4).

2. SCHEME OF INTEGRATION

To integrate the equations, we first observe that, from the trace of (1.4), we have $\alpha,_{\xi\eta} = 0$. Two independent solutions are

$$\begin{aligned}\alpha &= a(\xi) + b(\eta) \\ \beta &= a(\xi) - b(\eta)\end{aligned} \tag{2.1}$$

The gauge freedom is seen in one arbitrary function on ξ and one arbitrary function on η. Using this, we can choose $a(\xi)$ and $b(\eta)$ in any desirable form.

The basic eqs. (1.4) and (1.5) are equivalent to the following system:

$$\begin{aligned}&A,_{\eta} - B,_{\xi} = 0 \\ &A,_{\eta} + B,_{\xi} + \alpha^{-1}[AB] - \alpha,_{\eta}\alpha^{-1}A - \alpha,_{\xi}\alpha^{-1}B = 0\end{aligned} \tag{2.2}$$

Now follows the main step, from system (2.2) to a Schrödinger-type system for which the equations (2.2) will be the conditions of selfconsistency. Introduce operators:

$$D_1 = \partial_{\xi} - \frac{2\alpha,_{\xi}\lambda}{\lambda-\alpha}\partial_{\lambda} \quad ; \qquad D_2 = \partial_{\eta} + \frac{2\alpha,_{\eta}\lambda}{\lambda+\alpha}\partial_{\lambda} \tag{2.3}$$

$$([D_1 D_2] = 0 \quad \text{if and only if } \alpha,_{\xi\eta} = 0)$$

The Schrödinger type equations (for complex wave matrics Ψ) are

$$D_1\Psi = \frac{A}{\lambda-\alpha}\Psi \quad ; \qquad D_2\Psi = \frac{B}{\lambda+\alpha}\Psi \tag{2.4}$$

where λ is a complex spectral parameter.

The solution of (2.4) will give us matrices (λ,ξ,η) and $A(\xi,\eta)$, $B(\xi,\eta)$. At the point $\lambda=0$ eq's (2.4) coincide with definitions of A,B (1.3). Consequently:

$$g(\xi,\eta) = \Psi(\lambda,\xi,\eta)\big|_{\lambda=0} = \Psi(0,\xi,\eta) \tag{2.5}$$

For integration we need at least one particular solution g_o, A_o, B_o and Ψ_o. Instead of $\Psi(\lambda,\xi,\eta)$ let us introduce the matrix $\chi(\lambda,\xi,\eta)$:

$$\Psi = \chi\Psi_o \tag{2.6}$$

Equations for χ:

$$D_1\chi = \frac{1}{\lambda-\alpha}(A\chi - \chi A_o) \ ; \qquad D_2\chi = \frac{1}{\lambda+\alpha}(B\chi - \chi B_o) \tag{2.7}$$

These equations have symmetry. Using it we can put on χ the following additional conditions, which are equivalent to the demand of symmetry ($g_{12} = g_{21}$) and reality of the metric tensor g:

$$\overline{\chi}(\overline{\lambda}) = \chi(\lambda) \ ; \qquad g = \chi\left(\frac{\alpha^2}{\lambda}\right) g_o\tilde{\chi}(\lambda) \ ; \qquad \chi(\infty) = I \tag{2.8}$$

(these constraints are consistent with (2.7))

$$\text{Now:} \qquad g = \Psi(0) = \chi(0)\, g_o \tag{2.9}$$

3. SOLITONIC SOLUTIONS

In the general case the solution of our problem for χ is equivalent to the solution of so-called "Riemann problem" in the theory of functions of complex variables. The general solution for χ contains solitonic and nonsolitonic parts. In this section we will investigate the pure solitonic case, when the non-solitonic part vanish.

To find solutions for χ we should know the analytical structure of the matrix χ in the complex plane λ (this structure has a close connection with scattering data for the wave matrix Ψ). Solitons in $g(\xi,\eta)$ will appear, when in the λ-plane there exist points of degeneracy of the matrix χ (and χ^{-1}) where $\det\chi$ (or $\det\chi^{-1}$) tends to zero. This means (from additional constraints for χ) that χ (and χ^{-1}) should have poles in the λ-plane. From the constraints (2.8) one can see, that n poles in χ, at the points $\lambda = \mu_K(\xi,\eta)$ $(K = 1,2,\ldots n)$, will generate n poles in χ^{-1} at the points $\lambda = \nu_K(\xi,\eta)$, where $\nu_K = \alpha^2/\mu_K$. All poles are located either on the real axis of the λ-plane, or will be distributed by pairs: each pole μ_K (or ν_K) generating a complex conjugate pole $\overline{\mu}_K$ (or $\overline{\nu}_K$).

Let us consider the situation when there is no coincidence of poles, all poles are complex and simple (the case for real poles we can obtain by a limiting transition. In this case the matrix χ will be:

$$\chi = I + \sum_{\kappa=1}^{n} \left(\frac{R_\kappa}{\lambda - \mu_\kappa} + \frac{\overline{R}_\kappa}{\lambda - \overline{\mu}_\kappa} \right)$$

$$\chi^{-1} = I + \sum_{k=1}^{n} \left(\frac{S_\kappa}{\lambda - \nu_\kappa} + \frac{\overline{S}_\kappa}{\lambda - \overline{\nu}_\kappa} \right) \tag{3.1}$$

$$(R_\kappa = R_\kappa(\xi,\eta) \ , \quad S_\kappa = S_\kappa(\xi,\eta) \ , \quad \nu_\kappa = \alpha^2/\mu_\kappa)$$

$$g(\xi,\eta) = \chi(0) \cdot g_0 = \left[I - \sum_{\kappa=1}^{n} \left(\frac{R_\kappa}{\mu_\kappa} + \frac{\overline{R}_\kappa}{\overline{\mu}_\kappa} \right) \right] g_0 \tag{3.2}$$

Now to find $g(\xi,\eta)$ we need to find functions $R_\kappa(\xi,\eta)$ and $\mu_\kappa(\xi,\eta)$. From eq. (2.7) (in poles second order $(\lambda-\mu_\kappa)^{-2}$) one can obtain:

$$\mu_{\kappa,\xi} = \frac{2\alpha_{,\xi}\mu_\kappa}{\alpha - \mu_\kappa} \ ; \quad \mu_{\kappa,\eta} = \frac{2\alpha_{,\eta}\mu_\kappa}{\alpha + \mu_\kappa} \tag{3.3}$$

(the eq. for ν_κ will be the same). Solutions of (3.3) are the roots of the algebraic equation

$$\alpha^2/\lambda + 2\beta + \lambda = 2W_\kappa \qquad (W_\kappa = \text{const}) \tag{3.4}$$

$$\mu_\kappa = W_\kappa - \beta - \sqrt{(W_\kappa - \beta)^2 - \alpha^2} \ ; \quad \nu_\kappa = W_\kappa - \beta + \sqrt{(W_\kappa - \beta)^2 - \alpha^2}$$

Now let us rewrite eq. (2.7) in the following manner:

$$\frac{A}{\lambda - \alpha} = (D_1\chi)\chi^{-1} + \chi\frac{A_0}{\lambda - \alpha}\chi^{-1} \ ; \quad \frac{B}{\lambda + \alpha} = (D_2\chi)\chi^{-1} + \chi\frac{B_0}{\lambda + \alpha}\chi^{-1} \tag{3.5}$$

Only first-order poles, and only on the right sides now exist. This leads us to algebraic eq's. for the matrices R_κ, which depend only on known functions μ_κ, α, A_0, B_0 (the unknown matrices A and B will be out). The result of the calculation is:

$$(R_\kappa)_{ab} = n_a^{(\kappa)} m_b^{(\kappa)} \ ; \quad m_b^{(\kappa)} = m_{o\,c}^{(\kappa)} (M_\kappa)_{cb} \ ; \quad M_\kappa = \Psi_0^{-1}(\mu_\kappa, \xi, \eta)$$

$$\sum_{\ell=1}^{n} \frac{m_b^{(\ell)} m_c^{(\kappa)} (g_0)_{cb}}{\overline{\nu}_\kappa - \mu_\ell} n_a^{(\ell)} + \sum_{\ell=1}^{n} \frac{\overline{m}_b^{(\ell)} m_c^{(\kappa)} (g_0)_{cb}}{\nu_\kappa - \overline{\mu}_\ell} \overline{n}_a^{(\ell)} = -m_c^{(\kappa)} (g_0)_{ca} \tag{3.6}$$

$$\sum_{\ell=1}^{n} \frac{m_b^{(\ell)} \overline{m}_c^{(\kappa)} (g_0)_{cb}}{\overline{\nu}_\kappa - \mu_\ell} n_a^{(\ell)} + \sum_{\ell=1}^{n} \frac{\overline{m}_b^{(\ell)} \overline{m}_c^{(\kappa)} (g_0)_{cb}}{\overline{\nu}_\kappa - \overline{\mu}_\ell} \overline{n}_a^{(\ell)} = -\overline{m}_c^{(\kappa)} (g_0)_{ca}$$

After this from (3.5) one can obtain matrices A and B (in poles $\lambda=\alpha$, $\lambda=-\alpha$):

$$A = \left\{ -2\alpha\alpha,_{\xi}\frac{\partial\chi}{\partial\lambda}\chi^{-1} + \chi A_o\chi^{-1} \right\}_{\lambda=\alpha}$$
$$B = \left\{ -2\alpha\alpha,_{\eta}\frac{\partial\chi}{\partial\lambda}\chi^{-1} + \chi B_o\chi^{-1} \right\}_{\lambda=\alpha} \tag{3.7}$$

If we now calculate $Tr(A^2)$ and $Tr(B^2)$ and substitute into (1.5), we can obtain the metric coefficient $f(\xi,\eta)$ by quadratures.

4. SIMPLE SOLITONS (ONE POLE)

$$\lambda = \mu(\xi,\eta) \; ; \quad \mu = W - \beta - \sqrt{(W-\beta)^2 - \alpha^2}$$
$$W \text{ - real}, \quad \mu \text{ - real } (W-\beta)^2 \geq \alpha^2 \tag{4.1}$$

From the previous section we can calculate the matrices χ and $g(\xi,\eta)$:

$$g = \left(\frac{\mu}{\alpha}I - \frac{\mu^2-\alpha^2}{\alpha\mu}P\right) g_o \tag{4.2}$$

where

$$P = \frac{m_c(g_o)_{ca} m_b}{m_c m_d (g_o)_{cd}} \quad (p^2 = P \; ; \quad \det P = 0 \; ; \quad TrP = 1) \tag{4.3}$$

$$m_a = \underset{o}{m_b} M_{ba} \; ; \quad M = (\Psi_o^{-1})_{\lambda=\mu} \tag{4.4}$$

Calculating A,B, then TrA^2, TrB^2, substituting into (1.5), the integral in $f(\xi,\eta)$ is trivial, and the result is:

$$f = \frac{C\mu m_a m_b (g_o)_{ab}}{\sqrt{\alpha(W-2a)(W+2b)}} f_o \tag{4.5}$$

C is an arbitrary constant, a and b are functions from (2.1); f_o is the particular solution for f which corresponds to the particular solution g_o.

A concrete example is that of a Kasner solution.

$$g_o = \begin{pmatrix} \alpha^{2S_1} & 0 \\ 0 & \alpha^{2S_2} \end{pmatrix} \; ; \quad f_o = C_o\alpha,_{\xi}\alpha,_{\eta}\alpha^{S_1^2+S_2^2-1} \tag{4.6}$$

$(S_1 + S_2 = 1)$, $S_1 = \frac{1}{2} + q$; $S_2 = \frac{1}{2} - q$

q is an arbitrary constant

For $\Psi_o(\lambda,\xi,\eta)$ we have:

$$\Psi_o = \begin{pmatrix} (\alpha^2 + 2\beta\lambda + \lambda^2)^{s_1} & 0 \\ 0 & (\alpha^2 + 2\beta\lambda + \lambda^2)^{s_2} \end{pmatrix} \tag{4.7}$$

If we will choose coordinates

$$a(\xi) = \xi + W/2 \qquad b(\xi) = -\eta - \frac{W}{2}$$
$$\text{i.e.}\quad \alpha = \xi - \eta = t\ , \qquad \beta = \xi + \eta + W = z + W \tag{4.8}$$

the metric becomes

$$-ds^2 = \frac{c_1 t^{2q^2} \operatorname{ch}(qr + C_2)}{\sqrt{z^2 - t^2}} (-dt^2 + dz^2) +$$
$$+ \frac{\operatorname{ch}(s_1 r + C_2)}{\operatorname{ch}(qr + C_2)} t^{2S_1} dx^2 + \frac{\operatorname{ch}(s_2 r - C_2)}{\operatorname{ch}(qr + C_2)} t^{2S_2} dy^2 - \tag{4.9}$$
$$- \frac{2\operatorname{sh}(r/2)}{\operatorname{ch}(qr + C_2)} t\, dxdy$$

C_1, C_2 are arbititrary constants and

$$e^r = 2\frac{z^2}{t^2} - 1 - 2\sqrt{\frac{z^2}{t^2}\left(\frac{z^2}{t^2} - 1\right)} \tag{4.10}$$

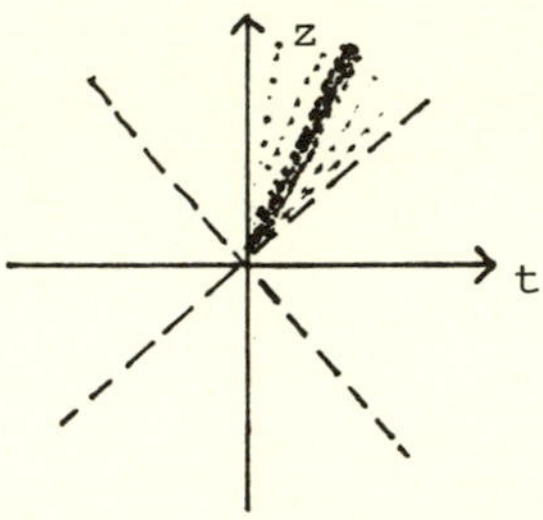

(g_{11}); $C_1 > 0$; t-time

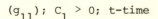

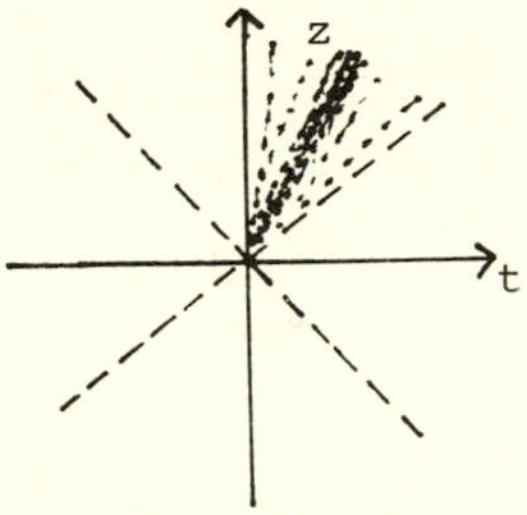

(g_{11}); $c_1 < 0$; z-time

On the light cone $r \to 0$ and $g \to g_o$. In region $z^2 < t^2$ we can also define our solution by analytical continuation. If $z^2 > t^2$, μ is complex:

$$\mu = W - \beta - i\sqrt{\alpha^2 - (W - \beta)^2}$$

The solution in this region will correspond to the case with two poles $\lambda = \mu$ and $\lambda = \bar{\mu}$, but in this situation we have $|\mu|^2 = \alpha^2$ and the poles are located on the circle $|\lambda|^2 = \alpha^2$. On this circle the matrix χ tends to I (see below) and $g \to g_o$ (a general result for the poles). On the light cone $(W - \beta)^2 = \alpha^2$ (this equation yields the pair of straight lines ξ = const, η = const) the solution is continuous, but discontinuity will arise in the first derivatives. Such phenomenon will not arise for complex poles.

5. TWO-SOLITON SOLUTIONS

Consider one complex pole $\lambda = \mu$. Consequently there are two poles in χ.

$$\chi = I + \frac{R}{\lambda - \mu} + \frac{\bar{R}}{\lambda - \bar{\mu}}$$

$$(R)_{ab} = n_a m_b$$

$m_b = m_{oc} M_{cb}$, $M = (\Psi_o^{-1})_{\lambda=\mu}$. We can calculate the vector n_a from the algebraic eq's. (3.6):

$$n_a = \frac{1}{\Delta}\left\{\frac{m_c \bar{m}_d (g_o)_{cd}}{\nu - \bar{\mu}}\, \bar{m}_b (g_o)_{ba} - \frac{\bar{m}_c \bar{m}_d (g_o)_{cd}}{\bar{\nu} - \bar{\mu}}\, m_b (g_o)_{ba}\right\}$$

$$\Delta = \frac{|m_a m_b (g_o)_{ab}|^2}{|\nu - \mu|^2} - \frac{|m_a \bar{m}_b (g_o)_{ab}|^2}{|\bar{\nu} - \mu|^2} \tag{5.1}$$

$\mu(\xi,\eta)$ comes from the equation

$$\alpha^2/\lambda + 2\beta + \lambda = 2W \tag{5.2}$$

and $W = W_1 - iW_2$

For the modules $|\mu| = \rho$ and phase ϕ ($\mu = \rho e^{i\phi}$) one has

$$\cos\phi = \frac{(2W_1 - 2\beta)\rho}{\alpha^2 + \rho^2}\, ; \qquad \sin\rho = \frac{2W_2\rho}{\alpha^2 - \rho^2} \tag{5.3}$$

If $W_2 \neq 0$ the poles are located either inside the circle

$|\lambda|^2 = \alpha^2$ $(\rho^2 < \alpha^2)$ or outside $(\rho^2 > \alpha^2)$. Let us consider $\rho^2 \leq \alpha^2$.

From (5.1 - 2) we can see that on the circle $\rho^2 \to \alpha^2$, $\frac{1}{\Delta} \to (\rho^2 - \alpha^2)^2 \to 0$ and $n_a \to \rho^2 - \alpha^2 \to 0$. Consequently on the circle $|\lambda|^2 = \alpha^2$, $R \to 0$, $\chi \to I$ and $g \to g_o$.

For illustration of the results let us now consider the same particular solution (4.7 - 4.6) (Kasner). We will investigate only two cases:

1) isotropic "background" $s_1 = s_2 = \frac{1}{2}$ (5.4)

2) flat space "background" $s_1 = 0$, $s_2 = 1$ (5.5)

For $s_1 = s_2 = \frac{1}{2}$ the result is:

$$
\begin{aligned}
- ds^2 &= C_1 \alpha^{3/2} \delta^{-1} Q(-dt^2 + dz^2) + \\
&+ \alpha Q^{-1} \Big\{ \Big[p_1^2 H - (1 - \delta)^2 \cos 2\phi + 2p_1 (1 - \delta^2) \sin^2\phi \Big] dx^2 + \\
&+ \Big[p_1^2 H - (1 - \delta)^2 \cos 2\phi - 2p_1 (1 - \delta^2) \sin^2\phi \Big] dy^2 \\
&- 2p_2 (1 - \delta)^2 \sin 2\phi \, dx \, dy \Big\}
\end{aligned}
$$

where

$$Q = p_1^2 H - (1 - \delta)^2; \; H = 1 + \delta^2 - 2 \cos 2\phi \; ; \; \delta = \rho^2/\alpha^2$$

C_1, p_1, p_2 are constants and

$$p_1^2 - p_2^2 = 1 \tag{5.6}$$

The functions ρ, ϕ are obtained from (5.3).

This solution describes the propagation and interaction of two localised "perturbations" on the isotropic background (not necessarily Friedman). The general features can be seen from the behavior of $g_{ab}(\xi, \eta)$ in space at any fixed moment of time t. If $\alpha = t$, $W_1 - \beta = Z$ we will have:

1) in regions $Z \to \pm\infty$ $\quad g \to g_o = \mathrm{diag}(t, t)$

2) everywhere (in Z) $\quad g_{11} > t$ and $g_{22} < t$

3) each components g_{ab} has (in Z) two extremal points

4) In asymptotic region $t \to \infty$ $(t \gg W_2)$ these extremums are

concentrated near the light cone $z^2=t^2$

5) During evolution (as time t decrease) the distances between extremums decrease and the amplitudes of the solitons increase.

By "amplitudes" we mean the absolute values of the extremums (in Z) of the components of the matrix

$$g = (g-g_o)g_o^{-1} = \begin{pmatrix} \frac{g_{11}-t}{t} & \frac{g_{12}}{t} \\ \frac{g_{12}}{t} & \frac{g_{22}-t}{t} \end{pmatrix} \tag{5.7}$$

at any fixed moment of time t.

6) In the components g_{11}, g_{22} both perturbations, in the limit $t\to 0$ (cosmological type singularity), approach each other and coincide (at t=0). The amplitudes are finite.

7) In the component g_{12} both perturbations reach some minimal distance ($=2W_2$) (there is no coincidence)

8) In the region $t\to\infty$ the amplitudes tend to zero. In region $t\to 0$ they tend to some finite values; the asymptotic form for g_{ab} at $t\to 0$ is:

$$g = \frac{1}{t}\begin{pmatrix} \frac{Z^2+s^2W_2^2}{Z^2+W_2^2} & \frac{1-s^2}{s}\,\frac{ZW_2}{Z^2+W_2^2} \\ \frac{1-s^2}{s}\,\frac{ZW_2}{Z^2+W_2^2} & \frac{s^2Z^2+W_2^2}{s^2(Z^2+W_2^2)} \end{pmatrix} \tag{5.8}$$

where $s = (1+p_1)/p_2$.

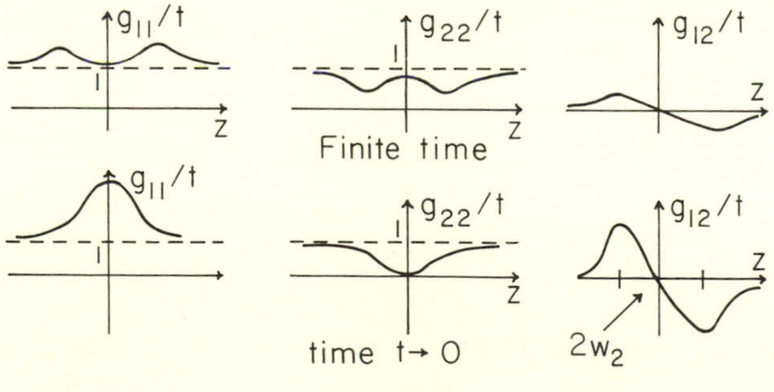

Now let us consider the second case, when $s_1=0$, $s_2=1$ (flat space). Let us choose coords:

$$\alpha = W_2 \text{sh}Z \text{ ch}t \; ; \qquad \beta = W_1 + W_2 \text{ ch}Z \text{ sh}t \tag{5.9}$$

For modulus ρ and phase ϕ ($\mu = \rho e^{i\phi}$) this yields:

$$\sin^2\phi = 1/\text{ch}^2 t \; ; \quad \cos^2\phi = \text{th}^2 t \; ; \quad \rho^2/\alpha^2 = \text{th}^2 \frac{Z}{2} \tag{5.10}$$

The resulting metric is

$$\begin{aligned} -ds^2 &= \omega(-dt^2+dZ^2) + \omega^{-1}(\gamma+a_1^2\text{sh}^2Z)dx^2 + \\ &+ \omega^{-1}\left[\gamma(2b_1\text{ch}Z-a_1\text{sh}^2Z)^2 + \text{sh}^2Z(r^2+a_1^2+b_1^2)^2\right]dy^2 - \\ &- 2\omega^{-1}\left[\gamma(2b_1\text{ch}Z-a_1\text{sh}^2Z) + a_1\text{sh}^2Z(r^2+a_1^2+b_1^2)\right]dxdy \end{aligned} \tag{5.11}$$

where:

$$\begin{aligned} \omega = r^2 + (b_1-a_1\text{ch}Z)^2 \; ; \quad \gamma = (a_1^2-b_1^2-m_1^2)\text{ch}^2t \\ r = m_1 + \sqrt{a_1^2-b_1^2-m_1^2} \quad \text{sh}t \end{aligned} \tag{5.12}$$

$a_1, b_., m_.$ - arbitrary constants ($a_1^2 \geq b_1^2+m_1^2$) and $W_2^2=a_1^2-b_1^2-m_1^2$.

This solution can in fact be obtained from the Kerr-NUT metric after complex coordinates transformation:

$$\theta = iZ \; ; \quad r = m_1 + \sqrt{a_1^2-b_1^2-m_1^2} \quad \text{sh}t \; ; \quad \tau = x \; ; \; \phi = y$$

where θ, r, ϕ, τ are Boyer-Lindquist coordinates. If $b_1=0$ in these coords we will have Kerr-metric with angular momentum parameter a_1 and mass m_1. In this case solution (5.11) corresponds to $a_1 \geq m_1$. This means, that (for the stationary analog of our original metric) the method of ISP generate the Kerr-solution from stationary double solitons.

6. GENERAL CASE

$$W = \left(1/2 \, \frac{\alpha^2}{\lambda} + 2\beta + \lambda\right) ; \quad D_1 W \equiv 0 \; ; \quad D_2 W \equiv 0 \tag{6.1}$$

$G_o(\lambda,\xi,\eta)$ is a matrix on the circle $|\alpha^2| = \alpha^2$, and

$$G_o = G_o(W). \tag{6.2}$$

W is real. The constraints are

$$G_o(\infty) = G_o(-\infty) = I \tag{6.3}$$

$$\overline{G}_o(\lambda) = G_o(\lambda) \ ; \quad G_o = \tilde{G}_o \tag{6.4}$$

Introduce a new matrix G:

$$G(\lambda,\xi,\eta) = \Psi_o G_o \Psi_o^{-1} \tag{6.5}$$

Ψ_o is a particular solution. Since $D_1 W \equiv D_2 W \equiv 0$, we have

$$D_1 G = \frac{1}{\lambda-\alpha}(A_o G - GA_o) \ ; \quad D_2 G = \frac{1}{\lambda+\alpha}(B_o G - GB_o) \tag{6.6}$$

Again we have a Riemann problem: χ_1 is analytical outside the circle $|\lambda|^2 = \alpha^2$, and χ_2 is analytical inside. On the circle

$$\chi_1 = \chi_2 G \quad (\chi_2(\infty) = I) \tag{6.7}$$

This problem has a unique solution if the matrices χ_1, χ_2 have no poles (points of degeneracy).

From (6.7):

$$(D_1\chi_1 + \frac{1}{\lambda-\alpha}\chi_1 A_o)\chi_1^{-1} = (D_1\chi_2 + \frac{1}{\lambda-\alpha}\chi_2 A_o) = \frac{1}{\chi-\alpha}A$$

$$(D_2\chi_1 + \frac{1}{\lambda+\alpha}\chi_1 B_o)\chi_1^{-1} = (D_2\chi_2 + \frac{1}{\lambda+\alpha}\chi_2 B_o)\chi_2^{-1} = \frac{1}{\chi+\alpha}B \tag{6.8}$$

$$g = \chi_2(0) g_o$$

The Eq's for χ_1, χ_2 are:

$$\chi_1 = I + \frac{1}{\pi i}\int_\Gamma \frac{\rho(Z)}{\lambda - Z + i0}\, dZ \qquad \Gamma \text{ denotes the circle } |\lambda|^2 = \alpha^2$$

$$\chi_2 = I + \frac{1}{\pi i}\int_\Gamma \frac{\rho(Z)}{\lambda - Z - i0}\, dZ \tag{6.9}$$

From (6.7) we now have

$$\rho(Z) + T(\lambda,\xi,\eta)\left(I + \frac{1}{\pi i}\Gamma\frac{\rho(Z')}{Z-Z'}\, dZ'\right) = 0$$

$$T = (I-G)(I+G)^{-1} \tag{6.10}$$

$G=I$ is the solitonic case.

In the asymptotic regions $\beta\to\pm\infty$ (or $\alpha\to\pm\infty$), and $G\to I$. This means that in the general case the asymptotics will be of solitonic type.

THE HYDRODYNAMICS OF PRIMORDIAL BLACK HOLE FORMATION AND THE DEPENDENCE OF THE PROCESS ON THE EQUATION OF STATE

I. D. Novikov, A. G. Polnarev

Institute for Space Research, Moscow

1. INTRODUCTION

Our paper deals with the problem of primordial black holes (PBH), with the numerical computations of their formation and their subsequent evolution. There is a connection between the problem of PBH formation and the problem of sources of the gravitational waves. On the one hand, the process of PBH formation is connected with the origination of primordial gravitational waves, on the other hand PBHs can be the source of the gravitational waves in the contemporary Universe. But we will not discuss these problems in this paper and will discuss here only the problem of the numerical computations of the PBH's formation.

The possibility of primordial black holes formation was mentioned in 1966 by Ya. B. Zeldovich, and I. D. Novikov (1966, 1967, 1968) and in 1971 by S. Hawking (1971). Later the PBH problem was treated in numerous papers [Hawking (1974), Carr and Hawking (1974), Carr (1975, 1976, 1977), Lin, Carr and Fall (1976), Chapline (1976), Page (1976), Zeldovich and Starobinsky (1976), Polnarev (1977), Nadejin, Novikov, and Polnarev (1977, 1978), Byalko (1977, 1978), Bicknell and Henriksen (1978)]. It is rather easy to construct a cosmological model with the formation of PBH's in the case when the pressure is much smaller than the energy density ε: $p \ll \varepsilon$, see Polnarev (1977). Indeed, if in the perturbed region ε is somewhat higher than the average density $\langle\varepsilon\rangle$, or the rate of expansion is lower than outside of the perturbed region, then the gravitational forces may be able to stop the expansion of this fraction of matter, contraction will begin and the PBH will be formed.

Thus in the case of the soft equation of state $p \ll \varepsilon$ the PBH can be formed even when the amplitudes of the perturbations are small in the sense of all parameters: $|\delta\varepsilon|/\varepsilon \ll 1$, $|\upsilon_{peculiar}/C| \ll 1$, gravitational perturbation $|h_i^K| \ll 1$. The situation changes drastically in the case of the relativistic equation of state $p = \varepsilon/3$. Lifshitz (1946) has demonstrated that for $p = \varepsilon/3$ the small initial perturbations cannot grow to large amplitude and so cannot turn into the PBH's. Thus at least the initial gravitational perturbations must be great enough (of order of unity) to form the PBH's when the equation of state is $p = \varepsilon/3$. The same conclusion is valid for the most stiff equation of state $p = \varepsilon$. In the very vicinity of the cosmological singularity the equation of state of the matter is unknown. Thus the PBH's which arise at the very beginning of the expansion of the Universe could be formed when the equation of state differs from $p = \varepsilon/3$. This is the reason why we investigate the dependence of the process of PBH formation on variations in the equation of state.

The following three questions are basic for the problem of PBH formation.

1) What are the deviations from the Friedmann cosmological model at the beginning of the expansion which result in PBH formation?

2) How does the accretion of surrounding matter proceed on the PBH already formed?

3) What is the dependence of the process of the PBH formation on the equation of state?

These problems had been analysed in the first papers on PBH [Zeldovich and Novikov (1966, 1967, 1968)] and subsequent papers [Hawking (1971), Carr and Hawking (1974), Carr (1975), Nadejin, Novikov, and Polnarev (1978), Bicknell and Henriksen (1978)]. It turned out that only numerical computations can give an exhaustive answer to all three questions.

Appropriate calculations carried out by Nadejin, Novikov and Polnarev (1978) demonstrate the hydrodynamic picture of PBH formation and the subsequent non-steady fluid accretion under the simplest assumption of spherical symmetry of the processes under consideration. We chose the case of spherical symmetry because it is the simplest case from the mathematical point of view. But on the other hand the spherical perturbation is the most favourable to the PBH formation. Thus using this approximation we gave the absolute lower limit on the amplitude of the perturbations which results in PBH formation. Note that in the case of the stiff equation of state ($p \sim \varepsilon$) the spherical symmetry is a good approximation for qualitative description of the evolution of the perturbations with the small deviations from sphericity. But in the case $p = 0$, even small initial deviations from sphericity could lead to the completely different picture of the evolution. In this last case the two-dimensional singularities - the so-called pancakes [see Zeldovich (1970)] could arise instead of the PBH's.

2. THE BASIC EQUATIONS AND FORMULATION OF THE PROBLEM

In this section we shall recall briefly the basic equations and formulation of the numerical problem. For details see our previous work [Nadejin, Novikov, and Polnarev (1978)]. As we have mentioned above the spherically symmetric problem will be considered here. In this case the line element can be written in the following form:

$$ds^2 = c^2 e^{\delta} dt^2 - e^{\omega} dR^2 - r^2(d\theta^2 + \sin^2\theta d\phi^2) \qquad (1)$$

From the very beginning let us use dimensionless variables. Introduce some characteristic mass (the mass of the perturbation) corresponding to gravitational radius r_g. Then the dimensionless variables (denoted by a tilde) are the following: Euler radius $r = r_g\tilde{r}$ (the surface area of spherical shell is equal to $4\pi r^2$), Lagrange radius $R = r_g\tilde{R}$, time $t = r_g\tilde{t}/c$, a mass contained inside the Lagrange radius R, $m = c^2\tilde{m}/(2Gr_g)$, energy density $\varepsilon = 3c^4\tilde{\varepsilon}/(8\pi G r_g^2)$. Below we shall work only with dimensionless variables, so it is convenient to omit the tilde. We also introduce a new variable u, related to $\dot{r} \equiv \frac{\partial r}{\partial t}$ by the equation $u^2 = e^{-\delta}\dot{r}^2$. This variable u means physical velocity of expansion or contraction of the surface of R = const relative to the comoving system of reference, which is chosen in this text. We shall denote the partial derivative with respect to R by a "prime". Finally, dimensionless pressure will be denoted by p.

The comoving system of reference is chosen. Then Einstein equations for the line element (1) are equivalent to the following system of equations [see e.g. Misner and Sharp (1964)] :

$$\dot{u} = -\frac{1}{2} e^{\delta/2} \left(\frac{2e^{-\omega} p' r'}{p + \varepsilon} + \frac{m}{r^2} + 3pr \right), \qquad (2)$$

$$\dot{r} = e^{\delta/2} u \qquad (3)$$

$$\dot{m} = -3pr^2\dot{r} \qquad (4)$$

$$\varepsilon = \frac{\partial m}{\partial r^3} \qquad (5)$$

Write the equation of state as

$$p = (\gamma - 1)\varepsilon, \quad \gamma = \text{const.} \qquad (6)$$

The first integral of the system (2)-(5) has the form [see Podurets (1964)]

$$u^2 + 1 - e^{-\omega} r'^2 - \frac{m}{r} = 0. \qquad (7)$$

This first integral can be used to check the accuracy of our numerical calculations. The above system of equations was used first in numerical calculations for the problem of collapse by May and White (1967) and Schwartz (1967). For numerical integration of the system (2)-(6) it is necessary to set initial conditions at the slice t = const.

Our goal is to investigate the PBH-formation in an expanding Universe and to determine how this process depends on the deviations of initial conditions from the Friedmann model. Zeldovich (1970) has demonstrated that the quantum effects of particle creation near the singularity result in Friedmann-like beginning of the expansion of the Universe. According to his result when $t \to 0$ (the singularity corresponds to $t=0$), the solution of the Einstein equations has the following form:

$$t \to 0, \quad dS^2 \approx dt^2 - a^2(t)\, g_{\alpha\beta}(x^1, x^2, x^3)\, dx^\alpha dx^\beta$$

ε does not depend on the spatial coordinates. Here $g_{\alpha\beta}$ are arbitrary functions of x^1, x^2, x^3. For example for $p = \varepsilon/3$, $a(t) \sim t^{1/2}$, $\varepsilon \approx 1/(4t^2)$ (in our notations, see above).

Only during the subsequent evolution does ε become different in different places of 3-space. In the case of spherical symmetry the Friedmann-like type of line element is

$$dS^2 = dt^2 - a^2(t)\,[dR^2 + \Phi^2(R)\,(d\theta^2 + \sin^2\theta d\phi^2)] \tag{8}$$

Here we consider the spherical perturbed region with the center located at the origin of the coordinate system. It is assumed also that outside the perturbed region the solution is exactly that of the Friedmann model. Hence the perturbation is such that the total mass of matter inside the perturbed region is exactly the same as the mass which would have been in this region in the case of the unperturbed Friedmann model. This means, that outside the perturbed region the line element is just the same as for Friedmann model and we assume also that this Friedmann model has a flat comoving 3-space:

$$dS^2 = dt^2 - a^2(t)\,[dR^2 + R^2(d\theta^2 + \sin^2\theta d\phi^2)] \tag{9}$$

To describe the perturbation we have to choose the function $\phi(R)$ in (8) near the origin of the coordinate system. Any perturbation can be characterized by two principal parameters: the perturbation amplitude and the steepness of the slope. The simplest choice of the perturbed region is the following. The deviation near the singularity is assigned as a spherical region with a comoving 3-space of constant positive curvature, i.e. the perturbed region corresponds to some part of the closed Friedmann model:

$$dS^2 = dt^2 - a^2(t)[dR^2 + \sin^2 R(d\theta^2 + \sin^2\theta d\phi^2)] \qquad (10)$$

The deviation amplitude can be characterized by a number measuring the fraction of the closed space with the constant positive curvature cut out. R_- is the value of R on the boundary of the perturbed region of the 3-space with positive constant curvature. Further the transition region ($R_+ - R_- = \Delta$) is where the solution gradually matches an external non-perturbed region with a perturbed inner region. The width Δ of this transition region characterizes the steepness of the slope of the perturbation. This is the second important parameter of the problem. The development of the process strongly depends on the width Δ. In fact, if Δ is small steep pressure gradients and violent hydrodynamic phenomena develop as the density perturbations $\delta\varepsilon$ grow. If Δ is large pressure gradients are small.

For a detailed description of the choice of the boundary and the initial conditions, the choice of the moment of the beginning of the computations, and details of the techniques of the computations see Nadejin, Novikov, and Polnarev (1978).

3. RESULTS OF NUMERICAL SIMULATIONS AND CONCLUSIONS

First of all we investigate the dependence of the process of PBH formation on the amplitude of the perturbation R_- and the width Δ in the case of the equation of state $p = \varepsilon/3$. Fig. 1 - 5 show the results of these calculations. The different figures correspond to different initial conditions: to the different amplitudes of a perturbation R_- and the different widths Δ of transition region. In Figures 1,2 the energy density ε is depicted versus the Lagrange radius R. The different curves correspond to different moments of time. At the beginning the density is nearly uniform over the entire space as it must be at the beginning of the Friedmann-like expansion. Only in the transition region is there a slight rarefaction related with the fact that the energy density is somewhat higher in the region of closed universe, $R<R_-$, than in the region of spatially flat Friedmann universe, $R>R_+$. Rarefaction occurs because the total mass contained in the region $R \leq R_+$ should be the same, as mentioned above, as the mass of the spatially flat Friedmann universe at the same Lagrange radius R. The general decreasing of the density with time is due to cosmological expansion. At the next stage the matter begins to contract at the central region of the perturbation.

If the initial perturbation of metric is not great enough (R_- is not great, Fig. 1) it turns into an acoustic wave propagated to infinity: $R_- = 0.75\ R_{max}$ (where $R_{max} = \frac{\pi}{2}$), and the black hole does not form. For larger $R_- = 0.80\ R_{max}$ density perturbations are large, but still no PBH forms, the perturbation spreads out as a wave package. For $R_- = 0.9\ R_{max}$, $\Delta = 0.5R_-$

(Fig. 2) a PBH forms.

To determine the moment of time when the black hole forms, and to find its boundary, we note at first that, as it is well known, there is no local criterion of the event horizon location. The integration of null geodesic curves should be carried out to find the event horizon. The position of the event horizon depends significantly on the behavior of matter, which can fall into black hole in the future. So it is more reasonable from the point of view of an observer near the black hole to determine the black hole boundary as an apparent horizon, i.e. to find the Lagrange radius corresponding to that mass which is within it's gravitational radius at given moment of time. In this work as well as in the previous one [Nadejin, Novikov, and Polnarev (1978)] we use this definition for the black hole boundary. This makes it possible to determine its position using only local calculations. One can see [Misner and Sharp (1964), May and White (1966, 1967), Carr and Hawking (1974), Nadejin, Novikov, and Polnarev (1978)], that some point with Lagrange radius R lies within its gravitational radius if and only if the following two conditions are fulfilled: $m/r>1$ and $u<0$. Note, the event horizon always lies outside the apparent horizon.

The moment of time when the black hole forms can easily be found with the help of Fig. 3, where the ratio m/r is depicted versus Lagrange radius R. In the regions where expansion occurs r increases and m decreases because of the work done by pressure gradients. That is why the ratio m/r decreases with expansion. Obviously when contraction replaces expansion the ratio m/r must increase. If the value m/r is increasing and it reaches unity before pressure gradients have time to halt the contraction, then both conditions of the criterion of black hole formation are fulfilled.

An interesting question arises now: what are the geometrical properties of space-like slices at the moment of black hole formation? In Fig. 5 the embedding diagrams are shown [see, e.g. Landau and Lifshitz (1962), Misner, Thorne, and Wheeler (1973)]. The surfaces obtained by revolution of curves depicted on Fig. 5 around a vertical-axis are geometrically identical to spacelike slices in equatorial plane at the different moments of time. One can see [Landau and Lifshitz (1962)] that the space geometry with line element (1) corresponds to the surface of revolution with

$$Z(r) = \int \sqrt{1 + e^{-\omega}(r')^{-2}}\, dr \tag{11}$$

The dependence of Z on r is depicted in Fig. 5. One can see from this figure, that the moment of black hole formation is not apparent from a geometrical point of view. It is interesting to note also that the formation of a black hole does not require occurrence of a semi-closed configuration [Zeldovich and Novikov (1971)], when some Lagrange radius exists for which $r' = 0$.

In Fig. 4 the curve is drawn that shows for which R_- and Δ the primordial black hole forms, and for which it does not and perturbations become acoustic waves. The following conclusions can be made. PBH's form only for very large deviations from the Friedmann model which correspond to $R_- \approx 0.85 - 0.9\ R_{max}$. The width of the transition region has a strong effect on the PBH formation. The narrower Δ, the greater is the role of pressure gradients that hinder PBH formation.

Before numerical calculations were performed, attempts were made to estimate the importance of pressure in PBH-formation by developing steady-state or self-similar solutions. An assumption was made that pressure could contribute to gas accretion by PBH's in the process of their formation and significantly enlarge their masses. Carr and Hawking (1974) showed that there is no self-similar solution that results in a catastrophic accretion of matter by a PBH, with its size growing as fast as the cosmological horizon. Our calculations show, that in fact pressure strongly hinders PBH formation, making their masses smaller than they would have been with the same initial perturbation but with no pressure, $p = 0$. Indeed, near the singularity, within the spatial cross-section $t = const$, the energy density in the perturbed region $R<R_-$ is higher than far out in the flat Friedmann model, and the outward pressure gradient at R_- tends to throw away the matter. In the transition region Δ the density ε is minimum, and on its outer boundary R_+ the inward pressure-gradient gives rise to accretion. However, this effect is less significant in PBH formation than the above mentioned gradient at R_- which results in the outflow of matter from the perturbed region.

Hence, the mass of the PBH which actually forms is 0.2 to 0.3 that of the PBH which would have formed with no outflow at all, it should be emphasized that the size of the PBH just after its formation is much smaller than the cosmological horizon. When a PBH forms, its mass is about 0.01 to 0.06 of that trapped within a sphere with a radius equal to the cosmological horizon. Under such conditions the accretion on a PBH is slow and it only slightly increases the mass of the PBH in the course of the subsequent evolution. Our calculations show this clearly. This conclusion was proved for self-similar solutions in Carr and Hawking (1974) (and was mentioned as one of the possibilities in Zeldovich and Novikov (1966)).

Let us fix now the initial conditions of the perturbation, say, $R_- = 0.95 \cdot R_{max}$ and $\Delta = 0.5 \cdot R_-$, and consider the process of PBH formation, varying the equation of state. The basic quantitative characteristics of the process as a whole are listed in Table 1 for seven different equations of state (for seven values of γ in an equation of state $p = (\gamma - 1)\varepsilon$). Besides this we present here the figures which illustrate in more detail the processes of PBH formation for two important cases of equation of state (1) $\gamma=4/3$ (the case of relativistic gas), and (2) $\gamma=2$ (the stiffest equation of state). In order to make a comparison between these space-times it is convenient to present the dependence of energy density ε (Fig. 6,7) and the ratio m/r (Fig. 8,9) on Lagrange radius R for various moments of time using logarithmic coordinates. If one compares the processes of PBH formation in the case $p=\varepsilon/3$ (Fig. 6 and Fig. 8) with a case of a softer equation of state $p=\varepsilon/5$ ($\gamma=1.2$ not shown), one can see that qualitatively the results are quite similar but quantitative difference takes place: the black hole forms earlier (see Table 1) and it's mass is larger in the case of $\gamma=1.2$, than in the case $\gamma=4/3$. Fig. 7 and Fig. 9 show the process of PBH formation in the case of the stiffest equation of state $p=\varepsilon$. One can see that qualitatively new feature of the process arises. The process up to the moment of black hole formation is qualitatively similar to the case $\gamma=4/3$ but this moment occurs later and the mass of black hole is smaller compared with the case $\gamma=4/3$. But subsequent evolution of the matter around black hole differs qualitatively from previous cases. One can see (Fig. 7) that discontinuity of matter flow arises just at the black hole horizon (at the boundary of black hole in the sense of above determination). The accretion of fluid into black hole seems to stop and the outer part of matter goes to infinity (to large Euler radii r), while the inside matter undergoes unlimited-contraction. Unfortunately, the sharp gradient prevents the continuation of numerical simulations, but the picture seems to be well understood without additional computations. This picture including black hole formation, subsequent accretion of matter into black hole and cessation of accretion, differs drastically from the similarity solution constructed in Lin, Carr and Fall (1976) and also in the case $p=\varepsilon$ (see also Bicknell and Henriksen (1978)).

One can suppose, that in general three characteristic times arise in the hydrodynamics of PBH formation. First, there is time of black hole formation (denote it as t_{BH}), second, there is the time when a physical singularity occurs at the center of the perturbation (energy density goes to infinity), denote this moment as t_s, and, third, there is the time, when discontinuity of matter flow occurs, denote it as t_{df}. As our computations demonstrate $t_{BH} < t_s < t_{df}$ for γ which is equal to 1.2, 4/3 and 1.4. Another unequality $t_{BM} < t_{df} < t_s$ is valid for γ which is equal to 1.6, 1.8, 1.9 and 2. As one can see from Fig. 7 an additional region of low density arises outside but near to black hole horizon. This region approaches the horizon and evolves to a discontinuity of matter flow. It is interesting to note that in the case $\gamma=1.6$ (in some sense intermediate between soft and stiff equations of state) the discontinuity also arises, but inside the horizon. See Table 1, where all these results are summarized.

In conclusion we represent the dependence of black hole mass on time (Fig. 10). One can see from Fig. 10, that accretion into the black hole is more intensive for softer equations of state, than for stiff equations of state. It is interesting to note, that for large γ, the dependence of black hole mass at the moment of it's formation is rather weak, while the time of the black hole formation depends crucially on γ. So one can conclude that catastrophical accretion (when the black hole grows approximately at the same rate as cosmological horizon), seems to be impossible. This time delay of the black hole formation is due to action of pressure gradients. As a result the ratio of black hole mass to the mass contained inside cosmological horizon is the lesser the larger is γ (Fig. 11). Besides this, as mentioned above, pressure gradients throw out some part of matter initially contained in the perturbed region. Thus, one can conclude that it is more difficult for a black hole to form the stiffer is the equation of state.

4. ACKNOWLEDGEMENTS

The authors would like to express their gratitude to Ya. B. Zeldovich for helpful discussions and D. K. Nadejin for the help in the carrying out of the numerical computations.

REFERENCES

Bicknell, G.V. & Henriksen, R.N. (1978) Self-similar growth of primordial black holes. I. stiff equation of state. Astrophys. J., 219, pp.1043-1057.

Byalko, A.Y. (1977, 1978). Preprints of Research Inst. for Theor. Phys., Hirosima Univ., USSR. Preprint numbers RRK 77-11, RRK 78-13.

Carr, B.J. (1975). The primordial black hole mass spectrum. Astrophys. J., 201, pp.1-19.

Carr, B.J. (1976). Some cosmological consequences of primordial black hole evaporations. Astrophys. J., 206, pp.8-25.

Carr, B.J. (1977). Black hole and galaxy formation in a cold early universe. M.N. Roy. Astro. Soc., 181, pp.293-309.

Carr, B.J. & Hawking, S.W. (1974). Black holes in the early universe. M.N. Roy. Astro. Soc., 168, pp.399-415.

Chapline, G.F. (1976). Quarks in the early universe. Nature, 261, pp.550-551.

Hawking, S.W. (1971). Gravitationally collapsed objects of very low mass. M.N. Roy. Astro. Soc., pp.75-78.

Hawking, S.W. (1974). Black hole explosions? Nature, 248, pp.30-31.

Landau, L.D. & Lifshitz, E.M. (1962). The Classical Theory of Fields, 2nd Ed., Addison-Wesley, Reading Mass.

Lifshitz, E.M. (1946). In Russian., JETP, 16, pp.587.

Lin, D.N.C., Carr, B.J., & Fall, S.M. (1976). The growth of primordial black holes in a universe with a stiff equation of state. M.N. Roy. Astro. Soc., 177, pp.51-64.

May, M.M. & White, R.H. (1966). Hydrodynamic calculations of general-relativistic collapse. Phys. Rev., 141.

May, M.M. & White, R.H. (1967). Stellar dynamics and gravitational collapse. In Methods in Computational Physics, vol. 7, ed. by B. Alder, S. Fernbach, & M. Rotenberg. Academic Press, New York.

Misner, C.W. & Sharp, D.H. (1964). Relativistic equations for adiabatic, spherically symmetric gravitational collapse. Phys. Rev., 136, pp.B571-576.

Misner, C.W., Thorne, K.G., & Wheeler, J.A. (1973). Gravitation. Freeman, San Francisco.

Nadejin, D.K., Novikov, I.D., & Polnarev, A.G. (1977). Hydrodynamics of primordial black hole formation. Abstracts of Contributed Papers GR8, pp.382.

Nadejin, D.K., Novikov, I.D., & Polnarev, A.G. (1978). In Russian. Astr. Z., 55, pp.216.

Page, D.N. & Hawking, S.W. (1976). Gamma rays from primordial black holes. Astrophys. J., 206, pp.1-7.

Podurets, M.A. (1964). On one form of Einstein's equations for a spherically symmetrical motion of a continuous medium. Sov. Astron. AJ, 8, pp.19-22. Translated from Astron. Z., 41, pp.28-32 (1964).

Polnarev, A.G. (1977). In Russian. Astrophysika, 13, pp.376.

Schwartz, R.A. (1967). Gravitational collapse, neutrinos and supernovae. Annals of Physics, 43, pp.42-73.

Zeldovich, Ya. B. (1970a). Particle production in cosmology. JETP Lett., 12, pp.307-311.

Zeldovich, Ya. B. (1970b). In Russian. Astrophysika, 6, pp.119.

Zeldovich, Ya. B. & Novikov, I.D. (1967). The hypothesis of cores retarded during expansion and the hot cosmological model. Sov. Astron-AJ, 10, pp.602-603. Translated from Astron. Z., 43, pp.758-760 (1966).

Zeldovich, Ya. B. & Novikov, I.D. (1968). Proceeding IAU Symposium N. 29, Byurakan, USSR.

Zeldovich, Ya. B. & Novikov, I.D. (1971). Relativistic Astrophysics. University of Chicago Press.

Zeldovich, Ya. B. & Starobinsky, A.A. (1976). Possibility of a cold cosmological singularity in the spectrum of primordial black holes. JETP Lett., 24, pp.571-572.

Table 1. The basic quantitative characteristics of PBH formation and evolution for various equations of state $P = (\gamma - 1) E$.

γ	1.2	4/3	1.4	1.6	1.8	1.9	2.0
t_{BH}	1.237	1.500	1.641	2.140	2.765	3.126	3.526
m_{BH} at $t=t_{BH}$	.428	.374	.351	.305	.289	.285	.282
t_{df}	-	-	-	4.349 within horizon	4.999	4.796	4.645
m_{BH} at $t=t_{df}$	-	-	-	450	364	340	325
t_s	1.85	2.58	3.161	-	-	-	-
m_{BH} at $t=t_s$	.582	.572	.558	-	-	-	-

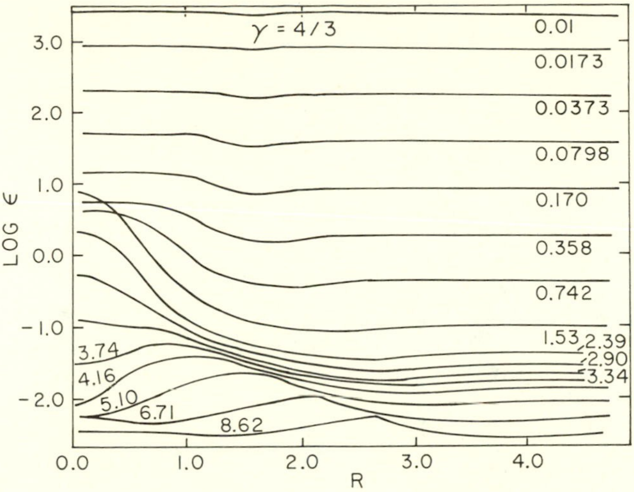

Figure 1 The energy density log ε versus Lagrange radius R. Various curves correspond to various moments of time. No black hole forms. The perturbation turns into an acoustic wave. The picture corresponds to $R_=0.75\ R_{max}$, $\Delta=0.5\ R_$ and $\gamma=4/3$.

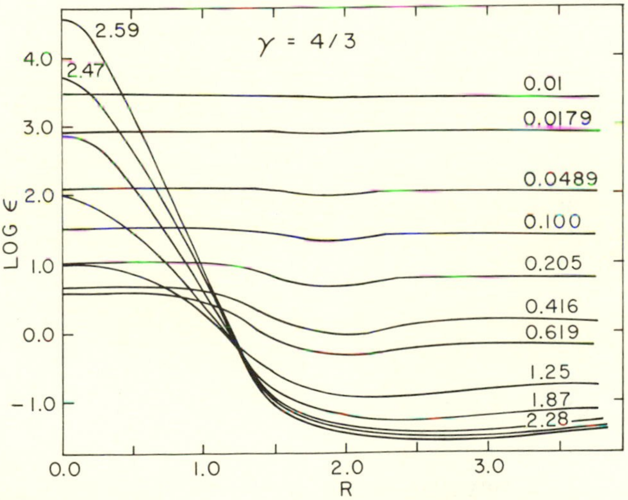

Figure 2 The energy density log ε versus Lagrange radius R. This picture describes black hole formation and corresponds to $R_=0.9\ R_{max}$, $\Delta=0.5\ R_$ and $\gamma=4/3$.

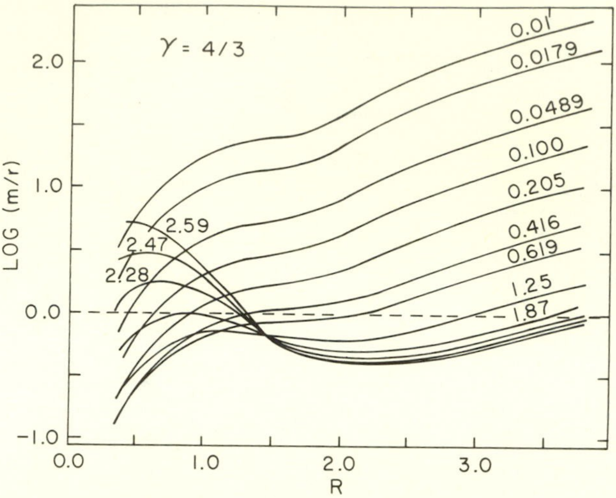

Figure 3 The ratio log m/r versus Lagrange radius R. The moment of black hole formation corresponds to the curve which is tangent to dotted line $m/r=1, R_-=0.9\ R_{max}$, $\Delta=0.5R_-$ and $\gamma=4/3$.

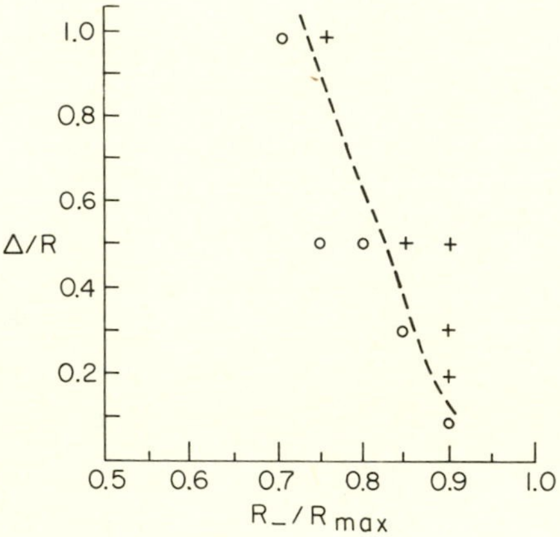

Figure 4 The conditions under which PBH form for $\gamma=4/3$ Results from various computations are plotted on the plane of initial conditions $(R_-/R_{max}, \Delta/R_-)$. The crosses correspond to formation of black hole while circles correspond to the generation of acoustic waves. The dotted line is a rough boundary of the region of PBH formation in the $(R_-/R_{max}, \Delta/R_-)$-plane.

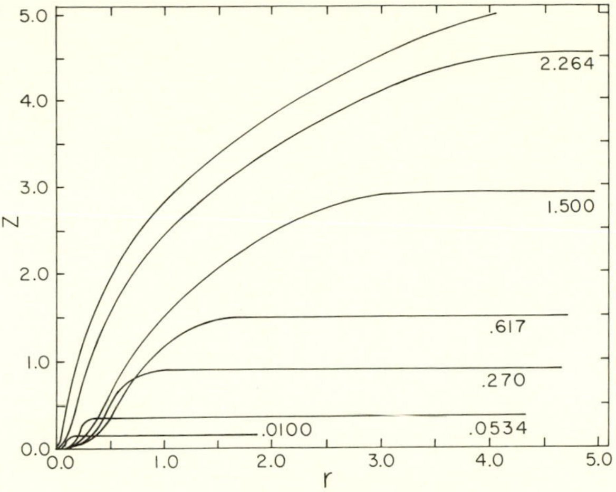

Figure 5 Embedding diagrams describing the space geometry at various moments of time. The properties of space geometry considered in our work are identical to the geometrical properties of the surfaces of revolution obtained from the curves in the figure by revolution of these curves around Z-axis. On the abscisa Euler radius r is plotted. The figure corresponds to $\gamma=4/3$, $R_-=0.95\ R_{max}$, and $\Delta=0.5\ R_-$.

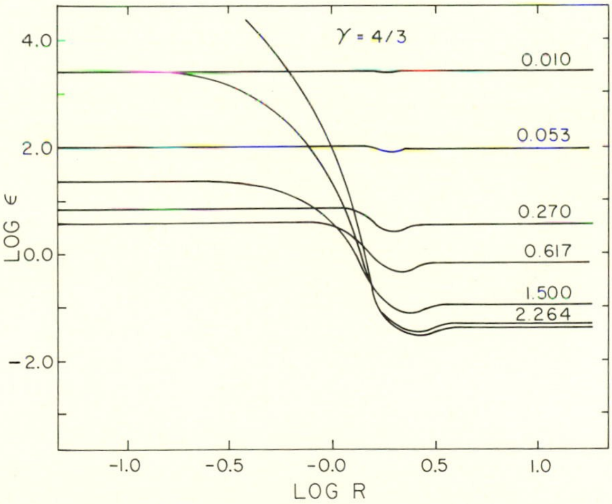

Figure 6 Log ε versus Log R for $\gamma=4/3$, $R_-=0.95\ R_{max}$, and $\Delta=0.5\cdot R_-$.

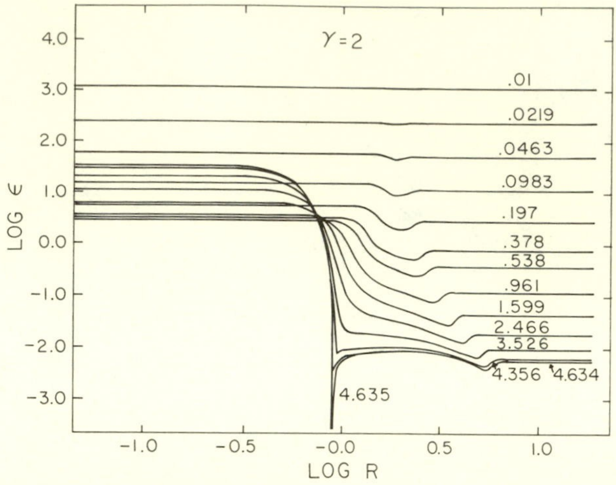

Figure 7 Log ε versus Log R for γ =2 (the stiffest equation of state), $R_{-}=0.95 \cdot R_{max}$, $\Delta=0.5\ R_{-}$. One can see here the development of flow discontinuity.

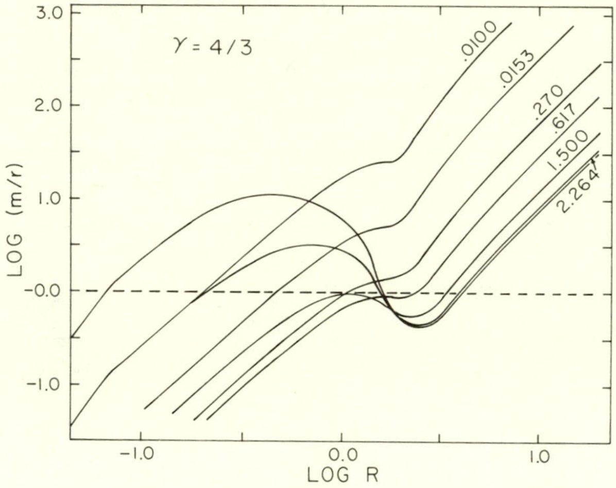

Figure 8 Log (m/τ) versus Log R, γ=4/3, $R_{-}=0.95 \cdot R_{max}$, $\Delta=0.5R_{-}$.

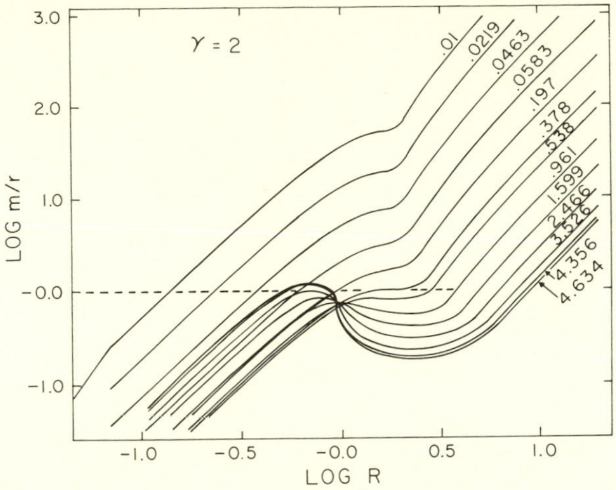

Figure 9 Log (m/γ) versus Log R. $\gamma=2$, $R_-=0.95\cdot R_{max}$, $\Delta=0.5\cdot R_-$.

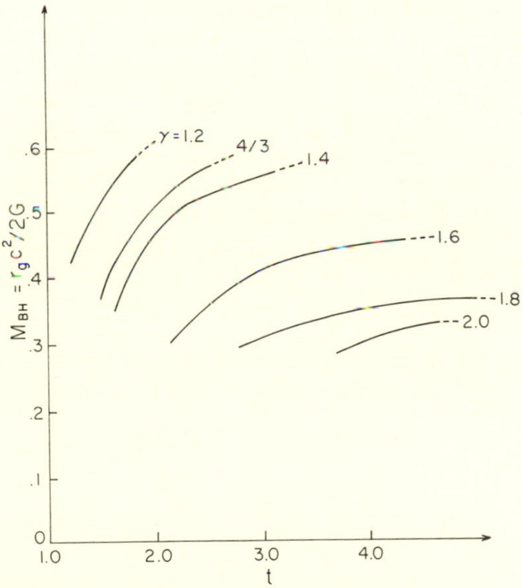

Figure 10 The dependence of the black hole mass m_{BH} on time t for various equations of state. $R_-=0.95\cdot R_{max}$, $\Delta=0.5\cdot R_-$.

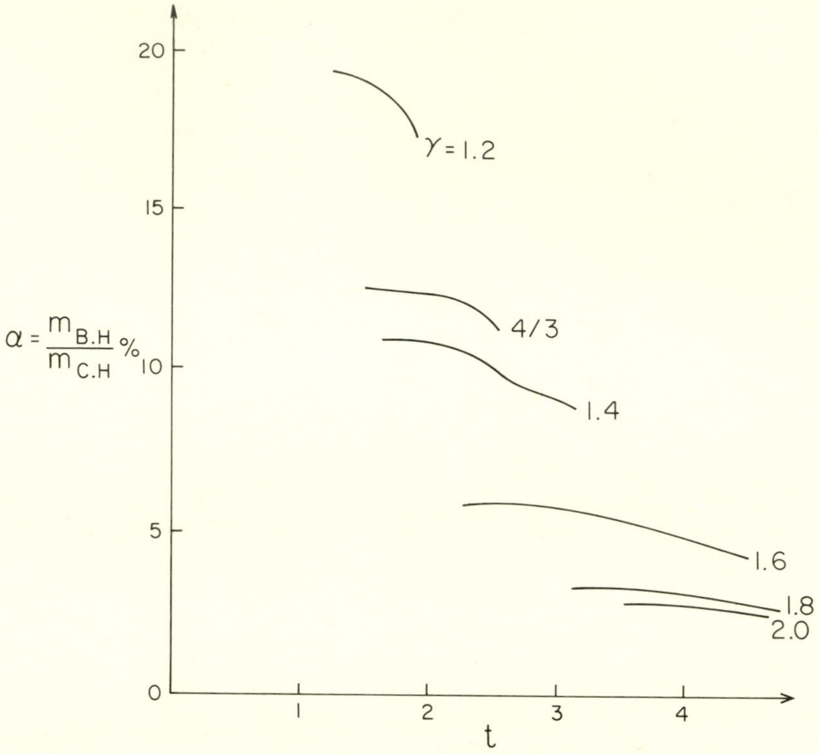

Figure 11 The dependence of the ratio $\alpha = \frac{m_{BH}}{m_{CH}}$ on time t for various equations of state, where m_{CH} is the mass of matter contained withing cosmological horizon.

MASSIVE BLACK HOLES AND GRAVITATIONAL RADIATION

R. D. Blandford

130-33 California Institute of Technology,
Pasadena, California 91125

INTRODUCTION

The search for black holes in galactic nuclei (and elsewhere) bears some superficial resemblance to the quest for extraterrestrial intelligence. It is difficult, potentially expensive, and controversial but the scientific rewards of success may be considerable. There is a further similarity in that it is virtually impossible to quantify the probability of discovery. That large gas masses always be able to avoid gravitational collapse and that we should be alone in the universe both seem to be very unlikely propositions, and yet if we try to set up the simplest type of rate equation we find that all our conclusions are predicated by blind guesses. On the optimistic side, almost every galaxy may contain in its nucleus a black hole of sizeable ($\gtrsim 10^6\ M_\odot$) mass and yet to be pessimistic only in the most active quasars and radio galaxies is there any evidence for extremely compact structure and even here the black hole interpretation is far from obligatory.

Unfortunately, unlike for the case of a distant civilization, a signal from the vicinity of a black hole is extremely difficult to recognize unambiguously. This of course is familiar from studies of galactic X-ray sources. Just as one cannot persuade a sceptic of the presence of a black hole in Cyg X-1, so we must expect that it will be hard to be confident that massive black holes are present in quasars or elsewhere.

In these two talks, I would like to fulfil three objectives: to describe in broad outline the physical picture that is emerging from observations of quasars and active galaxies, to present the case for the presence of massive black holes within these objects and elsewhere and finally to speculate upon the radiation of gravitational waves during black hole formation or interaction. I shall make no attempt to quantify the intensity and spectrum of the observed radiation.

R.D. BLANDFORD

OBSERVATIONS OF QUASARS AND ACTIVE GALACTIC NUCLEI

The observations and interpretation of quasars and active galaxies are fortunately well described in three recent conference proceedings (Hazard & Mitton, 1979; Ulfbeck, 1978 and Wolfe, 1978), and I shall only present a brief summary.

Taxonomy

There is, regrettably, a cumbersome and by now rather outmoded classification of extragalactic objects. Only when the various subsets can be clearly associated with particular physical characteristics does it seem likely that a sensible classification scheme, paralleling that in use for stars, will emerge. QUASARS are large redshift ($z \lesssim 3.5$) objects that appear star-like in optical appearance and generally have optical luminosities in excess of those associated with normal galaxies. A small fraction (roughly five percent are RADIO-LOUD and until recently most well-studied objects were in this class. The remainder are RADIO-QUIET. There are two types of radio source associated with quasars: flat spectrum COMPACT RADIO SOURCES ($\lesssim$ 10pc in size) and steep spectrum DOUBLE RADIO SOURCES ($\sim$ 30kpc to $\sim$6Mpc in size). Compact sources are frequently variable. Double sources show the characteristic twin radio lobe geometry associated with radio galaxies and almost always in addition a fainter, compact radio component associated with the optical identification (e.g. DeYoung, 1974).

Quasars generally show broad emission line spectra with the total widths of permitted lines like Hα of order $\Delta\lambda/\lambda \sim 0.03$. In addition the more distant objects with emission redshifts $z_e \gtrsim 1$ show narrow ($\Delta\lambda/\lambda \lesssim 3 \times 10^{-4}$) absorption lines identified with ground state transitions in a wide variety of ionization levels of the more abundant elements. A few objects are known to exhibit 0.21 m hydrogen absorption lines. The absorption redshift z_a is always less than $\sim z_e + 0.01$ and often less than $0.8z_e$. The origin of these lines is still undecided but the balance of current argument does seem to favour an extrinsic origin for the majority of them. However even if the lines are formed intrinsically in material ejected at high speed from the quasar, fairly strong arguments can be given to show that this material must be $\gtrsim$ 10 kpc from the continuum-producing region and therefore not directly related to the central powerhouse.

The optical continuum radiation is often observed to vary and in a few percent of cases (e.g. 3C 446, 279), the quasars are OPTICALLY VIOLENTLY VARIABLE and can change their flux levels by a factor of two in less than a week. OVV quasars can also display high ($\sim$ 10 percent) degrees of linear polarization which can on occasion vary even more rapidly. Most other quasars show $\lesssim$ 1 percent polarization (Stockman & Angel, 1978). Rapid variability and high polarization are also exhibited by another class called LACERTIDS after the eponymous variable 'star' BL Lacertae. These objects which are also variable flat spectrum radio sources are characterized by weak or undetectable emission lines. Several low redshift Lacertids appear under close examination to be surrounded by a faint optical nebulosity that seems to be a normal elliptical galaxy.

By contrast, SEYFERT GALAXIES (Seyfert, 1943) appear to be associated with spiral galaxies. They have bright nuclei and like the quasars have strong, broad emission lines, although they are typically only $\sim$ 0.01 as powerful as an average quasar. Seyferts are now subdivided into types 1 and 2 (Weedman, 1976). Type 1 Seyferts show both broad permitted lines and relatively narrow forbidden lines similar to quasars whereas type 2 Seyferts show only the narrow components. The spectra of type 2 Seyferts frequently exhibits evidence for dust and radio emission is more common in type 2 than type 1 (de Bruyn & Wilson, 1978).

Type 1 and type 2 Seyferts have a parallel amongst elliptical galaxies associated with radio sources in the two classes BROAD LINE RADIO GALAXIES and NARROW LINE RADIO GALAXIES respectively (Osterbrock, 1978). Many BLRG's are also N GALAXIES which are characterized by their brilliant star-like nuclei. N galaxies are generally somewhat less luminous than quasars and show sufficient nebulosity that the association with elliptical galaxies seems secure.

Radio Source Morphology

The typical extended double radio source shows two regions of strong, non-thermal, polarized emission, displaced typically $\sim$ 100 kpc on either side of the optical identification. However, the total linear sizes range all the way up to 6 Mpc (3C236; Willis, Strom & Wilson, 1975). The more powerful sources like Cygnus A (Hargrave & Ryle, 1974) are somewhat smaller and contain bright, apparently active hot spots in the radio lobes and on the basis of these features it was argued that energy had to be supplied continuously to the emitting regions probably directly from the nucleus of the associated galaxy or quasar. Support for this view has come with the discovery in a few closeby sources of radio jets (e.g. Turland, 1975; van Breughel & Miley, 1976; Waggett, Warner & Baldwin, 1977; Readhead, Cohen & Blandford, 1978) that presumably delineate the channel along which the supply occurs. In addition, observations using Very Long Baseline Interferometry have revealed the presence of parsec-size radio structure in the nuclei aligned with larger scale radio features.

A significant fraction of the most powerful compact radio sources have been shown by VLBI to display apparent 'superluminal' expansion. That is to say the sources are resolved into two or more components whose separation appears to increase at a rate faster than the speed of light, if we place these sources at the 'cosmological' distance indicated by their redshifts (Cohen *et al.*, 1977). This need not be incompatible with special relativity and several distinct explanations of this phenomenon have been proposed (reviewed in Blandford, McKee & Rees, 1977). It seems most likely that the explanation is linked to the existence of small scale radio jets and that the emitting regions move at almost the speed of light. From observations of the linear polarisation of some compact sources it is possible to infer that the plasma in the radio-emitting regions of rapidly variable sources must have a relativistic or nearly relativistic equation of state (Wardle, 1978). This is also consistent with relativistic expansion.

X-ray Observations

To date, 19 type 1 Seyferts, 3 quasars and 2 Lacertids and several active galaxies have been found to be powerful X-ray sources with luminosities up to 2×10^{38} W. In several objects, most of the radiated power is produced at X-ray energies although it is not yet clear whether or not this is by thermal or non-thermal processes. HEAO-B should detect many more quasars.

Rapid Variability

Barring relativistic effects and special geometries, the observation of a significant change of flux in a time t_{var} limits the size of the emitting region to $\lesssim ct_{var}$. The shortest variability time scales that have been reported are at optical and X-ray energies and range from hours to days (Véron, 1975; Martin, Angel & Maza, 1976; Tananbaum *et al.*, 1978). One light day is only $\sim$ 100 Schwarzschild radii for 10^8 $M_\odot$, which is about the smallest mass invoked in quasar models.

Dynamical Evidence

Recently Young *et al.* (1978) and Sargent *et al.* (1978) have carried out a high resolution photometric and spectroscopic investigation of the nearby elliptical galaxy M87 (famous for its optical jet). They have shown that in the central $\sim$ 200 pc of the nucleus, there is a cusp in the light distribution and that the velocity dispersion of the stars appears to increase towards the centre of this cusp. In most other elliptical galaxies, the surface brightness and velocity dispersion flatten off in a way that can be explained by a simple 'isothermal sphere' stellar model. The conclusion that has been drawn is that there is a large ($\sim 3 \times 10^9 M_\odot$), comparatively dark mass present in the central 100 pc with a 'mass to light' ratio more than 10 times that of the galaxy and that this is responsible for the stellar density cusp. Of course this is a long way from resolving a black hole and there is no objection of principle to associating this mass with say a dense cluster of unmagnetised neutron stars. However there is also a compact, unresolved radio source in the nucleus (Kellerman *et al.*, 1973) and this has a linear size $\lesssim 0.03$ pc which again is ~ 100 Schwarzschild radii for $3 \times 10^9 M_\odot$. Improved resolution may be achieved by searching for interstellar radio scintillations.

The nucleus of our galaxy exhibits evidence for much non-stellar activity and in particular another compact radio source, this time with a linear size of $\sim 10^{12}$ m has been detected (Kellerman et al. 1977). However there is also a dynamical upper limit of $\sim 4 \times 10^6 M_\odot$ on the mass of a putative central hole (Wollman et al. 1977) based on the observed width of an infra-red line seen within the central 0.1 pc. Once again there is a factor ~ 100 relating the suggested hole size to the linear resolution.

Evolutionary Considerations

Quasars and strong radio sources evolve quite dramatically with redshift. There are roughly a thousand times as many objects per unit co-moving volume at $z \sim 2$. It is probably too soon to know how far back in time this evolution continues although some authors have argued that quasar activity peaks at $z \sim 2.5$ (e.g. Schmidt, 1972).

For the purposes of computing possible gravitational wave event rates we list in Table 1 a set of estimates of the local space density of various types of extragalactic objects together with a description of their apparent density and luminosity evolution. Several features should be noted.

i) Quasars associated with flat spectrum compact radio sources appear to evolve less rapidly with redshift than those associated with steep spectrum, extended sources (Schmidt, 1976).

ii) The strong density evolution apparent in quasars and radio sources is consistent with the view that most of the nearby bright galaxies were once quasars.

iii) Roughly two percent of spirals are Seyfert galaxies now and of order five percent of giant ellipticals are strong radio galaxies. On the assumption that every galaxy has an active phase, this allows an estimate of $\sim 10^{8-9}$ yr for the duration of non-thermal activity within a galaxy.

iv) Only a small fraction of optically selected quasars ($\sim$ five percent) are radio sources.

THEORETICAL MODELS OF QUASARS AND ACTIVE GALACTIC NUCLEI

There is no generally accepted physical model of quasars although there is an emerging consensus that massive black hole models first proposed by Zel'dovich and Salpeter in 1964 are

Table 1. Local space densities ϕ_0 of extragalactic objects (after Schmidt, 1972, 1978; Huchra & Sargent, 1973). Density evolution is described by assuming that the co-moving density obeys a law $\phi \propto e^{c\tau}$, where τ is the 'look-back' time expressed as a fraction of the age of the universe. For a low-density, Friedmann universe, $\tau \sim z/(1+z)$. Most of the entries in this table are extremely rough estimates.

Object	Luminosity (W)	ϕ_0 (Gpc^{-3})	c
Radio quiet	10^{41} (optical)	0.02	10
Quasar	10^{40} (")	1	10
	10^{39} (")	70	10
Giant Elliptical Galaxies	10^{38} (")	1.5×10^4	-
Seyfert Galaxies	10^{38} (")	100	-
	10^{37} (")	3×10^4	-
Field Galaxies	10^{37} (")	3×10^6	-
	10^{36} (")	2×10^7	-
	10^{35} (")	10^8	-
Quasars with compact radio sources	10^{41} (radio)	0.01	3
	10^{40} (")	0.05	3
	10^{39} (")	0.3	3
Radio Galaxies	10^{38} (")	0.03	20
	10^{37} (")	10	8
	10^{36} (")	300	4
	10^{35} (")	2000	1

best able to account for the data. The arguments for this change of emphasis away from alternative 'star cluster' (e.g. Colgate, 1977; Arons et al., 1975) and 'superstar' (e.g. Hoyle & Fowler, 1963; Ozernoi & Usov, 1973) models are not very convincing. They include:-

(i) apparent absence of evidence for a quantum of outburst energy $\sim 10^{52}$ erg such as might be associated with the destruction of a stellar mass object,

(ii) the unstable and possibly violently explosive character of massive ($\gtrsim 10^6 M_\odot$) stars,

(iii) rapid variability in some objects,

(iv) prevalence of linear radio structure suggesting the presence of a good gyroscope that can point in the same direction in space for the lifetime $\sim 10^8$ yr of a strong radio source and yet is able to achieve and maintain the bifurcation of the energy supply within the central few parsecs of the nucleus.

It is most natural to associate the fixed direction of the radio structure with the spin axis of the hole. In addition, as originally discussed by Bardeen & Petterson (1975) any material that possesses sufficient angular momentum to form an accretion disc around the hole (Lynden-Bell, 1969) will, as a consequence of the Lense-Thirring precession, fall into the equatorial plane when the precession period becomes comparable with the radial infall time, typically at 100 Schwarzschild radii, and this large surface can further help define the collimation. In fact in order to channel a beam into a tight cone of opening angle of order a few degrees, the jet emerging along the spin axis must presumably still be in pressure equilibrium with a surrounding gas cloud at quite large distances from the hole.

The most compelling argument for a black hole model, which we expand further below, is that the inevitable endpoint of most other quasar models is a central black hole and that the radiative efficiency of this black hole phase is likely to be much greater than that of earlier phases (Begelman & Rees, 1978). Provided that the holes can be fuelled at an appreciable rate, black hole objects should dominate any observed sample of active nuclei.

How is the binding energy of the accreting gas liberated near the horizon? One thing that is very certain is that it is not liberated as black-body radiation. The black-body temperature required to radiate $\sim 10^{39}$ W from a disc about 6 Schwarzschild radii across around a $10^8 M_\odot$ hole is $\sim 3 \times 10^5$K, much less than the brightness temperature of the optical and infrared radiation. One promising possibility generalising upon models for Cygnus X-1 is to rely on gas dynamical process to transport a large energy flux into a hot corona and there upscatter the photons in energy using the hot electrons (e.g. Lightman, Giacconi & Tananbaum, 1978). Another idea is to rely on electromagnetic or hydromagnetic torques to extract energy and angular momentum from the disc and the hole (e.g. Blandford, 1976; Lovelace, 1976; Shields & Wheeler, 1976; Blandford & Znajek, 1977). This requires that there be a component of field normal to the disc presumably convected inwards by the accreting material. In both of these types of model, the net result may be a supersonic and possibly relativistic wind directed parallel to the spin axis. For sources that do not exhibit a preferred axis, dissipation of chaotic gas motions through turbulence and shocks has been proposed (Fabian et al. 1976; Meszaros & Silk, 1977).

All of these dissipation mechanisms can lead to the production of relativistic electrons which are needed to account for the synchrotron emission believed to be seen in polarised OVV objects. In these sources, it can be shown that the cooling lengths of the radiating electrons are smaller than the source sizes. This therefore implies that energy must be transported across the emitting region in a form suitable for efficient particle acceleration (Hoyle, Burbidge & Sargent, 1966).

A critical limiting accretion rate is given by assuming that the Eddington luminosity ($4\pi G M_H m_p c/\sigma_T$) is radiated with a typical efficiency of 0.1 (Salpeter, 1964). This gives

$$\dot{M}_H \sim 2\, M_{H8}\, M_{\odot} yr^{-1} \qquad (1)$$

and a corresponding growth time for the hole (mass M_H) of $\sim 5 \times 10^7$ yr, independent of mass. This is gratifyingly close to the lifetimes estimated above on observational grounds. After the black hole has grown by digesting all the fuel left over from the original nucleus, it will subsequently radiate at a power limited by the supply of gas. If M87 is any guide (in which the integrated radiated power is $\sim 10^{-5}$ of the Eddington luminosity) then this can be extremely sub-critical and maybe sources like this are dead or dying quasars. This gas can be supplied by normal stellar processes in the surrounding galaxy (Mathews & Baker, 1971; Gisler, 1976), stellar collisions and tidal disruption in a density cusp that may form around the hole (e.g. Frank, 1978 and references therein) or a passing galaxy (e.g. Gunn, 1978).

There is an important caveat that applies to all of these models. Most of the evidence is drawn from observations of the one or two percent of the most active and interesting objects. It is a definite assumption that all quasars contain a common powerhouse able to reproduce the features so far observed in only a small minority.

EVOLUTION OF GALACTIC NUCLEI AND THE FORMATION OF MASSIVE BLACK HOLES

A newly formed galaxy is likely to develop a dense central star cluster or gas cloud in its nucleus that may be able to evolve towards a black hole as a result of combined gravitational and dissipative processes. Various stages in the evolution of the nucleus have, as we mentioned in the previous section, been associated with the quasar phenomenon. The question that we must address is what is the likelihood of radiating a significant fraction of the black hole mass as gravitational waves? There are two opportunities for doing this effectively - during the formation of the hole if this is sufficienctly violent and during the coalescence of two massive holes.

Ideas about evolution of a galactic nucleus have been reviewed by Saslaw (1973) and Begelman & Rees (1978). I reproduce in Fig. 1 a 'metabolic pathway' taken from Rees (1978) that describes some of the many options open to a galactic nucleus. Provided that stars are formed, the cluster will initially relax in a time of order, the 'central reference time' (Spitzer & Schwarzschild, 1951)

$$t_R \simeq 8 \times 10^5 \, N^{\frac{1}{2}} \left(\frac{R}{1\text{pc}}\right)^{3/2} \left(\frac{m}{1M_\odot}\right)^{-\frac{1}{2}} \log^{-1}(N/2) \text{ yr.} \qquad (2)$$

Here N is the number of stars of mass m in a cluster of radius R and velocity disperion v. (This is $\sim$ N/logN times the time for a star to cross the cluster.) Initially a cluster will contract by evaporating stars with essentially zero total energy on an evaporation time scale $\sim 90t_R$ (Spitzer & Thuan, 1972) so that $N \propto R^{1/2} \propto v^{-2}$. This will continue until the density is so high that stellar collisions are important which happens when the random velocities become comparable with the stellar surface escape velocities ($\sim$ 600 km s^{-1} for the sun). Then either coalescence (Colgate, 1967) or disruption (Spitzer & Stone, 1967) becomes important. Coalescence leads to massive stars, supernovae and compact remnants which may be able to sink towards the centre if they are magnetised and there is enough interstellar gas, (Bisnovatyi-Kogan & Sunyaev, 1972). Disruption will probably lead to the formation of a dense rotating central gas cloud in which further star formation can take place. Only if this occurs is it possible to produce large pulses of radiation when the central cloud collapses to form a hole. (The collapse could in principle be initiated by a single stellar mass black hole accreting super-critically (Begelman, 1978).) Probably the best yield of radiation will occur if a barmode instability develops into two comparable mass holes which then radiate away their binding energy liberating a few percent of their rest mass as gravitational waves. Fragmentation into more than two components may lead to ejection from the nucleus (cf. Saslaw, Valtonen & Aarseth, 1974).

It is interesting that when the holes are produced from both single and binary systems as well, they may be able to recoil from the system by radiating a significant fraction $\gtrsim$ of their energy (in units of c) as linear momentum (Bekenstein, 1973; Rees, private communication). This may allow the hole to escape from the central potential well. Thus there is clearly a danger that the apparent presence of black holes in the centres of galaxies

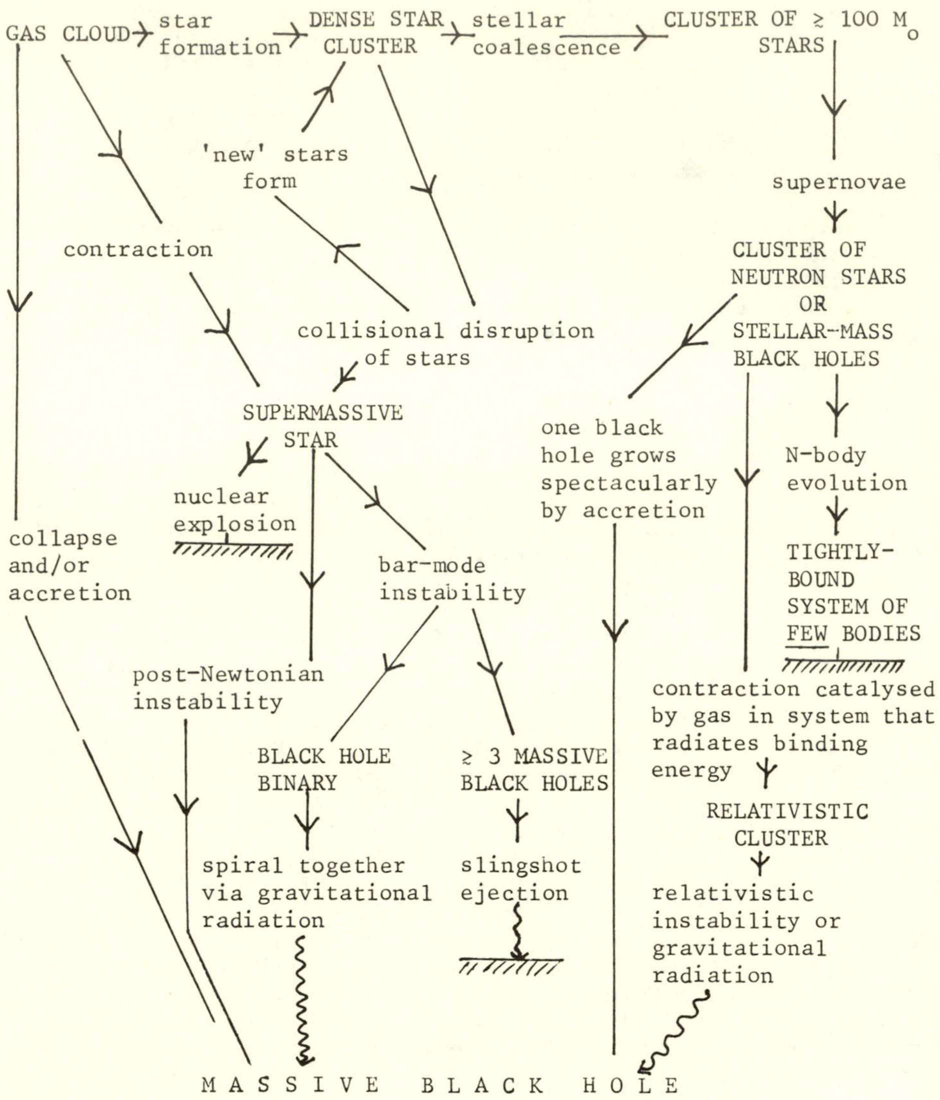

Fig.1. How a black hole may form in a galactic nucleus (from Rees 1978). Many of the stages in these different possible evolutions may be associated with violent activity but as long as there is an adequate supply of fuel, the black hole stage should be the most luminous. Black holes that form by the three routes (⁓⁓⁓›) are most likely to generate gravitational radiation efficiently.

rules out the possibility of efficient radiation production usually occurring during the hole formation. Better estimates of recoil velocities from post-Newtonian and fully non-linear numerical calculations should settle this matter.

If instead the hole grows from, say, a single 10 $M_\odot$ hole by accretion and capture of surrounding gas and stars then clearly there is no prospect of detecting pulses of radiation from quasars. The feasibility of this type of model is a problem that has attracted much theoretical interest in recent years (e.g. Young, 1977; Frank, 1978). (It is interesting as pointed out by Hills (1975) that solar mass stars can be swallowed whole by holes of mass $\gtrsim 10^8\ M_\odot$, and that this may provide a mechanism for stopping quasar activity after $\sim 10^8$ yr.) At present it seems doubtful that black holes massive enough to power the brightest quasars can grow from small seeds within a Hubble time solely by swallowing stars and stellar debris.

A more promising possibility is to build up the holes by agglomeration of fewer, larger fragments. Globular clusters mass M, may be able to spiral inwards under dynamical friction from a radius r with background density ρ on a timescale

$$t_{df} \sim 3 \times 10^9 \left(\frac{\rho}{1M_\odot pc^{-3}}\right)^{\frac{1}{2}} \left(\frac{r}{1kpc}\right)^3 \left(\frac{M}{10^6 M_\odot}\right)^{-1} \text{yr.} \qquad (3)$$

(Tremaine, Ostriker & Spitzer, 1975; Tremaine, 1976) where we have assumed an isothermal sphere distribution of background stars. It has been suggested that many globular clusters may contain 100-1000 $M_\odot$ holes in their cores (e.g. Bahcall & Ostriker, 1975; Lightman & Fall, 1977) because the characteristic relaxation times for many clusters are much less than the age of the universe. However, in the absence of suitable dissipation, only a small fraction of the initial cluster mass can form the black hole because the binding energy so released can evaporate most of the remainder of the cluster. In addition only clusters of mass $M \gtrsim 3 \times 10^6\ (r/1\ \text{kpc})^2\ M_\odot$ can spiral in from radius r in a Hubble time. This therefore does not seem to be a very efficient way of supplying large black holes to the nucleus, although as ground-based laser systems are most sensitive to waves from 100-1000 $M_\odot$ objects, somewhat less mass is needed in this form to ensure detectability.

A larger scale version of this process can occur in clusters and groups of galaxies. Large central giant elliptical galaxies can 'cannibalize' satellite galaxies as they spiral inwards, again under dynamical friction in the background medium (Ostriker & Tremaine, 1975; Ostriker & Hausman, 1977; White, 1978). From the expression for t_{df}, we see that larger mass galaxies will be preferentially ingested and if these too contain $10^8\ M_\odot$ holes it is of interest to ask if they can collect in the giant galaxy's nucleus? In general it seems that the answer is yes because the central core densities of the small galaxies are generally larger than the central densities of the cannibal, and tidal stripping to the bare central black hole should not occur until the nucleus is reached. From here the hole can spiral inwards on a comparatively short timescale. After the captured hole (mass M_c) has spiralled into a distance $r_H \sim (M_H/\rho)^{1/3}$ of the original (presumably larger) hole, it will be effectively bound in a binary orbit. However the orbit must shrink to a radius

$$r_{GR} \sim 10^{15}\left(\frac{M_H}{10^8 M_\odot}\right)^{1/4}\left(\frac{M_C}{10^8 M_\odot}\right)^{1/4}\left(\frac{M_H + M_C}{10^8 M_\odot}\right)^{1/4} \text{ m} \quad (4)$$

from where it can contract under gravitational radiation reaction in a reasonable fraction of the Hubble time. If $r_H > r_{GR}$, then the process of dynamical friction will be modified for $r_H \gtrsim r \gtrsim r_{GR}$, as the holes will orbit with a velocity in excess of the central stellar velocity dispersion. The binary will only lose energy to those stars that come within $\sim r$ and instead of giving them small deflections, the holes will generally impart sufficient velocity to escape from the cluster. Small angle diffusion in the outer parts of the cluster will repopulate the loss cone and simple estimates suggest that there is no problem in getting down to a binary radius $\lesssim r_{GR}$ and thus ensuring coalescence of the two holes.

A typical massive elliptical may eat several smaller galaxies during its lifetime, especially if massive halos (e.g. Ostriker, Peebles & Yahil, 1974) are a common feature of galaxies. Cannibalism may also be important in small groups where the galaxy velocity dispersions are somewhat lower than in clusters and capture is consequently easier.

FREQUENCY OF GRAVITATIONAL WAVE BURSTS FROM MASSIVE BLACK HOLES

As I emphasised in the Introduction, it is impossible, granted present understanding, to provide a reliable estimate of the rate of reception of gravitational wave pulses of a given amplitude and duration. Nevertheless it is important to try to answer this question in principle, because if these pulses are ever detected, they would clearly be telling us something quantitative about the births and lives of quasars.

This problem has been addressed by Thorne & Braginsky (1976) whose treatment I modify. Let us assume that every one of a set of once active objects of present density ϕ_o^H forms a black hole of mass M_H and gravitational wave energy release $\varepsilon_H M_H c^2$ and that in its subsequent evolution it coalesces with N captured holes of mass M_c and energy release $\varepsilon_c M_c c^2$. The presently observed event rate becomes $E = 4\pi(R^2\phi_o^H\ (1 + N))c$ where R is the circumferential radius (Misner, Thorne & Wheeler, 1973) or proper motion distance (Weinberg, 1972). For the low density ($\Omega_0 \sim 0.1$), older ($H_0 \sim 60$ km s^{-1} Mpc^{-1}) universe currently favoured by many cosmologists, $R \sim 5z(1+0.45z)/(1+z)$Gpc for redshifts $z \lesssim 5$. The corresponding age of the universe at the time of emission is $t_e \sim 17/(1+z)$Gyr.

The gravitational pulses will be redshifted and, as the luminosity distance is (1+z)R, the mean observed values of the dimensionless amplitude h_0 and frequency ν_0 will be given by

$$h_0 \sim 2 \times 10^{-15} \ (\varepsilon/0.1)^{\frac{1}{2}} (M_c/10^8 M_\odot)(R/1 \ \mathrm{Gpc})^{-1}$$

$$\nu_0 \sim 100(M_H/10^8 M_\odot)^{-1} \ (1+z)^{-1} \ \mu\mathrm{Hz}$$

(e.g. Thorne, 1978).

Optimistic estimates of detectable amplitudes from second generation Doppler tracking experiments give $h_0 \simeq 10^{-16}$ in this frequency range for pulses. We do not know the mass function ϕ_0^H (M) but it is reasonable to suppose that the event rate is dominated by fairly massive holes that are perhaps associated with the more luminous of galaxies. For illustrative purposes, we can assume that $\varepsilon \sim 0.1$, $N \sim 5$, $M_H \sim M_c \sim 10^8 M_\odot$, $h_0 \sim 2 \times 10^{-16}$. Then we can see out beyond $z \sim 2$ and in order to get an event rate greater than 1 yr^{-1} we need $\phi_0^H \gtrsim 10^6$ which from Table 1 is just possible if every spiral and large elliptical with $L \gtrsim 10^{37}$ W was at one stage a Seyfert or quasar. As concluded by Thorne & Braginsky there is therefore some chance that these sources can be detected, although to do so generally necessitates making fairly optimistic assumptions about the source efficiencies and event frequency.

A possible contraint on the burst frequency can arise if we expect a substantial electromagnetic pulse to accompany the gravitational pulse (e.g. Eichler & Solinger, 1976). One way by which this may be achieved is through shock heating of the surrounding gas by the strong, time-dependent tidal forces associated with the formation of the hole. Indeed, the coalescence of the holes may be controlled by gas-dynamical dissipation rather than gravitational radiation if the mass of the gas is high enough. The total power associated with a pulse from a hole forming with $\varepsilon \sim 0.1$ is ~ 0.03 $c^5/G \sim 10^{51}$ W, independent of mass. It is unlikely that more than 10^{-6} of this power can have been radiated at optical wavelengths more than once every ten years say, out to a redshift $z \sim 1$ from a hole of mass $\gtrsim 10^7 M_\odot$, without it having been noticed. More careful scrutiny of the whole electromagnetic frequency-pulse duration plane for bursts seems as well-motivated as the quest for gravitational waves.

Of course failure to observe these electromagnetic effects does not preclude the possible existence of detectable gravitational pulses. Even if a large fraction of the released energy is deposited as heat in the surrounding gas, and this gas can then radiate, the optical depth may be so large that the gas must expand by a large factor until the radiation can escape. As an example, $10^8 M_\odot$ of gas in a galactic nucleus must fill a region of size $\sim$ 100pc before it becomes optically thin to Thomson scattering. The 'pulse' would then be lengthened to $\gtrsim$ 1000 year, and the radiative efficiency would depend on how well the bulk kinetic energy of the expanding ejecta could be converted into photons at this distance.

PREGALACTIC BLACK HOLES

It is also possible that much of the matter in the universe, in particular the dark mass required to bind clusters, be in the form of massive black holes. If these holes formed, or if they somehow combined sufficiently recently, then in principle it is possible for their gravitational radiation to be detected. Of course this requires rather specialized and contrived conditions, especially for the large mass holes considered here. (A more convincing case can be made for the existence of 1-100 $M_\odot$ holes and if, for example, many of these are made in close binary pairs then a significant fraction should spiral together in a Hubble time.)

There are however some contraints that exist on the density of massive black holes (e.g. Rees, 1977; Carr, 1978; Bertotti & Carr, 1978 and references therein). The most reliable is derived from the apparent absence of gravitational lens images of distant quasars (Press & Gunn, 1973). From optical observations, we can probably conclude that no more than 0.02 of the critical density is contained in holes of mass $\sim 10^{12} - 10^{14}\ M_0$. Dynamical (e.g. van den Bergh, 1969; Carr, 1978) and radiative (e.g. Dahlbacka et al., 1974; Carr, 1978; Ipser & Price, 1977) arguments are somewhat less reliable but can in principle impose similar limits on a wider mass range.

It is an interesting question to ask whether or not pulses or a continuous background of gravitational radiation will be seen. Let us suppose that a population of holes of mass M_H is formed at a redshift z_f, liberating a fraction ε of their final rest mass as pulses of duration $\tau \sim 3\ GM_H/c^3$. Then these pulses will overlap if the product $E\tau$ exceeds unity. Equivalently for $z_f \gg \Omega_0^{-1}$, this requires that the present density of holes (in units of the critical density) satisfy $\Omega_H \gtrsim 0.05\ \Omega_0^2\ z_f^{-1}$, independent of mass. The corresponding dimensionless amplitude h_0 in a broad-band interval around $\nu_0 \sim 10(M_H/1\ M_\odot)\ z_f^{-1}$ kHz is $h_0 \sim H_0 \nu_0^{-1}(\varepsilon\Omega_H/z_f)^{\frac{1}{2}}$. It should be easier to detect the presence of a universal background than individual pulses simply by integrating the signal over a long period of time. For illustration purposes, if $\sim 10^{-3}$ of the mass of the universe forms $10^5 M_\odot$ black holes, corresponding to the

Jeans' mass (Peebles & Dicke, 1968), at recombination ($z_f \sim 10^4$), and $\varepsilon \sim 0.1$, then a continuous background will be produced with rms amplitude $h_0 \simeq 10^{-17}$ at a frequency $\nu_0 \sim 10$ μHz. Again, it is with higher frequencies associated with smaller holes that a better case for a continuous background can be made.

CONCLUSION

In summary, the case for the existence of massive black holes is being developed but is not as yet strong and may never be well-founded observationally. The possibility of detecting gravitation wave pulses associated with the formation and interaction of these holes certainly exists but only if fairly hopeful assumptions are made about the source density, wave generation efficiency and the development of the appropriate Doppler tracking techniques. This is clearly not a field for pessimists.

ACKNOWLEDGEMENTS

I thank many colleagues especially J. Baldwin, B. Carr, C. Caves, J. Ostriker, M. Rees and K. Thorne for informative discussions during the preparation of these talks and their comments on the manuscript. I also acknowledge support from a National Science Foundation grant (AST 76 20375).

REFERENCES

Arons, J., Kulsrud, R.M. & Ostriker, J.P. (1975). A multiple pulsar model for quasi-stellar objects and active galactic nuclei. Astrophys. J., 198, pp.687-707.

Bahcall, J.N. & Ostriker, J.P. (1975). Massive black holes in globular clusters? Nature, Lond., 256, pp.23-24.

Bardeen, J.M. & Petterson, J.A. (1975). The Lense-Thirring effect and accretion disks around Kerr black holes. Astrophys.J.Lett., 195, L65-67.

Bekenstein, J.D. (1973). Gravitational-radiation recoil and runaway black holes. Astrophys. J., 183, pp.657-664.

Begelman, M.C. (1978). Black holes in radiation-dominated gas: an analogue of the Bondi accretion problem. Mon. Not. R. astr. Soc., 184, pp.53-67.

Begelman, M.C. & Rees, M.J. (1978). The fate of dense stellar systems. Mon. Not. R. astr. Soc., in press.

Bertotti, B. & Carr, B.J. (1978). On the detectability of a cosmological background of gravitational radiation. Preprint.

Bisnovatyi-Kogan, G.S. & Sunyaev, R.A. (1972). Quasars and the nuclei of galaxies: A single object or a star cluster? Soviet Astron. A. J., 16, pp. 201-208.

Blandford, R.D. (1976). Accretion disc electrodynamics - a model for double radio sources. Mon. Not. R. astr. Soc., 176, pp. 465-481.

Blandford, R.D., McKee, C.F. & Rees, M.J. (1977). Super-luminal expansion in extragalactic radio sources. Nature, Lond., 267, pp. 211-216.

Blandford, R.D. & Znajek, R.L. (1977). Electromagnetic extraction of energy from Kerr black holes. Mon. Not. R. astr. Soc., 179, pp. 433-456.

Carr, B.J. (1975). The primordial black hole mass spectrum. Astrophys. J., 201, pp. 1-19.

Carr, B.J. (1978). On the Cosmological Density of Black Holes. Comments Astrophys. Space Sci., 7, pp. 161-173.

Cohen, M.H. et al. (1977). Radio sources with superluminal velocities. Nature, Lond., 268, pp.405-409.

Colgate, S.A. (1967). Stellar coalescence and the multiple supernova interpretation of quasi-stellar sources. Astrophys. J., 150, pp. 163-192.

Dahlbacka, G.H., Chapline, G.F. & Weaver, J.A. (1974). Gamma rays from black holes. Nature, Lond., 250, pp.36-37.

de Bruyn, A.G. & Wilson, A.S. (1978). The Radio Properties of Seyfert Galaxies. Astron. Astrophys., 64, pp. 433-444.

De Young, D.S. (1976). Extended extragalactic radio sources. Ann. Rev. Astron. Astrophys., 14, pp. 447-474.

Eichler, D. & Solinger, A. (1976). The electromagnetic background: limitations on models of unseen matter. Astrophys. J., 203, pp. 1-5.

Fabian, A.C., Maccagni, D., Rees, M.J. & Stoeger, W.R. (1976). The nucleus of Centaurus A. Nature, Lond., 260, pp.683-685.

Frank, J. (1978). Tidal disruption by a massive black hole and collisions in galactic nuclei. Mon. Not. R. astr. Soc., 184, pp.87-99.

MASSIVE BLACK HOLES

Gisler, G.R. (1976). The Fate of Gas in Elliptical Galaxies and the Density Evolution of Radio Sources. Astron. Astrophys., 51, pp.137-150.

Gunn, J.E. (1978). Feeding the monster. In Quasars and Active Nuclei, ed. C. Hazard and S. Mitton. Cambridge Univ. Press.

Hargrave, P.J. & Ryle, M. (1974). Observations of Cygnus A with the 5-km radio telescope. Mon. Not. R. astr. Soc., 166, pp. 305-327.

Hazard, C. & Mitton, S.A. (1978). Eds. Quasars and Active Nuclei. Cambridge University Press.

Hills, J.G. (1975). Possible power source of Seyfert galaxies and QSOs. Nature, Lond., 254, pp.295-298.

Hoyle, F., Burbidge, G.R. & Sargent, W.L.W. (1966). On the nature of the quasi-stellar sources. Nature, Lond., 209, pp.751-753.

Hoyle, F. & Fowler, W.A. (1963). Nature of strong radio sources. Nature, Lond., 197, pp.533-535.

Huchra, J. & Sargent, W.L.W. (1973). The space density of the Markarian Galaxies including a region of the south galactic hemisphere. Astrophys. J., 186, pp.433-443.

Ipser, J.R. & Price, R.M. (1977). Accretion onto pregalactic black holes. Astrophys. J., 216, pp.578-590.

Kellerman, K.I. et al. (1973). Absence of variations in the nucleus of Virgo A. Astrophys. J. Lett., 179, pp.141-144.

Kellerman, K.I. et al. (1977). The small radio source at the galactic center. Astrophys. J. Lett., 214, pp.61-62.

Lightman, A.P. & Fall, M. (1978). An approximate theory for the core collapse of two-component gravitating systems. Astrophys. J., 221, pp.567.579.

Lightman, A.P., Giacconi, R. & Tananbaum, J. (1978). X-ray flares in NGC 4151: A thermal model and constraints on a central black hole. Astrophys. J., 221, pp.567-579.

Lovelace, R.V.E. (1976). Dynamo model of double radio sources. Nature, Lond., 262, pp. 649-652.

Lynden-Bell, D. (1969). Galactic Nuclei as Collapsed Old Quasars. Nature, Lond., 223, pp.690-694.

Martin, P.G., Angel, J.R.P. & Maza, J. (1976). Night-to-night variations in the optical polarisation of the nucleus of NGC 1275. Astrophys. J. Lett., 209, pp. 21-23.

Mathews, W.G. & Baker, J.C. (1971). Galactic winds. Astrophys. J., 170, pp.241-259.

Mészáros, P. & Silk, J.I. (1977). X-rays from Massive Black Holes. Astron. Astrophys., 55, pp.289-293.

Misner, C.W., Thorne, K.S. & Wheeler, J.A. (1973). Gravitation. W.H. Freeman, San Francisco.

Osterbrock, D.E. (1978). Optical Emission-Line Spectra of Seyfert Galaxies and Radio Galaxies. Phys. Scripta, 17, pp.137-143.

Ostriker, J.P. & Hausman, M.A. (1977). Cannibalism among the galaxies: Dynamically produced evolution of cluster luminosity functions. Astrophys. J. Lett., 217, pp.125-129.

Ostriker, J.P., Peebles, P.J.E. & Yahil, A. (1974). The size and mass of galaxies, and the mass of the universe. Astrophys. J. Lett., 193, pp.1-4.

Ostriker, J.P. & Tremaine, S.D. (1975). Another evolutionary correction to the luminosity of giant galaxies. Astrophys. J. Lett., 202, pp. 113-117.

Ozernoi, L.M. & Usov, V.V. (1973). The supermassive oblique rotator: electrodynamics, evolution, observational tests. Astrophys. Space Sci., 25, pp.149-194.

Peebles, P.J.E. & Dicke, R.H. (1968). Origin of the globular star cluster. Astrophys. J., 154, pp.891-908.

Press, W.H. & Gunn, J.E. (1973). Method for detecting a cosmological density of condensed objects. Astrophys. J., 185, pp.397-41

Readhead, A.C.S., Cohen, M.H. & Blandford, R.D. (1978). A jet in the nucleus of NGC 6251. Nature, Lond., 272, pp.131-138.

Rees, M.J. (1977). In Experimental Gravitation, pp. 423-429. (Academia Nazionale dei Lincei, Rome).

Rees, M.J. (1978). In Structure and Properties of Nearby Galaxies. Eds. Berkhuijsen & Wiekbinskii. Reidel, Dordrecht.

Salpeter, E.E. (1964). Accretion of interstellar matter by massive objects. Astrophys. J., 140, pp.796-800.

Sargent, W.L.W. et al. (1978). Dynamical evidence for a central mass concentration in the galaxy M87. Astrophys. J., 221, pp. 731-744.

Saslaw, W.C. (1973). The dynamics of dense stellar systems. Publ. Astr. Soc. Pacific, 85, pp.5-23.

Saslaw, W.C., Valtonen, M. & Aarseth, S.J. (1974). The gravitational slingshot and the structure of extragalactic radio sources. Astrophys.J., 190, pp.253-270.

Schmidt, M. (1972). Statistical Studies of the evolution of extragalactic radio sources. II. Radio galaxies. Astrophys. J., 176, pp.289-301.

Schmidt, M. (1976). On the apparent absence of evolution of quasi-stellar radio sources with flat radio spectra. Astrophys. J. Lett., 209, pp.55-56.

Schmidt, M. (1978). The Local Space Density of Quasars and Active Nuclei. Phys. Scripta, 17, pp.135-136.

Seyfert, C. (1943). Nuclear emission in spiral nebulae. Astrophys. J., 97, pp.28-40.

Shields, G.A. & Wheeler, J.C. (1976). Magnetised Accretion Disks and the radio outbursts of 3C120 and Cygnus X-3. Astrophys. Lett., 17, pp.69-76.

Spitzer, L., Jr. & Schwarzschild, M. (1951). The possible influence of interstellar clouds on stellar velocities. Astrophys. J., 114, pp. 385-397.

Spitzer, L., Jr. & Stone, M.E. (1967). On the evolution of galactic nuclei. II. Astrophys. J., 147, pp.519-528.

Spitzer, L., Jr. & Thuan, T.X. (1972). Random gravitational encounters and the evolution of spherical systems. IV. Isolated systems of identical stars. Astrophys. J., 175, pp. 31-61.

Stockman, H.S. & Angel, J.R.P. (1978). A linear polarisation survey of bright QSOs. Astrophys. J. Lett., 220, pp.67-71.

Tananbaum, H. et al. (1978). Uhuru observations of X-ray emission from Seyfert galaxies. Astrophys. J., 223, pp.74-81.

Thorne, K.S. (1978). In Theoretical Principles in Astrophysics and Relativity. Eds. Lebovitz, Reid & Vandervoot, Chicago University Press.

Thorne, K.S. & Braginsky, V.B. (1976). Gravitational Wave bursts from the nuclei of distant galaxies and quasars: proposal for detection using Doppler tracking of interplanetary spacecraft. Astrophys. J. Lett., 204, pp.1-6.

Tremaine, S.D. (1976). The formation of the nuclei of galaxies. II. The local group. Astrophys. J., 203, pp.345-351.

Turland, B. (1975). 3C219: A double radio source with a jet. Mon. Not. R. astr. Soc., 172, pp.181-189.

Ulfbeck, O. (1978). Quasars and Active Nuclei of Galaxies. Ed. Phys. Scripta, 17, No. 3.

van Breughel, W.J.M. & Miley, G.K. (1977). Radio 'jets'. Nature, Lond., 265, pp.315-318.

Véron, M.P. (1975). New Photographic Observations of BL Lacertae. Astron. Astrophys., 41, pp.423-430.

Waggett, P.C., Warner, P.J. & Baldwin, J.E. (1977). NGC 6251, a very large radio galaxy with an exceptional jet. Mon. Not. R. astr. Soc., 181, pp.465-474.

Wardle, J.F.C. (1977). Upper limits on the Faraday rotation in variable radio sources. Nature, Lond., 269, pp. 563-566.

Weedman, D.W. (1976). Seyfert Galaxies, Quasars and Redshifts. Q. Jl. R. astr. Soc., 17, pp.227-262.

Weinberg, S. (1972). In Gravitation and Cosmology. Wiley, New York.

White, S. (1978). Simulations of merging galaxies. Mon. Not. R. astr. Soc., 184, pp.185-203.

Willis, A.G., Strom, R.G. & Wilson, A.S. (1974). 3C236, DA 240; the largest radio sources known. Nature, Lond., 250, pp.625-630.

Wolfe, A. (1978). Ed. Pittsburgh Conference on BL Lac Objects. Pittsburgh.

Wollman, E.R. et al. (1977). NeII 12.8 Micron emission from the galactic center. II. Astrophys. J. Lett., 218, pp. 103-107.

Young, P.J. (1977). The black tide model of QSOs. II. Destruction in an isothermal sphere. Astrophys. J., 215, pp.36-52.

Young, P.J. et al. (1978). Evidence for a supermassive object in the nucleus of the galaxy M87 from SIT and CCD area photometry. Astrophys.J., 221, pp.721-730.

BLACK HOLES AND GRAVITATIONAL WAVES: PERTURBATION ANALYSIS

S. L. Detweiler

Yale University

1. INTRODUCTION

Today a few outstanding questions in gravitational wave theory remain unanswered. Just how efficient are the most efficient sources of gravitational waves? What are the general characteristics of these gravitational waves? In these questions the phrase "most efficient sources" is perhaps a euphemism for "black holes". It seems eminently reasonable that the best gravitational wave sources involve the strongest gravitational fields and hence involve black holes.

Discussions of black holes and of gravitational waves form the majority of this volume. In this review I consider the gravitational waves and their sources as small perturbations of the geometry of a black hole. All mathematical techniques are discussed but only in a schematic manner. To improve readibility almost none of the complicated formulae which abound in this subject are contained herein.

Over twenty years ago Regge & Wheeler (1957) pioneered in the use of perturbation methods to study black hole dynamics. I find it remarkable that after all of this time the subject is still yielding interesting information and has ample opportunities for research projects.

2. PERTURBATION METHODS - SCHEMATICALLY

The general theory of relativity is a complicated, non-linear theory. A perturbation treatment results in substantial simplification; nevertheless long and messy algebraic expressions and differential equations remain. To simplify this section (and perhaps to keep the attention of an unenthusiastic reader) the formal perturbation analysis is outlined with a pretend theory of gravity based upon the field equation

$$\nabla^2\Phi - \partial^2\Phi/\partial t^2 = 4\pi\rho \ . \qquad (1)$$

rather than with Einstein's theory of gravity. Of course throughout the discussion references are supplied which allow the enthusiastic reader to fill in the correct formulae.

A more detailed review of the methods of this subject is given by Breuer (1975).

The Schwarzschild (1916) and the Kerr (1963) metrics are stationary, vacuum solutions to Einstein's field equations; in the pretend theory they are represented by a solution Φ_0 to the equation

$$\nabla^2\Phi_0 = 0 \tag{2}$$

and form the background geometry for the dynamics of the perturbation.

To this background field is now added an infinitesimal perturbation in the form of a small piece of matter, $\delta\rho(\vec{x},t)$, or a small gravitational wave, $\delta\Phi(\vec{x},t)$. By virtue of (1) and (2) the perturbation must satisfy

$$\nabla^2\delta\Phi - \partial^2\delta\Phi/\partial t^2 = 4\pi\delta\rho. \tag{3}$$

This is the fundamental perturbation equation in the pretend theory.

In the analysis of the perturbations of a black hole the situation is much more complicated. Einstein's equations must be used; and metric and matter perturbations are of second rank tensors not of scalar fields. The non-linearity of Einstein's equations complicates the transition from (1) to (3) wherein all terms of zeroth order in the perturbation drop out by virtue of the background geometry satisfying Einstein's equations, and all terms of second and higher order are discarded because they are much smaller than the first order terms. Typically at the stage of (3) the perturbed Einstein equations form a system of coupled partial differential equations inhomogeneous and linear in the metric perturbations. Regge & Wheeler first calculated these equations for perturbations of the Schwarzschild metric. And Chandrasekhar and Friedman (1972) did so for the limited class of axisymmetric perturbations of the Kerr metric.

At this stage a rather surprising simplification occurs. For the Schwarzschild metric Regge & Wheeler and Zerilli (1970) find that specific linear combinations of the components of the perturbed metric tensor satisfy uncoupled partial differential equations very similar to the pretend-theory equation (3). And in a novel approach, based upon perturbing the Newman-Penrose (1962) equations, Bardeen & Press (1973, with the Schwarzschild metric) and Teukolsky (1973, with the Kerr metric) neatly bypass the metric perturbations and instead find a single, decoupled partial differential equation not too unlike (3).

Return to the pretend theory. Equation (3) is reduced to an ordinary differential equation by the separation of the variables. The quantity $\delta\Phi$ may be Fourier analyzed in time and decomposed in terms of spehrical harmonics,

$$\delta\Phi(\vec{x},t) = \sum_{\ell,m} \int_{-\infty}^{+\infty} \frac{e^{i\sigma t}}{r} Y_{\ell m}(\theta,\phi)\ \psi_{\sigma\ell m}(r)\ d\sigma \tag{4}$$

and $\delta\rho$ may be handled similarly. Then equation (3) reduces to

$$d^2\psi_{\sigma\ell m}/dr^2 + [\sigma^2-\ell(\ell+1)/r^2]\psi_{\sigma\ell m} = T_{\sigma\ell m}; \tag{5}$$

the quantity $T_{\sigma\ell m}$ is just the σ, ℓ, m component of $\delta\rho$.

In perturbing Einstein's equations the separation goes through but is more complicated. For the metric perturbations by Regge & Wheeler the tensors must be decomposed in terms of tensorial spherical harmonics. For the Newman-Penrose approach the decomposition of the perturbation for the Kerr metric is made in terms of spin-weighted spheroidal harmonics of Teukolsky (1973).

In the study of the Schwarzschild metric one ultimately arrives at what are commonly called the Regge-Wheeler equation and the Zerilli equation which describe the odd parity and the even parity perturbations respectively and are similar to (5). Given a solution, $\psi(r)$, to one of these equations the metric perturbations may be constructed from a linear combination of ψ and its first derivative. For the Kerr metric it is easiest for computational purposes to work with a version of the Teukolsky equation. Again the metric perturbations may be constructed from a solution to this equation (cf. Chrzanowski 1975, Chandrasekhar 1978a,b,c or Wald 1978). Chandrasekhar (1975) and Chandrasekhar & Detweiler (1975a, b) discuss the relationships among these varied methods and verify consistency in the proper limits.

All of the named equations above are of the form

$$d^2\psi_{\sigma\ell m}/dr_*^2 + [\sigma^2-V_{\sigma\ell m}(r)]\ \psi_{\sigma\ell m} = T_{\sigma\ell m}, \tag{6}$$

where $T_{\sigma\ell m}$ is a combination of the components of the stress-energy tensor which is the source of the perturbation in the gravitational field. Also r_* is a radial coordinate defined by

$$dr_*/dr = (r^2+a^2)/(r^2-2Mr+a^2), \tag{7}$$

with the quantity a being the Kerr angular momentum parameter.

The quantity r_* then runs from $-\infty$ to $+\infty$ as r runs from the event horizon to infinity. For the case of the Regge-Wheeler equation

$$V_{\sigma\ell m}(r) = (1-2M/r)\left[\ell(\ell+1) - 6M/r\right]/r^2; \tag{8}$$

the other potential functions are more complicated but do not differ qualitatively from this one. Even the Teukolsky equation may be written in a form where $V_{\sigma\ell m}$ is a purely real function of σ, ℓ, m and r and where the solutions are well behaved asymptotically (cf. Chandrasekhar & Detweiler 1976).

To solve the inhomogeneous form of (6) it is convenient to define two independent homogeneous solutions, ψ_∞ and ψ_{r+}, which behave as

$$\psi_\infty \sim e^{-i\sigma r^*} \qquad \text{as } r_* \to +\infty \tag{9}$$

$$\sim B_{in}\, e^{ikr^*} + B_{out}\, e^{-ikr^*} \qquad \text{as } r_* \to -\infty \tag{10}$$

$$\psi_{r+} \sim A_{in}\, e^{i\sigma r^*} + A_{out}\, e^{-i\sigma r^*} \qquad \text{as } r_* \to +\infty \tag{11}$$

$$\sim e^{ikr^*} \qquad \text{as } r_* \to -\infty. \tag{12}$$

The quantity k is just $\sigma + am/2Mr_+$.

The physically interesting boundary conditions which go with (6) are that no radiation comes out of the hole nor comes in from infinity. The formal solution in terms of a Green function is thus

$$\psi_{\sigma\ell m}(r_*) = -\frac{\psi_\infty(r_*)}{2i\sigma A_{in}} \int_{-\infty}^{r_*} \psi_{r+}(r_*')\, T_{\sigma\ell m}(r_*')\, dr_*'$$

$$-\frac{\psi_{r+}(r_*)}{2i\sigma A_{in}} \int_{r_*}^{+\infty} \psi_\infty(r_*')\, T_{\sigma\ell m}(r_*')\, dr_*'. \tag{13}$$

The most interesting part of this solution is the amplitude of the radiation outgoing at infinity,

$$\psi_{\sigma\ell m}(r_* \to \infty) = -\frac{e^{-i\sigma r^*}}{2i\sigma A_{in}} \int_{-\infty}^{+\infty} \psi_{r+}(r_*')\, T_{\sigma\ell m}(r_*')\, dr_*'. \tag{14}$$

The total energy radiated to infinity for a given σ, ℓ, and m in the Regge-Wheeler development is now

$$\frac{d^2E}{d\sigma\, d\Omega} = \sigma^2 \, |\psi_{\sigma\ell m}(r_* \to \infty)|^2 \, |Y_{\ell m}(\theta,\phi)|^2. \tag{15}$$

In the Teukolsky development the angular dependence must be replaced by a suitably normalized spin weighted spheroidal harmonic.

The method of calculating the gravitational radiation emitted by, say, a small particle falling into a black hole is now complete. In the pretend theory $\delta\rho(x,t)$ is a Dirac δ-function along the trajectory of the particle; the quantity $T_{\sigma\ell m}(r)$ is found usually with the aid of a computer, and $\psi_{\sigma\ell m}(r_* \to \infty)$ comes from (14). Then the energy spectrum may be calculated from (15) or the wave form reconstructed via (4). The only complication (at the numerical level) introduced by starting with Einstein's equations rather than equation (1) is that $T_{\sigma\ell m}(r)$ is more difficult to find and involves the perturbation in the stress-energy tensor.

3. THE FREE OSCILLATIONS OF A BLACK HOLE

As a drummer beats his drum a sound is emitted which is easily distinguishable from that of a piano. A clever listener makes this distinction by mentally taking the Fourier transform of the sound; the result is a specific mixture of tones and overtones. As the drummer continues the relative amplitudes of the tones change but their frequencies remain fixed and are characteristics of the drum, not of the drummer. A very clever listener might even be able to use this information to deduce the size, tension and density of the drumhead.

With a similar analysis a black hole may someday be identified by the characteristic frequencies of the gravitational waves it emits in an interaction. At this time there have been a number of studies, using the methods outlined in the last section, of the gravitational waves emitted by various sources in the vicinity of a black hole. Typically, after a short burst, the wave form settles into a superposition of damped sinusoidal components, whose complex frequencies are characteristic of the free oscillations as calculated by Chandrasekhar & Detweiler (1975a) and by Detweiler (1977, 1979). For a description of this effect caused by gravitational collapse see Moncrief et.al. in this volume.

3.1 Description

The free oscillations of a black hole have been variously described as normal modes, quasi-normal modes and eigenmodes. Difficulties with these descriptive terms are that the wavefunctions of the free oscillations are not square-integrable, nor orthogonal, nor do they form a complete set in any sense.

Consider the transmission amplitude, $1/A_{in}$, of a wave, normalized by (11) and (12), as a complex function over the complex frequency plane of σ. The frequency of a free oscillation, a resonant frequency, is defined as the location of a pole of $1/A_{in}$ or equivalently as a zero of A_{in}. Thus if a wave of resonant frequency is sent in from infinity, then it will be infinitely amplified upon transmission down the hole and upon scattering off the hole (if this seems counter-intuitive see Price & Thorne 1969). It is apparent that the black hole is free to oscillate at its resonant frequency with no external stimulus. The conservation of energy and the stability of black holes imply that the imaginary part of σ is positive so that the oscillation does decay exponentially (cf. Press & Teukolsky 1973).

A different interpretation of a free oscillation is given by Detweiler (1977). Let the perturbing source of a wave be located close to the event horizon. If the source is broad-band in frequency, then the integral in (14) is a relatively smooth function of σ. And if the resonant frequency is close to the real axis, then the energy flux from (15) will be proportional to $1/|A_{in}|^2$ and exhibit the characteristic Lorentzian line shape of a resonance with the half-width at half-maximum giving the imaginary part of the resonant frequency.

3.2 How to find the frequencies of free oscillations

A natural prescription for finding a resonant frequency begins with a choice of some frequency σ. Then the initial conditions at $r_* \to -\infty$ for $\psi_{r+}(r_*)$ are determined in (12), and the homogeneous form of (6) is integrated from a large negative value of r to a to a large positive value. At large r_*, ψ_{r+} can be decomposed into ingoing and outgoing pieces so that A_{in} and A_{out} are evaluated. This process is repeated for different complex values of σ until a zero of A_{in} is found. This corresponds to a free oscillation.

While the above process works in principle it is not satisfactory in practice. All resonant frequencies seem to be stable, so interest is mainly centered upon the $\mathrm{Im}(\sigma) \geq 0$ region of the complex plane. As the numerical integration proceeds out to large values of r_* the two independent solutions of (6) behave as $\exp(i\sigma r_*)$ and $\exp(-i\sigma r_*)$. But if $\mathrm{Im}(\sigma) > 0$ then the amplitude of the outgoing solution is growing exponentially while the ingoing dies exponentially. Eventually the outgoing piece dominates the ingoing so much that the latter is lost in the inaccuracy of any numerical computation. And A_{in} cannot be evaluated. A similar problem exists near the event horizon.

But this is a difficulty in practice not in principle and may be circumvented in a number of different ways. The most straightforward response to the problem is to ignore it. Brute strength along with small step size may be used in the numerical work. Experience shows that as long as $\mathrm{Im}(\sigma)$ is not too large when compared with $\mathrm{Re}(\sigma)$ then the decomposition into ingoing and outgoing pieces can take place at a value of r_* large enough that (11), supplemented with some additional terms in the asymptotic expansion, is accurate but small enough that the ingoing piece is not dominated by the outgoing. This is essentially the method used by Chandrasekhar & Detweiler (1975a), who first calculated some of the resonant frequencies of a Schwarzschild black hole. But generally evaluating A_{in} for complex values of σ is an arduous task and for examining rotating black holes proves insurmountable (at least to me).

If a free oscillation of frequency σ_r is particularly long lived so that $\mathrm{Im}(\sigma_r)$ is small compared with $\mathrm{Re}(\sigma_r)$ then it is easy to determine σ_r. It seems safe to assume that A_{in} is an analytic function of σ at least in the region near σ_r. Thus

$$A_{in}(\sigma \approx \sigma_r) = \mathrm{const}^{-\frac{1}{2}}\,(\sigma - \sigma_r); \tag{16}$$

so a plot of

$$|A_{in}|^{-2} = \mathrm{const}/\{[\sigma - \mathrm{Re}(\sigma_r)]^2 + [\mathrm{Im}(\sigma_r)]^2\} \tag{17}$$

for real values of σ will have the standard Lorentzian line shape near a resonant frequency. This method is used by Detweiler (1977) to find the resonant frequencies of rotating black holes with $a \gtrsim 0.9M$ - a region where some of the zeros of A_{in} are close to the real axis.

Recently it has been possible to find many of the resonant frequencies for black holes with all values of a. The method is essentially an extension of the previous one. The location of some of the resonances of a Schwarzschild hole are known; for a Kerr hole if a is small then the resonances should not be displaced too far from their Schwarzschild locations. The procedure then is first to evaluate A_{in} for some number N of real values of σ which are close to the Schwarzschild value of $\mathrm{Re}(\sigma_r)$. Then an N-1 order polynomial is fit to $A_{in}(\sigma)$. A zero of this polynomial close to the Schwarzschild value of $\sigma_r(a=0)$ is σ_r (small a). Continuing in this manner one can follow the motion of a resonant frequency about the complex plane as a varies from 0 to M.

In summary there are numerical difficulties in finding the complex resonant frequencies; however the difficulties are in practice not in principle. Generally the larger the imaginary part of σ_r the more difficult it is to locate σ_r. Fortunately, as we see in the next section, the most interesting resonant frequencies, those with the smallest imaginary parts, are amenable to numerical methods.

3.3 The resonant frequencies of black holes

3.3.1 Schwarzschild black holes.

For Schwarzschild black holes some of the resonances were first found by Chandrasekhar & Detweiler (1975). They found 2 or 3 frequencies for each value of $\ell = 2,3$, and 4; from the spherical symmetry the resonant frequencies must be independent of the choice of m. For each ℓ there is one resonance whose imaginary part is much smaller than that of the others; this is the fundamental resonance. For reasons mentioned in section 3.2 this is the easiest to find; and as we discuss in section 4 this is also the most interesting one. Thus we focus attention only on these fundamental resonances.

For large values of ℓ the Regge-Wheeler potential, equation (8), acts as a high pass filter with the cutoff at $\sigma = \ell/3\sqrt{3}$ $(=0.19\ell)$. And Press (1971) noted that for large ℓ an arbitrary initial perturbation in the region near the black hole evolves into a more-or-less damped sinusoidal oscillation with a frequency again of $\ell/3\sqrt{3}$. Goebel (1972) interprets this result as a high frequency wave which is caught in the null circular orbit at r=3M but which slowly decays away.

Figure 1 shows the location in the complex plane of the resonant frequencies for different values of ℓ, m and a; for the moment look only at the locations for a=0. It is amazing how well the estimate of $\ell/3\sqrt{3}$ pinpoints the real part of the frequencies. The imaginary part is $\sim 0.95M^{-1}$ and nearly independent of ℓ for

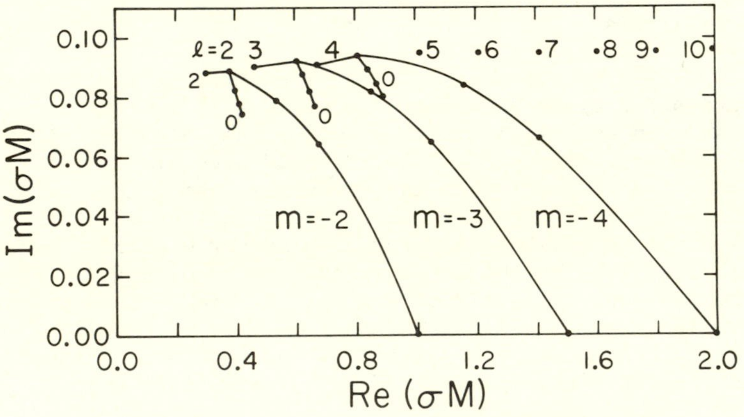

Fig. 1 Illustrating the location in the complex plane of the fundamental resonant frequencies. Each curve is labeled by its value of m; only $\ell = \pm m, 0$ are shown. The curves are parametrized by a/M and the dots denote a/M = 0, 0.7, 0.9 and 1.0 except for m = ℓ when a/M = 0 and 1.

$\ell \geq 4$. As ℓ increases the Q of the resonance also increases but as we see in section 4, as was already seen by the people who studied synchrotron gravitational radiation, and as predicted by Goebel, it is increasingly difficult to excite the resonances as ℓ increases, and thus the lowest values of ℓ are the most interesting.

3.3.2 <u>Black holes with moderate values of a.</u> Figure 1 also shows the motion of the resonant frequencies as the Kerr parameter a increases. The splitting of the azimuthal degeneracy for small values of a is at first linear in a and rather smooth. But as a passes 0.9M and approaches 1.0M the modes for large negative m take a nose dive toward the real axis. And as $a \to M$ the imaginary part of the frequency goes to zero and the real part goes to $-\frac{1}{2}m/M$.

This variation of the resonant frequency with a has a physical interpretation. For a = 0 the free oscillations resemble the Kelvin oscillations of an incompressible fluid sphere. The period of the oscillation is comparable to the dynamical time scale, $1/\sqrt{\rho}$ for the sphere and M for the black hole. And the oscillations are degenerate over the spherical harmonic index m because of the spherical symmetry. As the black hole spins up the symmetry is broken and the splitting occurs. The hole is acting like a Maclaurin spheroid. The frequency of the oscillation is comparable to the (dynamical time scale)$^{-1}$ modified by a term of the order of the rotational frequency. Finally when $a \to M$ the perturbed black hole resembles a Jacobi ellipsoid. The dragging of the inertial frame pulls the perturbation along with the hole and the radiation comes out at -m times the rotational frequency of the event horizon, $-ma/2Mr_+$, which approaches $-m/2M$ as $a \to M$.

3.3.3 <u>Analytic treatment for $a \approx M$.</u> For the case of $a \approx M$ and $\sigma \approx -am/2Mr_+$ Teukolsky's equation is amenable to analytic treatment as shown by Starobinsky and Churilov (1973) and by Teukolsky & Press (1974). The latter note that for fixed a very close to M, as σ approaches and passes $-am/2Mr_+$ the reflection amplitude executes a finite number of oscillations. And the closer a is to M, the more oscillations there are. With a minor extension of this work it may be seen that these oscillations are caused by a number of resonant frequencies (all for the same value of ℓ and m) which are piling up on the stable side of $\sigma = -am/2Mr_+$. As a gets closer to M more resonances pile up; and for a = M there is an infinite sequence of resonances with $-am/2Mr_+$ as a limit point. It appears then that in some sense the maximal Kerr black hole, with a = M is marginally unstable.

Teukolsky and Press note however that the oscillations in the reflection amplitude do not appear until a is at least within 1 part in 10^7 of M. It seems then that the piling up may be unimportant except for a possible role in keeping a black hole from passing the a = M limit and consequently losing its event horizon.

4. THE GRAVITATIONAL WAVES EMITTED BY A TEST PARTICLE FALLING INTO A SCHWARZSCHILD BLACK HOLE

As a small test particle falls into a black hole its gravitational field becomes a perturbation of the geometry of the black hole and radiates out to infinity as a gravitational wave. This process may be examined in a straightforward (although sometimes numerically tedious and difficult) manner via the methods outlined in section 2. In this section only the results of such analyses are presented.

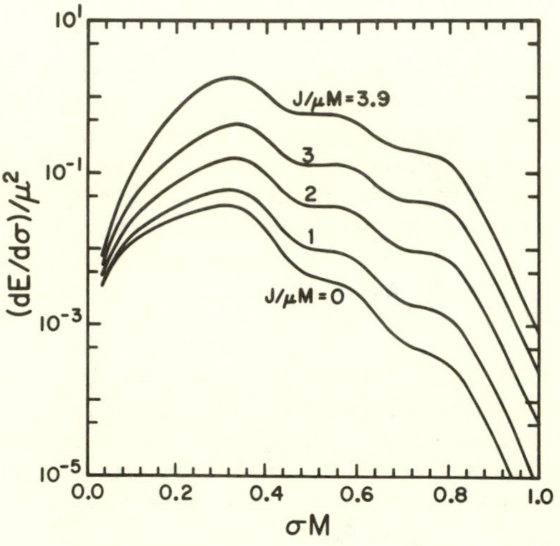

Fig. 2 The energy spectrum of the gravitational radiation emitted by a particle, with angular momentum J, falling into a Schwarzschild black hole.

Davis, Ruffini, Press & Price (1971) first calculated the energy spectrum from a particle falling radially into a Schwarzschild black hole; some of their results are presented in figure 2. The $\ell = 2$ radiation dominates the spectrum and is peaked at about $\sigma = 0.32M^{-1}$ - a bit below the fundamental resonance. At the low frequency end the spectrum goes to zero as $\sigma^{4/3}$; this is caused by the early time behavior of the source and may be accurately described by the quadrupole moment formalism of Landau & Lifshitz (1975).

When integrated over all frequencies the $\ell = 2$ radiation yields an amount of energy $\Delta E(\ell = 2) = 0.0092\mu(\mu/M)$ where μ is the mass of the test particle. The higher multipoles are responsible for the plateaus in figure 2, and the energy loss falls off by about a factor of 10 for each step in ℓ.

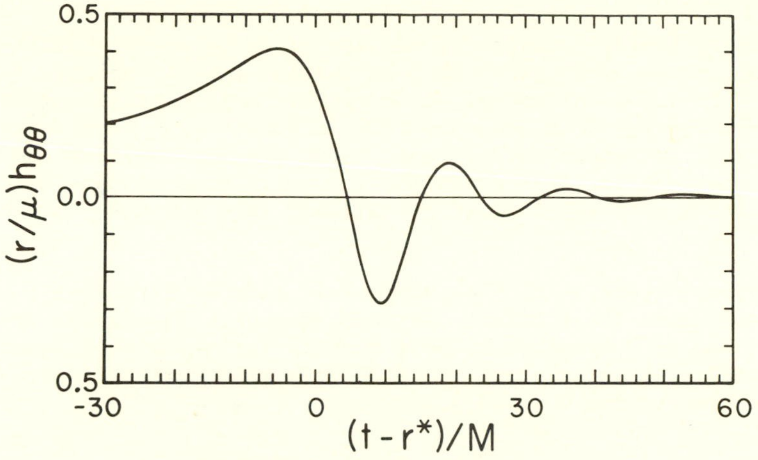

Fig. 3 The gravitational wave form on the equator for a particle falling radially into a Schwarzschild black hole along the z-axis.

From dimensional analysis it follows quite generally that for any test particle of mass μ falling along some geodesic into or around a black hole, rotating or not, the energy of the emitted gravitational radiation is of the form

$$\Delta E = \frac{\mu^2}{M} f(\vec{J}/\mu M,\ E_\infty/\mu,\ a/M) \tag{18}$$

where f is some dimensionless number dependent upon the characteristics of the black hole and of the trajectory of the particle; $\vec{J}$ is the angular momentum of the particle, E_∞ its total energy at infinity. For a particle falling radially into a Schwarzschild hole

$$f(\vec{J}=0,\ E_\infty/\mu=1,\ a/M=0) = 0.01 \tag{19}$$

Davis, Ruffini & Tiomno (1972) Fourier transform the above spectrum to calculate the waveform of the metric perturbation. Detweiler & Szedenits (1979) repeat this calculation and their results are presented in figure 3. As noted by Davis et al. (1972) the waveform has three distinct parts (i) a precursor, (ii) a sharp pulse, and (iii) an oscillatory tail. The precursor is adequately described by the quadrupole moment formalism at least at very early times. The sharp pulse carries much of the energy of the wave and lasts for about 10M. But quickly the waveform settles into a damped oscillation at precisely the ℓ = 2 fundamental resonant frequency given in figure 1. This effect alone is firm justification for the detailed treatment given in section 3. The contributions from the higher multipoles are essentially negligible in this waveform.

Ruffini (1973) extends the results of Davis et al. (1971) by allowing the incident particle some initial kinetic energy at infinity. If the total energy of the particle is twice its rest mass then the quantity f in equation (18) is about 0.25. Typically the enhancement increases rapidly with E_∞ as the test particle becomes ultra relativistic. Unfortunately the usefulness of this mechanism for building astrophysical, gravitational wave sources is questionable.

Another extension of this subject is made by Detweiler & Szedenits who allow the incident particle some orbital angular momentum J. Figure 2 summarizes their results for the energy spectrum integrated over all angles. Considerable enhancement is seen as $J/\mu M$ increases from 0 up to 3.9.

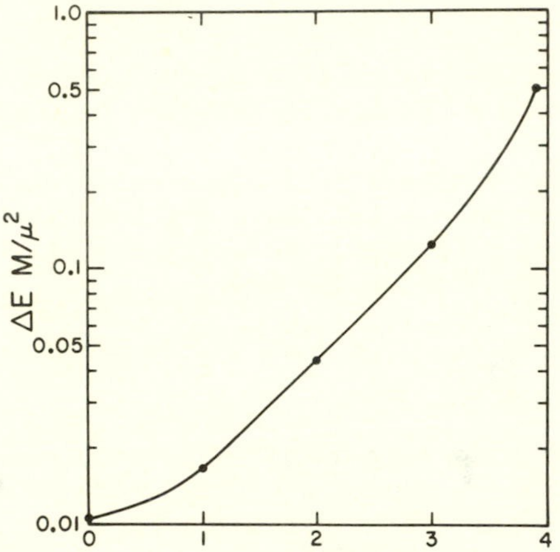

Fig. 4 Illustrating the total energy of the gravitational waves emitted by a particle with angular momentum J falling into a non-rotating black hole.

The choice of examining the $J/\mu M = 3.9$ case is significant. At $J/\mu M = 4$ and $E_\infty/\mu = 1$ the particle has just the proper amount of angular momentum and energy to come in from infinity, spiral around the hole and asymptotically approach the unstable, marginally bound, circular orbit at $r = 4M$. Thus this particular geodesic emits an infinite amount of energy over an infinite amount of time, at least with the approximations made in this chapter. Clearly this result is unphysical. Instead of evaluating the case $J/\mu M = 4$ we choose 3.9 which allows the particle enough angular momentum to circle the hole once before crossing over the event horizon.

Figure 4 shows that with the addition of a modest amount of angular momentum the total energy emitted increases by a factor of 50. It seems to me that the lack of axial symmetry is primarily responsible for this increase. This subject is returned to in section 6.

5. THE GRAVITATIONAL WAVES EMITTED BY A TEST PARTICLE IN A CIRCULAR ORBIT ABOUT A BLACK HOLE

Davis, Ruffini, Tiomno & Zerilli (1972) are the first to have numerically examined the gravitational radiation from the circular orbits of a Schwarzschild black hole. Their main concern is to examine the possibility of synchrotron gravitational radiation. They show that as the radius of the orbit decreases, the energy flux goes up dramatically, and, while the higher multipoles are enhanced for $3M < r_o < 6M$ more radiation comes out in a lower ℓ value than in a higher. Detweiler (1978) has extended those calculations to include rotating black holes.

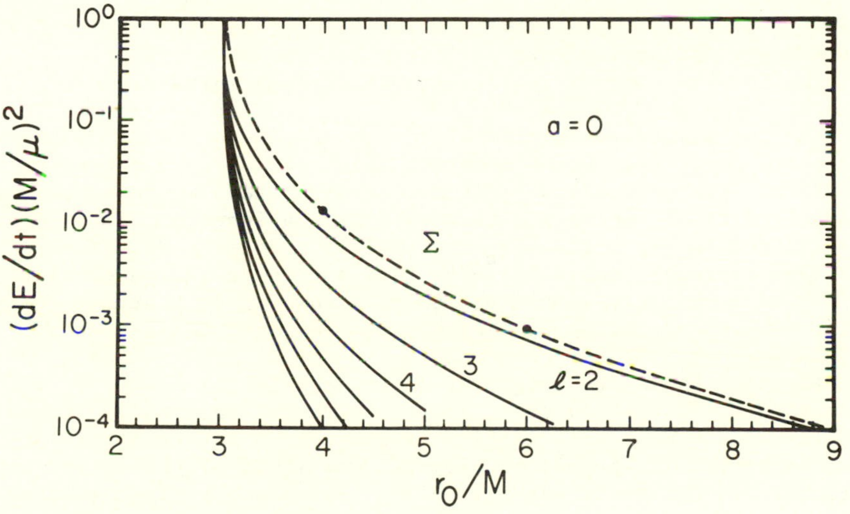

Fig. 5 The gravitational wave luminosity as a function of orbital radius for a Schwarzschild black hole. The luminosity summed over ℓ is denoted by Σ, and the two dots signify the marginally bound and the marginally stable circular orbits.

Figure 5 illustrates the gravitational wave luminosity as a function of orbital radius, r_o, for a Schwarzschild black hole. Only the radiation with $m = -\ell$ is shown; for all other m the flux is down by a couple of orders of magnitude. The flux diverges as the particle becomes ultra-relativistic at the circular photon at $r = 3M$.

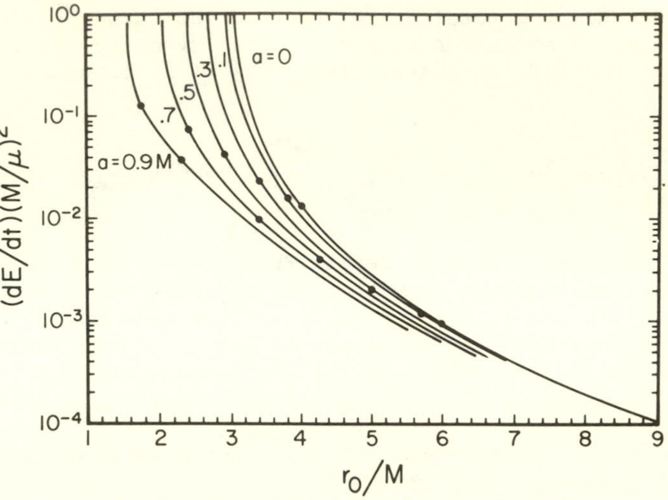

Fig. 6 The total luminosity as a function of orbital radius for black holes of varying Kerr parameter a. The dots signify the marginally bound and marginally stable circular orbits.

The luminosity calculations are summarized in figure 6. A few facts are apparent. For a constant value of r_o, as a increases the total flux decreases. This may be understood in terms of the linearized theory where the flux is a strong function of frequency. As a increases for a given r_o the orbital frequency decreases, thus the flux decreases too. But also as a increases the marginally bound and marginally stable orbits move inward, so that the luminosity from these important orbits increases with increasing a.

Figure 7 shows the waveform, in the equatorial plane, of the radiation for a typical orbit just outside the marginally stable orbit. Only the $h_{\theta\theta}$ polarization is present on the equator. As for the spiralling infall problem the magnitude of the metric perturbation is typically μ/r for an orbit reasonably close to the black hole.

The speculation in the next section uses figure 8 which shows how much energy is radiated per orbit for a particle in the marginally stable and marginally bound orbits. In this manner a dimensionless quantity f similar to that defined in (18) may be introduced,

$$\Delta E(\text{per orbit}) = \frac{\mu^2}{M} f(\text{orbital parameters}) \tag{20}$$

Then f is greater than 1 for some stable orbits with a $\gtrsim$ 0.9M and for some bound but unstable orbits with a $\gtrsim$ 0.25M.

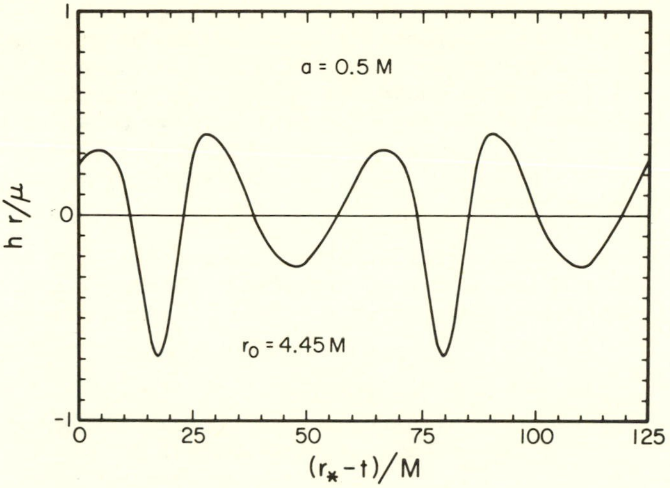

Fig. 7 The metric perturbation wave form as it might be detected far from the black hole for a circular orbit.

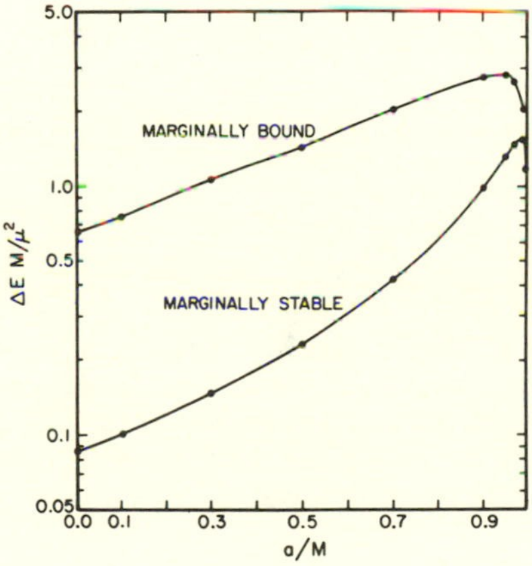

Fig. 8 The total energy emitted per orbit by a particle in the marginally bound or the marginally stable circular orbit, versus the angular momentum parameter a of the black hole.

6. UNSOLVED PROBLEMS, CONJECTURES AND SPECULATIONS

Much of the research presented in the previous two sections is leading toward evaluating the dimensionless quantity f, defined in equation (18), for all geodesics about all black holes. Since f effectively reflects the possible efficiencies of gravitational wave emission, large values of f should be good news to experimenters. So far f has been found for spiralling trajectories into nonrotating holes and for circular orbits about rotating holes. It remains to find f for all spiralling trajectories into rotating holes and for elliptic, parabolic or retrograde-circular orbits about rotating holes.

Only a small sample of all possible geodesics have been examined thus far. Nonetheless a bit of intuition and conjecture can stretch our knowledge considerably. Table 1 summarized our conclusions.

TABLE 1
The energy radiated by particles in various trajectories about various black holes.

Geodesic	a/M	$f=\Delta E(M/\mu^2)$
radial infall	0	0.01
one orbit infall	0	0.50
marginally bound orbit	0	0.65
marginally bound (one orbit infall?)	0.25	1.0
"	0.70	2.0
"	0.95	2.8

For a particle falling radially into a nonrotating black hole $f \simeq 0.01$ - a small number which has led to discouragement among experimenters. However see figure 4; if a modest amount of angular momentum is given to the particle then f increases by a factor of fifty for the particle which makes one circuit before plunging into the hole.

Compare this value of f to the value f = 0.65 from figure 8 for the marginally bound circular orbit about a nonrotating hole. That these two values of f are comparable is no coincidence. The particle with $J/\mu M = 3.9$ momentarily hangs up near the marginally bound circular orbit for a long enough time to complete one revolution before falling into the hole. Also the angular momentum for the marginally bound orbit is $J/\mu M = 4$ and the total energy is $E_\infty/\mu = 1$. Thus the spiralling orbit is quite similar to one revolution of the marginally bound orbit.

Such a comparison should go through for rotating black holes also. A particle which spirals into a rotating hole should emit about as much energy as a particle emits in one revolution in the marginally bound orbit. The marginally bound curve in figure 8 now takes on the new significance of being comparable to the energy emitted by particles spiralling into rotating black holes; and it is clear that f is typically greater than 1 for $a \gtrsim 0.25M$ and greater than 2 for $a \gtrsim 0.7M$.

All in all the black hole capture of a small mass is considerably more efficient for gravitational wave generation than was first believed based upon the results of Davis et al. (1971).

An approach based upon Newtonian trajectories coupled with the Landau & Lifshitz (1975) quadrupole moment formalism (cf. Zeldovich & Novikov, 1971, or Smarr, 1977) indicates that the energy emitted in a head on collision of two black holes might be

$$\Delta E \approx 0.02\ \mu^2/M, \tag{21}$$

where μ is the reduced mass and M the total mass of the system. Remarkably this back of the envelope calculation is consistent with the fully relativistic treatment not only in the limit when μ is small, equation (19), but also in the limit when the black holes are of equal mass (cf. Smarr, this volume). In the latter case $\mu = M/4$ and

$$\Delta E \approx 10^{-3}\ M. \tag{22}$$

The results of sections 4 and 5 indicate that the axisymmetric, head on collision of two black holes is probably the weakest gravitational wave generator of all black hole interactions. But until the four-dimensional problem of spiralling capture is solved we are left to conjecture upon this more general problem. It seems reasonable (to me) that the spiralling infall problem will scale in the same manner as the direct infall problem. If this is so, then the earlier results of this section and Table 1 imply that

$$1\ \mu^2/M \lessapprox \Delta E \lessapprox 3\ \mu^2/M. \tag{23}$$

Or after a substitution for the reduced mass

$$0.06\ M \lessapprox \Delta E \lessapprox 0.20\ M. \tag{24}$$

One can only conclude that violent black hole interactions may indeed be efficient sources of gravitational waves.

This work was supported in part by a grant with Yale University from the National Science Foundation.

REFERENCES

Bardeen, J. M. & Press, W. H. (1973), "Radiation fields in the Schwarzschild background", Journ. Math. Phys. 14, 7 - 19.

Breuer, R. A. (1975), Gravitational Perturbation Theory and Synchrotron Radiation (Springer-Verlag, New York).

Chandrasekhar, S. (1975), "On the equations governing the perturbations of the Schwarzschild black hole," Proc. R. Soc. Lond. A., 343, 289 - 298.

Chandrasekhar, S. (1978a), "The gravitational perturbations of the Kerr black hole. I. The perturbations in the quantities which vanish in the stationary state," Proc. R. Soc. Lond. A., 358, 421-439.

Chandrasekhar, S. (1978b), "The gravitational perturbations of the Kerr black hole. II. The perturbations in the quantities which are finite in the stationary state," Proc. R. Soc. Lond. A., 358, 441-465.

Chandrasekhar, S. (1978c), "The gravitational perturbations of the Kerr black hole. III. Further amplifications," in press.

Chandrasekhar, S. & Detweiler, S. (1975a), "The quasi-normal modes of the Schwarzschild black hole," Proc. R. Soc. Lond. A., 344, 441 - 452.

Chandrasekhar, S. & Detweiler, S. (1975b), "On the equations governing the axisymmetric perturbations of the Kerr black hole," Proc. R. Soc. Lond. A., 345, 145 - 167.

Chandrasekhar, S. & Detweiler, S. (1976), "On the equations governing the gravitational perturbations of the Kerr black hole," Proc. R. Soc. Lond. A., 350, 165 - 174.

Chandrasekhar, S. & Friedman, J. L. (1972), "On the stability of axisymmetric systems to axisymmetric perturbations in general relativity. I. The equations governing nonstationary, stationary, and perturbed systems," Astrophys. J., 175, 379 - 405.

Chrzanowski, P. L. (1975), "Vector potential and metric perturbations of a rotating black hole," Phys. Rev. D., 11, 2042 - 2062.

Chrzanowski, P. L. (1976), "Applications of metric perturbations of a rotating black hole: Distortion of the event horizon," Phys. Rev. D., 13, 806 - 818.

Davis, M., Ruffini, R., Press, W. H. & Price, R. H. (1971), "Gravitational radiation from a particle falling radially into a Schwarzschile black hole," Phys. Rev. Lett., 27, 1466 - 1469.

Davis, M. Ruffini, R. & Tiomno, J. (1972), "Pulses of gravitational radiation of a particle falling radially into a Schwarzschild black hole," Phys. Rev. D., 5, 2932 - 2935.

Davis, M., Ruffini, R., Tiomno, J. & Zerilli, F. (1972), "Can synchrotron gravitational radiation exist?" Phys. Rev. Lett., 28, 1352 - 1355.

Detweiler, S. (1977), "On resonant oscillations of a rapidly rotating black hole," Proc. R. Soc. Lond. A., 352, 381 - 395.

Detweiler, S. (1978), "Black holes and gravitational waves. I. Circular orbits about a rotating hole," Astrophys. J., 225.

Detweiler, S. (1979), "Black holes and gravitational waves. III. Free oscillations of a rotating black hole," Astrophys. J., in press.

Detweiler, S. & Isper, J. R. (1973), "The stability of scalar perturbations of a Kerr-metric black hole," Astrophys. J., 185, 675-683.

Detweiler, S. & Szedenits, E. J. (1979), "Black holes and gravitational waves. II. Spiralling geodesics into a Schwarzschild black hole," Astrophys. J., in press.

Goebel, C. J. (1972), "Comments on the vibrations of a black hole" Astrophys. J., 172, L95 - L96.

Hartle, J. B. & Wilkins, D. C. (1974), "Analytic properties of the Teukolsky equation," Commun. Math. Physics, 38, 47 - 63.

Kerr, R. P. (1963), "Gravitational field of a spinning mass as an example of algebraically special metrics," Phys. Rev. Lett., 11, 237 - 238.

Landau, L. D. & Lifshitz, E. M. (1975), The Classical Theory of Fields (Pergamon Press, Oxford).

Newman, E. T. & Penrose, R. (1962), "An approach to gravitational radiation by a method of spin coefficients," J. Math. Phys., 3, 566 - 578.

Press, W. H. (1971), "Long wave trains of gravitational waves from a vibrating black hole," Astrophys. J., 170, L105 - L108.

Press, W. H. & Teukolsky, S. A. (1973), "Perturbations of a rotating black hole. II. Dynamical stability of the Kerr metric," Astrophys. J., 185, 649 - 673.

Price, R. & Thorne, K. S. (1969), "Non-radial pulsation of general relativistic stellar models. II. Properties of the gravitational wave," Astrophys. J., 155, 163 - 182.

Regge, T. & Wheeler, J. A. (1957), "Stability of a Schwarzschild singularity," Phys. Rev., 108, 1063 - 1069.

Ruffini, R. (1973), "Gravitational radiation from a mass projected into a Schwarzschild black hole," Physical Rev. D7, 972 - 976.

Schwarzschild, K. (1916),"Über das Gravitationsfeld eines Massenpunktes nach der Einsteinschen Theorie," Sitzber. Deut. Akad. Wiss. Berlin, Kl. Math.-Phys. Tech., 189 - 196.

Smarr, L. (1977), "Space-time generated by computers: Black holes with gravitational radiation," Ann. N. Y. Acad. Sci., 302, 569.

Starobinski, A. A. & Churilov, S. M. (1974), "Amplification of electromagnetic and gravitational waves scattered by a rotating black hole," Sov. Phys. JETP, 38, 1 - 5.

Teukolsky, S. A. (1973), "Perturbations of a rotating black hole. I. Fundamental equations for gravitational, electromagnetic, and neutrino-field perturbations," Astrophys. J., 185, 635 - 647.

Teukolsky, S. A. & Press, W. H. (1974), "Perturbations of a rotating black hole. III. Interaction of the hole with gravitational and electromagnetic radiation," Astrophys. J., 193, 443 - 461.

Wald, R. M. (1978), "Construction of solutions of gravitational, electromagnetic, or other perturbation equations from solutions of decoupled equations," Phys. Rev. Lett., 41, 203 - 206.

Zeldovich, Ya. B. & Novikov, I. D. (1971), Stars and Relativity (University of Chicago Press, Chicago).

Zerilli, F. J. (1970), "Effective potential for even-parity Regge-Wheeler gravitational perturbation equations," Phys. Rev. Lett., 24, 737 - 738.

RADIATION FROM SLIGHTLY NONSPHERICAL MODELS OF GRAVITATIONAL COLLAPSE

Vincent Moncrief
Department of Physics, Yale University

Christopher T. Cunningham and Richard H. Price
Department of Physics, University of Utah

INTRODUCTION

There are two current techniques for computing the properties of gravitational radiation emitted during the formation of a black hole:

(i) linearized and higher order perturbations of known (non-radiative) exact solutions and

(ii) full scale numerical codes for solving the Einstein equations for axially symmetric (and ultimately non-symmetric) systems.

In this paper we shall discuss the first method and some results we have obtained in applying it to perturb a specific, simple model of gravitational collapse.

Our program has been to study the first and second order perturbations of the Oppenheimer-Snyder model of spherical, pressureless collapse to a Schwarzschild or slowly rotating Kerr black hole. Our idea was to formulate the first and second order perturbation problems in terms of suitable gauge invariant quantities and to solve the relevant perturbation equations numerically for some reasonable initial conditions. Our aim was to compute the waveforms, spectra and total energies emitted for a variety of representative collapses.

The main advantage to our approach is that we reduce the perturbation equations, by an expansion in tensor harmonics, to a partial differential system in only two independent variables and achieve, through the use of gauge invariants, a considerable decoupling of this system. Our emphasis on computing gauge invariants also helps us to isolate the physically relevant features of the perturbations. The same techniques would be applicable to perturbing any spherically symmetric background. We achieve significantly greater numerical accuracy with this approach than is currently attainable with the non-perturbative codes. The main limitation of our technique is of course that one can treat only small departures from spherical symmetry.

In the following sections we shall sketch the formulation and reduction of the perturbation equations and discuss our numerical results and their significance in relation to general questions about the radiation emitted during the formation of a black hole. Specifically we discuss the dominance of quasi-normal ringing in our radiation outputs, the occurence of (predicted) power law tails in our late-time waveforms and the total energies emitted in the models we have studied.

The full details of the methods used and results described here appear elsewhere (see Cunningham, Price & Moncrief, 1978, 1979). These details are too lengthy to bear repetition here. For that reason we shall only sketch the methods and describe their application to the Oppenheimer-Snyder problem.

PERTURBATION METHODS

General techniques

The procedure for generating the linearized and higher order perturbation equations is well known. Suppose that $\overset{o}{g}_{\mu\nu}$, $\overset{o}{u}_{\mu}$, $\overset{o}{\rho}$,... are the metric, velocity, density, etc. of some known solution of Einstein's equations (units: G=c=1):

$$G_{\mu\nu}(g) = 8\pi T_{\mu\nu}(g,u,\rho,\ldots),$$

One expands

$$g_{\mu\nu}(\varepsilon) = \overset{o}{g}_{\mu\nu} + \varepsilon\delta g_{\mu\nu} + \frac{\varepsilon^2}{2}\ \delta^{(2)}g_{\mu\nu} + \ldots,$$

etc., substitutes these expansions into the field equations and requires these equations to hold for arbitrary small values of the expansion parameter ε. Designating the background fields $\{\overset{o}{g}_{\mu\nu},\overset{o}{u}_{\mu},\overset{o}{\rho}\}$ collectively by $\{\overset{o}{b}\}$ and the perturbations $\{\delta g_{\mu\nu},\delta u_{\mu},\delta\rho\}$ by $\{\delta b\}$ we can write the linearized equations as

$$DG_{\mu\nu}(\overset{o}{b})\cdot\delta b - 8\pi DT_{\mu\nu}(\overset{o}{b})\cdot\delta b = 0$$

where $DG_{\mu\nu}(\overset{o}{b})\cdot\delta b$ represents the first variation of $G_{\mu\nu}$ about $\overset{o}{b}$ in the direction δb, etc. Similarly the n-th order equations have the form

$$DG_{\mu\nu}(\overset{o}{b})\cdot\delta^{(n)}b - 8\pi DT_{\mu\nu}(\overset{o}{b})\cdot\delta^{(n)}b$$
$$= S_{\mu\nu}(\overset{o}{b},\delta b,\ldots\delta^{(n-1)}b)$$

where $S_{\mu\nu}$ is a source term constructed from all the lower order perturbations. Thus each order of perturbation theory beyond the first requires the solution of a linear inhomogeneous system of equations with an inhomogeniety determined from the chosen solutions of all the preceeding orders. The linear operator acting on the n-th order perturbations is always identical to that encountered at the first order. Thus all the techniques of separating variables and decoupling equations which work at first order also apply to the n-th order problem.

The first of these reduction techniques is simply to expand the various perturbation fields $\delta g_{\mu\nu}$, etc., in tensor harmonics (see Regge and Wheeler 1957). This accomplishes the usual separation of angle variables always possible on a spherically symmetric background. The second main reduction technique is to define an appropriate set of gauge invariant linear functions of the perturbation fields $\delta g_{\mu\nu}$, δu_{μ}, $\delta\rho$ and to reexpress the perturbation equations in terms of these quantities. Any equation which determines the evolution of a gauge invariant quantity must necessarily be expressible purely in terms of gauge invariants. Thus expressing the perturbation equations in terms of invariants automatically leads to a significant decoupling of the physically interesting (i.e., invariant) quantities from the remaining gauge dependent ones.

That one can always define a complete, non-redundant set of gauge invariant perturbation functions is assured (for vacuum backgrounds with compact Cauchy surfaces) by a decomposition result of Moncrief (1975). The asymptotically flat case, allowing non-vacuum spacetimes, can probably be handled similarly with suitable care for the asymptotic conditions. This decomposition approach has the advantage of generality (since no symmetry of the background is assumed) but the disadvantage of non-locality. Fortunately, when the background spacetime has enough symmetry, the expansion in harmonics often simplifies the perturbation problem sufficiently that the construction of a complete set of invariants can be given explicitly. This is always true in the case of a spherically symmetric background. For details of this construction in the Schwarzshild exterior see Moncrief (1974). The corresponding construction for the Friedmann interior and the matching of the interior and exterior perturbations is given by Cunningham, et al. (1978, 1979).

The most familiar technique of simplifying the perturbation equations is that of choosing a special gauge. However any such choice is irrelevant to the determination of the invariant quantities which are of most interest. Furthermore some foresight is necessary in choosing gauge conditions for the interior and exterior regions in order to allow a regular matching of the perturbations across the boundary. For these reasons we have refrained from imposing gauge conditions except at the stellar boundary. We found it convenient to require that the boundary surface be a surface of constant radial coordinate even in the perturbed spacetime where its geometry is no longer spherical.

The Oppenheimer-Snyder Model

The Oppenheimer-Snyder (OS) model consists of a spherical piece of a k=+1 Friedmann cosmological model matched to a Schwarzschild exterior. The k=+1 model was chosen because it has a hypersurface of time symmetry (τ=0) which provides a natural initial data surface for the perturbation problem. We can imagine that prior to τ=0 the star was a static Schwarzschild model perturbed into slow rotation or otherwise deformed: The model loses its pressure at τ=0 and its collapse proceeds as a small deformation of the OS model.

There are several classes of perturbations which must be distinguished. All of the various tensor harmonics (labeled by integers ℓ,m with $|m|\leq\ell$) decouple from one another and further split into even (electric) and odd (magnetic) parity classes. Only the modes with $\ell\geq2$ admit gravitational radiation. For simplicity we shall describe only the axisymmetric (m=0) case for which the various modes of perturbation have the following physical significance.

Odd (Magnetic) Parity Modes:

ℓ=0: empty,

ℓ=1: azimuthal velocity perturbations which reduce uniquely in the exterior to the Kerr perturbation of the Schwarzschild metric,

$\ell\geq2$: differential, azimuthal velocity perturbations which couple, through the stellar boundary, to gravitational waves.

Even (Electric) Parity Modes:

ℓ=0: purely spherical (Bondi-Tolman) perturbations of the background,

ℓ=1: dipole-type perturbations of the fluid interior which reduce to pure gauge (translational) perturbations in the Schwarzschild exterior,

$\ell\geq2$: convective velocity perturbations and density perturbations which couple, through the stellar boundary, to gravitational waves.

The quadrupole (ℓ=2) and higher multipole matter perturbations are all capable of driving gravitational waves (at first order) but do so only in a somewhat subtle way. It is well known from the work of Lifshitz and Khalatnikov (1963) that first order matter perturbations do not drive gravitational waves in a complete Friedman cosmological model. In our problem the coupling of matter disturbances to gravitational waves arises only because of the occurence of the stellar boundary across which the perturbations must match. When initial conditions are chosen to exclude radiation on the initial surface the radiation which subsequently appears is effectively produced by "sources" localized along the boundary surface. This peculiarity is an artifact of our use of the homogeneous (Friedmann) interior and, even in the Friedmann case, disappears at second order where genuine interior sources arise.

None of the radiative ($\ell \geq 2$) modes described above carry a net angular momentum. To perturb the collapsing star into slow rotation with non-vanishing angular momentum requires the $\ell=1$, odd parity mode which is non-radiative. To see the radiation induced by this mode one must continue to second order where the $\ell=1$, odd mode drives an $\ell=0$, even and an $\ell=2$, even mode. The latter is therefore the fundamental quadrupole mode excited in the formation of a Kerr black hole. It is convenient to split the second order exterior metric perturbation $\delta^{(2)} g_{\mu\nu}$ into a pure Kerr part and a part describing the radiation. The former gives a particular solution to the exterior second order perturbation equations and thus reduces the exterior problem to the homogeneous case. The resulting exterior metric has the form

$$g_{\mu\nu}(\varepsilon) = \overset{o}{g}_{\mu\nu} + \varepsilon \delta g_{\mu\nu}^{\text{Kerr}(1)} + \frac{\varepsilon^2}{2} [\delta^{(2)} g_{\mu\nu}^{\text{Kerr}(2)} + \delta^{(2)} g_{\mu\nu}^{\text{radiation}}]$$

which, after the radiation field dies out, reduces to that of a slowly rotating Kerr black hole to second order in the Kerr parameter $\varepsilon = a/M$.

We have studied several classes of perturbations numerically, the $\ell=2$ and $\ell=3$, odd and the $\ell=2$, even perturbations which radiate at first order and the second order problem ($\ell=1$, odd driving $\ell=2$, even) discussed above. In each case we chose initial data to represent initially quiescent (i.e., radiation-free) conditions at the onset of collapse. Some of our results are discussed in the following section.

NUMERICAL RESULTS AND CONCLUSIONS

The formalism described above leads to evolution equations for interior and exterior gauge invariant wave functions and to matching conditions which relate these functions across the collapsing stars boundary. After specifying suitable initial conditions for the wave functions and source (i.e., the density and velocity perturbations) we can integrate the evolution equations numerically to determine the gravitational radiation.

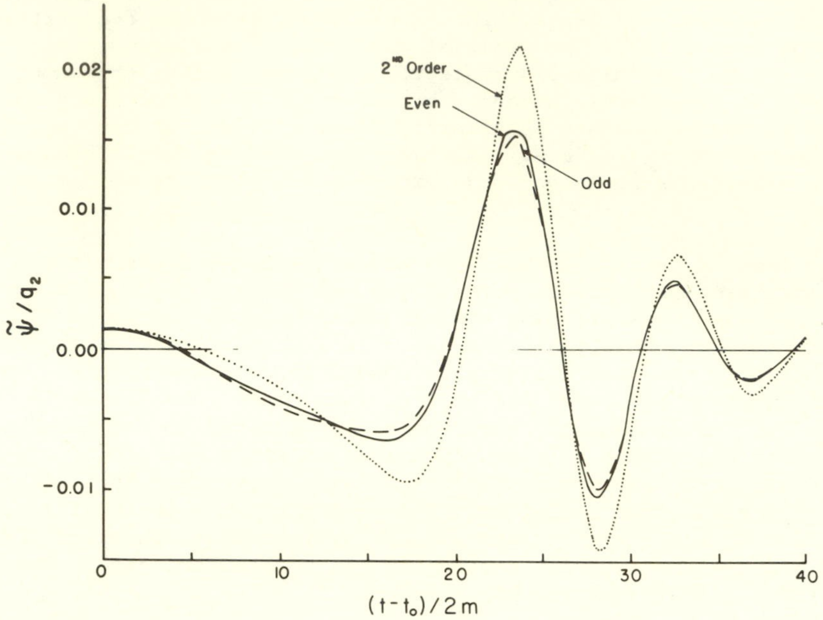

Fig. 1 Waveforms for the exterior quadrupole gravitational wave functions $\tilde{\psi}$ for collapse from an initial radius $r_0 = 8M$. The waveforms at a large constant radius are plotted as functions of $t-t_0$, the time after the passage of the first outgoing ray signaling the onset of collapse. The normalizations of $\tilde{\psi}$ are such that

$$d(\text{Energy})/dt = [d\tilde{\psi}/dt]^2/(384\pi).$$

For each case shown the initially stationary field has the asymptotic form

$$(\tilde{\psi}/q_2) \to (2M/r)^2 \quad \text{as } r \to \infty .$$

For the second order results q_2 represents only the radiative part of the total quadrupole moment.

We chose our initial conditions so that the fields were momentarily stationary on the initial hypersurface. In particular the exterior perturbations were chosen to be stationary multipoles with suitably defined moments q_ℓ. The character of the interior fields was somewhat more arbitrary since the physics of the star prior to its collapse was not modeled in detail. Fortunately we found that the evolution of the exterior field, and hence the radiation, was almost unaffected (variations of a few percent) by a significant variation of the interior initial data. We therefore settled on the (mathematically) simplest choice of "non-

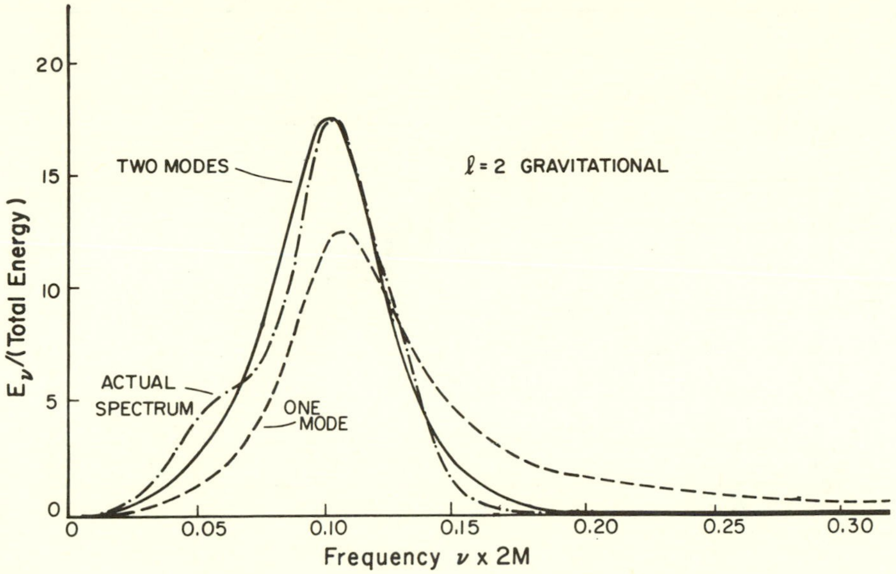

Fig. 2 Gravitational quadrupole spectrum for collapse from r_o=8M, compared with spectra for one (ω_1) and for the two quasi-normal modes:

$$\omega_1 = (0.64734 + 0.17792i)/2M \text{ and}$$

$$\omega_2 = (0.69688 + 0.5493i)/2M.$$

A good fit to the actual spectrum is achieved by the two mode spectrum plotted,

$$\tilde{\psi} \propto \text{Re}\,[(1-0.3i)\exp(i\omega_1 t) - (1 + 0.5i)\exp(i\omega_2 t)]$$

corresponding to roughly equal excitation of the two modes.

dynamical" interior initial data. This data evolves adiabatically until information signaling the onset of collapse has propagated inward from the stellar boundary.

A principal result of our numerical studies concerns quasinormal ringing, the damped sinusoidal oscillations (with discrete, complex frequencies) which characterize a perturbed black hole. We find in our models that quasinormal ringing overwhelmingly dominates the evolution of the exterior wave function (see Figs. 1 and 2). The fundamental frequency of this ringing agrees to an accuracy of 0.5% with that calculated directly by Chandrasekhar

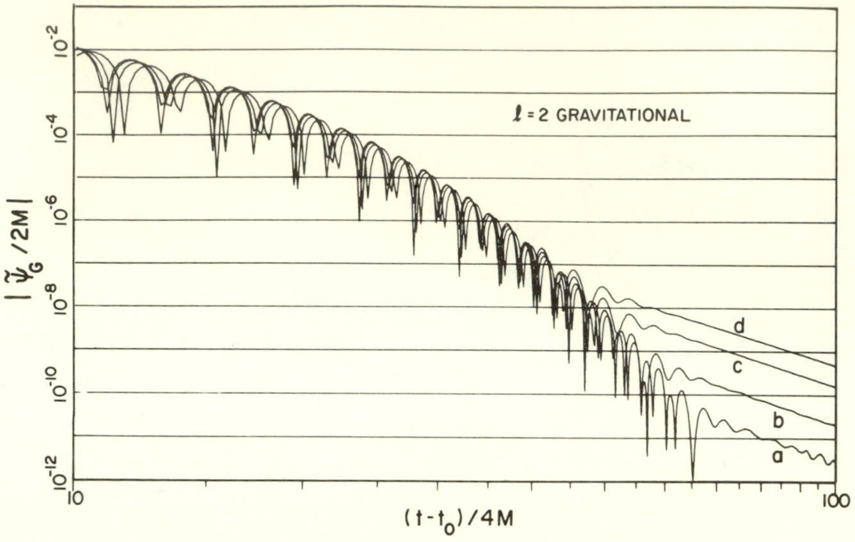

Fig. 3 Odd-parity gravitational wave amplitudes as functions of time, showing development of the predicted $t^{-(2\ell+2)}$ tails. Plots are for collapse from r_0=4M, with unit initial quadrupole moment (q_2/2M = 1), computed at four values of r/2M: (a) 3.1419, (b) 6.2491, (c) 13.3970, (d) 20.9241. The oscillations appearing in the tail of the wave function at r/2M = 3.1419 are caused by round-off errors in the initial stationary solution and are not of physical significance.

and Detweiler (1975).

Quasinormal ringing does not, of course, give a complete description of the exterior field. Soon after the onset of collapse the field goes through an initial phase which depends on the detailed nature of the collapsing star. The amount of radiation energy involved in this initial phase is however very small. At late retarded times, at any finite radius, the exponentially damped quasinormal ringing of the exterior field gives way to a power law tail caused by the backscattering of outgoing radiation (Price, 1972). Both the fall off in time ($t^{-(2\ell+2)}$) and the predicted amplitude of this tail have been verified to considerable accuracy in our models (See Fig. 3).

The quasinormal mode dominance, the relative independence of the exterior field on the details of the interior and the nearly identical results for even and odd-parity wave forms (see Fig. 1) strongly suggest that the exterior radiation is a phenomenon associated with the black hole geometry itself. The main effect

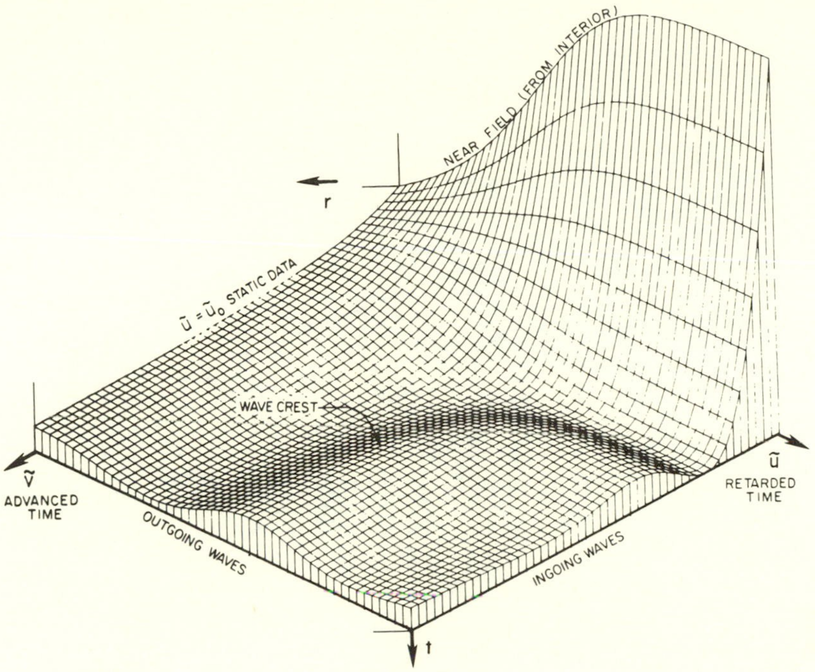

Fig. 4 A picture of the radiation being formed in the exterior. Height above the base plane represents the amplitude of the exterior field $\tilde{\psi}$. Note that the wave crest corresponds to outgoing radiation at large r and ingoing radiation near the event horizon.

of the collapse has been to excite the bell ringing modes of the black hole which remains behind. This interpretation is supported by the picture of the development of the exterior field in Fig. 4. In this figure it is evident that the field near the surface of the collapsing star is non-oscillatory. The ripples which describe the ingoing and outgoing radiation develop slightly outside the boundary surface of the collapsing star. These ripples represent the initiation of the quasinormal ringing.

The excitation of the quasinormal modes seems to depend primarily upon the behavior of the near field when the star's surface radius is small and this in turn is rather independent of the collapse details. We can in fact give a simple and reasonably accurate prediction of the excitation of the quasinormal modes in terms of the velocity of the collapsing surface in the neighborhood of r=4M. Since, for a given initial multipole moment, even and odd parity fields develop almost identically, this prediction applies equally to both parity classes. If we assume that the radiated energy is mostly in the form of quasinormal ringing we

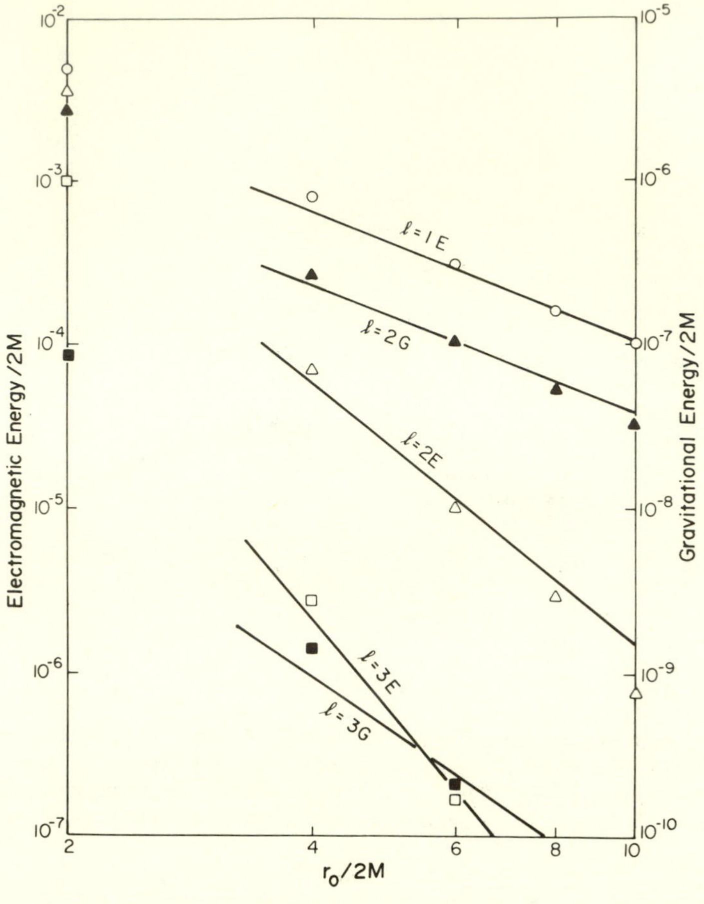

Fig. 5 The total odd-parity energies radiated compared to predicted energies based on quasinormal mode dominance (solid curves labeled E for electromagnetic and G for gravitational radiation respectively). In each case the initial field was taken to be an initially stationary multipole of unit multipole moment $(q_\ell/2M)=1$, which implies that $(\tilde{\psi}/2M)\rightarrow(2M/r)^\ell$ as $r\rightarrow\infty$. The exterior gravitational fields are normalized such that

$$d(\text{Energy})/dt=(16\pi)^{-1}[(\ell-2)!/(\ell+2)!]\,[d\tilde{\psi}/dt]^2$$

while the electromagnetic normalization corresponds to

$$d(\text{Energy})/dt = (4\pi)^{-1}[(\ell+1)!/(\ell-1)!][d\tilde{\psi}/dt]^2.$$

The first order even-parity energies, with the same initial moments, differ at most by a few percent from the odd-parity results. r_0 is the initial radius from which the collapse occurs.

can furthermore arrive at a simple formula for the total energy radiated in the collapse. The predictions of this formula (discussed in detail in Cunningham, et al., 1978) are compared with actual numerical results in Fig. 5.

In our first order perturbation formalism the exterior field is, of course, proportial to the initial multipole moment q_ℓ, and hence the radiated energy is proportional to $q_\ell{}^2$. To arrive at reasonable estimates of the total radiated energy we must give estimates of q_ℓ. There are, however, upper bounds upon the magnitudes of these initial moments which are implied by the consistent use of perturbation theory. Initial values exceeding these bounds would require a non-perturbative treatment. In the odd-parity case we have estimated these bounds by requiring that the rotations giving rise to q_ℓ be sufficiently small that no mass shedding occurs before the star has collapsed through its horizon. This leads to an energy limit for a star collapsing from a large (>>M) initial radius, in the case ℓ=2

$$E_{2,odd} \lesssim 1.2 \times 10^{-3} M.$$

For (first order) even parity perturbations we can estimate the bound on the initial quadrupole moment by considering the collapse of Newtonian dust spheroids to pancake singularities. We require that the initial moment be small enough that the singularity is reached only after the matter has fallen through its Schwarzschild radius. This leads to a smaller quadrupole moment than in the odd-parity case and to an energy limit

$$E_{2,even} \lesssim 2.2 \times 10^{-4} M,$$

about one fifth as large. It is important to remember that these limits are rough estimates of the limit of applicablity of perturbation theory and do not represent physical bounds upon the total energy that could be radiated in a highly non-spherical collapse.

In formulating the second order perturbation problem we specify the (ℓ=1, odd) velocity sources which determine the value of the Kerr parameter ε= a/M and associated angular momentum $J = aM = \varepsilon M^2$. The values of ε,M and the initial radius r_0 determine, at least roughly, the value of the initial quadrupole moment q_2. For convenience q_2 is here defined as the difference between the actual moment and the Kerr moment for the given choice of J and M since this difference represents the radiative part of the total initial moment. We estimate values for q_2 by computing the moments of Maclaurin spheroids with the same values of M,J and r_0.

The second order waveform for a collapse from r_0=8M is shown in Fig. 1 where it may be compared with odd and even parity first order waveforms generated by the same initial quadrupole moment q_2. The occurence of quasinormal ringing, with a somewhat enhanced amplitude, is readily apparent in these preliminary results. To estimate the total energies attainable in the second order calculations we must again consider the limits upon q_2 imposed by the validity of perturbation theory. To do this we consider the collapse of rotating Newtonian dust spheroids which start collapsing from momentarily stationary configurations with the appropriate values of M,J and r_o. We again require that the pancake singularlity be reached only after the matter has fallen through its Schwarzschild radius. This imposes a restriction upon the initial value of J for a given r_o and M. Our preliminary results give, for collapses from initial radii from 8M to 16M, the approximate energy limit (corresponding to $a/M \lesssim .32$)

$$E_{2nd\ order} \approx (a/M)^4 \times 1.9 \times 10^{-2}\ M \lesssim 2.0 \times 10^{-4}\ M$$

which is almost identical to that of the first order problem. While these second order results are still somewhat preliminary they provide the first direct estimates of the energies produced in the formation of a Kerr black hole from a slowly rotating collapsing cloud of matter.

REFERENCES

Chandrasekhar, S. & Detweiler, S. (1975). The quasi-normal modes of the Schwarzschild black hole. Proc. R. Soc. Lond. A., 344, pp. 441 - 452.

Cunningham, C., Price, R. & Moncrief, V. (1978). Radiation from collapsing relativistic stars. I. Linearized odd-parity radiation. Ap. J., 224, pp. 643 - 667.

Cunningham, C., Price, R., & Moncrief, V. (1979). Radiation from collapsing relativistic stars. II. Linearized even-parity radiation. (In preparation).

Lifshitz, E. M. & Khalatnikov, I. M. (1963). Investigations in relativistic cosmology. Advances in Physics, 12, pp. 185 - 249.

Moncrief, V. (1974). Gravitational perturbations of spherically symmetric systems. I. The exterior problem. Ann. Phys. (N.Y.), 88, No. 2, pp. 323 - 342.

Moncrief, V. (1975). Decompositions of gravitational perturbations. J. Math. Phys., 16, No.8, pp. 1556 - 1560.

Price, R. (1972). Nonspherical perturbations of relativistic gravitational collapse, I: Scalar and gravitational perturbations. Phys. Rev., D5, pp. 2419 - 2438.

Regge, T. & Wheeler, J. A. (1957). Stability of a Schwarzschild singularity. Phys. Rev., 108, pp. 1063 - 1069.

The research reported here has been supported by National Science Foundation grants to Yale University and the University of Utah.

GAUGE CONDITIONS, RADIATION FORMULAE AND THE TWO BLACK HOLE COLLISION

Larry Smarr

Harvard-Smithsonian Center for Astrophysics and
Lyman Laboratory of Physics
Harvard University

1 INTRODUCTION

In this paper I discuss several topics in axisymmetric numerical relativity. First, I review the variety of gauge conditions which can be imposed on axisymmetric line elements. This discussion will define a variety of specific computational frames using the general 3+1 machinery of lapse functions and shift vectors discussed by York in this volume. Next I will discuss how, given a frame, gravitational radiation information can be read out of a spacetime. These two sections are meant as a general preface to the articles by Eppley, Piran, Wilson and myself on particular numerically generated spacetimes.

The last section is a summary of the results of the two black hole collision calculation carried out by Eppley and myself. This first example of a numerically constructed asymptotically-flat spacetime containing black holes producing gravitational radiation is now fairly complete. The full nonlinear interaction turns out to share remarkable similarities with perturbation calculations described by Detweiler and Moncrief in this volume.

2 AXISYMMETRIC GAUGE CONDITIONS

2.1 Kinematics and Dynamics

As reviewed by York in this volume, the spacetime metric can be split into kinematic and dynamic parts. Geometrically the kinematics describes spacetime as being foliated by a family of spacelike time slices which are threaded by a congruence of curves. The time coordinate is that scalar function which is constant on each time slice. The spatial coordinates are 3 scalar functions constant along each curve of the congruence. The lapse function (α) specifies the normal separation of the time slices while the shift vector (β^i, i=1,2,3) specifies the boost velocity between the time slice normals and the coordinate trajectories (see York figure 1). In a coordinate representation $\alpha=(g^{oo})^{-1/2}$ and $\beta_i=g_{oi}$.

Thus, a particular choice of (α,β^i) fixes a computational frame or gauge in which the 3-metric (γ_{ij}) is evolved by the Einstein equations [York equations (35) and (39)]. Note that these equations cleanly separate between kinematic terms involving spatial derivatives of (α,β^i) and dynamic terms involving spatial derivatives of γ_{ij}. While in principle it makes no difference how many components of β^i or γ_{ij} are nonzero, in practice it makes a great difference because of the nonlinearity of R_{ij} in the Einstein equations [York equation (39)]. When one has to write out the Einstein equations for a particular line element the number of terms in R_{ij} grows very quickly with the number of nonzero γ_{ij}. For this reason, one often prefers to zero some components of γ_{ij} in exchange for some nonzero components of β^i. That is, one adjusts the coordinate trajectories in such a way as to keep certain submatrices of γ_{ij} orthogonal.

However, there are sometimes other motivations for choosing β^i. For instance, one might wish to make the coordinate congruence comove with matter or globally minimize the distortion of the grid. Both of these options, discussed by York herein, leave the maximum number of components γ_{ij} nonzero. Finally, one might simply desire normal coordinate trajectories ($\beta^i=0$).

In all these possibilities, I have assumed the lapse function α was left free to determine the nature of the time slicing. As also discussed by Eardley, York, and Wilson in this volume, a geometric slicing condition, such as maximal time slicing, is required if a global nonsingular foliation is to be constructed. In many previous investigations of the axisymmetric Einstein equations this requirement was not considered. As a result there are several well known gauges which will almost surely fail to globally cover the spacetime regions of interest in strong field problems.

Because the lapse-shift formulation gives a framework to describe all these gauges, I use it to classify the variety of coordinate systems used in studies of axisymmetric spacetimes. Most of these studies have involved analytic solutions which typically have extra symmetries (Killing vector fields) besides axial symmetry. I wish to separate them out for discussion because this extra symmetry automatically zeroes certain components of γ_{ij}. The following chart accomplishes this separation:

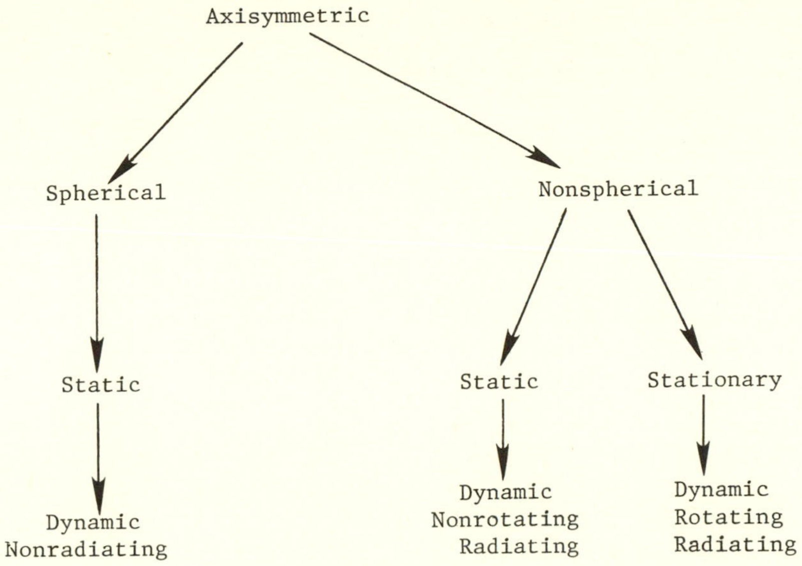

An important subclass of the nonspherical systems are the cylindrical spacetimes which are reviewed by Piran in this volume. Since he comprehensively discusses the possible cylindrical gauges, I shall not repeat these here.

In the following discussion an important question is what are the minimum number of components of γ_{ij} needed to carry the true dynamical degrees of freedom of the gravitational field. As York discusses in his article (cf. his section 8), one of the six independent components of γ_{ij} describes the length scale of space which is fixed by the Hamiltonian constraint (York equation 70). This leaves the five components of the conformal 3-metric $\tilde{\gamma}_{ij}$. The use of the three components of β^i then cut the freedom in $\tilde{\gamma}_{ij}$ down to two - the two gravitational degrees of freedom or polarization states. In spherical systems both degrees of freedom vanish. In nonrotating dynamic axisymmetric systems only one is present. Rotation allows for both.

2.2 Spherical Spacetimes

The high symmetry of such a spacetime implies that the most general 3-metric can be written as:

$$^{(3)}ds^2 = A\,dr^2 + Br^2(d\theta^2 + \sin^2\theta d\phi^2) \tag{1}$$

where A and B are functions of (t,r) only. The kinematic variables needed are at most (α, β^r), also functions of (t,r). From the preceding paragraph, one can say that only one metric function is needed to describe the scale of the space; none is needed for radiation information. We will classify all the possible gauge choices by specifying the conditions placed on β^r.

2.2.1 Eulerian gauge. The most obvious choice is to set $\beta^r = 0$. I will term this a pure Eulerian gauge since the spatial coordinates are constant along the normal or Eulerian trajectories [see Smarr and York (1978) or York, section 2, this volume]. The lapse α is left free to define the slicing while both A and B are evolved. In certain restricted cases, even greater simplification occurs. For instance, in static spherical spacetime (e.g. the standard Schwarzschild time slicing) $\alpha = 1 - \frac{2M}{r}$ is maximal (and time independent). One can use different radial coordinates to label the normal trajectories, for example, the Schwarzschild radial coordinate

$$A = A(r) = (1 - \frac{2M}{r})^{-1} , \; B = 1 \tag{2}$$

or the isotropic radial coordinate

$$A = B = (1 + \frac{M}{2r})^4 \tag{3}$$

Note that only one time independent metric function is required. In homogeneous spherical spacetime (e.g. Robertson-Walker spacetime), $\alpha = 1$ is a geodesic or, in this case, also a slicing of constant mean extrinsic curvature (see York this volume). Here $A = B = A(t)$ only.

However, in general, for a slicing which is not carried into itself by a time Killing fector field, such as the maximal slicing of Kruskal spacetime by Estabrook, et al. (1973), both A and B must be evolved if $\beta^r = 0$. This was the procedure used by the numerical code in Estabrook, et al. (1973) while the analytic solution presented therein had $\beta^r \neq 0$ (see below). This Eulerian gauge results in "spatial coordinates being sucked into the black hole" [Estabrook, et al. (1973)] since the shift vector cannot be used to push them outward.

2.2.2 Lagrangian gauge. The general theory of the comoving or Lagrangian gauge has been worked out by Taub (1978). This can be applied whenever matter fills the spacetime and is not restricted to spherical symmetry. The idea is to choose the coordinate trajectories to lie along the matter worldlines so that the matter is at rest with respect to the spatial coordinates. From the 3 + 1 point of view the obvious way to do this is to leave α free to determine the slicing and use β^i to locally boost from the time slice normals to the matter 4-velocity. Thus, one uses the shift vector to simplify the matter description instead of the gravitational field description. In spherical symmetry this leaves A and B to be evolved. This method was used by Chrzanowski (1977) for collapse calculations with pressure and by Eardley and Smarr (1978) for inhomogeneous dust collapse (Tolman-Bondi spacetimes). The lapse is used for maximal slicing so that the collapse singularity can be avoided (this does not always work in extreme examples - see Eardley and Smarr (1978)).

However, most work done on spherical systems in comoving gauge have not followed this natural approach. Instead β^r has been set to zero and the lapse function α has been used to bend the time slice so the matter is at rest in the slice. This is the gauge used for dust by Tolman (1934), Bondi (1947), and Oppenheimer and Snyder (1939); for pressure by Misner and Sharp (1964) and May and White (1966). This procedure can only work in spherical systems where the matter flow is irrotational and therefore always hypersurface orthogonal. This violation of the kinematical approach to Einstein's equations leads to the time slice becoming singular within a very short time of a black hole forming [see e.g. May and White (1966)]. This is the classic example of the utility of a kinematical analysis of gauge conditions.

2.2.3 Mixed Eulerian-Lagrangian Gauge. Finally, we come to the set of gauges in which the shift vector freedom is used to simplify the gravitational field. The computational frame which results is neither normal (Eulerian) nor comoving (Lagrangian). I will assume α is left free to specify the slicing. There are then three choices. The first is a generalization of the Schwarzschild like radial coordinate equation (2) achieved by setting $B = 1$. An example of this is the analytic solution of maximal slicing of Kruskal in Estabrook, et al. (1973). The second generalizes isotropic coordinates equation (3) by setting $A = B$. This is the gauge used by Wilson and myself in studying spherical collapse numerically (see Wilson this volume). In this case one can consider A as the conformal factor. The fact that the 3-metric is then conformally flat shows explicitly that no gravitational degrees of freedom are present. This gauge is a special case of the minimum distortion shift vector [Smarr and York (1978a,1978b); York, this volume]. Finally, one can set $A = 1$ which implies r is the proper radial distance in the 3-space. These three gauges have all been used by Wilson and myself on model spherical collapse with maximal slicing. Even with the spacetime and slicing fixed, the time development of β^r and A or B is quite different in different gauges. It is not yet clear which of these choices is to be preferred. In any case the important point of principle is that by allowing $\beta^r \neq 0$, the 3-metric can be simplified down to the single function A or B needed to describe the gravitational physics. It should be remarked in passing that by using a grid velocity (see Wilson-this volume), the mixed Eulerian-Lagrangian schemes can be made more or less Lagrangian by letting the grid points move relative to the spatial coordinates.

2.3 Nonspherical Axisymmetric Spacetimes

The key step in the development of numerical relativity was the success in solving the nonspherical Einstein equations (Smarr 1977). This is because only nonspherical systems allow gravitational radiation to be emitted. Thus, only with this step did one actually begin solving a full dynamical field theory. One might therefore imagine that the line elements will become much more complicated than in the spherical case. However, from the kinematical arguments presented by York in this volume, one can see that only two 3-metric functions are needed for a nonrotating axisymmetric system (one for scale, one for the + polarization waves) and only three for the rotating case (one more for the x polarization). That this can be obtained with the use of a natural gauge choice, while still retaining the lapse freedom for time slicing was only recently recognized by Wilson (see his discussion-this volume). This "isothermal" gauge will be outlined below.

Besides allowing radiation degrees of freedom, nonspherical flows have a tendency to form intersecting fluid worldlines, with complicated shock fronts and self intersecting swirls (see Wilson-this volume). For that reason Lagrangian schemes are not well suited to nonspherical systems. Therefore I will concentrate on discussing the classes of gauges most often met in the general relativistic literature.

2.3.1 Static and Stationary Gauges. Let us begin with the simplest case: the static, nonrotating, vacuum axisymmetric line element [Weyl (1917); Levi-Civita (1917)]. In 3 + 1 terms the lapse α is maximal (the time slices are time symmetric), the shift vector β^i vanishes, and the 3-metric γ_{ij} has only one time independent component:

$$ds^2 = A^2(dR^2 + dZ^2) + R^2 \alpha^{-2} d\phi^2 \tag{4}$$

The essential feature of this gauge is that $\gamma_{RR} = \gamma_{ZZ}$. This is termed the "isothermal" form (Synge 1960).

The relation $g_{\phi\phi} = R^2\alpha^{-2}$ holds only for vacuum (actually only $R^\phi_\phi + R^t_t = 0$ is required - Synge 1960). In the general case with matter (but all other restrictions holding), $g_{\phi\phi} = R^2B^2$ where B is an independent metric function, but of course it is time independent. Lewis (1932) showed that one can easily generalize the static line element to the stationary case. In the 3 + 1 language we can say that the generalization obtains by leaving the 3-metric in the Weyl "isothermal" form (1) and adding a shift vector with only a ϕ-component (β^ϕ). As in the static case, $g_{\phi\phi} = \rho^2\alpha^{-2}$ in vacuum or $g_{\phi\phi} = \rho^2B^2$ if matter is present. The existence of the time Killing vector implies that the natural gauge is maximal slicing (α) and minimal distortion

(β^i)(Smarr and York 1978).

2.3.2. General Time Dependent Gauge. It is perhaps easier to classify various gauge specializations if we start by discussing the general case first. In a nonrotating axisymmetric spacetime we lose no generality by setting $\beta^\phi = 0$ and using a 3-metric:

$$ds^2 = AdR^2 + BdZ^2 + 2CdRdZ + DR^2d\phi^2 \tag{5}$$

with four functions of $(t,R,Z)(\gamma_{R\phi} = \gamma_{Z\phi} = 0)$. This leaves (α,β^R,β^Z) free for kinematic chores. If the spacetime is rotating then all ten components of the four metric are nonzero prior to any gauge choice. Two examples of gauge choices which require this generality would be to demand comoving coordinates with matter (Taub 1978) to simplify the hydrodynamics equations or to demand minimal distortion shift vectors (Smarr and York 1978a,b; York-this volume) to reduce unphysical shearing of the coordinates. Both of these schemes have many applications to questions of pure theory, but neither has been successfully used numerically (see, however, Eppley 1975) because of the extreme complexity of the equations. Also, it is clear that both have some drawbacks computationally, as remarked above.

For nonrotating spacetimes, the pure Eulerian scheme of normal gauge ($\beta^R = \beta^Z = 0$) has been used frequently. The lapse is left free for slicing and the Einstein equations are used to evolve the four functions A,B,C,D in equation (5). In particular, this scheme has been used exclusively in the two black hole collision problem. Hahn and Lindquist (1963) used geodesic slicing ($\alpha = 1$) which leads to singular time slices in a freefall time scale (Smarr and York 1978). Building on work of Smarr, Čadež, DeWitt, and Eppley (1975), maximal time slicing was finally successfully used for the vacuum Einstein equations in the two black hole collision problem (Smarr 1977, Eppley and Smarr 1978a,b; this article). A bad result of using normal coordinates is that 1) large kinematic shear builds up in the 3-metric and 2) the Einstein equations have a very large number of terms in them through the 3-dimensional Ricci tensor built out of four independent metric functions.

For rotating spacetimes, a natural extension of the normal gauge is to require $\beta^R = \beta^Z = 0$ and use $\beta^\phi \neq 0$ to take out the rotational motion in some manner. This would follow the method Lewis (1932) used in going from static to stationary spacetime. Unfortunately, whereas in the stationary case the coordinate trajectory aligns itself with the time Killing vector field, in the dynamic case there is no unique specification of β^ϕ. One possibility would be to apply the minimal distortion shift vector equation [York equation (126)-this volume] only to β^ϕ. Another attempt is that of Pachner (1975) where he demands comoving coordinates but also $\beta^R = \beta^Z = 0$. This forces him to give up the time slicing freedom in α. Indeed his condition on the lapse is just the rotating version of the one used by May and White (1965) and therefore will fail in strong fields for the same reasons as mentioned in section 2.2.2.

2.3.3 Diagonal Gauge. We now move to the gauges which kill off particular components of the 3-metric. Historically, of course, most researchers have not thought in terms of 3 + 1, but rather in terms of the full 4-metric. Therefore, the philosophy was to zero as many 4-metric components as possible and then impose the full Einstein equations $G_{\mu\nu} = T_{\mu\nu}$. This leads to the least number of terms in the field equations. Unfortunately, even in the axisymmetric nonrotating case the equations were too complicated to solve analytically. Thus, even the researchers who followed this approach (Rosen and Shamir 1957; Voorhees 1971; Chandrasekhar and Friedman 1972; Cooperstock 1974) were forced to study only the linearized equations of the following "diagonal gauge" form:

$$^{(4)}ds^2 = -\alpha^2 dt^2 + AdR^2 + BdZ^2 + DR^2 d\phi^2 \tag{6}$$

Evidently, by comparison with the general nonrotating gauge equation (5), the lapse function α must be fixed by:

$$C \equiv \gamma_{RZ} = \partial_t \gamma_{RZ} = 0 \tag{7}$$

since the shift vector freedom has been thrown away $\beta^R = \beta^Z = 0$. Thus, in the diagonal gauge the time slice must bend itself in such a way that the intrinsic 3-metric of the slice stays diagonal. This makes the lapse function α do the work which is naturally reserved for the spatial coordinate congruence (through β_i). Worst of all, this gauge does not allow the time slicing to be freely specified and therefore is unlikely to have sufficient flexibility to avoid hitting singularities in strong field regions.

The generalization to the rotating case of the diagonal gauge was carried out by Chandrasekhar and Friedman (1972). The submatrix of $g_{\mu\nu}$ not involving ϕ is still diagonal (i.e. $\gamma^{Z\rho} = \beta^Z = \beta^\rho = 0$), but the three new quantities $(\beta^\phi, \gamma^{Z\phi}, \gamma^{R\phi})$ are added. The problems with the lapse function being determined by $\partial_t \gamma_{ZR} = 0$ are the same as in the nonrotating case. Furthermore, one has given up the kinematic freedom in β^i without simplifying either the number of 3-metric components or their shear.

2.3.4 Isothermal Gauge. Obviously, what one would like is to leave α free for slicing and still eliminate as many metric functions as possible. As noted in section 2.3, one needs only two functions in the 3-metric for the nonrotating case and three functions for the rotating case. Wilson (this volume) has recently investigated such a gauge. Here the kinematic freedom in

(β_R, β_Z) is used to keep the dynamic 3-metric in the same form as the static Weyl 3-metric, i.e.

$$ds^2 = A^2(dR^2 + dZ^2) + B^2R^2d\phi^2 \tag{8}$$

That is, the two shift vector components (β_R, β_Z) are fixed by the demand (7) as well as:

$$\gamma_{RR} - \gamma_{ZZ} = \partial_t(\gamma_{RR} - \gamma_{ZZ}) = 0 \tag{9}$$

The resulting simplification of the Einstein equations is enormous (Wilson and Eppley this volume) over the normal coordinate gauge (Eppley 1975) even though in both cases there are five 4-metric functions. As mentioned above, this is because of the clean geometric split between kinematic variables (α, β_i) and dynamic variables (γ_{ij}) which occurs in the 3 + 1 approach (York-this volume). It is a general lesson that if one can trade dynamic variables for kinematic ones, then the resulting Einstein equations are much simpler. One might think that the non-3 + 1 diagonal gauge (6) which requires one less 4-metric function than the Wilson gauge (8) would have even simpler field equations. However, a comparison of the equations (Chandrasekhar and Friedman, 1972, with $\omega = q_2 = q_3 = 0$ and Wilson, this volume), shows otherwise. We note in passing that in the special case of spherical symmetry, the Wilson isothermal gauge and the minimum distortion gauge A = B are the same.

The natural generalization of the isothermal gauge to the rotating case was suggested by Wilson and Smarr (1978). Since we must add $(\beta^\phi, \gamma_{R\phi}, \gamma_{Z\phi})$, we choose β^ϕ by demanding

$$\gamma_{R\phi} - \gamma_{Z\phi} = \partial_t(\gamma_{R\phi} - \gamma_{Z\phi}) = 0 \tag{10}$$

This leaves the 3-metric with only three functions:

$$ds^2 = [A(dR + dZ) + 2Ed\phi](dR + dZ) + DR^2d\phi^2 \tag{11}$$

which treat (R,Z) on equal footing.

3 GRAVITATIONAL RADIATION FORMULAE

3.1 Calculating Radiation

Assuming one chooses some gauge for an axisymmetric spacetime, as described in section 2, one then builds a spacetime numerically as described by Eppley, Piran, Wilson, or myself in this volume. One of the major problems is how to extract information about the gravitational radiation from that numerical spacetime. Preliminary results were discussed by Smarr (1977). Since then a systematic approach of testing many different methods in the literature has been carried out by Eppley, Smarr, and Teukolsky (1979), hereafter referred to as EST. In this section I will briefly summarize the theoretical results of that study, while Eppley (this volume) discusses the numerical experiments on linearized waves. This section will be useful as background for the full nonlinear calculations described in my section 4 and by Piran and Wilson in their articles.

3.2. Curvature Formulae

I shall divide the radiation formulae into two broad classes: The curvature formulae and the connection formulae. The first measures the gravitational wave energy flux by time integration of the Riemann curvature tensor, while the second directly combines connection quantities such as Γ^i_{jk} and K_{ij}, the 3-dimensional Christoffel symbols and the extrinsic curvature, to yield an energy flux. These classes are each subdivided into formulae obtained by considering spacetime split into spacelike hypersurfaces and by those formulae derived from the null formation. As shown in Figure 1 of Smarr (1977), one can consider a 2-sphere S, surrounding the radiating system, to lie either in a spacelike hypersurface or in a null hypersurface. Thus, to leading order in $1/r$, one can write each formula as a flux vector through a 2-sphere of area $4\pi r^2$ with surface element $d\Omega$:

$$d\varepsilon/dt = \frac{1}{4\pi} \oint_S P^r \, r^2 d\Omega \tag{12}$$

where it is assumed that the flux is normal to the 2-sphere. Here $\varepsilon(t)$ is the energy carried through the 2-sphere by a gravitational wave. The local time dependence $\varepsilon(t)$ must be integrated in time to get a physical quantity.

In vacuum, where most radiation measurements are made, the Riemann tensor is identical to the Weyl tensor C_{abcd}, which has ten independent components. This can be split into "pieces" by the introduction at each event of either a timelike or null vector. Since our approach is based on the 3 + 1 split (see York-this volume) I consider the unit timelike vector n^a, normal to the time slices as fundamental. This vector field splits the Weyl tensor into its electric and magnetic parts (Matte 1952; for notation see Smarr 1977):

$$\begin{aligned}
E_{ab} &= n^c C_{cabcd}\, n^d \\
B_{ab} &= \tfrac{1}{2}\, \varepsilon_a^{\ cd}\, n^e \gamma_b^f\, C_{efcd} \\
E^a_a &= B^a_a = E_{ab}\, n^b = B_{ab}\, n^b = 0 \\
E_{[ab]} &= B_{[ab]} = 0
\end{aligned} \tag{13}$$

Note that E_{ab} and B_{ab} each contain five of the ten components of C_{abcd}.

Alternatively, this vector n^a can be combined with a spacelike unit vector r^a (Smarr 1977) to yield ingoing (k^a) and outgoing (ℓ^a) null vector fields:

$$k^a = \frac{1}{2}(n^a - r^a)$$
$$\ell^a = n^a + r^a \tag{14}$$

which are normal to the spacelike 2-sphere S spanned by the complex null vectors m^a, $\overline{m}^a$. The Weyl tensor can then be split into the five complex Newman-Penrose curvature quantities ψ_0, ψ_1, ψ_2, ψ_3, ψ_4, again yielding ten independent pieces of C_{abcd}. I will only need ψ_4, the curvature quantity which falls off as 1/r in the wave zone:

$$\psi_4 = - R_{abcd}\, k^a\overline{m}^b k^b\overline{m}^d$$
$$= \frac{1}{4}\left[\overline{m}^a E_{ab}\overline{m}^b - 2\overline{m}^d B_{de}\varepsilon^{cab} r_a \overline{m}_b - r^a m^b \varepsilon_{abr} E^{mn} \varepsilon_{ncd} r^c \overline{m}^d\right] \tag{15}$$

These curvature tensors can then be used to measure energy loss by gravitational radiation. The method used by many perturbation calculations of black holes (see e.g. Teukolsky 1973) is to take $\psi_4(u,\theta,\phi)$, here built out of E_{ab}, B_{ab} by equation (15), and integrate in time fixed $r(u \equiv t-r)$.

$$P^r_{PSI} \equiv \left[\int_0^t dt'\psi_4(t' - r,\theta,\phi)\right]^2 \tag{16}$$

Since at large distances from the source

$$\psi_4 \to R_{\hat{t}\hat{\theta}\hat{t}\hat{\theta}} + i\, R_{\hat{t}\hat{\theta}\hat{t}\hat{\phi}}$$
$$\to E_{\hat{\theta}\hat{\theta}} + i\, E_{\hat{\theta}\hat{\phi}} \tag{17}$$

(where carets denote orthonormal tetrad components). one could (see e.g. Gibbons and Hawking 1971) use $E_{\hat{\theta}\hat{\theta}}$ directly as

$$P^r_{TID} \equiv \left[\int_0^t dt'(E_{\hat{\theta}\hat{\theta}} + i\, E_{\hat{\theta}\hat{\phi}})\right]^2 \tag{18}$$

Since in the far wave zone $E_{\hat{\theta}\hat{\theta}} = -E_{\hat{\theta}\hat{\theta}}$, it may be more accurate to use $\frac{1}{2}$ $(E_{\hat{\theta}\hat{\theta}} - E_{\hat{\phi}\hat{\phi}})$ in (18) instead of $E_{\hat{\theta}\hat{\theta}}$. Note that the ψ_4 formula (15) uses both E_{ab} and B_{ab} and thus effectively averages over more curvature components to find the mass loss than using (18) with only E_{ab}.

As pointed out in Smarr (1977), the time integral in equations (16) and (18) can be replaced by

$$\int_0^t \rightarrow - i/\omega \tag{19}$$

for a monochromatic linearized plane wave with time dependence $e^{i\omega t}$. One can formally extend this to a wave packet of diverging wave fronts by defining Poynting vectors:

$$P^r_{PSI\omega} \equiv |\psi_4|^2/\omega^2 \tag{20}$$

$$P^r_{TID\omega} \equiv |E_{\hat{\theta}\hat{\theta}} + i\, E_{\hat{\theta}\hat{\phi}}|^2/\omega^2. \tag{21}$$

The errors made by this approximation will be discussed below. The other curvature measure of energy loss, introduced in Smarr (1977), is the frequency weighted Bel-Robinson vector

$$P^r_{BR\omega} \equiv E_{ab}\, \varepsilon^{brd}\, B^a_d/(2\omega^2) \tag{22}$$

(This very useful approach is due in large measure to Eppley). Note that this uses both E_{ab} and B_{ab}, like the ψ_4 formula (15), but in the form E·B rather than $(E + B)^2 \sim E^2 + 2E{\cdot}B + B^2$.

3.3 Connection Formulae

Since all the curvature formulae require a time integration to obtain energy loss, it is not surprising that one could also use directly those quantities which when differentiated yield curvature tensors. These are the connection formulae. The null approach uses the connections in the form of complex spin coefficients derived from the null vectors ℓ^a, k^a defined by equation (14). The two we need are:

$$\sigma \equiv m^a m^b \nabla_a \ell_b \tag{23}$$

$$\tilde{\sigma} = m^a m^b \nabla_a k_b \tag{24}$$

which measure the shear of the outgoing and ingoing null vector respectively. As shown in Smarr (1977) equation (27), σ can be related to the spacelike connections Γ^i_{ij}, K_{ij}. To get $\tilde{\sigma}$ one changes the relative sign between Γ^i_{jk} and K_{ij} in equation (27) and replaces the overall factor of 1/2 by 1/4 (D. Eardley - private communication).

These connections are then used in the Bondi et al. (1962) mass loss formula, which can be written as equation (12) with

$$P^r_{B\sigma} = r^2 |\partial\sigma/\partial t|^2 \tag{25}$$

assuming we are at fixed r, or with $\tilde{\sigma}$:

$$P^r_{B\tilde{\sigma}} = |\tilde{\sigma}|^2 \tag{26}$$

Since both σ and $\tilde{\sigma}$ are obtained from K_{ij}, Γ^i_{jk} by virtually identical formulae (see remarks above), it is clear that using $\tilde{\sigma}$ is much simpler than using σ which must be differentiated. (This was pointed out to me by D. Eardley-private communication).

The spacelike connection formulae include the energy-momentum-stress pseudotensor (Landau and Lifshitz 1962) and the Arnowitt, Deser, and Misner (1962) Poynting vector. These all agree if Isaacson averaged (Misner, Thorne, and Wheeler 1973) over several wavelengths. However, if not averaged, which has not been practical in numerical relativity, they are quite gauge dependent. We choose the ADM Poynting vector as representative:

$$P^r_{ADM} = -\frac{1}{8} f^{ij} f^{\ell n} f^{mp} [(\partial_t h_{mp})(2D_m h_{j\ell} - D_j h_{\ell m})] \tag{27}$$

where we assume the 3-metric γ_{ij} can be split into

$$\gamma_{ij} = f_{ij} + h_{ij} \tag{28}$$

where f_{ij} is a background, usually flat, plus a perturbation. Then D_j is the covarient derivative with respect to f_{ij}. Note that this spacelike formula has the form $P \sim K \cdot \Gamma$ whereas the null formula (26) has the form $P \sim (K \pm \Gamma)^2 \sim K^2 \pm 2K \cdot \Gamma + \Gamma^2$. This suggests that one might use connection formulae with P built out of just K^2 or Γ^2. Such a formula has been discovered by Eppley (this volume) and Wilson (this volume). However, this formulae holds only in the isothermal gauge discussed in section 2.3.4 above. Note that K^2 is a sort of energy density since it appears in the Hamiltonian constraint equation (York-this volume-eqn(23)). Penrose (1966) discusses using $\sigma^2 \sim (K \pm \Gamma)^2$ as an energy density of gravitational waves from the null point-of-view.

3.4 A Linearized Solution

As described by Eppley in this volume, and in more detail in EST, one can find an analytic solution to the linearized Einstein equations which represents a diverging pure quadrupole gravitational wave in transverse traceless gauge ($\alpha = 1$, $\beta^i = 0$). Keeping only the leading terms in $1/r$ from Eppley's equation (38) one has

$$h_{\theta\theta} = r^2 \zeta \sin^2\theta \qquad h_{\phi\phi} = -r^2 \zeta \sin^2\theta$$

$$h_{r\theta} = h_{rr} = 0$$

$$\zeta \equiv \frac{3}{4}\left[\frac{F^{(4)}}{r} + \frac{2F^{(3)}}{r^2}\right] \qquad F = F(t-r) \tag{29}$$

where h_{ij} is the perturbation away from Minkowski spacetime written in spherical coordinates.

In EST we evaluate all the Poynting vectors discussed above

for the full metric given in Eppley (this volume). Here I will summarize the results for the leading 1/r term in Table 1. One notes the clear distinction between curvature formulae which involve second derivatives of ζ and the connection formulae which involve only first derivatives of ζ. All the curvature formulae require a time integration or division by ω. Furthermore, one sees that all the formulae require the square of the derivatives of ζ. It is interesting that the null expressions are the square of the sum or difference of derivatives, while the spacelike expressions are direct products of derivatives. Finally, one can see that the forms requiring division by ω^2 are $\pi/2$ out of phase with the others [$F^{(6)}$ vs. $F^{(5)}$].

Table 1. Gravitational Radiation Formulae. $d\varepsilon/dt = \frac{1}{4\pi} \oint P^r r^2 d\Omega$

For each Poynting vector defined by equation number in column 2, we list whether it is a curvature (CR) or connection (CN) formula (column 3) and whether it is a spacelike (S) or null (N) quantity.

Name	Eqn.	CR or CN	S or N	P^r linearized wave	$\frac{128}{9} * d\varepsilon/dt$
P^r_{PSI}	(16)	CR	N	$\frac{1}{16}[\int_0^t d\tilde{t}\ (\ddot{\zeta}-\dot{\zeta}')]^2 \sin^4\theta$	$[F^{(5)}]^2$
$P^r_{PSI\omega}$	(20)			$\frac{1}{16\omega^2}(\ddot{\zeta}-\dot{\zeta}')^2 \sin^4\theta$	$\frac{1}{\omega^2}[F^{(6)}]^2$
P^r_{TID}	(18)	CR	S	$\frac{1}{4}[\int_0^t d\tilde{t}\ \ddot{\zeta}]^2 \sin^4\theta$	$[F^{(5)}]^2$
$P^r_{TID\omega}$	(21)			$\frac{1}{4\omega^2}(\ddot{\zeta})^2 \sin^4\theta$	$\frac{1}{\omega^2}[F^{(6)}]^2$
$P^r_{BR\omega}$	(22)	CR	S	$-\frac{1}{2\omega^2}\ddot{\zeta}\ \dot{\zeta}' \sin^4\theta$	$\frac{1}{\omega^2}[F^{(6)}]^2$
$P^r_{B\sigma}$	(25)	CN	N	$\frac{r^2}{4}[\partial_u(\dot{\zeta}+\zeta')]^2 \sin^4\theta$	$[F^{(5)}]^2$
$P^r_{B\tilde{\sigma}}$	(26)	CN	N	$\frac{1}{16}(\dot{\zeta}-\zeta')^2 \sin^4\theta$	$[F^{(5)}]^2$
ADM	(27)	CN	S	$-\frac{1}{4}\dot{\zeta}\ \zeta' \sin^4\theta$	$[F^{(5)}]^2$

3.5 Numerical Problems

I believe the previous section clearly illustrates the unexpected variety in which the various radiation formulae extract energy loss information from even the simplest linearized wave. Of course, when ζ (equation 29) is substituted in analytically, these all agree exactly. However, when one attempts to perform numerically the operations on ζ listed in Table 1, one finds some formulae are far superior to others. I will briefly discuss some of the major problems we have discovered.

First, for the curvature formulae, one must decide whether to time integrate or to divide by some frequency ω. We have found the latter method to be more useful even though the waves generated by collisions or collapse are not monochromatic. The reason is that time integration of an oscillating function is effectively subtracting large quantities (peaks and troughs) to find a small quantity (the total area under the curve). Thus, for the shape of the waveform or details of near zone - far zone transition (see section 4. - below), we find that using ψ_4/ω or P^r_{BR} yields more detail than is seen in the time integrated quantities. As for total energy loss, the absolute numerical value may be slightly in error if the Fourier spectrum of the wave is not peaked about the value of ω chosen. However, for both the linearized and Brill waves (Eppley-this volume) and perturbations of black holes (Detweiler and Moncrief-this volume), one finds that simply measuring the distance between wave crests in ψ_4 to define λ and then using $\omega = 2\pi c/\lambda$ works very well. For black hole perturbations, this is because most of the energy comes out in nearly monochromatic normal modes. (See discussion by Detweiler this volume). On the other hand, if time integration is performed rather than division by ω, the total energy loss diverges badly in all our strong field calculations. We have traced this to the numerical noncancellation mentioned above. Even more impossible is the double time integration to get the amplitude A_+ as suggested by Thorne (1978). These are difficulties which may be overcome with more sophisticated numerical techniques.

Second, we now understand why the outgoing Bondi formula (25), discussed by Smarr (1977), did not work very well. If one constructs $\dot{\zeta} + \zeta'$ from equation (29), one sees that the 1/r terms cancel exactly leaving the energy flux to be determined from the $1/r^2$ terms. This is why only the $P^r_{B\sigma}$ Poynting vector has a r^2 multiplying it. The fact that it is a higher order formulae is disastrous numerically. The numerical noncancellation of the 1/r parts of $\dot{\zeta}$ and ζ' will in general dominate the $1/r^2$ piece. The use of the ingoing Bondi $P^r_{B\tilde{\sigma}}$ avoids this problem. The geometric reason for this problem was pointed out to me by D. Eardley. Since the shearing gravitational waves are propagating outward along ℓ^a, the effect of the shear on ℓ^a is higher order.

However, since the waves cross the k^a congruence, the effect on their shear, $\tilde{\sigma}$, is direct.

Finally, we have found that the curvature measures are more reliable for measuring energy loss (see Eppley, this volume, for discussion) than the connection measure. I suspect this is simply a reflection of the fact that curvature is more gauge invariant than connection. This shows up in at least two ways. One is that our radial coordinate grid moves in and out during the calculations (see below). This coordinate velocity may show up much more strongly in first derivatives than in second. The other reason is that ψ_4 or $P^r_{\tilde{B}R}$ are gauge invariant in the sense that they vanish in spherical backgrounds. In fact, $P^r_{\tilde{B}R}$ also vanishes on a time-symmetric initial data slice. Since calculations, such as the two black hole collision, start with time symmetric conditions and evolve to spherical states with outgoing radiation perturbations, $P^r_{\tilde{B}R}$ has no zeroth order background piece to hide the first order radiating piece. For similar reasons, the use of gauge invariant quantities has been found to be essential in perturbation problems (Moncrief and Detweiler-this volume) so it is not surprising that they are even more so in the full theory.

A more detailed discussion of the success and failure of various radiation formulae is contained in EST and in Eppley and Smarr (1978b). Much more work needs to be done in this area: 1) to find ways in which the connection formulae can be used and 2) to investigate rigorously the relation between time integration and division by ω.

4 THE TWO BLACK HOLE COLLISION

As an illustration of the use of the concepts discussed above, I will describe the results of the two black hole collision problem. Since this subject has been extensively reviewed [Smarr, Čadež, DeWitt, and Eppley (1976); Smarr (1977)], I will focus on the work done since these articles. A more detailed analysis will appear soon [Eppley and Smarr (1978a,b)]. The major reason for discussing this subject is that it is the only complete calculation to date of an asymptotically flat, highly nonspherical, radiating solution of the full nonlinear Einstein equations. As such, it is a laboratory for learning about the strong field sources of gravitational radiation which one someday hopes to observe (Weiss-this volume). This work was done in collaboration with Kenneth Eppley at Lawrence Livermore Laboratories.

4.1 The Spherical Limit

In Table 2, I list a comparison of the one and two black hole problems. The vacuum black holes are represented by one

Table 2. A comparison of the one and two black hole problems.

Property	One Black Hole	Two Black Holes
Initial Data	$K_{ij} = 0 \qquad \gamma_{ij} = \psi^4 \delta_{ij}$	$K_{ij} = 0 \qquad \gamma_{ij} = \psi^4 \delta_{ij}$
Constraint Equation	$\Delta\psi = 0 \qquad \psi = 1 + \frac{M}{2r}$	$\Delta\psi = 0 \quad \psi = 1 + \Sigma \operatorname{csch}(n\mu_o)[\frac{1}{+r_n} + \frac{1}{-r_n}]$ $\pm r_n = \{R^2 + [Z + \coth(n\mu_o)]^2\}^{\frac{1}{2}}$
Parameters	Total Mass M	M , $L/M = f(\mu_o)$ = proper distance between throats
Topology	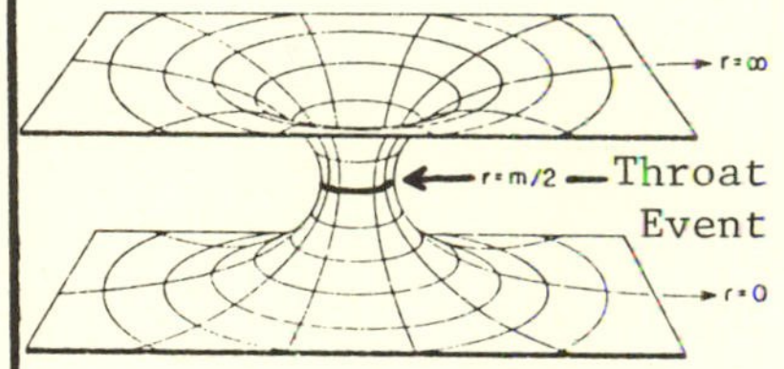	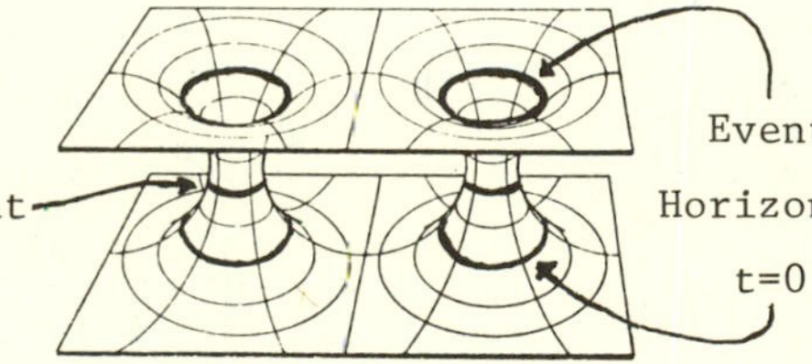
Lapse Function	$\alpha \to 1$ as $r \to \infty$ $\alpha = 0$ on throat → static slicing $\partial_i\alpha = 0$ on throat → dynamic slicing	$\alpha \to 1$ as $r \to \infty$ $\alpha = 0$ on throat } dynamic $\partial_i\alpha = 0$ on saddle } slicing
Spacetime Metric	$ds^2 = -\alpha^2 dt^2 + \psi^4(Adr^2 + Br^2 d\Omega^2)$	$ds^2 = -\alpha^2 dt^2 + \psi^4(AdZ^2 + BdR^2)$ $+ 2CdRdZ + DR^2 d\phi^2$

or two Einstein - Rosen (1935) bridges between two asymptotically flat spaces. [The figures are adapted from Hahn and Lindquist (1963)]. The initial data is formally the same: time symmetric and conformally flat. The conformal factors are simple solutions to the flat space Laplacian. Being spherical, the one black hole has only a scale-the total mass of the hole. The two black hole initial data has in addition a dimensionless distortion parameter μ_o which, for instance, gives the proper separation of the throats L/M at t = 0, in units of the total mass of the system. At t = 0, the throat and the event horizon coincide for one black hole, while the throat lies inside the event horizon (which may or may not be disjoint) for two black holes.

If we evolve these initial data using maximal time slicing and normal coordinates (zero shift), then we find the line elements given by (1) and (5) discussed in section 2 above. These are rewritten in Table 2 as they were used in the computation, with the conformal factor taken out front. The reason we do this is that the determinant of the 3-metric will be time independent in this gauge [equation (2.26) of Smarr and York (1978)]. Since the overall conformal factor can be taken to be $(\det\gamma)^{-1/3}$ [York-this volume, section 7], this implies that $\psi^4 = (\det\gamma)^{-1/3}$ for all time. Therefore the metric functions (A,B) for one black hole and (A,B,C,D) for two black holes, are just the components of the conformal 3-geometry discussed by York in this volume.

The maximal time slicing depends on the boundary conditions placed on the lapse function α on the throat. In the one black hole, setting $\alpha = 0$ there leads to the standard static time slices, while using the Neumann condition, $\partial_i\alpha = 0$ on the throat, leads to the dynamic evolution of Estabrook, et al. (1973). (See York section 8.5 for more details). In the two black holes there is a mixture since $\alpha = 0$ on the throats, but $\partial_i\alpha = 0$ on the saddle point, halfway between the throats. Since at late times, the saddle point will be the "center" of the coalesced black hole, we might expect the late time metric evolution to be similar in the two spacetimes.

That they are was demonstrated in Eppley (1975) and Smarr (1977). Eppley's (1975) Figure 3 shows the time dependence of the functions (α,A,B) listed in Table 2. One sees that a large peak develops in the radial metric A just inside the horizon, while a dip occurs in B. The lapse function collapses to zero in the region inside the black hole [Smarr and York (1978); York-this volume, section 8.6]. It was reported in Smarr (1977) that this also happened for the two black holes. With the runs since then we can describe this process in more detail.

4.2 The Lapse Collapse

In our latest work on the two black holes, Eppley and I made long time runs ($t = 0 \rightarrow t = t_{final}$) for three initial values of μ_o or L/M (see Table 3). In Smarr (1977) only Runs I and II had been made, each for less time. The initial separations L/M as well as the Newtonian free-fall time t_{ff}/M [Smarr (1977) equation (39)] are also given in Table 3.

Table 3. Parameters of the three runs discussed in the text.

RUN	μ_o	$\frac{L}{M}$	$\frac{t_{ff}}{M}$	$\frac{t_{collapse}}{M}$	$\frac{t_{collision}}{M}$	$\frac{t_{bulge}}{M}$	$\frac{t_{final}}{M}$
I	2.00	3.9	6.9	14	7	6	66
II	2.75	6.6	17.3	24	17	21	80
III	3.25	9.6	31.4	40	33	34	80

The initial lapse function for run II is shown in Figure 1a. The vertical scale is 0 to 1. Only the innermost region of the grid ($r < 8M$) is shown. The edge of the full grid was at $r \sim 40M$. For Run(I,III) the lapse was (smaller, larger) in the saddle region. As the evolution proceeds the lapse drops in the central region until at $t \sim t_{collapse} \sim 24M$ (see Table 3), the value of α at the saddle becomes $\lesssim 0.05$. By $t \sim 36M$ the entire central region has $\alpha \simeq 0$ (Figure 1b) and looks very similar to the late time Schwarzschild lapse function. In Run III, the lapse first collapses in a region around each throat and only at $t \sim 40M$ finally collapses in the central region. Thus, as L/M increases the lapse shows that the holes act as individual black holes for a time and then coalesce.

To try and pinpoint the moment of collision, one can notice that in the maximal slicing of Schwarzschild $t_{collapse} \sim 7M$, even though the central region of the grid is always inside a black hole. As discussed by York in this volume, one expects the lapse to collapse only as curvature builds up in the central region. Thus, it is a good first guess to say that 7M before $t_{collapse}$ the saddle region came inside the final black hole. As shown in Table 3, this $t_{collision}$ agrees remarkably well with the Newtonian free-fall time t_{ff}. This gives one some confidence in the internal consistency of the solutions.

That the lapse falls to zero in time for the slices to avoid hitting the supposed spacelike singularity of this spacetime

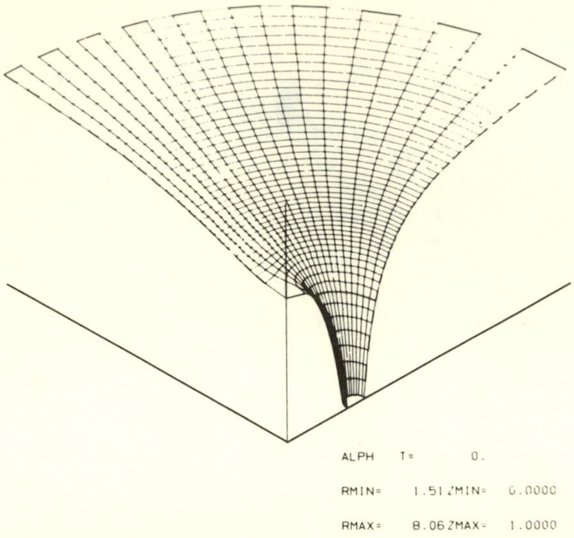

Figure 1a. The lapse function for the two black hole run II at t = 0. The equator (Z = 0) is to the left and the symmetry axis (R = 0) is to the right. The vertical scale is 0 to 1. The lapse is zero on the throat.

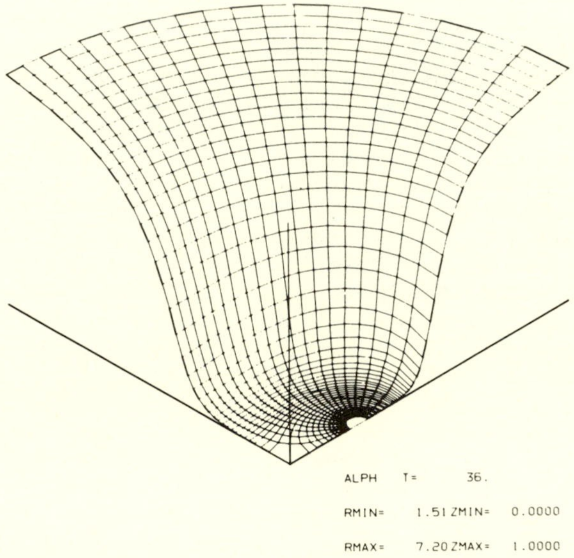

Figure 1b. The lapse function at t = 36M. The entire central region is now inside the final black hole.

gave strong support to the notion that maximal slicing always avoids singularities. However, as shown by Smarr and York (1978), York-this volume, Eardley and Smarr (1978), and Eardley-this volume, we now know that this is not necessarily true. Whether maximal slicing can fail in a vacuum spacetime is an open question.

4.3 Metric Shear

As was anticipated by Čadež (1975), the gauge choice of maximal slicing and zero shift vector leads to large shearing in the 3-metric. This is an example of the unnecessary coordinate shear that the minimal distortion shift vector [Smarr and York (1978); York this volume] was designed to get rid of. However, the minimal distortion shift vector equation [York equation (126)-this volume] becomes enormously complicated when there are four nonzero metric functions [see Eppley (1975)]. Furthermore, as York discusses one would need to impose the Neumann boundary condition on the throats to really minimize the coordinate shear and this would feed grid lines from the bottom sheet to the top sheet. This would complicate matters considerably, since at present we evolve only the upper sheet.

Therefore, we used a zero shift vector. In Smarr (1977), the conformal radial 3-metric for Run I at t = 20M is shown. It is clear that for this "coalescence spacetime" a spherical final state is reached very shortly after t = 0 ($t_{collision} \sim 7M$). The peak in the conformal radial metric is characteristic of the maximal slicing of a single black hole. In Figure 2, I show the radial metric for Run III at t = 22M. The outermost grid line has r $\sim$ 11M. It is clear that each hole is still acting

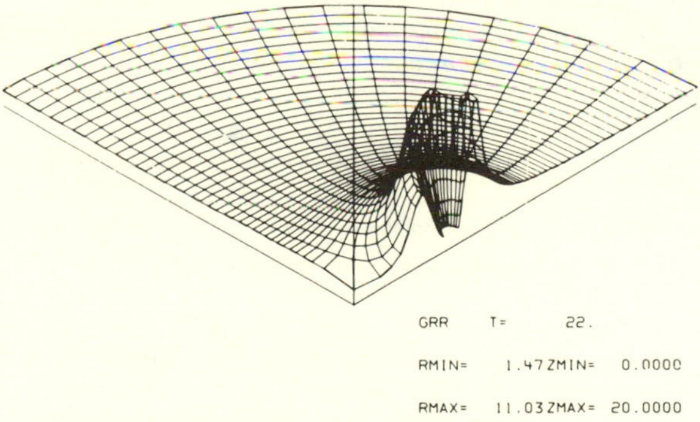

Figure 2. The conformal radial metric function $\tilde{\gamma}_{rr}$ constructed from $\tilde{\gamma}_{ZZ}$, $\tilde{\gamma}_{ZR}$, $\tilde{\gamma}_{RR}$ for Run III at t $\simeq$ 22M.

individually at this time, just as was shown by the lapse. In Table 3, I list t_{bulge} which is the time a bulge in the conformal radial metric first appears on the equator, signalling the coalescence (here $t_{bulge} \simeq 34M$). Note that this measure of the collision time agrees fairly well with that determined by the lapse.

It should be mentioned that this "stretching of the grid" near the horizon is difficult to follow accurately numerically, particularly when a small amount of artificial viscosity has been added to the numerical scheme (see my other article in this volume). As a result, the proper areas of 2-spheres near the horizon are not particularly accurate measures of where the exact horizon is. This effect gets worse as the initial distance between the holes is increased. However, Eppley and I feel that this problem does not seriously affect the radiation generated. [For more details see Eppley and Smarr (1978a)].

4.4 Gravitational Radiation

Whereas the previous sections (4.1, 4.2, 4.3) have shown how the gauge considerations of section 2 effect the metric evolution in a particular spacetime, I will now show how the gravitational radiation formulae discussed in section 3 are implemented in this spacetime.

Very preliminary results on the gravitational radiation from colliding black holes were reported in Smarr (1977). There has been a major advance in this portion of the work. Eppley and I have integrated nearly three times as far into the future for Run II as we had in Smarr (1977). As a result, we could afford to put our radiation 2-sphere at $r \sim 25M$ instead of $r \sim 8M$ as in Smarr (1977). This made measuring gravitational radiation much more reliable. Even though our 2-sphere was further out, we were able to get much more of the waveform out than previously. Also, we corrected a few coding errors and learned new methods for analyzing the generation process. I think the radiation results are now fairly solid and physically reasonable.

One thing which has <u>not</u> changed since the previous discussion is the usefulness of the Bel-Robinson vector for following gravitational radiation. As explained in section 3.4 above, we think this is because the P^r_{BR}, equation (22), is formed by <u>multiplying curvatures</u> and is automatically zero in a time symmetric ($t = 0$) time slice or in a spherically symmetric one (late times). We have also successfully used ψ_4, equation (15), to locate radiation. Using the $1/\omega$ rule discussed in section 3.2, we have used both $P^r_{PSI\omega}$ and $P^r_{BR\omega}$ to measure the energy radiated. These answers agree quite closely as described below.

One trick we used to make ψ_4 work was to drop all factors of the conformal factor ψ^4 appearing in E_{ab} and B_{ab}. This is because the background metric can be thought of as being completely determined by ψ^4, while the radiation should be determined by the conformal metric (A,B,C,D) in Table 2. Since analytically $\psi_4 = 0$ on the initial data slice, this procedure helped the problem of noncancellation of background terms in equation (15).

In Figure 3, is presented the value of $r \cdot \psi_4(\theta = \pi/2)$ versus time for Run II. The measurement is made on the 2-sphere at $r \sim 25M$. A comparison of the results for Runs I and III [Eppley and Smarr (1978b)] yield the following points. First, the amplitude of ψ_4 stays roughly the same. Second, the wavelength increases from $\lambda \sim 16M$ for Run I, to $\lambda \sim 22M$ for Runs II and III. Third, there is qualitatively very little difference as far as the shape of the waveform. This remarkable agreement on three runs with different grids and initial separations gives us a lot of confidence in the results.

Even more striking is the comparison between our ψ_4 and the ψ_4 calculated for a particle of mass μ falling radially into a black hole of mass M, with $\mu \ll M$. This is a perturbation calculation around a static Schwarzschild black hole background. This calculation was first performed by Davis, et al. (1971) and followed up by Davis, et al. (1972); Chung (1973); and Detweiler and Szedenits (1977). Already in Smarr (1977), we had noted the resemblence in amplitude and phase of the energy radiated, dE/dt, between the perturbation calculation and the full collision problem. In Figure 4 is shown ψ_4 for the perturbation calculation (see Detweiler-this volume) taken from Detweiler and Szedenits (1979). I have used the perturbation units on the left-hand side. On the right hand side, I have scaled-up the perturbation calculation by assuming μ is the reduced mass and M the total mass for two masses m_1 and m_2, so that if $m_1 = m_2$, $M^2/\mu \rightarrow 4M$ [Smarr (1975); Smarr (1977)]. This scale is the same as for Figure 3.

The agreement is so remarkable for such different calculations that something deep must be at work. For the present, we can say that it appears that the black hole perturbation calculations are reliable far beyond limits anyone expected. The other side of the coin is a confirmation of the conclusion in Smarr (1977) that the full nonlinear theory does not induce a significant amplification of gravitational radiation during the coalescence of event horizons.

To obtain the amount of energy radiated by the gravitational wave in Figure 3, we divided ψ_4^2 by $\omega^2 = (2\pi/\lambda)^2$ and integrated in time. We did the same for the Bel-Robinson vector. The total energy radiated is $\sim 0.2 \times 10^{-3}M$, $0.6 \sim 10^{-3}M$, and $1.0 \times 10^{-3}M$ for runs I, II, III, with an uncertainty of probably less than a factor

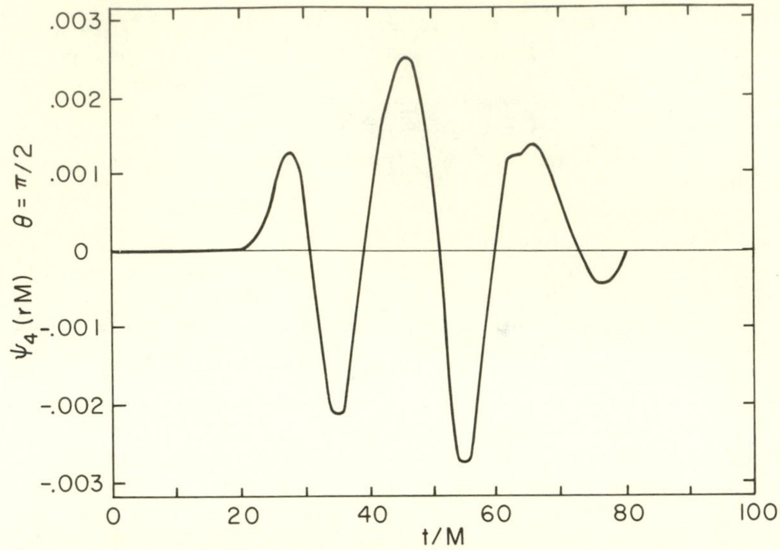

Figure 3. The curvature $\psi_4 \cdot rM$ in the equatorial plane crossing the 2-sphere at r = 25M as a function of time. This is for the two black hole collision Run II.

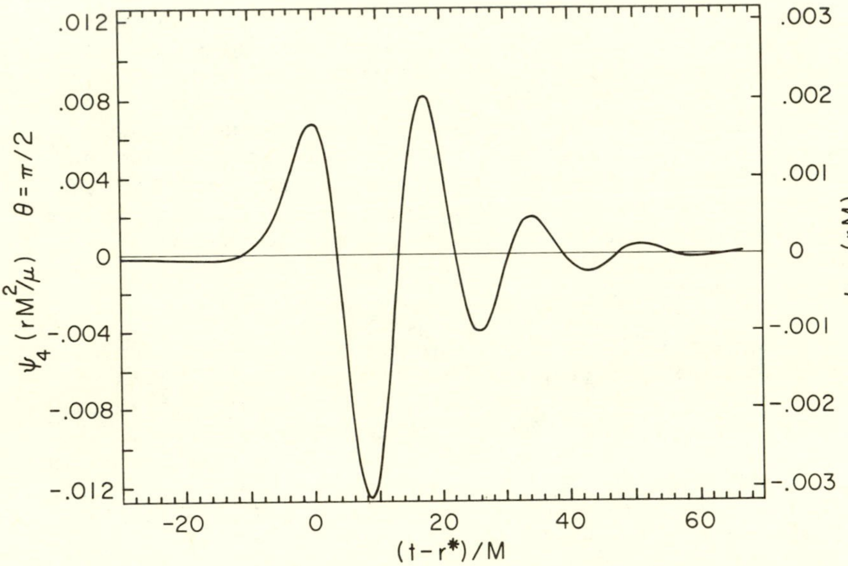

Figure 4. The same quantity as in Figure 3 except from the perturbation calculation of a particle of mass μ falling into a black hole of mass M. The abscissa is retarded time. The vertical scales are explained in the text. Only the quadrupole contribution is shown here.

of 2. For Run II the difference between the total energy loss measured by ψ_4 and Bel-Robinson was 10%. This seems to me to be very good agreement indeed, considering how differently ψ_4 and P_{BR} are constructed. These results fall in the shaded regions of Figure 18 of Smarr (1977) and are consistent with the picture described there. That is, there is no new gravitational radiation mechanism operating in the equal mass collision (done here) that does not operate in the large mass ratio collision [Davis, et al. (1971)].

This would suggest that the generation of the waveform in the strong field region is similar to that found in perturbation theory, i.e. that it is dominated by the "ringing region" near $r \simeq 3M$ [Press (1971); Goebel (1972)]. To investigate this we studied both P_{BR} and ψ_4 as functions of r on the equatorial plane. It is fascinating to watch the individual pulses form and move outward one by one. A full account will be given elsewhere [Eppley and Smarr (1978b)]. In figure 5 we show log $[r^2 \cdot |P^r_{BR}(\theta = \pi/2)|]$ versus radial zone number for Run II at t=48M. Across the top are the values of the areal radius r.

There are a number of interesting features one can see here. First, there are the five outgoing pulses A,B,C,D,E. The pulse A has already propagated off the outer edge of the grid at $r \sim 40M$. The pulses do not decay in amplitude as they move from $r \sim 15M$ to $r \sim 40M$ indicating that the artificial viscosity is small enough that its damping can be ignored. There is a rapid rise in the value of the plotted quantity between $r \sim 10M$ and $r \sim 4M$, indicating that P^r_{BR} is falling off faster than $1/r^{-2}$, as expected in this region. Between $r \sim 2M$ and $r \sim 4M$, there are a series of four more pulses being generated. Thus, we clearly see the "ringing region" where the later pulses are forming. Note that we have placed many more grid points per unit of r in this region in anticipation of the importance of this region. As discussed in more detail in Eppley and Smarr (1978b), this "ringing region" does not appear until after the holes have coalesced. This again is consistent with our overall physical picture. When watched in time one sees the direction of the Bel-Robinson vector flipping between ingoing and outgoing in this region. The Figures 14 and 15 of Smarr (1977) are qualitatively in agreement with our latest results. In particular pulses A and B come off before the collision (as in Figure 14) and the phase of the pulses A,B, and C agree almost exactly with the three pulses in Figure 15.

If one uses ψ_4 rather than P_{BR} versus r, one obtains almost exactly the same picture. It is all of this internal consistency coupled with the external consistency of the perturbation calculations that gives us confidence in our results. We have tried all of the other radiation formulae described in section 3. None of them work very well as detailed in Eppley

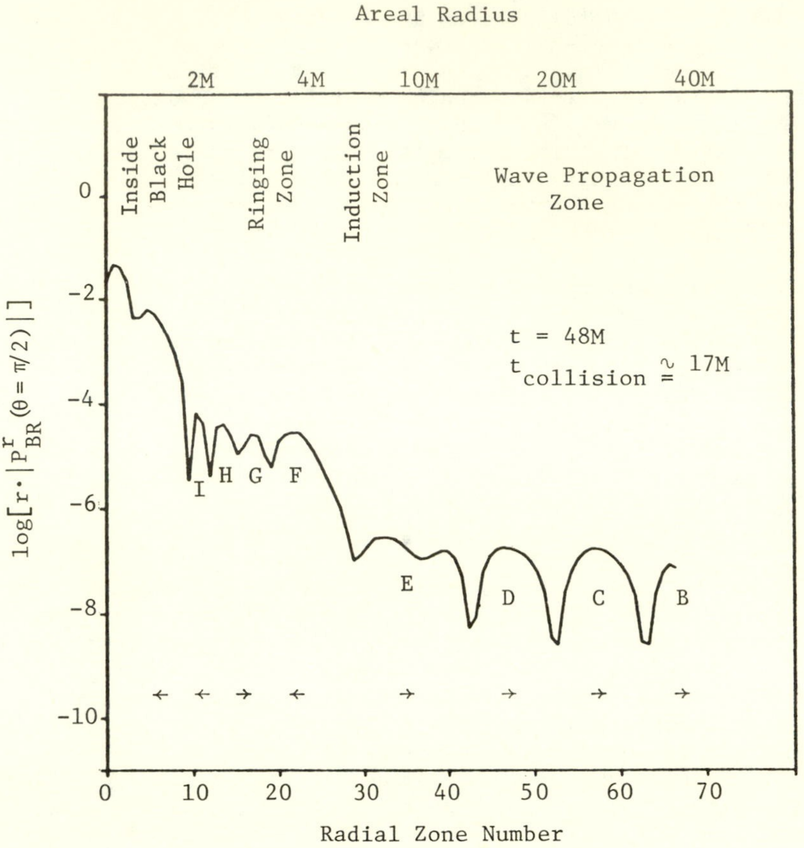

Figure 5. The generation and propagation of gravitational waves in the two black hole collision run II. Plotted is the log of the areal radius r^2 times the Bel-Robinson vector in the equatorial plane. The abscissa is the radial zone number. The pulses are labeled in order of appearance. The arrows give the direction of P^r_{BR}.

and Smarr (1978b). The time integrated ψ_4 formula (16) agrees with the energy loss for ψ_4/ω for the first few pulses, then gets very large due to the noncancellation of the peaks and troughs discussed in section 3.4. This area of energy loss formulae still requires more work for strong field sources. [For weak fields agreement is excellent - see Eppley, this volume].

4.5 Summary

The two black hole collision problem has been largely completed. This opened the door to many other strong field calculations such as discussed by Piran and Wilson, this volume. We find a large degree of internal consistency as well as remarkable agreement with the perturbation calculations of a small particle falling into a black hole. This latter result seems to be telling us that the nonlinearities of Einstein's equations may not be very important for generating gravitational radiation. The prelimimary results with matter spacetimes (see Wilson, this volume) are consistent with this view. Taken together with the perturbation results of Detweiler and Moncrief, et al., this volume, we can draw the conclusion that, at least for axisymmetric motions, black hole waveforms will be dominated by ringing modes which wash out most of the details of the black hole interaction. That is, the waveform tells us about the mass and angular momentum of the final hole, but not much about how it was formed. The agreement found here between the full nonlinear problem and the perturbation problem justifies a certain confidence in Detweiler's extrapolating his spiral perturbation calculations to the two black hole spiral collision. It would appear on the basis of those arguments that Hawking's upper limit for efficiency of 29% may be approached by a non-axisymmetric collision. However, to be sure, one would have to extend the techniques developed herein to a full 3 + 1 computer code capable of solving for the general globally hyperbolic solution of Einstein's equations.

REFERENCES

Arnowitt, R., Deser, S., & Misner, C.W. (1962). The dynamics of general relativity. In Gravitation: An Introduction to Current Research. Wiley, New York.

Bondi, H. (1947). Spherically symmetric models in general relativity. Mon. Not. Roy. Astro. Soc., 107, pp.410-425.

Bondi, H., van der Burg, M.G.J., & Metzner, A.W.K. (1962). Gravitational waves in general relativity VII. Waves from axi-symmetric isolated systems. Proc. R. Soc. Lond., A269, pp.21-52.

Čadež, A. (1975). Some remarks on the two-body-problem in geometrodynamics. Ann. Phys., 91, pp.58-74.

Chandrasekhar, S. & Friedman, J.L. (1972). On the stability of axisymmetric perturbations in general relativity I. The equations governing nonstationary, stationary, and perturbed systems. Astrophys. J., 175, pp.379-405.

Chrzanowski, P. (1977). Talk at the Yale Workshop on the Dynamical Construction of Spacetime (unpublished).

Chung, K.P. (1973). A four-dimensional Green's function approach to the calculation of gravitational radiation from a particle falling into a black hole. Nuovo Cim., 14B, pp.293-308.

Cooperstock, F.I. (1974). Axially symmetric two-body problem in general relativity. Phys. Rev., D10, pp.3171-3180.

Davis, M., Ruffini, R., Press, W.H., & Price, R.H. (1971). Gravitational radiation from a particle falling radially into a Schwarzschild black hole. Phys. Rev. Lett., 27, pp.1466-1469.

Davis, M., Ruffini, R., & Tiomno, J. (1972). Pulses of gravitational radiation of a particle falling radially into a Schwarzschild black hole. Phys. Rev., D5, pp.2932-2935.

Detweiler, S. & Szedenits, E., Jr. (1979). Black holes and gravitational waves II. Spiralling geodesics into a Schwarzschild black hole. Preprint.

Eardley, D.M. & Smarr, L. (1978). Time functions in numerical relativity I. Marginally bound dust collapse. Harvard Center for Astrophysics Preprint #1036.

Einstein, A. & Rosen, N. (1935). The particle problem in the general theory of relativity. Phys. Rev., 48, pp.73-77.

Eppley, K.R. (1975). The numerical evolution of the collision of two black holes. Ph.D. dissertation, Princeton University.

Eppley, K.R. & Smarr, L. (1978a). The collision of two black holes II. Evolution of metric functions. Preprint.

Eppley, K.R. & Smarr, L. (1978b). The collision of two black holes III. Gravitational radiation. Preprint.

Eppley, K.R. & Smarr, L. & Teukolsky, S. (1979). Evolution of time-symmetric gravitational waves II. Radiation formulae and an analytic weak-field solution. Preprint.

Estabrook, F., Wahlquist, H., Christensen, S., DeWitt, B., Smarr, L., & Tsiang, E. (1973). Maximally slicing a black hole. Phys. Rev., D7, pp.2814-2817.

Gibbons, G.W., & Hawking, S.W. (1971). Theory of the detection of short bursts of gravitational radiation. Phys. Rev., D4, pp.2191-2197.

Goebel, C.J. (1972). Comments on "vibrations" of a black hole. Astrophys. J. Lett., 172, pp.L95-L96.

Hahn, S.G., & Lindquist, R.W. (1964). The two body problem in geometrodynamics. Ann. Phys., 29, pp.304-331.

Landau, L.D. & Lifshitz, E.M. (1962). The Classical Theory of Fields, 2nd ed., Addison-Wesley, Reading, Mass.

Levi-Civita, T. (1919). Rend. Acc. Lincci, 28, p.3.

Lewis, T. (1932). Some special solutions of the equations of axially symmetric fields. Proc. Roy. Soc. Lond., A136, pp.176-185.

Matte, A. (1953). Can. J. Math., 5, p.1.

May, M.M., & White, R.H. (1966). Hydrodynamic calculations of general-relativistic collapse. Phys. Rev., 141, pp.1232-1241.

Misner, C.W. & Sharp, D.H. (1964). Relativistic equations for adiabatic, spherically symmetric, gravitational collapse. Phys. Rev., 136, pp.B571-576.

Misner, C.W., Throne, K.S., & Wheeler, J.A. (1973). Gravitation. Freeman, San Francisco.

Oppenheimer, J.R. and Snyder, H. (1939). On continued gravitational contraction. Phys. Rev., 56, pp.455-459.

Pachner, J. (1975). Numerical integration of exact time-dependent Einstein equations with axial symmetry. In General Relativity and Gravitation, ed. G. Shaviv and J. Rosen, Wiley, New York, pp.143-168.

Penrose, R. (1966). General-relativistic energy flux and elementary optics. In Perspectives in Geometry and Relativity. Indiana University Press.

Press, W.H. (1971). Long wave trains of gravitational waves from a vibrating black hole. Astrophys. J. Lett., 170, pp.L105-108.

Rosen, N. & Shamir, H. (1975). Gravitational field of an axially symmetric system in first approximation. Rev. Mod. Phys., 29, pp.429-431.

Smarr, L. (1975). The structure of general relativity with a numerical illustration: The collision of two black holes. Ph.D. Dissertation University of Texas at Austin.

Smarr, L., Čadež, A., DeWitt, B. & Eppley, K. (1976). Collision of two black holes: Theoretical framework. Phys. Rev., D14, pp.2443-2452.

Smarr, L. and York, J.W., Jr. (1978a). Radiation gauge in general relativity. Phys. Rev., D17, pp.1945-1956.

Smarr, L. & York, J.W., Jr. (1978b). Kinematical conditions in the construction of spacetime. Phys. Rev., D17, pp. 2529-2551.

Synge, J.L. (1960). Relativity: The General Theory. North-Holland, Amsterdam.

Taub, A.H. (1978). Relativistic fluid mechanics. Ann. Rev. Fluid Mech., 10, pp.301-32.

Teukolsky, S.A. (1973). Perturbations of a rotating black hole I. Fundamental equations for gravitational, electromagnetic and neutrino-field perturbations. Astrophys. J., 185, pp.635-647.

Thorne, K.S. (1977). On the mathematical description of gravitational waves. Preprint CRSR-662 Cornell University.

Tolman, R.C. (1934). Effect of inhomogeneity on cosmological models. Proc. Nat. Acad. Sci., 20, pp.169-176.

Voorhees, B.H. (1971). Axially symmetric distributions of matter in general relativity. Ph.D. Dissertation University of Texas at Austin.

Weyl, H. (1917). Ann. Physik., 54, pp.117.

Wilson, J.R. & Smarr, L. (1978). Numerical axially symmetric geometrohydrodynamics. Preprint.

PURE GRAVITATIONAL WAVES

Kenneth Eppley

Department of Physics and Astronomy
University of Maryland

1. INTRODUCTION

The computer generation of spacetimes has advanced from the first arduous attempts to simply perform a stable evolution for reasonable times to the level where the routine solution of realistic problems seems imminent. An important step in the process has been the study of pure gravitational radiation in solutions with axial symmetry. The primary purpose of that study was to learn how to calculate the energy flux of gravitational radiation in numerically constructed spacetimes. This aim has been largely accomplished through the study of the Brill (1959) waves and analytic linearized solutions. The Brill waves have also been interesting in their own right, since they are the first application of the general methods for the completely numerical construction and evolution of gravitational initial data. They have also shed light on the problem of accurately defining the mass of a numerically generated initial data set and on the question of whether any non-flat spacetime can remain forever non-singular. In this article we review the work that has been accomplished on vacuum axisymmetric spacetimes and discuss new results involving the application of various gauge conditions to the evolution. We will also discuss the generation and evolution of non-time-symmetric initial data.

2. TIME-SYMMETRIC WAVES

The initial data for the vacuum Einstein equations consists of the 3-metric γ_{ab} and the extrinsic curvature K^a_b. We first studied the case of time-symmetric data, i.e., $K^a_b=0$ at t=0, which reduces the Hamiltonian constraint (York-this volume, eqn. 23) to

$$^3R = 0 \tag{1}$$

Brill (1959) studied the solutions of (1) in detail for vacuum axisymmetric data. One can always find coordinates so that the metric takes the form

$$ds^2 = \psi^4 \left(e^q(d\rho^2+dz^2) + \rho^2\, d\phi^2\right) \tag{2}$$

where q is arbitrary subject to the restrictions:

$$\begin{aligned} &q = 0 \text{ when } \rho = 0 \\ &q_\rho = 0 \text{ when } \rho = 0 \\ &q_z = 0 \text{ when } z = 0 \\ &q \sim r^{-2} \text{ or faster asympotically.} \end{aligned} \tag{3}$$

Then (1) becomes

$$\nabla^2\psi = -\,1/8\ (q_{\rho\rho}+q_{zz})\ \psi \equiv \Phi\psi \tag{4}$$

where ∇^2 is the ordinary flat space Laplacian in three dimensions:

$$\nabla^2\psi \equiv \psi_{\rho\rho} + \psi_{zz} + \psi_\rho/\rho \tag{5}$$

One can write the mass of the initial data from the usual surface integral expression (Brill, 1959):

$$m = -\frac{1}{2\pi} \oint \vec{\nabla}\psi\cdot d\vec{s}\ \ (G=c=1) \tag{6}$$

which can be transformed to

$$m = -\frac{1}{2\pi} \int \nabla^2\psi d^3x \tag{7}$$

Alternately (Eppley, 1977) the mass can be rewritten as

$$m = \frac{1}{2\pi} \int \left(\frac{\nabla\psi}{\psi}\right)^2 d^3x \tag{8}$$

which is manifestly positive definite for any $q \neq 0$.

We solve (4) by numerical methods by casting it into a finite difference form and solving it iteratively, (Eppley, 1975) e.g. by successive overrelaxation (SOR). For example, one could write

$$\nabla^2\psi \cong \frac{1}{\Delta\rho^2} (\psi_{i,j+1} - 2\psi_{i,j} + \psi_{i,j-1}) + \frac{1}{2\rho\Delta\rho} (\psi_{i,j+1} - \psi_{i,j-1})$$

$$+ \frac{1}{\Delta z^2} (\psi_{i+1,j} - 2\psi_{i,j} + \psi_{i-1,j}) \tag{9}$$

which is accurate to second order. We found however, that using this difference expression the solution we obtain has a mass (6) which does not agree either with (7) or (8) by over an order of magnitude, even with rather fine mesh size. We found it necessary to use a fourth order accurate conservative form:

$$\nabla^2\psi \cong \frac{1}{24}\,\Delta\rho^2\ [(-1-d)\psi_{i,j+2} + (d-1)\psi_{i,j-2} + (28+26d)\ \psi_{i,j+1}$$

$$+ (28-26d)\ \psi_{i,j-1} - 54\psi_{i,j}] + \frac{1}{24}\,\Delta z^2\ [-\psi_{i+2,j} - \psi_{i-2,j}$$

$$+ 28\ (\psi_{i+1,j} - \psi_{i-1,j}) - 54\psi_{i,j}] \tag{10}$$

where $d \equiv \Delta\rho/2\rho$.

Using this expression gave a solution for which all the mass integrals agree reasonably well. (A conservative differencing scheme is one for which a macroscopic conservation law, e.g. a surface integral equals a volume integral, is identically satisfied by the difference equations as well as the differential equations.) The positive definite expression (8) gave approximately the same results for all the differencing schemes, and was fairly insensitive to the mesh size and number of points.

We applied this method to a particular choice of q:

$$q = A\rho^2/(1+(r/\lambda)^n) \qquad (A,\lambda \text{ constant}) \tag{11}$$

the simplest function we could devise satisfying the conditions (3). We chose $n = 5$, and $\lambda = 1$. As can be predicted (Wheeler, 1964) for small amplitude A the mass is proportional to A^2. We found

$$m/A^2 \cong 1.4 \times 10^{-2} \tag{12}$$

for this choice of q. We will describe below the time development of this initial data.

Note that ψ is the constrained metric function which gives static properties like the total mass. The unconstrained function q is the freely specifiable gravitational wave variable.

3. NON-TIME-SYMMETRIC WAVES

There has been some worry that the time symmetric condition was too restrictive, so we generalized Brill's discussion to the non-time-symmetric case. For an axisymmetric nonrotating spacetime, $K^a{}_b$ has four independent components (K^z_z, K^ρ_ρ, K^ρ_z, K^ϕ_ϕ). There are two non-trivial momentum constraints, which for a metric of form (2) become

$$\begin{aligned}
&K^\rho_{z,\rho} - K^\phi_{\phi,z} - K^\rho_{\rho,z} + K^z_\rho\ (2\Gamma^\rho_{\rho\rho} + \Gamma^\phi_{\rho\phi}) \\
&+ (K^z_z - K^\rho_\rho)\ \Gamma^z_{zz} + (K^z_z - K^\phi_\phi)\ \Gamma^\phi_{z\phi} = 0 \\
&K^z_{\rho,z} - K^\phi_{\phi,\rho} - K^z_{z,\rho} + K^\rho_z\ (2\Gamma^z_{zz} + \Gamma^\phi_{z\phi}) \\
&+ (K^\rho_\rho - K^z_z)\ \Gamma^\rho_{\rho\rho} + (K^\rho_\rho - K^\phi_\phi)\ \Gamma^\phi_{\rho\phi} = 0
\end{aligned} \tag{13}$$

On a maximal hypersurface (which forces $K^\phi_\phi = -K^z_z - K^\rho_\rho$) these equations can be rewritten as the two simple elliptic equations (equation 48) given by Wilson (with matter terms zeroed). This leaves one free function, say K^ϕ_ϕ, to complement q as a free gravitational wave variable.

The inclusion of $K^a{}_b$ alters the scale equation for ψ:

$$\psi_{zz} + \psi_{\rho\rho} + \psi_{\rho/\rho} = -\frac{1}{8}(q_{zz} + q_{\rho\rho})\,\psi - \frac{e^q\sigma}{8}\,\psi^5 \tag{14}$$

with $\sigma \equiv K^a{}_b K^b{}_a - K^2$.

Solutions of this equation do not necessarily exist for any σ. It is necessary to transform the momenta as some power of ψ to insure that a solution will exist. York's method is to take

$$\bar{K}^a{}_b = (\psi/\psi_o)^{-6} K^a{}_b \tag{15}$$

$$\bar{\sigma} = (\psi/\psi_o)^{-12}\,\sigma_o \text{ where } \psi_o \text{ is the previous value of } \psi.$$

This has the advantage of decoupling the momentum constraints from the scale equations. However (14) is still non-linear. While numerical techniques exist to solve such non-linear equations (Eppley, 1975), we chose to take

$$\begin{aligned}\bar{K}^a{}_b &= (\psi/\psi_o)^{-5/2}\, K^a{}_b \\ \bar{\sigma} &= (\psi/\psi_o)^{-5}\,\sigma_o\end{aligned} \tag{16}$$

Now in terms of the initial values of ψ_o, $K^a{}_b$ and σ_o, (14) becomes

$$\psi_{zz} + \psi_{\rho\rho} + \psi_\rho/\rho = -\frac{1}{8}(q_{\rho\rho} + q_{zz})\,\psi - \frac{e^q\sigma_o}{8}\,\psi_o{}^5 \tag{17}$$

ψ_o is the value of ψ before solving the equation and ψ is the new value. The constraints are still coupled so it is necessary to alternately iterate the momentum and Hamiltonian constraints (13) and (14), updating the source terms in each equation each time, until the solution converges. We found that the solutions seem to converge after a few repetitions. However, the requirements

$$\left.\begin{aligned}K^z_\rho &= 0 \\ K^\rho_\rho &= K^\phi_\phi\end{aligned}\right\} \quad \text{when } \rho = 0 \tag{18}$$

seem difficult to satisfy in a smooth fashion by this method. The first requirement follows from symmetry, the second is needed if we are to maintain q=0 for ρ=0. As particular choices of data we took

$$q = 0$$

$$K^{\phi}_{\phi} = - K_1 = A\rho^m/(1+r^n) \qquad (19)$$

We picked K^{ϕ}_{ϕ} to go to zero at ρ=0 so that the condition $K^z_z=K^{\rho}_{\rho}$ at ρ=0 could be imposed, which improved the behavior near the z axis.

As in the time-symmetric case, the mass is positive definite. We can generalize (8) to the form

$$m = \frac{1}{2\pi} \int d^3x \left[\left(\frac{\vec{\nabla}\psi}{\psi}\right)^2 + \frac{e^q\sigma\psi^4}{8} \right] \qquad (20)$$

where $(\vec{\nabla}\psi)^2 \equiv \psi_{\rho}^2 + \psi_z^2$.

This is in a sense a positive definite "local" energy density for any metric of the form (2), and may be useful for computing energy flux.

Wilson (eqn. 61) has constructed an energy flux from (20) by assuming that the relationship between the energy flux and the energy density is the same as for other massless fields, i.e. the flux normal to the wave vector is just proportional to the energy density, and taking (20) as this local energy density. Actually he used only the σ term in (20) because the $(\nabla\psi/\psi)^2$ term falls off faster (r^{-4}) than σ (r^{-2}). We have tested this formula for the Brill waves and will describe below how it compares with other methods. (As Smarr has pointed out, the correct Bondi formula is basically equivalent to this expression.)

4. THE EINSTEIN EVOLUTION EQUATIONS

The time development of the metrics γ_{ab} and the extrinsic curvatures $K^a{}_b$ are given by the Einstein evolution equations (York equations 35,39). As a choice of lapse function α we have used either geodesic slicing (α=1) or maximal slicing (York eqn. 104).(Geodesic slicing is only usable for very weak gravitational fields.) To perform maximal slicing we must solve the lapse equation (York eqn. 104) which is elliptic and can be solved numerically by relaxation. We note that in a numerical evolution it is better to evolve $K^a{}_b$ rather then K_{ab} when we do maximal slicing. That is because $\hat{K}$ is an algebraic sum of the $K^a{}_b$, while it is not such a simple expression in terms of K_{ab}. If we evolve the $K^a{}_b$, and solve the differenced version of the lapse equation exactly, then K remains identically zero. If we evolve K_{ab} there is always a small error, which however, will tend to grow in the presense of a strongly focusing singularity. Another useful technique is to add an extra term so that

$\dot{K} = -\lambda K$ so the lapse equation becomes (21)

$$D_a D^a \alpha = ({}^3R+K^2)\, \alpha-\lambda K \tag{22}$$

Then even if we do not solve the elliptic equation exactly small errors will tend to damp. This term allows the use of fewer iterations and thus saves time.

There is greater leeway in the choice of the shift vector β^a. We can generally choose $\beta^a=0$ without real difficulties. However, other choices of β^a may prove advantageous. Smarr and York (1978) have devised what they call the minimal distortion gauge (York eqn. 125). It is claimed that this gauge removes the non-radiative degrees of freedom so the metric will act like a radiation field.

This gauge has the disadvantage for numerical work of great complexity. Even in axial symmetry the equations for β^a are extremely long and involved (Eppley, 1975). It appears at present that for a fully general three-dimensional metric the minimal distortion gauge is too complicated to apply to a numerical code unless, perhaps, some symbol manipulation language is used to generate it.

Wilson (this volume) has devised a gauge for the purpose of simplifying the metric which has proved quite useful. His conditions are

$$\begin{aligned} B &= \dot{B} = 0 \\ C &= \dot{C} = 0 \end{aligned} \tag{23}$$

where

$$\begin{aligned} B &\equiv \gamma_{zz} - \gamma_{\rho\rho} \\ C &\equiv \gamma_{z\rho} \end{aligned} \tag{24}$$

This gauge is especially suited to the Brill waves, since it keeps the metric in the form (2) for all time. Since the metric is diagonal, great simplification occurs in the equations.

Wilson chooses to minimize the numbers of hyperbolic equations for the unknowns, $(q, \psi, K^\rho_\rho, K^\phi_\phi, K^z_\rho, K^z_z)$ by solving the constraints (Wilson, eqns. 47, 48). Here we give the full list of hyperbolic equations. We denote $K_1=-K^\phi_\phi$, $K_2=K^z_z-K^\rho_\rho$.

$$\begin{aligned} \dot{q} &= \alpha\, (2K^\phi_\phi-K_1) + \partial_z\, (q\beta^z) + \partial_\rho\, (q\beta^\rho) \\ &\quad - 2\beta^\rho/\rho + (1-q)\, (\beta^z{}_{,z}-\beta^\rho{}_{,\rho}) \end{aligned} \tag{25}$$

$$\dot{D} = (-2\alpha K^{\phi}_{\phi}+2\beta^{\rho}/\rho)\, D + \partial_{\rho}\, (\beta^{\rho}D) + \partial_{z}\, (\beta^{z}D) - D\, (\beta^{\rho}_{\ ,\rho}-\beta^{z}_{\ ,z}) \tag{26}$$

where $D \equiv \psi^4$

$$\dot{K}_1 = \alpha K K_1 + \frac{1}{A}\, [\alpha(R_{zz}+R_{\rho\rho}) - \alpha_{zz} - \alpha_{\rho\rho}] + \partial_z\, (\beta^z K_1) + \partial_{\rho}\, (\beta^{\rho}K_1) - K_1\, (\beta^z_{\ ,z} + \beta^{\rho}_{\ ,\rho}) \tag{27}$$

$$\dot{K}^{\phi}_{\phi} = \alpha(KK^{\phi}_{\phi} + R^{\phi}_{\phi}) + \gamma^{\phi\phi}\, (\Gamma^{z}_{\phi\phi}\alpha_z + \Gamma^{\rho}_{\phi\phi}\alpha_{\rho}) + \partial_z\, (\beta^z K^{\phi}_{\phi}) + \partial_{\rho}\, (\beta^{\rho}K^{\phi}_{\phi}) - K^{\phi}_{\phi}\, (\beta^z_{\ ,z}+\beta^{\rho}_{\ ,\rho}) \tag{28}$$

$$\dot{K}_2 = \alpha K K_2 + \frac{1}{A}\, [\alpha(R_{zz}-R_{\rho\rho}) - \alpha_{zz} + \alpha_{\rho\rho}] + 2\, (\Gamma^{z}_{zz}\alpha_z - \Gamma^{\rho}_{\rho\rho}\alpha_{\rho}) + \partial_z\, (\beta^z K_2) + \partial_{\rho}\, (\beta^{\rho}K_2) + 2K^z_{\rho}\, (\beta^{\rho}_{\ ,z}-\beta^{z}_{\ ,\rho}) - K_2\, (\beta^z_{\ ,z}+\beta^{\rho}_{\ ,\rho}) \tag{29}$$

$$\dot{K}^z_{\rho} = \alpha K K^z_{\rho} + \frac{1}{A}\, [\alpha R_{z\rho} - \alpha_{z\rho} + \Gamma^{\rho}_{\rho\rho}\alpha_z + \Gamma^{z}_{zz}\alpha_{\rho}] + \partial_z\, (\beta^z K^z_{\rho}) + \partial_{\rho}\, (\beta^{\rho}K^z_{\rho}) - K^z_{\rho}\, (\beta^z_{\ ,z}+\beta^{\rho}_{\ ,\rho}) + \frac{1}{2}\, K_2\, (\beta^z_{\ ,\rho}-\beta^{\rho}_{\ ,z}) \tag{30}$$

$$\gamma^{\phi\phi}\Gamma^{z}_{\phi\phi} = -\frac{2}{A}\, \frac{\psi_z}{\psi} \tag{31}$$

$$\gamma^{\phi\phi}\Gamma^{\rho}_{\phi\phi} = -\frac{1}{A}\, \left(\frac{2\psi_{\rho}}{\psi} + \frac{1}{\rho}\right) \tag{32}$$

$$R_{zz} + R_{\rho\rho} = - q_{zz} - q_{\rho\rho} - \frac{6}{\psi}\, (\psi_{zz}+\psi_{\rho\rho}) + \frac{2}{\psi^2}\, (\psi^2_z+\psi^2_{\rho}) - \frac{4}{\rho}\, \psi_{\rho}/\psi \tag{33}$$

$$R_{zz} - R_{\rho\rho} = \frac{2}{\psi}\, (\psi_{\rho\rho}-\psi_{zz}) + \frac{6}{\psi^2}\, (\psi^2_z-\psi^2_{\rho}) + \frac{2}{\psi}\, (q_z\psi_z - q_{\rho}\psi_{\rho}) - q_{\rho}/\rho \tag{34}$$

$$R^{\phi}_{\phi} = -\frac{2}{A}\, \left[\frac{\psi_{zz}+\psi_{\rho\rho}}{\psi} + \frac{\psi^2_z+\psi^2_{\rho}}{\psi^2} + \frac{2}{\rho}\, \psi_{\rho}/\psi\right] \tag{35}$$

$$^3R = \frac{1}{A}\, [-q_{zz}-q_{\rho\rho} - \frac{8}{\psi}\, (\psi_{zz}+\psi_{\rho\rho}+\psi_{\rho}/\rho)] \tag{36}$$

In the wave zone, aside from the β^a terms, these equations now look almost exactly like a scalar wave equation in two dimensions, i.e., dropping the β^a terms:

$$\ddot{D} \simeq D_{zz} + D_{\rho\rho}$$
$$\ddot{q} \simeq \frac{1}{2}\, (D_{zz} + D_{\rho\rho}) + q_{zz} + q_{\rho\rho} \tag{37}$$

D satisfies a pure wave equation and q obeys the equation of a wave whose source is also wavelike in character, so in a sense this can be considered to be a "radiation gauge". We will see that the metrics do show wave behavior in this gauge while they do not for $\beta^a=0$ (where the Einstein equations are more complicated).

5. AN ANALYTIC EXAMPLE - QUADRUPOLE SOLUTIONS OF THE LINEARIZED EQUATIONS

A very useful example of pure gravitational radiation is an analytic solution of the linearized Einstein equations (Eppley, Smarr and Teukolsky 1976). This is a pure quadrupole wave whose metric for all time is

$$\begin{aligned}
\gamma_{zz} &= 1+3d\cos^4\theta+6(b-c)\cos^3\theta + 3\ c-a \\
\gamma_{\rho\rho} &= 1+3\sin^2\theta\cos^2\theta\ d-a \\
\gamma_{z\rho} &= 3\sin\theta\cos\theta(d\cos^2\theta + b-c) \\
\gamma_{\phi\phi} &= \rho^2(1+3(a-c)\sin^2\theta-a)
\end{aligned} \tag{38}$$

$$\begin{aligned}
a &= 3[\frac{F^{(2)}}{r^3} + \frac{3F^{(1)}}{r^4} + \frac{3F}{r^5}] \\
b &= -\ [\frac{F^{(3)}}{r^2} + \frac{3F^{(2)}}{r^3} + \frac{6F^{(1)}}{r^4} + \frac{6F}{r^5}] \\
c &= 1/4\ [\frac{F^{(4)}}{r} + \frac{2F^{(3)}}{r^2} + \frac{9F^{(2)}}{r^3} + \frac{21F^{(1)}}{r^4} + \frac{21F}{r^5}] \\
d &= a+c-2b \\
F &\equiv F\ (t-r)
\end{aligned} \tag{39}$$

This solution is not time-symmetric. However, we can form a time-symmetric, non-singular solution by superposing solutions of $F(t+r) - F(t-r)$. Then at $t=0$ the metric is given by (38) using now:

$$\begin{aligned}
a &= 3\ [\frac{F^{(2)}(r)-F^{(2)}(-r)}{r^3} - \frac{3}{r^4}\ (F^{(1)}(r)+F^{(1)}(-r)) \\
&\quad + \frac{3}{r^5}\ (F(r)-F(-r))] \\
b &= -\ [\frac{-1}{r^2}\ (F^{(3)}(r)+F^{(3)}(-r)) + \frac{3}{r^3}\ (F^{(2)}(r)-F^{(2)}(-r)) \\
&\quad - \frac{6}{r^4}\ (F^{(1)}(r)+F^{(1)}(-r)) + \frac{6}{r^5}\ (F(r)-F(-r))] \\
c &= \frac{1}{4}\ [\frac{1}{r}(F^{(4)}(r)-F^{(4)}(-r)) - \frac{2}{r^2}\ (F^{(3)}(r) + F^{(3)}(-r)) \\
&\quad + \frac{9}{r^3}\ (F^{(2)}(r) - F^{(2)}(-r)) - \frac{21}{r^4}\ (F^{(1)}(r) + F^{(1)}(-r)) \\
&\quad + \frac{21}{r^5}\ (F(r) - F(-r))]
\end{aligned} \tag{40}$$

$d = a+c-2b$

A specific example would be

$$F(u) = ue^{-u^2} \tag{41}$$

Then at t=0

$$a=24e^{-r^2}$$
$$b=8(3-2r^2)e^{-r^2} \tag{42}$$
$$c=8(3-4r^2+r^4)e^{-r^2}$$

We can solve this initial data numerically using geodesic normal coordinates ($\alpha=1,\beta=0$) and compare the results to the analytic solution. We used the staggered leapfrog differencing method (Smarr, 1977). This code was able to reproduce the solution (38) to within a few per cent on a fairly course (20 x 20) grid for a typical case.

The time symmetry of these waves gives them an "imploding-exploding" character. That is (Figure 1) the waves are incoming at $t=-\infty$, with no singularities anywhere. (The situation is analogous to an electromagnetic field for which either $\vec{E}$ or $\vec{B}$ is exactly zero at t=0 (Wheeler, 1964)). Since the solution contains both incoming and outgoing waves at t=0, it will produce more than one pulse of radiation. The peak intensity will come from the radiation located at the origin at t=0, but the radiation which was outgoing already at t=0 will produce a precursor, while the radiation which was still incoming will produce a tail. If we observe the flux across a two-sphere at radius r_o, the maximum pulse will arrive at $t=r_o$, while the secondary pulses will be symmetric both before and after. (Of course the pattern also depends on what quantity we use to characterize the radiation field.) This is essentially what we observe in the numerical evolution (Figure 2).

From the analytic solution we can calculate the instantaneous energy flux over any two-sphere. For example, using the A.D.M. expression, (see Smarr, this volume) we obtain

$$\frac{dm}{du} = \frac{9}{128}\,[F^{(5)}(u)]^2 \tag{43}$$

for a wave of form (39), while for the time symmetric case (40) it is

$$\frac{dm}{du} = \frac{9}{128}\,[F^{(5)}(t+r) - F^{(5)}(t-r)]^2 \tag{44}$$

We will discuss later how well various mass-loss formulas

agreed with these results.

We observe from the evolution of the metrics of the Teukolsky data that they indeed represent pure radiation. Not only the radiation invariant ψ_4 (Smarr, 1977) looks like a pure gravitational quadrupole but the metrics as well look very similar, as is clear from the general form (38). This solution is transverse traceless, and there is no "gauge piece" to the metric.

6. EVOLUTION OF THE BRILL WAVES

The linearized waves (38) do not represent a self-consistent spacetime, since they do not have the correct asymptotic behavior of the metric (i.e., that ψ falls off as r^{-1} at infinity). But the Brill waves (2) are a true spacetime, and we wish to study the evolution of this initial data, to observe, for example, whether the metric remains always non-singular. Originally we used the geodesic normal slicing gauge.

When the initial data (2) is evolved in this gauge, the radiation pattern shown in ψ_4 is extremely similar to that produced by the linearized waves, but the metrics are quite different (Figure 3). The "imploding-exploding" character also occurs here, a result of the time-symmetry. The waves disperse to infinity, leaving the near region flat. However, the metrics do not display any wave character and do not become flat in the near zone even after all the radiation has departed. Rather, a "lump of gauge" remains in the center.

The radiation pattern is primarily a quadrupole (i.e. the amplitudes have a $\cos^2\theta$ angular dependence), although there are some higher multipoles present. Since the initial data was chosen with the simplest possible angular dependence, it is not surprising that the lowest multipole, the quadrupole, predominates.

It is straightforward to apply the Wilson gauge to the staggered leapfrog code. The complete equations are shown above ((25) to (36)). We must solve Wilson's shift equation for β^z and β^ρ. Although B and C are assumed to be zero in calculating the curvatures, it is useful to evolve them separately so we can modify the shift condition (23) to be

$$\dot{B} = -\lambda B \qquad \dot{C} = -\lambda C \tag{45}$$

Then the shift equations become

$$(\partial_{zz} + \partial_{\rho\rho})\ \beta^z = (F_1)_z + (F_2)_\rho$$
$$(\partial_{zz} + \partial_{\rho\rho})\ \beta^\rho = (F_2)_z - (F_1)_\rho \qquad (46)$$

where

$$F_1 = \alpha K_2 - \lambda B/(2A)$$
$$F_2 = 2\alpha K^z_\rho - \lambda C/A$$

Thus, errors in satisfying B=C=0 will tend to damp in time, allowing greater accuracy and also fewer iterations. Typically we take $\lambda \lesssim \Delta t^{-1}$.

When the Brill initial data are evolved in this gauge the invariant quantities such as ${}^4R_{ab}\ {}^4R^{ab}$ and ψ_4 are virtually unchanged from β^a=0 as they should be, except that some numerical errors near the z axis are reduced. But the metrics are quite altered. In the Wilson gauge they display wave behavior (Figure 4) and resemble the metric of the linearized waves quite closely. Further, when the wave disperses the metrics now become flat in the center.

Using the leapfrog method one may use either A and D as variables or q and ψ. The latter form has some advantages, especially if we wish to resolve the Hamiltonian constraint for ψ on each time level. That is, the hyperbolic equation for D (or ψ) is replaced by an elliptic equation. Since both are derived from Einstein's equations, both have equal accuracy. When we do geodesic slicing it does not matter especially that the Hamiltonian is not satisfied exactly. But when we do maximal slicing in strong fields the scalar curvature ${}^3R$ tends to become negative if this constraint is not satisfied and the lapse equation can then blow up. We compared a weak field evolution evolving ψ both hyperbolically and elliptically and found no significant difference in the evolution of the metric or the radiation pattern.

It is also possible to resolve the momentum constraints, i.e., replace the hyperbolic equations for K^z_ρ and K_2 by elliptic equations (Wilson equation 48). There are no real problems from the momentum constraints being unsatisfied. We do have a numerical difficulty in resolving these constraints, i.e., that $K^\phi_\phi - K^\rho_\rho$ does not go to zero smoothly on the z-axis, so q will not stay zero there. There are also problems in making K^z_ρ go to zero on the axis, both for the elliptic and hyperbolic solutions. The hyperbolic methods seem slightly better, especially for strong fields. One can also "smooth" the solution for K^z_ρ numerically near the axis if we evolve it hyperbolically but not if we solve for it elliptically.

When matter is present these conclusions may change. Wilson and Smarr (1978) have found it useful to resolve the momentum constraints in the collapsing star problem. There the equations for K^z_ρ and K_2 are the same except for the addition of source terms involving the matter velocities. In effect the matter acts as a source for the "gauge" part of the gravitational field while the "dynamic" part is evolved hyperbolically. If the evolution of the matter is more accurate (which may be the case since the equations of motion can be written in conservative form) then it may prove better to solve for K^z_ρ and K_2 elliptically. We notice that for a spherically symmetric collapse, the equations in this form reduce to a pure elliptic problem for the gravitational field, since q is zero for any spherically symmetric solution.

7. MEASURING ENERGY FLUX

Smarr (this volume) describes the various methods to calculate the energy flux carried by gravitational waves. Three formulas we have found useful are the A.D.M. "Poynting" vector, the mass loss calculated from the Bel-Robinson vector, and the Newmann-Penrose ψ_4 function.

For the linearized waves these three formulas agree with the analytic prediction to within 5 or 10 per cent for a typical evolution (Table 1). Of course the instantaneous values differ (Figure 2). For an ideal plane wave the ψ_4 expression is exactly 90° out of phase with the A.D.M. formula, but agrees with the Bel-Robinson result. The reason for the phase change is simply that the ψ_4 expression involves one more derivative of the metric than the A.D.M., and as a result is out of phase. Since the definition of local energy flux is completely observer dependent there is no physical significance to this instantaneous phase. The phase lag between the analytic result and the A.D.M. is due to an error of about 2 per cent in the effective speed of light.

We note that the Bondi-Sachs formula based on outgoing null rays gave very poor results, for reasons described by Smarr (this volume). However, the Bondi formula based on incoming null rays agreed with the correct result.

For the Brill waves the three expressions did not agree quite so well but still gave reasonable results. A comparison between the initial total mass and the mass loss calculated by the various formulas is given in Table 2 for the case $\beta^a=0$. These results were almost completely unchanged when the evolution was made with the Wilson gauge.

The Wilson mass loss expression was tried for the evolution using his gauge. With the complete expression (20) including the $(\nabla\psi/\psi)^2$ term the result was 1.6×10^{-7}, while the expression with only the σ terms gave 0.9×10^{-7}. Since the two-sphere was placed quite close to the source, it is not surprising that exact agreement is not obtained.

We have found generally for relatively weak fields (i.e., no black holes present), that gauge-dependent expressions (i.e., terms built from the four dimensional Christoffel symbols) such as the A.D.M. and the Wilson formulas gave close agreement to the manifestly invariant expressions (i.e. terms built from the curvatures) such as the Bel-Robinson vector and ψ_4. But when black holes are involved, only the invariants gave reasonable results. We suspect that the problem results from the fact that the coordinate system around a black hole is not static but is drawn into the hole at nearly the speed of light. It may be possible to rewrite some of the gauge dependent formulas to compensate for this coordinate motion, but no method has succeeded as yet.

8. EVOLUTION OF NON-TIME-SYMMETRIC INITIAL DATA

We evolved typical non-time-symmetric initial data, generated as described above. Although the momenta initially were quite different from any values they obtain during the Brill evolution, the evolution of both the metric and the radiation field are quite similar to the behavior of the Brill waves. Aside from some numerical problems near the z axis, ψ_4 looks almost indistinguishable from the time-symmetric case. Once again the wave is primarily a quadrupole. The most important conclusion is that the time-symmetric condition does not seem overly restrictive, nor does it seem to greatly alter the behavior of pure gravitational waves.

We also evaluated the mass loss for these waves. The results are shown in Table 2. Since the initial data were not calculated with a fourth order accurate scheme, but a second order accurate method on a non-uniform grid, we did not obtain as close agreement between the initial calculated mass and the radiation loss.

Table 1. Comparison of mass loss formulas for the pure quadrupole waves ($F(u) = 10^{-5}ue^{-u^2}$).

Analytic value	2.22×10^{-8}
A.D.M.	2.2×10^{-8}
ψ_4	2.0×10^{-8}
Bel-Robinson	2.0×10^{-8}
Bondi-Sachs (outgoing)	6.8×10^{-8}
Bondi-Sachs (incoming)	2.2×10^{-8}

Table 2. Comparison of mass loss for numerically constructed waves.

	Time-symmetric ($A=10^{-3}$)	Non-time-symmetric ($A=10^{-2}$)
Initial mass	1.4×10^{-7}	6.6×10^{-6}
A.D.M.	1.5×10^{-7}	9.9×10^{-6}
ψ_4	1.1×10^{-7}	3.7×10^{-6}
Bel-Robinson	1.2×10^{-7}	3.3×10^{-6}

Figure 1. Spacetime diagram showing the "imploding-exploding" nature of time-symmetric data such as the Teukolsky waves. The fact that both incoming and outgoing waves are present initially is shown in the radiation pattern at a fixed radius.

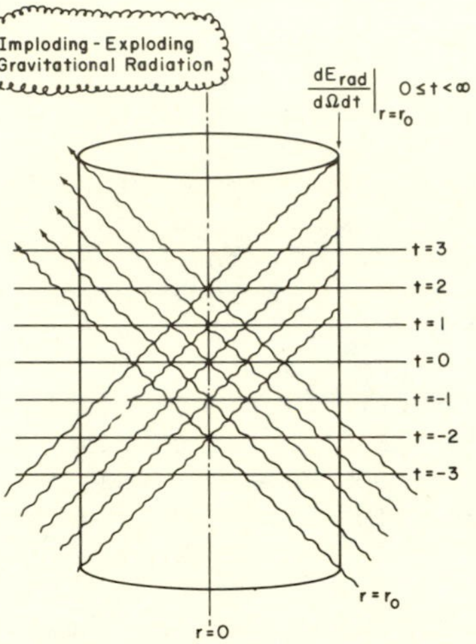

Figure 2. Instantaneous mass loss for the Teukolsky waves at a fixed radius for a typical numerical evolution. The analytic result is shown along with the results using the A.D.M. and Bel-Robinson expressions.

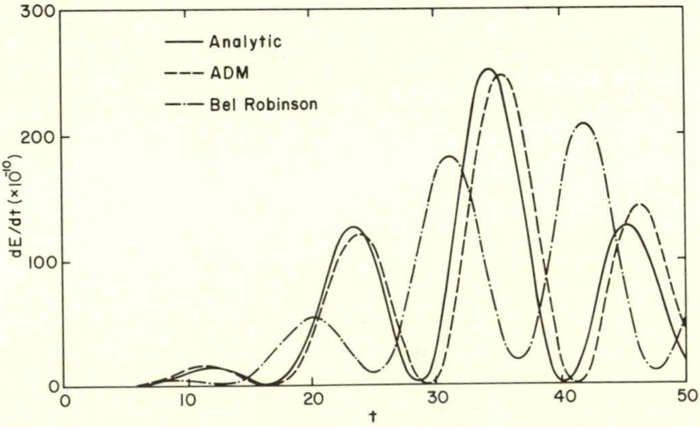

Figure 3. Evolution of Brill waves with $\alpha=1$, $\beta^a=0$, amplitude = 10^{-3}. The metric $A \equiv \gamma_{zz} - 1$ is plotted. (The z axis is to the right, the ρ axis to the left, and the **vertical** axis is the value of the function.)

3(a). The metric A at t = 0. ($-2.1\times10^{-4} \leq A \leq 1.9\times10^{-4}$)

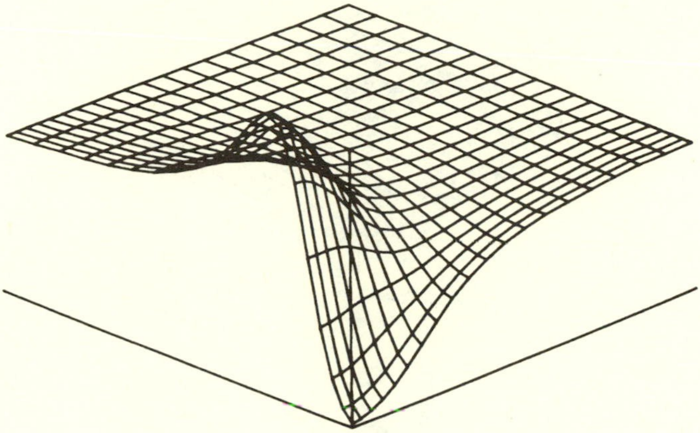

3(b). The metric A at t = 3. Note the non-flat behavior near the origin. ($-3.8\times10^{-4} \leq A \leq 1.0\times10^{-4}$)

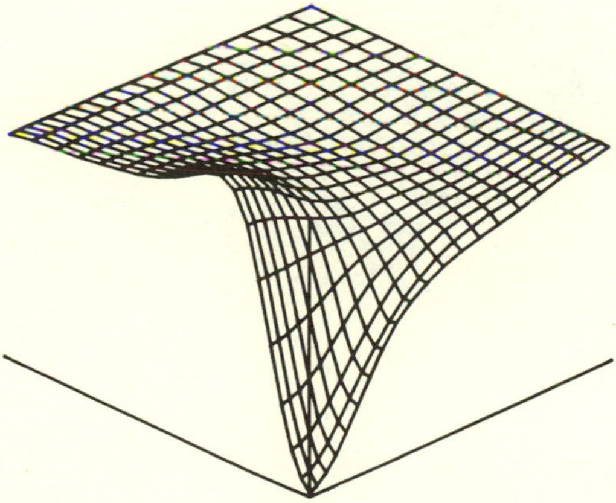

Figure 4. Evolution of the Brill waves of Figure 3 using the Wilson shift vector.

4(a). The metric q is plotted at t = 3. (For small amplitudes q becomes identical to A in the Wilson gauge.) Note that q displays wave behavior and becomes flat near the origin. ($-1.0\text{x}10^{-4} \leq A \leq 1.0\text{x}10^{-4}$)

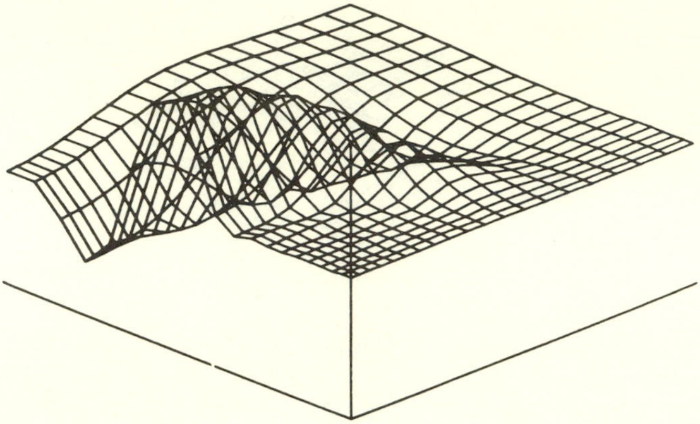

4(b). The ψ_4 radiation field at t = 3. Note the essentially quadrupole dependence ($\cos^2\theta$). ($-4.0\text{x}10^{-4} \leq \psi_4 \leq 3.9\text{x}10^{-4}$)

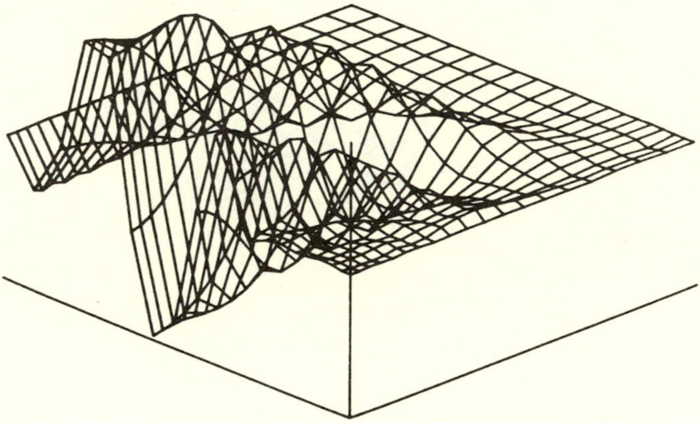

REFERENCES

Brill, D.R. (1959). On the positive definite mass of the Bondi-Weber-Wheeler time-symmetric gravitational waves. Ann. Phys., 7, pp. 466-483.

Eppley, K.R. (1975). The Numerical Evolution of the Collision of Two Black Holes. Ph.D. Thesis, Princeton University (unpublished).

Eppley, K.R. (1977). Evolution of time-symmetric gravitational waves: Initial data and apparent horizons. Phys. Rev. D, 16, 6, pp. 1609-1614.

Eppley, K.R., Smarr, L., and Teukolsky, S. (1978). Evolution of Time-Symmetric Gravitational Waves. II. Radiation Formulae and an Analytic Weak-Field Solution. Preprint.

Smarr, L. (1977). Space-time generated by computers: Black holes with gravitational radiation. In Eighth Texas Symposium on Relativistic Astrophysics, ed. M.D. Papagiannis, pp. 569-604. New York Academy of Sciences.

Smarr, L. and York, J.W. (1978). The radiation gauge in general relativity. Phys. Rev. D, 17, pp. 1945-1956.

Wheeler, J.A. (1964). Geometrodynamics and the issue of the final state. In Relativity, Groups and Topology, ed. C. DeWitt and B. DeWitt, pp. 317-520.

Wilson, J.R., and Smarr, L. (1978). Numerical axially symmetric geometrohydrodynamics. Unpublished preprint.

GRAVITATIONAL RADIATION FROM HYPERBOLIC ENCOUNTERS

P.D. D'Eath

Department of Applied Mathematics and Theoretical Physics,
Silver Street, Cambridge, England

0. INTRODUCTION

The aim of this lecture is to review the various perturbation methods developed in recent years for calculating gravitational radiation emitted in two-body encounters. Especial emphasis will be given to high-speed encounters, since these are the most efficient in generating waves. Section 1 will discuss techniques and results for distant encounters, starting with methods adapted to low velocities, then considering the work of Peters and of Thorne and Kovács, which spans the range from low to high speeds, and then turning to methods involving plane-fronted waves, specific to high speeds. Some implications of the plane-fronted wave approach for analytical radiation calculations in fully nonlinear strong-field collisions are described in Section 2.

1. GRAVITATIONAL RADIATION FROM DISTANT ENCOUNTERS

We begin by outlining the language used by Thorne (1977) in describing gravitational radiation. Complete information about the wave field in an asymptotically flat spacetime is carried by the transverse-traceless part of the spatial metric, h^{TT}_{jk} . The massless spin-2 field is resolved into its two polarization states by projection onto two orthogonal basis tensors:

$$h^{TT}_{jk}(t,r,\theta,\phi) = A_+ e^{+}_{jk} + A_\times e^{\times}_{jk} \quad . \qquad (1.1)$$

Here polar coordinates r,θ,ϕ are used in the asymptotic region, and the basis tensors are chosen to be

$$\underline{\underline{e}}^{+} = \underline{e}_\theta \otimes \underline{e}_\theta - \underline{e}_\phi \otimes \underline{e}_\phi \ , \quad \underline{\underline{e}}^{\times} = \underline{e}_\theta \otimes \underline{e}_\phi + \underline{e}_\phi \otimes \underline{e}_\theta \qquad (1.2)$$

where $\underline{e}_\theta$ and $\underline{e}_\phi$ are unit vectors in the θ and ϕ directions. Thorne (1977) has shown that A_+ and $A_\times$ are frame-independent scalar fields: if a boost is applied to h^{TT}_{jk} , which is then projected onto new spatial transverse-traceless basis tensors constructed from the boosted $e^{+}_{jk}, e^{\times}_{jk}$, one finds that the same A_+ and $A_\times$ result. In a given frame, the rate of energy output per unit solid angle in gravitational waves is then

$$\frac{dE}{d\Omega dt} = \frac{r^2}{16\pi}\left[\left(\frac{\partial A_+}{\partial t}\right)^2 + \left(\frac{\partial A_\times}{\partial t}\right)^2\right] , \tag{1.3}$$

using geometrical units with $c = G = 1$.

The perturbation methods developed for distant encounters fall essentially into three classes. The domains of validity depend on the encounter velocity and impact parameter, although there is some overlap, and the predictions for different régimes mesh smoothly together. The basic reference which summarizes information on the first two approaches, namely post-Newtonian and second-linearized methods, is Paper IV of Kovács & Thorne (1978), henceforth called KT IV; the basic reference for the third approach involving plane-fronted waves is D'Eath (1978).

(a) Post-Newtonian methods

Suppose that the two bodies concerned have masses m_A, m_B , and that they approach from infinity with relative velocity v and impact parameter b . Suppose further that typical radii r_A, r_B of the bodies are much smaller than the minimum separation between A and B , so that they interact as point particles. If the conditions

$$v \ll 1 , \quad (m_A+m_B) \ll bv \tag{1.4}$$

hold, then it is valid to treat the gravitational field around the bodies as a slow-motion perturbation of flat space, where time derivatives are much smaller than spatial derivatives, and the field equations become elliptic. The bodies move nearly on Newtonian orbits, and the radiation is given at leading order by the quadrupole-moment expression

$$h^{TT}_{jk} = (2/r)\ \ddot{I}^{TT}_{jk}(t-r) \tag{1.5}$$

derived in Secs.36.9,10 of Misner, Thorne & Wheeler (1973) (MTW).

If moreover the condition

$$(m_A+m_B) \ll bv^2 \tag{1.6}$$

holds, then the gravitational deflection produced by the encounter is small, i.e. the encounter is "distant". In describing the radiation, use a frame in which A starts at rest at the origin, while B moves in the xz plane, with initial velocity in the positive z-direction. Arrange that $t = 0$ at closest approach, and measure retarded time by defining

$$T = (t-r)v/b \quad , \tag{1.7}$$

corresponding to the typical timescale b/v of the radiation. Define

$$\ell(T) = (1+T^2)^{\frac{1}{2}} \quad , \tag{1.8}$$

$$\mathcal{A}_+ = (br/4m_A m_B)A_+ \quad , \quad \mathcal{A}_\times = (br/4m_A m_B)A_\times \quad . \tag{1.9}$$

Then (KT IV, Turner 1977) the quadrupole radiation has

$$\begin{aligned} \mathcal{A}_+ &= \tfrac{1}{2}(\ell^{-3}+\ell^{-1})\sin^2\theta + \tfrac{1}{2}\ell^{-3}(\sin^2\phi - \cos^2\theta\cos^2\phi) \\ &\quad + (T\ell^{-3}+T\ell^{-1}+1)\cos\theta\sin\theta\cos\phi \quad , \\ \mathcal{A}_\times &= \ell^{-3}\cos\theta\cos\phi\sin\phi - (T\ell^{-3}+T\ell^{-1}+1)\sin\theta\sin\phi \quad . \end{aligned} \tag{1.10}$$

The total energy radiated (Ruffini & Wheeler 1971, Hansen 1972) is

$$\Delta E = (37\pi/15)m_A^2 m_B^2 v/b^3 \quad . \tag{1.11}$$

These calculations can be refined to include higher-order terms in the post-Newtonian expansion (Epstein & Wagoner 1975, Wagoner & Will 1976). Such improvements applied to the distant encounter problem (Turner & Will 1978) yield results which remain accurate well into the régime covered by second-linearized theory, for speeds $v \lesssim 0.4$.

(b) Second-linearized methods

We next describe the method of Thorne and Kovács, used in analysing the bremsstrahlung problem. This again treats the geometry produced by the two interacting bodies as a perturbation of flat spacetime:

$$^{\mu\nu} = \eta^{\mu\nu} - \bar{h}^{\mu\nu} \quad , \quad |\bar{h}^{\mu\nu}| << 1 \quad , \tag{1.12}$$

where

$$^{\mu\nu} = (-g)^{\frac{1}{2}}g^{\mu\nu} \quad , \quad g = \det(g_{\mu\nu}) \tag{1.13}$$

and $\eta^{\mu\nu}$ is the Minkowski metric. The gauge condition

$$\bar{h}^{\mu\nu}{}_{,\nu} = 0 \tag{1.14}$$

is imposed, and the field equations are conveniently written in the form (Sec.20.3 of MTW)

$$H^{\mu\alpha\nu\beta}{}_{,\alpha\beta} = 16\pi(-g)(T^{\mu\nu}+t^{\mu\nu}_{L-L}) \quad , \tag{1.15}$$

$$H^{\mu\alpha\nu\beta} = \mathfrak{g}^{\mu\nu}\mathfrak{g}^{\alpha\beta} - \mathfrak{g}^{\alpha\nu}\mathfrak{g}^{\mu\beta} \quad , \tag{1.16}$$

where $T^{\mu\nu}$ is the energy-momentum tensor and $t^{\mu\nu}_{L-L}$ is the pseudotensor of Landau and Lifschitz (1962, Sec.100).

In the bremsstrahlung calculation, one builds up a sequence of approximations to $\bar{h}^{\mu\nu}$, in powers of the masses m_A, m_B. The first-order term ${}_1\bar{h}^{\mu\nu}$, is the sum of the linearized far-fields of A and B, where e.g. the contribution from A can be found by taking a "Coulomb" m_A/r linearized gravitational field in the gauge (1.14), and giving it a Lorentz boost up to A's velocity. There is no restriction on the relative velocity v, provided that the encounter is distant; time derivatives are treated on a par with spatial derivatives, and the field equations are regarded as hyperbolic. In particular, the second approximation ${}_2\bar{h}^{\mu\nu}$ satisfies

$$\eta^{\alpha\beta}\, {}_2\bar{h}^{\mu\nu}{}_{,\alpha\beta} = {}_2T^{\mu\nu} \quad , \tag{1.17}$$

where ${}_2T^{\mu\nu}$ is a source containing matter terms plus a contribution quadratic in ${}_1\bar{h}^{\mu\nu}$ and its derivatives. The resulting terms in ${}_2\bar{h}^{\mu\nu}$ proportional to $(m_A)^2$ and $(m_B)^2$ arise from the self-fields of A and B, and are non-radiative. The radiation field is contained in the part of ${}_2\bar{h}^{\mu\nu}$ produced by the interaction, proportional to $m_A m_B$. Thorne & Kovács (1975) show how to calculate this field using Green's functions in the spacetime which is weakly curved by the perturbation ${}_1\bar{h}^{\mu\nu}$. The details of the application to a distant encounter are given by Kovács & Thorne (1977). The original bremsstrahlung calculation of Peters (1970) gives an alternative derivation of the same radiation field. Peters assumes $m_B \ll m_A$, and considers perturbations about the Schwarzschild far-field of A; however the $m_A m_B$ part of the metric which he calculates is actually valid whatever the ratio m_B/m_A.

The second-linearized radiation calculation is accurate for encounters sufficiently distant that the condition

$$(m_A+m_B) \ll bv^2(1-v^2)^{\frac{1}{2}} \tag{1.18}$$

holds. At low velocities the condition is just (1.6), and the method reproduces the results of Sec.1(a). The restriction (1.18) at high speeds will be explained in Sec.1(c). Such encounters produce only small orbital deflections.

As before, let B approach with speed v in the positive z-direction in the rest-frame S of A, with the orbits lying in the xz plane and x = +b initially for B. Define

$$\gamma = (1-v^2)^{-\frac{1}{2}} \tag{1.19}$$

It is also convenient to consider the centre-of-velocity frame $\tilde{S}$, in which A and B move initially with velocities $-\tilde{v}\underline{e}_z$, $+\tilde{v}\underline{e}_z$, where

$$\tilde{v} = \gamma v/(\gamma+1) , \quad \tilde{\gamma} \equiv (1-\tilde{v}^2)^{-\frac{1}{2}} = [(\gamma+1)/2]^{\frac{1}{2}} . \tag{1.20}$$

The spatial origin of $\tilde{S}$ is chosen to lie midway between the two trajectories. Now the quadratic radiation field depends only on the product $m_A m_B$, and hence can be found by considering an equal-mass encounter, which has complete forward-backward symmetry in frame $\tilde{S}$. This leads (KT IV) to the symmetry relations

$$\begin{aligned} A_+(\tilde{t},\tilde{r},\tilde{\theta},\tilde{\phi}) &= A_+(\tilde{t},\tilde{r},\pi-\tilde{\theta},\tilde{\phi}+\pi) , \\ A_\times(\tilde{t},\tilde{r},\tilde{\theta},\tilde{\phi}) &= -A_\times(\tilde{t},\tilde{r},\pi-\tilde{\theta},\tilde{\phi}+\pi) . \end{aligned} \tag{1.21}$$

The wave-forms for arbitrary v are given in KT IV, but are too long to be written down here. Provided that v is only moderately relativistic, they describe radiation with the same general time-scale b/v and amplitude $A_{+,\times} \sim m_A m_B/br$ as in the low-velocity limit. However, when v is close to 1, Kovács and Thorne distinguish two characteristic timescales. First, in frame S, A sees B's gravitational field Lorentz-contracted, and undergoes accelerations over a timescale $b/(v\gamma)$. The appropriate retarded-time quantity for measuring the resulting radiation is then

$$T_A = v\gamma(t-r)/b . \tag{1.22}$$

Second, B accelerates in its initial rest frame, with timescale $b/(v\gamma)$ in that frame, and corresponding wave generation. The relevant timescale in frame S is found by Lorentz transformation, and hence depends on the angular coordinates (θ,ϕ) of the distant observer. The natural retarded-time parameter with which to view this radiation in frame S is

$$T_B = \frac{T_A + \beta v\gamma}{\gamma(1-\alpha v)} , \tag{1.23}$$

where $\alpha = \cos\theta$, $\beta = \sin\theta \cos\phi$.

At high speeds, most of the radiated energy is thrown forward into a narrow cone with a half-angle $\sim \gamma^{-1}$ in S , and its angular structure is best viewed by using the variable

$$\psi = \theta\gamma \quad , \tag{1.24}$$

which is of order unity in the beam. The basic timescale in this "forward region" is $b/(v\gamma)$; Kovács & Thorne (KT IV) are led to describe the leading radiation field using the time variable

$$\hat{T} = T_A + \psi \cos\phi \quad . \tag{1.25}$$

With the definitions

$$c = \cos\phi \quad , \qquad s = \sin\phi \quad , \tag{1.26}$$

$$\ell_A = [1+(\hat{T}-\psi c)^2]^{\frac{1}{2}} \quad , \qquad S^2 = 1 - \frac{4\psi c}{(1+\psi^2)}\hat{T} + \frac{4\psi^2}{(1+\psi^2)^2}\hat{T}^2 \quad , \tag{1.27,28}$$

in (1.10), the leading "forward region" field is given by

$$\begin{aligned} A_+ = \frac{4\gamma^2}{(1+\psi^2)^2 S^2} \{ & \psi c S^2 + \left(\frac{1+\psi^4}{1+\psi^2}\right) |\hat{T}| \left(2c^2 - 1 - \frac{2\psi c}{1+\psi^2}\hat{T}\right) \\ & + \frac{1}{\ell_A}\Big[\frac{1}{2}\langle(1-2c^2)-(1+2c^2)\psi^2-2c^2\psi^4\rangle + \psi c \langle 1+2c^2+2(1+c^2)\psi^2\rangle \hat{T} \\ & + \langle(1-2c^2)-(1+4c^2)\psi^2\rangle \hat{T}^2 + 2c\psi\hat{T}^3\Big]\} \quad , \end{aligned} \tag{1.29}$$

$$\begin{aligned} A_\times = \frac{4\gamma^2 s}{(1+\psi^2)^2 S^2} \{ & -\psi(1+\psi^2)S^2 + 2(1-\psi^2)|\hat{T}| \left(-c + \frac{\psi}{1+\psi^2}\hat{T}\right) \\ & + \frac{1}{\ell_A}\Big[c(1+\psi^2) - 2\langle(1+c^2)+c^2\psi^2\rangle\psi\hat{T} + 2c(1+2\psi^2)\hat{T}^2 - 2\psi\hat{T}^3\Big]\} \quad . \end{aligned}$$

There is an apparent discontinuity in the time derivatives at $\hat{T} = 0$, produced by the $|\hat{T}|$ terms, but the wave-forms are in fact smooth when viewed on the shorter timescale $\Delta T_B \sim 1$ $[\Delta\hat{T} \sim (1+\psi^2)/\gamma]$ near $\hat{T} = 0$.

The wave-forms (1.29) are accurate in the forward region:

$$\theta \ll \gamma^{-\frac{1}{2}} \quad . \tag{1.30}$$

In the centre-of-velocity frame $\tilde{S}$, they describe a beam at angles $\tilde{\theta} \ll \frac{\pi}{2}$. By the symmetry properties (1.21), there is a corresponding beam in $\tilde{S}$ in the backward direction, where $\pi - \tilde{\theta} \ll \frac{\pi}{2}$. Hence, by making Lorentz transformations and using (1.21,29), one can find the wave-forms in the "backward region" in S :

$$\theta \gg \gamma^{-\frac{1}{2}} \quad . \tag{1.31}$$

Overlapping with both forward and backward regions is the "intermediate region":

$$\gamma^{-1} \ll \theta \ll \frac{\pi}{2} \quad , \tag{1.32}$$

(i.e. $\tilde{\gamma}^{-1} \ll \tilde{\theta}$, $\tilde{\gamma}^{-1} \ll \pi-\tilde{\theta}$) , where the radiation changes smoothly in character as θ varies from forward to backward. The typical intermediate-region radiation timescale is $b\theta$, and the waveforms there are very simple [KT IV, Eqs.(4.14a-c)], being found most easily using plane-fronted wave methods.

The characteristic power/solid angle emitted into the forward region at high speeds is

$$\frac{dE}{d\Omega dt} \sim \frac{\gamma^6 (m_A m_B)^2}{b^4} \tag{1.33}$$

for $\theta \lesssim \gamma^{-1}$. Since $\Delta t \sim \gamma^{-1} b$ and $\Delta\Omega \sim \gamma^{-2}$ in the forward region, and since nearly all the energy is beamed forward in S , one can estimate the total energy ΔE radiated in S . Numerical results provide an estimate of the relevant coefficient (Peters 1970, KT IV):

$$\Delta E = (20.0 \pm 0.3)(m_A m_B)^2 \gamma^3 b^{-3} \quad . \tag{1.34}$$

This estimate holds whenever γ is large, provided that $\gamma \ll b/(m_A+m_B)$ [Eq.(1.18)].

(c) Bremsstrahlung calculations using plane-fronted waves

The approaches outlined so far have in effect relied on treating the masses m_A, m_B as small parameters, in order to construct perturbation expansions. However, if one is specifically interested in high-speed encounters, then another small parameter is available, namely γ^{-1} . In this Section we show how large-γ methods can be used to extend the results of Sec.1(b) down to impact parameters b which are comparable with $(m_A+m_B)\gamma$.

Suppose one takes a body of mass m and boosts it up to high speed v in the positive z-direction, with $\gamma \gg 1$. Define $\mu = m\gamma$, and fix attention on the gravitational field over length-scales of order μ in transverse x,y directions, but over scales of order m in the direction of motion. Then (D'Eath 1978) the metric, written in suitable coordinates, can be given the asymptotic expansion

$$g_{\mu\nu}(t,x,y,z,\gamma) = \eta_{\mu\nu} + \gamma^{-1}h^{(1)}_{\mu\nu}[\gamma^{-1}x,\gamma^{-1}y,z-vt] + \gamma^{-2}h^{(2)}_{\mu\nu}[\gamma^{-1}x,\gamma^{-1}y,z-vt] + \ldots , \tag{1.35}$$

where the only non-zero components of $h^{(1)}_{\mu\nu}$ are

$$h^{(1)}_{xx} = -h^{(1)}_{yy} = [(\gamma^{-1}y)^2-(\gamma^{-1}x)^2]\ f[\gamma^{-1}\rho,z-vt] ,$$
$$h^{(1)}_{xy} = h^{(1)}_{yx} = -2(\gamma^{-1}x)(\gamma^{-1}y)\ f[\gamma^{-1}\rho,z-vt] , \tag{1.36}$$

where

$$\rho^2 = x^2 + y^2 \tag{1.37}$$

and the function f is defined by

$$f(p,q) = 4m[q-(p^2+q^2)^{\frac{1}{2}}]p^{-4} + 2m(p^2+q^2)^{-\frac{1}{2}}p^{-2} . \tag{1.38}$$

This expansion is valid as $\gamma \to \infty$ with $\gamma^{-1}x, \gamma^{-1}y$, $(z-vt) \to$ constants. The leading term $h^{(1)}_{\mu\nu}$ depends only on the Coulombic far-field of the body, and not on any more detailed information. The Lorentz-contracted shock wave structure of (1.35,36) was first appreciated by Pirani (1959).

If one examines this high-velocity gravitational field using "blurred spectacles" which cannot resolve the detailed shock structure, or if one takes the limit $\gamma \to \infty$ (while scaling m down to keep μ finite), one sees an impulsive axisymmetric plane-fronted shock wave (Penrose 1972, Ehlers & Kundt 1962). In one coordinate system (Aichelburg & Sexl 1971) the impulsive shock has metric

$$ds^2 = dudv + dx^2 + dy^2 - 4\mu\ \ell n(x^2+y^2)\delta(u)du^2 . \tag{1.39}$$

This represents two pieces of Minkowski space joined together with a warp across the null surface $u = 0$. The warp involves identifying coordinates by sliding up or down the null generators:

$$x_+ = x_- ,\quad y_+ = y_- ,\quad v_+ = v_- + 4\mu\ \ell n(x^2+y^2) , \tag{1.40}$$

where $\pm$ refers to $u \gtrless 0$.

Now consider a high-speed encounter of two bodies with comparable masses m_A, m_B , using the conventions of Sec.1(b), and working in the centre-of-velocity frame $\tilde{S}$. Here, to describe encounters sufficiently close that b is comparable with $(m_A+m_B)\gamma$, write

$$d = b/(m_A+m_B)\gamma \tag{1.41}$$

and regard d as a parameter of order unity. (For a more formal treatment see Sec.VI of D'Eath 1978). In describing the incoming fields of A and B, concentrate now on length-scales of order b or $(m_A+m_B)\tilde{\gamma}^2$ in transverse directions, and scales of order $b\tilde{\gamma}^{-1}$ in the z-direction, and arrive at two asymptotic expansions analogous to (1.35). E.g. B's field viewed in this way has the form

$$g_{\mu\nu}(t,x,y,z,\tilde{\gamma}) = \eta_{\mu\nu} + \tilde{\gamma}^{-2} j^{(2)}_{\mu\nu}[\tilde{\gamma}^{-2}x,\tilde{\gamma}^{-2}y,\tilde{\gamma}^{-1}(z-\tilde{v}t)] + \ldots \tag{1.42}$$

before collision. When the fields collide, B's shock feels A's gravitational field effectively as an impulsive shock, and is subjected to delays of the type (1.40). The leading metric perturbation which describes the resulting structure of shock B just after passing through shock A is still transverse, given by $\tilde{\gamma}^{-2} g^{(2)}_{\mu\nu}$, where

$$g^{(2)}_{xx} + i g^{(2)}_{xy} = -(\bar{X}+iY)^2 f[(\bar{X}^2+Y^2)^{\frac{1}{2}}, \tilde{\gamma}^{-1}(z-t) - 4m_A \ln(X^2+Y^2) - 8m_A \ln\tilde{\gamma}],$$

$$g^{(2)}_{yy} = -g^{(2)}_{xx} , \tag{1.43}$$

$$X = \tilde{\gamma}^{-2}x , \quad \bar{X} = X - 2(m_A+m_B)d , \quad Y = \tilde{\gamma}^{-2}y ,$$

$$m \to m_B \quad \text{in Eq.(1.38)} . \tag{1.44}$$

The incoming "equilibrium" shock structure of (1.35,36) has been disturbed by the logarithmic time-delay in the second argument of f, and as shock B continues to propagate forward thereafter, it has to re-arrange itself, thereby producing a beam of radiation.

The radiative evolution of shock B after collision is described using the expansion

$$g_{\mu\nu}(t,x,y,z,\tilde{\gamma}) = \eta_{\mu\nu} + \tilde{\gamma}^{-2} g^{(2)}_{\mu\nu}(U,X,Y,V) + \ldots \tag{1.45}$$

$$U = \tilde{\gamma}^{-3}t , \quad V = \tilde{\gamma}^{-1}(z-t) - 8m_A \ln\tilde{\gamma} . \tag{1.46}$$

Here one again considers the shock structure over scales comparable with b in transverse directions $[\Delta X, \Delta Y \sim (m_A+m_B)]$, and over scales $b\tilde{\gamma}^{-1}$ in the z-direction (taking a convenient origin for V), but one must also allow for slow evolution of the shock over a timescale $\Delta U \sim (m_A+m_B)$. This natural timescale is understood most easily by examining the encounter in B's rest-frame. There one is viewing the disruption of B's Coulomb far-field, caused by shock A, over scales of order b in all space-time directions; using two null and two transverse coordinates, then boosting back to frame $\tilde{S}$, one arrives at suitable coordinates $\tilde{\gamma}^{-1}(z-t)+\text{const.}$, $\tilde{\gamma}^{-3}(z+t), \tilde{\gamma}^{-2}x, \tilde{\gamma}^{-2}y$. Then note that $z \sim t$ in shock B, to see that $U = \tilde{\gamma}^{-3}t$ is a suitable time coordinate. In these coordinates, the evolution of shock B for $U > 0$ is governed by wave equations:

$$Lg^{(2)}_{xx} = 0\ , \quad Lg^{(2)}_{xy} = 0\ , \quad Lg^{(2)}_{yy} = 0\ , \tag{1.47}$$

$$L = 2\partial^2/\partial U\partial V + \partial^2/\partial X^2 + \partial^2/\partial Y^2 \quad . \tag{1.48}$$

These should be solved with the boundary conditions (1.43,44) on the "null surface" $U = 0$, and the requirement that $g^{(2)}_{\mu\nu} \to 0$ rapidly in directions ahead of shock B. Care must be taken to separate out the boosted Coulomb field of B, after its small-angle deflection by A, from the radiative part of $g^{(2)}_{\mu\nu}$.

The resulting "forward region" radiation is beamed with half-angle $\sim \tilde{\gamma}^{-1}$ in $\tilde{S}$, and is best described using the angular coordinate

$$\psi = \tilde{\gamma}\tilde{\theta} \quad , \tag{1.49}$$

where $\tilde{\theta}, \tilde{\phi} = \phi$ are polar angles in $\tilde{S}$. Taking a convenient origin for retarded time τ, the time-derivatives of the wave amplitudes in the beam are

$$\frac{\partial}{\partial\tau}(A_+ + iA_\times) = \frac{3m_B}{\pi r}\int_{-\infty}^{\infty}\int_{-\infty}^{\infty} dXdY[X-2(m_A+m_B)d + iY]^2 \cdot$$
$$\cdot\{[X-2(m_A+m_B)d]^2+Y^2+[\tilde{\gamma}^{-1}\tau+X\psi\cos\phi+Y\psi\sin\phi+4m_A\ln(X^2+Y^2)]^2\}^{-5/2} \quad . \tag{1.50}$$

Provided $d \sim 1$ and $m_A \sim m_B$ this describes radiation with strong-field intensity in the beam (power/solid angle $\sim c^5/4\pi G \sim 3.6\times 10^{59}/4\pi$ erg/sec/steradian). The radiation timescale is then $\sim (m_A+m_B)\tilde{\gamma}$ or $\sim b/\tilde{\gamma}$, corresponding to $\Delta T_A \sim 1$ in Sec.1(b). There will be a similar beam in the backward direction, and the total energy emitted in such a close encounter is a fraction $\sim \tilde{\gamma}^{-2}$ of the incoming energy $(m_A+m_B)\tilde{\gamma}$ in $\tilde{S}$.

If d is increased to values $d >> 1$, the effects of the delay term in (1.50) become small, and the overall size of the integral (1.50) scales as d^{-2}, with dependence $\propto m_A m_B$ on the masses. The radiation for $d >> 1$ can be found in closed form (D'Eath 1978), and agrees with the large-γ results in Sec.1(b) (KT IV). However the overlap with Sec.1(b) extends no further - second-linearized methods cannot deal with the nonlinear mass-dependence found here, once b is comparable with $(m_A+m_B)\gamma$.

If instead d is reduced to values $d << 1$, (1.50) still describes the radiation near the forward direction, although there will no longer be a beam for $d \lesssim \tilde{\gamma}^{-1}$ [i.e. $b \lesssim (m_A+m_B)\tilde{\gamma}$] - see Sec.2. Numerical computation should show the angular width of the beam gradually increasing as d is reduced, until eventually (1.50) only describes some detailed structure near $\tilde{\theta} = 0$ in a more extensive radiation pattern.

As with (1.29), so the function (1.50) has a discontinuity at $\tau = -2\tilde{\gamma}(m_A+m_B)d\psi\cos\phi - 8m_A\tilde{\gamma}\ \ln[2(m_A+m_B)d]$, the retarded time at which the body B is seen by the distant observer to pass through shock A. Again, the radiation is in fact smooth when viewed on shorter timescales $\Delta T_B \sim 1$. This behaviour is missed by looking at the gravitational field over transverse distances $\sim b$, but could be found using a variant of the "method of virtual quanta" devised by Matzner & Nutku (1974). There one approximates the wave generation near B by working in B's rest frame, sufficiently close to B that shock A can be treated as an exact plane wave which scatters as a weak perturbation off B's Schwarzschild far-field. This method will not yield the dominant part of the radiation, as calculated above, since it ignores the logarithmic transverse variations of the shocks. Rather, it complements other methods by describing the high-frequency domain ($\omega >> \tilde{\gamma}/b$ in frame $\tilde{S}$).

Radiation in the intermediate region can also be found using plane-fronted waves. The timescale for this radiation is sufficiently long that one can ignore the detailed shock structure, adopting the "blurred spectacles" approach mentioned earlier, to see only the collision of two weak impulsive plane-fronted shocks, travelling at the speed of light. The radiation is then found by a second-linearized calculation (a simplified version of the Thorne-Kovács treatment) as carried out by Curtis (1975).

2. HIGH-SPEED COLLISIONS

When objects A and B with comparable masses approach at

high speed, as viewed in $\tilde{S}$, their characteristic strong-field radii in transverse directions are $m_A\tilde{\gamma}, m_B\tilde{\gamma}$ ("active gravitational radii"). If the impact parameter b is comparable with $(m_A+m_B)\tilde{\gamma}$, or smaller, then a highly nonlinear collision occurs, and only small parts of the resulting space-time can readily be found by perturbation methods. An interaction region strongly curves on length- and time-scales $\sim (m_A+m_B)\tilde{\gamma}$ develops after the incoming shocks cross, and the collision is expected to produce a black hole of mass $\sim (m_A+m_B)\tilde{\gamma}$ if the ratio $b/(m_A+m_B)\tilde{\gamma}$ is sufficiently small. The interaction region is bounded by curved shocks, found by continuing shocks A and B after they cross; the curved shocks can be described in perturbation theory, leading to analytical results for wave generation.

For simplicity, consider an equal-mass head-on collision, where the space-time is axisymmetric. Again work in frame $\tilde{S}$, writing $\tilde{\mu} = m_A\tilde{\gamma} = m_B\tilde{\gamma}$. The methods of Sec.1(c) can still be used to analyse the structure of shock B on transverse length scales $\sim \tilde{\mu}\tilde{\gamma}$ and length scales $\sim \tilde{\mu}$ in the z-direction. The characteristic initial data at $U = 0$, just after shock B has passed through shock A , are given by (1.43,44) with $d = 0$. Note that B's gravitational field suffers a logarithmically infinite time delay as $X,Y \to 0$ in this case, so that the initial data are regular - the singularity present at $\bar{X} = Y = 0$ when $d > 0$ has been pushed off to infinity by the delay. Solution of the wave equations (1.47,48) gives the gravitational radiation which escapes out to future null infinity at angles $\tilde{\theta}$ of order $\tilde{\gamma}^{-1}$ in the forward region, travelling within shock B ahead of the strong-field interaction region. Describe these waves using the news function (Bondi, van der Burg & Metzner 1962)

$$c_\tau(\tau, \tilde{\theta} = \psi\tilde{\gamma}^{-1}) = \frac{r}{2}\frac{\partial A_+}{\partial \tau} \tag{2.1}$$

at retarded time τ , where only the + polarization is present since the spacetime is invariant under $\phi \leftrightarrow -\phi$. Then (1.50) gives the forward-region news function

$$c_\tau(\tau, \tilde{\theta} = \psi\tilde{\gamma}^{-1}) = \frac{3\tilde{\mu}}{2\pi}\int_0^\infty dP \int_0^{2\pi} d\phi\, P^3 \cos 2\phi[P^2+(\psi P\cos\phi+\tau+8\tilde{\mu}\ell nP)^2]^{-5/2} \tag{2.2}$$

apart from small higher-order corrections in $\tilde{\gamma}^{-1}$. Again, this describes strong-field radiation, with timescale $\sim \tilde{\mu}$.

This news function is plotted as a function of τ for selected

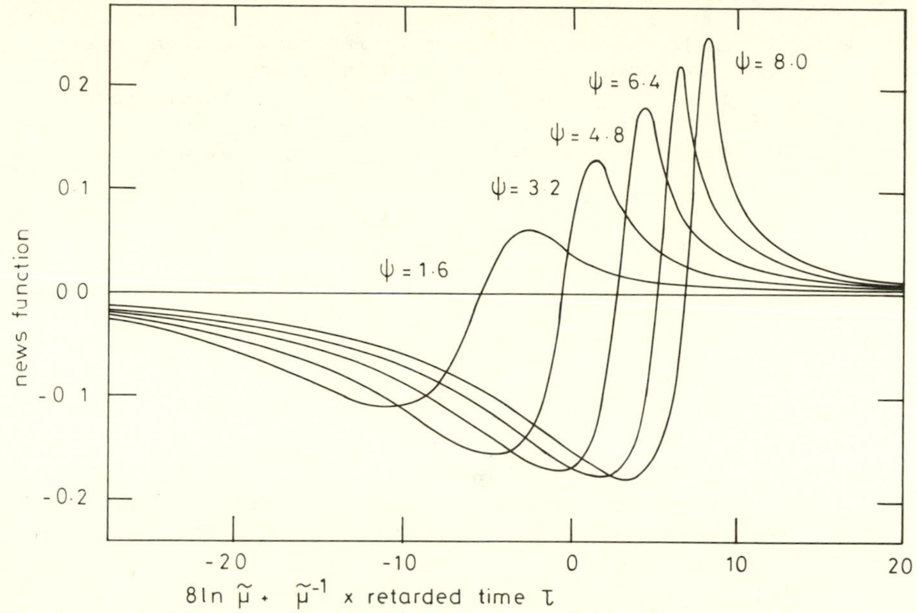

Fig.1. The news function produced by a high-speed head-on collision, at selected angles $\tilde{\theta} = \psi\tilde{\gamma}^{-1}$ close to the forward direction.

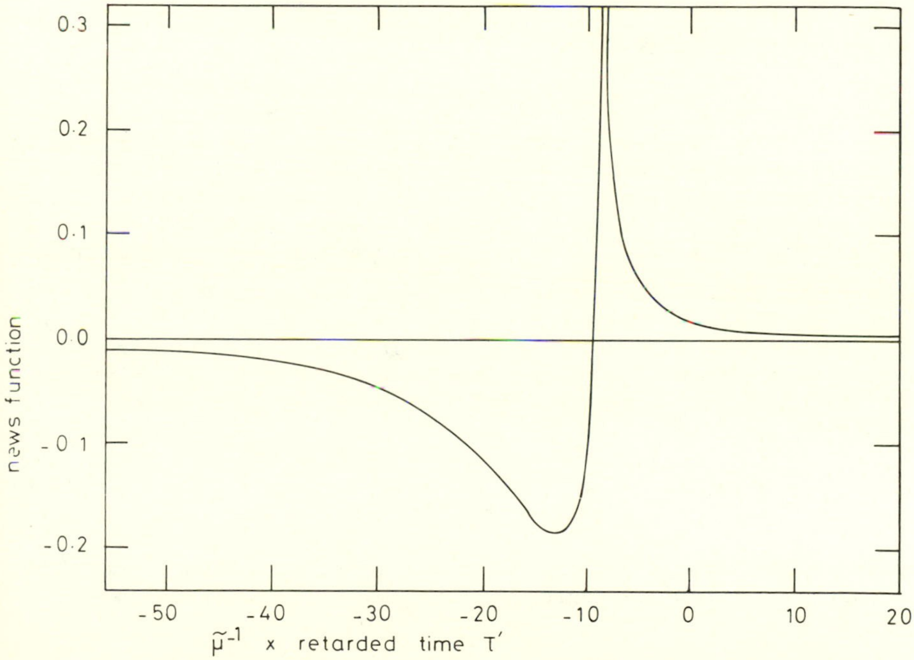

Fig.2. The news function at angles $\tilde{\gamma}^{-1} << \tilde{\theta} << 1$, produced by a high-speed head-on collision.

ψ-values in Fig.1. As one continues to increase ψ, the news function does not decrease - i.e. there is no beaming. This behaviour is essentially caused by the infinite spreading-out of B's shock field by the logarithmic delay at shock A. When $\psi \to \infty$, the integral (2.2) tends to a limiting form, centred on later and later τ-values, best viewed by making the change of origin (super-translation):

$$\tau' = \tau - 8\tilde{\mu}\,\ell n(\psi/\tilde{\mu}) \quad . \tag{2.3}$$

Then as $\psi \to \infty$ with τ' fixed, the integral (2.2) tends to the function

$$a_o(\tau') = \frac{4}{\pi}\int_{\mathcal{D}} \frac{dP}{P^2}\left[\frac{2(8\ell nP+\tau'/\tilde{\mu})^2}{P^2}-1\right]\left[1-\frac{(8\ell nP+\tau'/\tilde{\mu})^2}{P^2}\right]^{-\frac{1}{2}} \tag{2.4}$$

where $\mathcal{D}$ is the domain $\{P: 8\,\ell n\,P-P+\tau'/\tilde{\mu} < 0 < 8\,\ell n\,P+P+\tau'/\tilde{\mu}\}$. This function is plotted in Fig.2.; note that the secont peak visible in Fig.1 has turned into a mild but infinite logarithmic spike at $\tau'/\tilde{\mu} = 8-8\,\ell n\,8$.

The limiting news function $a_o(\tau')$ describes the forward radiation accurately at angles satisfying $\tilde{\gamma}^{-1} \ll \tilde{\theta} \ll 1$. The total energy/solid angle radiated at these angles is $\tilde{\mu}/8\pi$, since numerical calculation (Chapman 1978) shows that

$$\int_{-\infty}^{\infty} d\tau'[a_o(\tau')]^2 = \frac{\tilde{\mu}}{2} \tag{2.5}$$

to high accuracy. Thus, if the radiation were distributed isotropically with this news function, a quantity $\frac{1}{2}\tilde{\mu}$ of energy would be radiated out of the incoming $2\tilde{\mu}$, giving an efficiency of 25%. Fourier transformation of $a_o(\tau')$ gives the energy radiated per unit frequency per steradian $dE/d\omega d\Omega$ plotted in Fig.3 (Chapman 1978). This drops rapidly as ω is increased from zero; a large part of the energy is contained in frequencies $\omega \lesssim 0.5\,\tilde{\mu}^{-1}$.

Perturbation theory can in principle be carried further to obtain more information on the angular distribution of radiation. The integral (2.2) only gives the first term in an asymptotic expansion

$$c_\tau(\tau, \tilde{\theta} = \psi\tilde{\gamma}^{-1}) = \sum_{n=0}^{\infty} \tilde{\gamma}^{-2n}\, Q_{2n}(\tau,\psi) \tag{2.6}$$

valid in the forward region. Similarly, the radiation outside the forward (and backward) region will be given by an expansion

$$c_{\tau'}(\tau',\tilde{\theta}) = \sum_{n=0}^{\infty} \tilde{\gamma}^{-n}\, S_n(\tau',\tilde{\theta}) \quad . \tag{2.7}$$

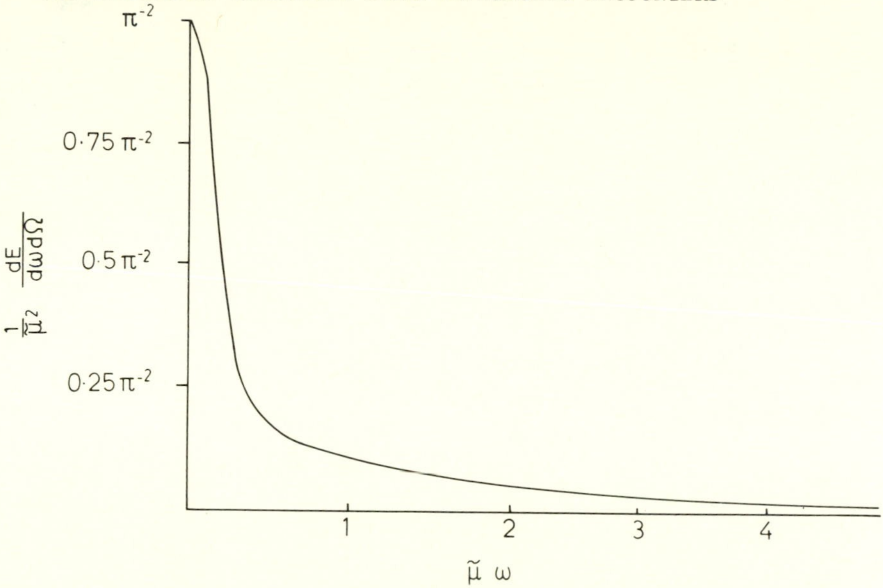

Fig.3. The energy spectrum for the news function shown in Fig.2.

So far, only $S_o(\tau', \tilde{\theta} = 0) = a_o(\tau')$ is known by matching these expansions, but higher-order calculations in the forward region followed by matching out to larger angles will reconstruct the full high-speed news function $S_o(\tau', \tilde{\theta})$ as a Taylor series about $\tilde{\theta} = 0$ in powers of $\tilde{\theta}^2$.

Particularly interesting is the possibility that $S_o(\tau', \tilde{\theta})$ is isotropic, i.e. that $S_o(\tau', \tilde{\theta}) \equiv a_o(\tau')$. This is suggested by the zero-frequency limit (ZFL) results of Smarr (1977). The ZFL method allows one to calculate the radiation emitted at low frequencies $\omega \simeq 0$ in problems where the initial and final states are known, such as the distant encounters of Sec.1, or equal-mass head-on collisions. If h^{TT}_{jk} is known in the far field both initially and finally, one can take the difference to find $A_{+f} - A_{+i}$ and $A_{\times f} - A_{\times i}$, the zero-frequency parts of $\dot{A}_+$ and $\dot{A}_\times$. Applied to the forward region in our high-speed collision, assuming that the final state is a Schwarzschild black hole at rest, this gives

$$\left.\frac{dE}{d\omega d\Omega}\right|_{\omega=0,\, \tilde{\theta}=\psi\tilde{\gamma}^{-1}} \rightarrow \frac{\tilde{\mu}^2 \psi^4}{\pi^2 (1+\psi^2)^2} \quad , \quad \tilde{\gamma} \rightarrow \infty \quad , \tag{2.8}$$

which can be confirmed analytically using (2.2). Away from the forward and backward regions, it gives the isotropic behaviour

$$\left.\frac{dE}{d\omega d\Omega}\right|_{\omega=0} \to \frac{\tilde{\mu}^2}{\pi^2} \quad , \quad \tilde{\gamma} \to \infty \quad , \tag{2.9}$$

or equivalently (with regard for signs)

$$\int_{-\infty}^{\infty} d\tau' \; S_o(\tau',\tilde{\theta}) = -2\tilde{\mu} \quad , \tag{2.10}$$

for all $\tilde{\theta}$. This allows some optimism concerning the isotropy of the complete news function $S_o(\tau',\tilde{\theta})$.

REFERENCES

Aichelburg, P.C. & Sexl, R.U. (1971). Gen.Relativ. and Gravitation 2, 303.

Bondi, H., van der Burg, M.G.J. & Metzner, A.W.K. (1962). Proc. Roy.Soc. (London) A 269, 21.

Chapman, G.I.A. (1978). Personal communication.

Curtis, G.E. (1975). D.Phil. thesis, University of Oxford.

D'Eath, P.D. (1978). Phys.Rev. D, in press.

Ehlers, J. & Kundt, W. (1962). In Gravitation, ed. L. Witten. Wiley, New York.

Epstein, R. & Wagoner, R.V. (1975). Ap.J. 197, 717.

Hansen, R.O. (1972). Phys.Rev. D 5, 1021.

Kovács, S.J. & Thorne, K.S. (1977). Ap.J. 217, 252.

Kovács, S.J. & Thorne, K.S. (1978). Ap.J., in press; cited in text as KT IV.

Landau, L.D. & Lifschitz, E.M. (1962). The Classical Theory of Fields. Pergamon Press, London.

Matzner, R.A. & Nutku, Y. (1974). Proc.Roy.Soc. (London) A 336, 285.

Misner, C.W., Thorne, K.S. & Wheeler, J.A. (1973). Gravitation. Freeman, San Francisco.

Penrose, R. (1972). In General Relativity: papers in honour of J.L. Synge, ed. L. O'Raifeartaigh. Oxford University Press, Oxford.

Peters, P.C. (1970). Phys.Rev. D 1, 1559.

Pirani, F.A.E. (1959). Proc.Roy.Soc. (London) A 252, 96.

Ruffini, R. & Wheeler, J.A. (1971). In Relativistic Cosmology and Space Platforms, Proceedings of the Conference on Space Physics, E.S.R.O., Paris.

Smarr, L. (1977). Phys.Rev. D 15, 2069.

Thorne, K.S. (1977). Cornell University preprint CRSR 662.

Thorne, K.S. & Kovács, S.J. (1975). Ap.J. 200, 245.

Turner, M. (1977). Stanford University preprint.

Turner, M. & Will, C.M. (1978). Ap.J. 220, 1107.

Wagoner, R.V. & Will, C.M. (1976). Ap.J. 210, 764.

GRAVITATIONAL COLLAPSE OF EVOLVED STARS AS A PROBLEM IN PHYSICS

W. David Arnett

Enrico Fermi Institute
University of Chicago

1. INTRODUCTION

These lectures will discuss the underlying physical processes which currently dominate our descriptions of the gravitational collapse of evolved massive stars. Emphasis will be placed on model-independent aspects of the problem. In the following section the physics of core collapse will be examined; new results on neutrino trapping and adiabatic, General Relativistic hydrodynamics and their implications for gravitational wave generation will be presented. Then the nature of pre-collapse configurations, and reasons for their tendency to converge to a common core structure, will be summarized. Finally, the event rates proposed for core collapse will be critically discussed.

2. PHYSICS OF CORE COLLAPSE

2.1 Overview

A discussion of the physics of core collapse can be divided into two parts: (1) a general examination of the problem, featuring the microscopic physics, and (2) an illustrative special case that has now begun to be explored in some detail, that of spherically-symmetric collapse (called "1D"). From the conceptual point of view it is essential to understand the spherical case before attempting the more general problem in two and three space dimensions. From the computational point of view it is important to use simplifications developed in 1D analysis to make solution of the 2 and 3D cases feasible.

In another sense the problem divides into two parts, depending upon whether the matter is transparent or opaque to neutrinos. In the early, transparent stages the deviations from spherical symmetry should be smaller, but these should grow, perhaps to become significant, during collapse to super-nuclear densities in the opaque stage. Different physical processes dominate in these two stages.

2.2 The Pre-Collapse State

Massive stars tend to evolve toward a common state of their inner regions as they approach hydrodynamic collapse (see the following section). The evolution from hydrostatic contraction to hydrodynamic collapse is smooth and continuous but one can define the onset of instability as the point at which the pressure-weighted adiabatic exponent, Γ, becomes less than a critical value,

$$\Gamma_{crit} = 4/3 + \mathcal{O}(Gm/rc^2),$$

where the second term is due to GTR. At central densities of $\rho_c \simeq 10^9$, 10^{12}, and 10^{15} g cm^{-3}, the GTR corrections are roughly 0.1, 1, and 10%. In this sense the onset of instability occurs at a central density of

$$\rho_c \simeq (3 - 6) \times 10^9 \text{ g cm}^{-3}$$

and a central temperature of

$$T_c \simeq 6 \times 10^9 \text{°K} = 0.5 \text{ MeV}.$$

The "core" mass, that is, the mass interior to the silicon-burning shell (see below), is close to the Chandrasekhar maximum mass for degenerate electron configurations in hydrostatic equilibrium:

$$M_{core} \gtrsim M_{Ch} = 1.45\ (2Y_e)^2\ M_\odot$$

where Y_e is the number of electrons per nucleon. Some exceptions are very massive stars ($M > 100\ M_\odot$) or "unusually" mixed stars. This core material (silicon burning "ashes") consists of neutron-rich iron-group nuclei, such as ^{58}Fe, ^{60}Fe, ^{62}Ni, etc.

2.3 Nuclear Statistical Equilibrium

At such high temperature and density, reactions involving strong and electromagnetic interactions are so rapid as to approach equilibrium with their inverses:

$$(Z,A) + p \rightleftarrows (Z+1,A+1) + \gamma$$

$$(Z,A) + n \rightleftarrows (Z,A+1) + \gamma$$

and so on for reactions like (α,γ), (α,n), (α,p), (p,n), etc. In general for a system in "chemical" (i.e., reaction) equilibrium,

$$\sum_i \nu_i \mu_i = 0$$

where the ν_i are the stoichiometric constants and the

$$\mu_i = u_i + m_i c^2$$

are the chemical potentials of the ith component, and m_i is its rest mass. Two special cases are simple and useful here. For a degenerate Fermi-Dirac gas, u_i is equal to the kinetic Fermi energy. For a Maxwell-Boltzmann (MB) gas,

$$\frac{u_i}{kT} = \ln\left\{\frac{Y_i}{z_i} \; A\,\rho\left(\frac{2\pi h^2}{m_i kT}\right)^{3/2}\right\}$$

where A is Avogadro's number, z_i the partition function, and $Y_i = n_i/\rho A$ the mole fraction (number of component i particles per nucleon). The former is useful for electrons; the latter for nuclei and nucleons (sometimes). In general more complex relations must be used. For nuclei, using a recurrsion procedure,

$$\mu_{AZ} = Z\mu_p + N\mu_n$$

or (for a MB gas)

$$Y_{AZ} = Y_p^{\,Z}\, Y_n^{\,N}\, G_{AZ}(T,\rho,A,Z,B_{AZ}),$$

which is the law of mass action. Note that G_{AZ} depends on the difference of masses through the nuclear binding energy B_{AZ}.

We need two more equations to specify Y_p and Y_n, the mole fractions of free protons and neutrons. They are nucleon conservation,

$$\sum_i Y_i A_i = 1$$

and charge conservation,

$$\sum_i Y_i Z_i = Y_e.$$

There are several problems. Some of them are:

(1) For nuclei far from the valley of beta stability (experimentally unknown nuclei), it is difficult to know what to use for z_i and B_i.

(2) At high density, corrections for finite nuclear size and nuclear surface energy must be considered for the transition to nuclear matter.

(3) At high density, the effects of nucleon interactions and neutron degeneracy must be included.

Thus, if the three quantities ρ, T, and Y_e are specified, we can use nuclear statistical equilibrium (NSE) to determine the composition (Y_i's), pressure, specific energy and entropy (the equation of state).

A useful conceptual aid (Epstein & Arnett 1975) is to consider the NSE distribution as having only four components: free neutrons, free protons, alphas, and "nuclei." Here "nuclei" can be thought of as a nucleus having the properties of the ensemble average for $A > 4$. At high temperature and low density free nucleons dominate. At low T and high ρ the dominant components are nuclei (and perhaps free neutrons if neutron drip occurs for low Y_e). In an intermediate regime of ρ and T the alpha abundance has a maximum, which lies (roughly) at

$$t \simeq 0.16 + (0.115 + 0.035r)r$$

where $t = \log_{10}(T/10^{10}\,°K)$ and $r = \log_{10}(\rho/10^{10}\ \mathrm{g\ cm^{-3}})$. These features are relatively insensitive to Y_e. A typical evolutionary trajectory is roughly

$$t \simeq (0.168 + 0.037r)r;$$

this fits the results of Arnett (1977a, hereafter A77) and Mazurek (1978), while those of Wilson (1978) lie slightly below and Bruenn, Buchler and Yueh (quoted in Mazurek 1978) lie above. Over the whole path the abundance of nuclei relative to free nucleons is significant. As we shall see, this is important because it affects the equation of state and the rate of weak interaction processes.

2.4 Neutrino Emission

Neutrinos are produced by thermal processes and by neutronization. Thermal processes involve the annihilation of a real or virtual electron-positron (or muon-antimuon, etc.) pair, forming a neutrino-antineutrino pair. Lepton number is unchanged if the neutrino pair escapes. The quantity Y_e used in NSE (see above) is, strictly speaking, related to the net electron-positron charge:

$$Y_e = [n(e^-) - n(e^+)]/\rho A,$$

so that pair annihilation does not change Y_e. The most important thermal processes are:

(1) pair:

$$e^+ + e^- \to \nu + \overline{\nu},$$

(2) plasmon:

$$(\text{plasma excitation}) \to \nu + \overline{\nu},$$

(3) photo:

$e^- + \gamma \to e^- + \nu + \bar{\nu}$,

(4) bremsstrahlung:

$e^- + (Z,A) \to (Z,A) + e^- + \nu + \bar{\nu}$.

These processes dominate until the collapse really begins in earnest (see A77; Freedman, Schramm & Tubbs 1978).

Neutronization refers to electron capture driven by the high electron Fermi energy (i.e., high density). The most important processes are capture on:

(1) nuclei,

$e^- + (Z,A) \to \nu + (Z-1,A)$;

(2) free protons,

$e^- + p \to \nu + n$.

The latter is the inverse of neutron decay, so that the matrix element is well known; however this is not true for most neutron rich nuclei, especially as excited states may be involved in both target and product. Thus the capture rate on nuclei is more uncertain.

Nuclei resist electron capture because of the large threshold energies required as they become more neutron rich, and because only "valence" protons are likely to be able to undergo superallowed transitions. Electron capture on free protons is limited by the small abundance of free protons. These problems are eased by higher density (nuclei) and higher temperature (protons), so neutronization speeds up as collapse continues. Note that by making holes in the electron Fermi sea, and by leaving nuclei in excited states, electron capture _heats_ the matter even if the neutrinos freely escape.

Neutronization reduces Y_e, thus reducing the electron pressure and affecting NSE. Once collapse begins, neutronization becomes the dominant mode of neutrino production, overwhelming thermal processes, although thermal processes should eventually take over again as the net lepton number (per nucleon) becomes small.

2.5 Neutrino Opacity

The escape of neutrinos is impeded by the inverses of the processes of emission and by scattering. The most important processes are (see A77; Freedman _et al._ 1978):

(1) neutral current scattering,

$\nu + (Z,A) \rightarrow \nu + (Z,A)$,

$\nu + n \rightarrow \nu + n$, and similarly for antineutrinos;

(2) nucleon absorption,

$\nu + n \rightarrow p + e^-$,

$\bar{\nu} + p \rightarrow n + e^+$;

(3) electron-neutrino scattering,

$e^- + \nu \rightarrow e^- + \nu$,

$e^- + \bar{\nu} \rightarrow e^- + \bar{\nu}$.

The last process may be important because it modifies the neutrino energy as seen from the fluid frame comoving with the matter ("nonconservative" scattering), thus speeding thermalization of the neutrino gas.

Are the neutrinos formed early enough in the collapse to escape the star before they are dragged along with the matter? Apparently not (Mazurek 1975; A77). At densities $\rho \gtrsim 10^{12}$ g cm^{-3}, the neutrinos are trapped and move with the matter; in terms of timescales,

$$\tau_{diffusion}(\nu) >> \tau_{infall}.$$

The neutrino distribution function approaches a Fermi-Dirac form, so the microscopic system can be specified by three quantities again: T, ρ, Y_e. Because lepton number $Y_\ell = Y_e + Y_\nu$ and entropy S are conserved during further collapse (until shocks form which alter S), it is often useful to use S and Y_e instead of T and Y_e. Thus the transition from neutrino transparency to trapping may be simply thought of in terms of a change in entropy ΔS and in lepton number ΔY_e.

The effect of neutral currents is to increase the neutrino opacity without changing the emissivity much, thus inhibiting neutrino transport and enhancing neutrino trapping.

2.6 Neutronization Shell: Accretion or Explosion?

As the collapse continues there is a spherical shell in which neutronization produces neutrinos, some to be trapped and some to escape. Since Colgate and White (1966) most work has been based on the view that the neutrino luminosity of this shell must become large enough to cause an explosion of the overlying layers. This does not seem to happen using the physics thought to apply at such density ($\rho \sim 10^{12}$ g cm^{-3}); see A77 and Wilson (this volume) and references therein.

It is not clear that a strongly radiating neutronization shell would give an explosion however; it might give a standing accretion shock. To test this possibility, the behavior of a core collapse with a strongly radiating neutronization shell was calculated. The methods were those of A77, with minor improvements (for this problem) in treatment of e-ν scattering and nuclear partition functions. In order to keep core hydrodynamics from overwhelming the neutronization shell phenomena, GTR and equation of state corrections due to nucleon-nucleon interactions were not included. This gave a soft bounce at $\rho_c \simeq 3 \times 10^{13}$ g cm^{-3}. To get a high shell luminosity the NSE conditions were (carefully) modified so that the proton abundance Y_p increased (to about 1% in the shell). This allowed electron capture to occur on free protons primarily, and neutronization happened at lower density so that more neutrinos could escape (in the shell ΔY_e was about 0.3).

As more matter fell in, the neutronization shell became thinner ($r \sim 3 \times 10^6$ cm, $\Delta r \lesssim 0.1r$), and did radiate strongly. The dominant stress on the matter falling into the neutronization region was due to neutrinos, but there was not the vaguest hint of an explosion. Instead a steady-state accretion shock, with thickness determined by neutrino stress (not pseudo-viscosity), surrounded the hot neutron-star core which was quasi-hydrostatic. The neutron-star simply grew in mass as more matter fell in.

The failure of such a favorable case to give explosive behavior suggests that explosive mechanisms must be sought elsewhere.

2.7 Entropy and Lepton Number in the Core

The importance of considering the entropy explicitly was first emphasized by Bethe and Brown (see Bethe, Brown, Applegate & Lattimer 1978). The entropy change in going through neutronization and neutrino trapping seems to be of order unity, $\Delta S/k \simeq 1$, in both the "standard" case (A77) and for the high luminosity neutronization shell described in the previous section. An important, and perhaps dominant, source of entropy generation is downscattering of neutrinos.

For example, in the reaction

$$e^- + p \rightarrow n + \nu$$

the particle energies approximately satisfy:

$$\langle\varepsilon(e^-)\rangle \simeq \langle\varepsilon(\nu)\rangle_{form} \simeq \varepsilon_f(e^-),$$

so that these neutrinos are formed with essentially the electron Fermi energy

$$\varepsilon_f(e^-) \simeq 34\ \rho_{12}^{1/3}\ \text{MeV}.$$

Nonconservative scattering (and absorption-emission) tend to thermalize the neutrinos (at least initially the ν number density is low so that at the same temperature the mean energy per neutrino in thermal equilibrium is less than the mean energy per electron, hence the tendency for neutrino energy to decrease). The neutrinos which escape do so with lower energy than they were formed with,

$$<\varepsilon(\nu)_{escape}> \simeq 10\ \text{MeV} << \varepsilon_f(e^-).$$

Suppose a fraction f of the degradation of neutrino energy is due to irreversible processes (e-ν scattering for example). The entropy change per neutrino degraded is

$$\Delta S/\nu \simeq f[\varepsilon_f(e^-) - <\varepsilon(\nu)_{escape}>]/T$$

$$\simeq kf(30 - 10)\ \text{MeV}/3\ \text{MeV} = (20/3)fk.$$

The number of neutrinos so processed is then $\Delta Y_e \simeq 0.3$, so per nucleon we have

$$\Delta S \simeq 0.3\Delta S/\nu \simeq 2fk.$$

The initial entropy was $S(0) \simeq 0.5$ to 1 k, so the entropy upon trapping is of order

$$S_{trapp} \simeq (0.5 + 2f)k,$$

i.e., in the range 0.5 - 3 or so. This rough analysis can be made much more precise (and complex) but the general result seems to be that S_{trapp} is small (not, say, 5 or 10). Further, essentially the same result obtains in the "standard" case (A77), so we may already have the adiabat (S_{trapp}) for further collapse hydrodynamics well determined. Such a small range in entropy pins down the temperature fairly well for a given equation of state.

The lepton number varies more between these two cases; $Y_e = Y_e + Y_\nu$ is about 0.09 for the radiating case and 0.23 for A77. The difference is simply dependent on the degree of trapping; the latter value seems more plausible at present.

If further work confirms these ranges,

$$S_{trapp} < 3k,$$

and

$Y_e > 0.1$,

then the major uncertainty in the further evolution may lie not in the thermodynamic trajectory but in our ability to properly include the complex physics of the equation of state of a high density interacting gas.

2.8 Adiabatic Hydrodynamics of the Core

From the discussion above it is clear that a useful approximation is that of adiabatic hydrodynamics. Thus the major theoretical difficulty, neutrino transport, is not explicitly calculated. For densities below $\rho \sim 10^{12}$ g cm^{-3} caution must be used. During infall the neutronization and neutrino escape can be approximated by an "effective equation of state." If there is a core explosion, with a shock propagating outward, the nature of the shock may be altered at densities below $\rho \sim 10^{12}$ g cm^{-3} due to neutrino transport. Nevertheless this approximation may allow us to study the dynamics of the core, which is of primary importance for the generation of gravitational waves.

Adiabatic hydrodynamics allows the 1D problem to be thoroughly explored; and the 2 and 3D problems become tractable. Van Riper (1978) has examined the Newtonian 1D case in detail, and finds core explosions only for the adiabatic exponent Γ near the critical value (Γ_{crit} = 4/3 for Newtonian gravity). No satisfactory "back of the envelope" explanation is yet available, but some understanding of the result can be obtained as follows. For Γ near critical, the inner part of the core which falls as a unit encounters a weak restoring force and thus overshoots equilibrium quite far. Upon expansion the large amplitude means that we have an ordered outward motion with large momentum. For Γ suddenly exceeding the critical value so that the restoring force is strong, small amplitude overshoot results. This tends to give a standing shock around the quasi-static inner core, which converts ordered infall energy into disordered thermal energy. The "coherence" of the motion for $\Gamma \simeq \Gamma_{crit}$ seems crucial for a vigorous explosion to result. We note that for the entropy range given above the equation of state (at least for $\rho \lesssim 10^{13}$ g cm^{-3}) has $\Gamma \simeq \Gamma_{crit}$ = 4/3 as Van Riper's constraints require; for higher density the situation is less certain due to GTR and nucleon interactions.

It appears that collapse will continue until greater than nuclear density is attained. Either the increased gravitation due to GTR (Van Riper & Arnett 1978; Van Riper 1979), or the attractiveness of the nucleon-nucleon interaction at sub-nuclear density (Lattimer, Lamb, Pethick & Ravenhall 1978), can prevent the low density bounce found in calculations not including these effects. Because of the near balance between pressure gradient and gravitational forces, even as a perturbation rotation could affect the density for bounce. Certainly the neglect of rotation becomes more suspect as the bounce density increases, but we simply do not know what the pre-collapse rotational state is. Van Riper and Arnett (1978) found that massive cores ($M_{core} \gg M_{crit}$) were on such high adiabats that a subnuclear bounce occurred, but the adiabatic approximation needs to be verified for such conditions.

For core mass near critical (that is, near the Oppenheimer-Volkoff mass for the chosen equation of state), Van Riper and Arnett (1978) found large amplitude bounces and vigorous outgoing shocks (analogous to the $\Gamma \simeq 4/3$ Newtonian cases of Van Riper 1978). For slightly larger masses a black hole was formed directly. For lower masses a standing accretion shock developed around the dense inner core; it is not clear what the long term behavior would be. Energy in the explosions varied, with many values of the order of several times 10^{51} ergs, a range of values interestingly close to those inferred from supernova light curves, and from remnants.

This "reflected shock" mechanism of explosion had been found previously in the search of a neutrino transport explosion (e.g., Wilson 1971; Bruenn 1975), and may have confused some interpretation of the numerical results.

2.9 Problems and Prospects

These results may change our concepts and techniques for core collapse. Some implications are:

(1) At densities high enough for neutrino trapping, adiabatic hydrodynamics is the appropriate zeroth-order approximation. Without rotation the bounce occurs at super-nuclear density, so that (adiabatic) GTR hydrodynamics in 2D and 3D becomes a highly relevant problem. (see Wilson 1979)

(2) High bounce density enhances damping of core oscillations by hydrodynamics and gravitational waves relative to neutrino loss.

(3) It is of vital importance to improve the equation of state (at low entropy, $S/k \sim 1$), especially for densities $\rho \gtrsim \rho_{nuclear}$.

(4) A well-founded treatment of nuclear partition functions for $1\ \mathrm{MeV} \lesssim kT \lesssim 10\ \mathrm{MeV}$ is needed.

(5) Several independent investigations of the conditions in the neutronization shell are needed to pin down better the jump in entropy and lepton number. Because this occurs at $\rho \ll \rho_{nuclear}$, 1D calculations may be adequate, at least in most cases.

(6) As a "reflected" shock propagates to lower density, ν-transport can affect the shock structure (Bruenn, Buchler & Yueh

1978). This needs to be pursued.

(7) Because of the considerable effort expended by many groups in clarifying the physics of the core collapse, we now seem to be approaching a unified picture of the collapse through the first core bounce in the 1D case. A description of this core behavior is needed to estimate the gravitational radiation, so we may be on the verge of providing the answer required. However the supernova explosion uses only about 1% of the energy available; the explosion problem may be more difficult in that its mechanism can be a subtle perturbation on the general phenomenon of collapse.

(8) At present the "reflected" shock seems to be the best candidate for explosion mechanism, but Wilson (this volume) finds that at low densities it damps by ν_μ, $\bar{\nu}_\mu$ radiation. Other effects that require more effort are neutrino convection (Epstein 1977), thermonuclear shell detonation, mantle "re-implosion" and the mass of the remnant, nucleosynthesis in the mantle, and mantle accretion (?) by a black-hole core.

3. PHYSICS OF PRECOLLAPSE EVOLUTION

3.1 Thermonuclear Burning Stages

Stars in hydrostatic equilibrium have temperature gradients; this, taken with the high temperature sensitivity of thermonuclear reaction rates, gives stars a tendency to develop a heterogeneous composition. Unless mixing is complete (generally assumed to be a rare occurrence), the star will tend to develop a structure of spherical shells composed of material characteristic of the different stages of nuclear burning. These burning stages are determined by the temperature and the fuel available.

Table 1 summarizes their characteristics (see Arnett 1973 for detail and for references). The six stages fall naturally into three groups: (1) the photon-cooled stages which occupy 99% of the star's life; (2) the first two neutrino-cooled stages which have fairly complex reaction chains which are not in equilibrium; and (3) the last two neutrino-cooled stages which burn towards quasi-equilibrium (Woosley, Arnett & Clayton 1973). If neutrino cooling dominates over radiative diffusion, thermonuclear burning can be thermally stable only if a convective instability develops to carry away the excess heat (Arnett 1972); thus the latter four stages are always associated with convective regions if the burning is vigorous.

Table 1
Thermonuclear Burning Stages

Fuel	T(°K)	Ashes	Cooling*	q(erg/g)	Comments
H(CNO)	2×10^7	$^4He(^{14}N)$	γ's	7×10^{18}	$\sim$90% of star's life
$^4He(^{14}N)$	2×10^8	$^{12}C, ^{16}O(^{22}Ne)$	γ's	7×10^{17}	$\sim$10% of star's life
^{12}C	1×10^9	$^{20}Ne, ^{23}Na, Mg$	ν's	5×10^{17}	< 1% of star's life
^{20}Ne	1.3×10^9	$^{16}O, ^{24}Mg$	ν's	$\sim 1 \times 10^{17}$	weak
^{16}O	1.8×10^9	Si-Ca	ν's	5×10^{17}	quasi-equilibrium
Si	3.4×10^9	Sc-Ni (Fe group)	ν's	2×10^{17}	quasi-equilibrium

*photons $\equiv$ radiative diffusion of energy
neutrinos $\equiv$ free escape of rarely formed $\nu\bar{\nu}$ pair

3.2 Nuclear Reaction Networks

The sets of coupled nonlinear differential equations governing the rate of change of nuclear abundances are called "reaction networks." As nuclear evolution proceeds, higher temperatures allow more reactions to become important, requiring the use of more complex reaction networks. Finally nuclear statistical equilibrium is approached, allowing use of statistical physics (see §2.3 above). Just prior to this the quasi-equilibrium approximation (Bodansky, Clayton & Fowler 1968) can be used to ease the computational problem.

Not only are the nuclear reactions important to the thermal balance of the star, but the composition of the layers surrounding the core is of vital importance for nucleosynthesis and determination of abundances which form one of the fundamental probes of galactic and stellar evolution.

The first calculations involving networks coupled to stellar evolution used an "alpha-chain" of nuclei (^4He, ^{12}C, ^{16}O, ^{20}Ne . . . ^{56}Ni); see Arnett (1969), especially figures 10 and 12 for oxygen and silicon burning. These results gave a good indication of the qualitative nature of these advanced burning stages, but indicated a quantitative problem. At the densities involved electron capture on the products of oxygen and silicon burning occurred significantly, requiring that more neutron-rich nuclei be added to the network. This was not feasible at that time, so subsequent work (cf. Arnett 1973, 1977b) used cruder approximations.

In a major development Weaver, Zimmerman and Woosley (1978) have managed to do the evolution with a more general set of nuclei, following the electron capture. This is important not only in its own right, but also because it should give us enough insight to do future calculations in a far more efficient way (their computer usage was nontrivial).

3.3 Ignition Masses and the Chandrasekhar Mass

The Weaver et al. results are surprisingly similar to the earlier calculations of Arnett (1977b), with perhaps the major difference being the treatment of semi-convective mixing. This is probably due to the tendency of the physical processes driving the evolution to force all the calculations to a similar final state ("core convergence"). To better understand this effect, consider: What is the maximum temperature a star of given mass can attain? We will assume a simple (polytropic) structure; the cores of evolved stars can be so approximated. The answer will tell us if a core of given mass can directly ignite its fuel.

The equation of state is dominated by the electron contribution, so

$$\frac{P}{\rho} \simeq \mathcal{R} Y_e T + K_\gamma {Y_e}^\gamma \rho^{\gamma-1}$$

which is an interpolation formula between Maxwell-Boltzmann and degenerate behavior. Here Y_e is the number of electrons/nucleon (i.e., mole fraction) and $\mathcal{R}$ the gas constant. The quantity K_γ is constant for $\gamma = 5/3$ and $4/3$, the nonrelativistic and extreme relativistic limits, respectively.

For hydrostatic equilibrium and spherical symmetry,

$$\frac{dP}{dr} = -Gm\rho/r^2,$$

and

$$\frac{dm}{dr} = 4\pi r^2 \rho.$$

Thus, roughly,

$$P_c/\rho_c \simeq GM/R$$

and

$$M \simeq \frac{4\pi R^3}{3} \rho_c,$$

where the subscripts denote central values, or

$$\frac{P_c}{\rho_c} = fGM^{2/3}\rho_c^{1/3}$$

where f is a structure form factor.

Combining these two relations,

$$\mathcal{R}YT_c = fGM^{2/3}\rho_c^{1/3} - K_\gamma Y_e^\gamma \rho_c^{\gamma-1}.$$

For large mass M, the first term on the RHS dominates so $T \sim \rho^{1/3}$. Continued contraction gives even higher temperature, so any fuel will eventually ignite.

For small M and $\gamma > 4/3$, there is a zero in T at

$$\rho \simeq (fGM^{2/3}/K_\gamma Y_e^\gamma)^{\frac{1}{\gamma-4/3}}$$

For smaller ρ, $T \sim \rho^{1/3}M^{2/3}$ while larger ρ are unphysical ($T < 0$). Thus as the star contracts it passes through a maximum T, then cools to be supported entirely by electron degeneracy.

These two sorts of behavior are divided by the Chandrasekhar limiting mass,

$$M_{ch} = [K_{4/3}/fG]^{3/2}Y_e^2$$
$$\simeq 5.85\ Y_e^2\ M_\odot.$$

Silicon burning occurs at $kT \gtrsim 0.6\ m_e c^2$; to get such high maximum temperatures masses $M \simeq 0.99\ M_{ch}$ are required.

Using a more precise version of these arguments, the mass required to ignite a given fuel can be calculated. Note that $M_{ch} \sim Y_e^2$; electron capture reduces M_{ch}.

3.4 Core Convergence

The nuclear evolution is a sequence in which the ashes of one stage are the fuel for the next. Let us call the core mass that matter which is the ashes of the most recently completed burning stage. If $M_{core} < M_{ignition}$, ignition of the next fuel must wait until shell burning increases M_{core}. Thus the minimum core mass that can exhaust all fuels, that is burn silicon, is $M > 0.98\ M_{ch}$. The precise Y_e left after oxygen burning is not well known, but even after silicon burning $Y_e > 0.44$ (from NSE estimates and from Weaver et al. 1978), so

$$M_{core} \gtrsim 1.2\ M_\odot$$

for collapse.

If $M_{core} > M_{ignition}$, then hydrostatic contraction continues until nuclear energy generation balances energy losses. The effects of neutrino cooling and convection tend to reduce the mass in ashes, M_{ashes}, with each stage. Between burning stages, neutrino cooling in the core reduces entropy near the center (neutrino emission, and cooling, goes as a high power, $n \sim 6$ to 12, of the temperature). This gives a positive entropy gradient which inhibits the growth of the convective core during the subsequent burning stage.

Thus the net effect is for M_{core} to approach M_{ch}. This is the standard result. Perhaps large scale mixing (due perhaps to rotation or binary interaction) could break this tendency, and allow $M_{core} >> M_{ch}$ at collapse.

3.5 Stellar Death as a Function of Mass

Which stars collapse? Almost all stellar evolutionary investigations involve treating stars as isolated objects. Although a promising start has been made on the evolution of binaries, there remain serious questions about loss of mass and angular momentum which can drastically modify our picture of the evolution of such systems.

Table 2. Outcome of Stellar Evolution

Range of Mass	Result Expected
$M \lesssim M_\odot$	lifetime longer than age of universe
$1 \lesssim M/M_\odot \lesssim (3 \text{ to } 6)$	white dwarf + planetary nebula, mass loss
$(3 \text{ to } 6) \lesssim M/M_\odot \lesssim (5 \text{ to } 8)$	(a) degenerate ignition of $^{12}C + {}^{12}C$ (1) "fizzle" → core collapse, or (2) detonation and dispersal of core, or (3) deflagration and (?) (b) pulsationally driven mass loss to white dwarf
$(5 \text{ to } 8) \lesssim M/M_\odot \lesssim 100$	core collapses + supernova sometimes black hole (?)
$100 \lesssim M/M_\odot$	$e^\pm$ pair instability supernova

If we ignore these complexities for the moment, the nature of stellar death as a function of mass can be roughly categorized as in Table 2. The results for stars which have degenerate ignition of $^{12}C + ^{12}C$ is particularly uncertain. This table is far from being as reliable as we would like; clearly much remains to be done.

4. EVENT RATES

4.1 The Identity Crisis

In discussing event rates for supernovae (SN) and gravitational collapse, a problem of identity (or nonidentity) must be faced. Empirically speaking, supernovae fall into two groups: historical supernovae which do not occur (yet) in modern times but which happened in our own galaxy, and extragalactic supernovae. Some (5) supernova remnants (SNR) have been identified with historical supernovae. Only extragalactic supernovae have good type identification (type I, II) because spectra and a light curve are needed. Only two pulsars (PSR) are well identified with SNR, one of which (the Crab) was a historical SN. The relation of x-ray sources (XRS), which are thought to include some neutron stars and black holes, to the above objects is unclear, but probably important to this problem. The standard assumption, that most of these things correspond to each other, will be made in what follows: beware.

It should also be pointed out that the white dwarf formation rate is high, requiring that most stars die to form them, in contradiction to some very high collapse rates that appear in the literature.

4.2 Historical Supernovae

All historical supernovae fall in a 60 degree sector of the Galaxy which contains the sun. Table 3 gives the distance and height above the Galactic plane of each of these supernovae; Cas A is included although it does not seem to have been observed when it exploded. It is interesting to note that these are fairly high above the plane, contrary to conventional ideas. Following Tammann (1978), the Galactic supernova rate is

$$R \simeq n/f\Delta t = 1/28 \text{ years}$$

where $n = 6$ SN, $f = 60/360 = 1/6$ is the fraction of the Galaxy observed and $\Delta t = 10^3$ years is the interval of observation. If the Galaxy has a a disk area of $A = 7.1 \times 10^8\ \rho_c^{\ 2}$ (radius of 15 kpc), the areal rate is

$$\dot{\sigma} \simeq 5 \times 10^{-11} \text{ pc}^{-2} \text{ yr}^{-1}.$$

Table 3. Historical Supernovae

Supernova (date)	Distance (Kpc)	Height (pc)
1006	2.4	600
1054	2	200
1181	8	430
1572 (Tycho)	6	150
1604 (Kepler)	10	1200
. . .(Cas A)	3	110

4.3 Extragalactic Supernovae

There are less than a half-dozen well-observed extragalactic supernovae, and only about 30 SN II and 50 SN I observed at all! These small numbers give statistical problems. Most observed SN occur in Sc galaxies (Tammann 1978); the observed rate/Sc galaxy can be converted to a rate of stellar death as a function of stellar mass. Talbot and Arnett (1975) have used UBV photometry and their galactic disk models to do this, and find the ranges $M > 6\ M_\odot$ or $4 < M/M_\odot < 8$ both contain enough stars to explain the observed rate.

Note that apparently most SN are not seen by the observers:

$$R_{observed} \simeq 1/300 \text{ yr-Galaxy}$$

for a typical spiral, while Tammann infers (for the Milky Way)

$$R_T \simeq 1/30 \text{ yr-Galaxy},$$

(part of the increase is because the Galaxy is bigger than most). Van den Berg (1978) suggests that fewer SN are lost,

$$R_{VdB} \simeq 1/60 \text{ yr-Galaxy}.$$

These numbers should probably be regarded as in good agreement with the estimate above.

4.4 Supernova Remnants

There are about 10^2 supernova remnants known in the Galaxy, as well as some in the Magellanic Clouds. The two classical problems of rate estimation occur here: (1) the representativeness of the sample seen and (2) the correctness of the age assigned. To find the age one uses the Sedov solution for a blast wave in a homogeneous medium, but this is oversimplified. The medium is clumpy, and identification of shock velocity and radius corresponding to the Sedov case is not easy. Further, systematic variation in the mean density of the interstellar medium means that SNR slow faster and have higher surface brightness in more dense regions, giving an observational selection against SNR out of the Galactic plane (see 4.2 above!). Thus,

$$R_{SNR} > 1/150 \text{ yr-Galaxy}$$

or so (Clark 1978).

4.5 X-ray Binaries

Although these objects are obviously of importance for the rate question, fundamental uncertainties (e.g., angular momentum and mass loss) regarding their evolution make it difficult to infer a reliable rate (current ideas are consistent with the rates above; see Clark, this volume).

4.6 Pulsars

The pulsar rate is somewhat controversial (Taylor & Manchester 1977); it is consistent with the estimates above, but (for reasons given below) does not have the high accuracy sometimes claimed.

The pulsars form a flattened distribution in the Galaxy, so that the change in areal density can be written as

$$\dot{\sigma} \simeq \mathcal{B} - \mathcal{D},$$

where $\dot{\sigma} \simeq 0$ is a steady state. Further,

$$\mathcal{B} \simeq \mathcal{D} \simeq \langle\sigma/\tau\rangle$$

where τ is the age characteristic of the set of objects. If this set is homogeneous,

$$\mathcal{B} \simeq \langle\sigma\rangle/\langle\tau\rangle \ ,$$

which need not be the case for pulsars. Thus an appropriate "age" τ and surface density σ are needed. The latter is just

$$\sigma = \text{number/area sampled.}$$

Here the area goes as ℓ^2, where ℓ is the projected distance in the Galactic plane. This is usually inferred from the dispersion measure,

$$DM = \int_0^\ell n_e d\ell = \overline{n}_e \ell$$

which is determined from the dispersion in arrival time of pulses at different frequencies. It is found that $\overline{n}_e$ is not constant, and in fact its fluctuation is of the same order: $\delta\overline{n}_e \simeq \overline{n}_e$. Plotting DM versus distance inferred some other way (which is possible for a couple of dozen PSR's) gives $\overline{n}_e \simeq 0.035\ cm^{-3}$ with a factor of two error either way.

The age τ has been inferred from the characteristic age, $A_p = P/2\dot{P}$, which is constructed from the pulsar pulse period and its rate of change, and from the kinematic age,

$$A_k = \int_0^h dz/u_\perp \simeq \langle h/u_\perp \rangle,$$

where h is the height of the pulsar above the Galactic plane and $u_\perp$ its velocity perpendicular to the plane. To use the kinematic age it must be assumed that all pulsars are born in the plane, which is suspect observationally (see 4.2, esp. the Crab, where $h \simeq 200$ pc at birth). Further, in practice, $A_k \simeq \langle h \rangle / \langle u_\perp \rangle$, so that different sets of objects are used to estimate $\langle h \rangle$ and $\langle u_\perp \rangle$, a dangerous procedure if untested.

Simply plotting A_p versus h gives a scatter diagram. If A_p were a good measure of the age, this would tell us about kinematics but A_p has trouble too. If pulsars are formed spinning fast, slow down, and become undetectable at long periods, then the "flow" of pulsars through each age range should be constant (except for very short or long ages). For the five bins of $\Delta \log A_p = 0.5$ in the range $4 \leq \log A_p < 6.5$, this is true, very crudely (factor of ten in the rate). For larger ages the rate falls to low values. This failure seems to be a general characteristic of simple versions of pulsar theory. As complexity increases and alternative explanations appear, the reliability of our inferences must decrease correspondingly unless we can check the conceptual framework independently. In addition there is the question of pulsars unseen because they beam away from us. The net result of these different uncertain methods is a rate

$$\dot{\sigma} \simeq 4 \times 10^{-11\pm1}\ pc^{-2}\ yr^{-1}$$

or

$$R_{PSR} = 1/(35 \times 10^{\pm1}\ yr)$$

for the Galaxy. Note that the extremes of 3.5 and 350 years between events are getting unreasonable on other grounds, reflecting the severe uncertainties in the pulsar rate.

4.7 Death Rates and Stellar Masses

Although the rate of events per galaxy is the quantity needed to predict a rate of events in a detectable volume, to connect these rates to theoretical descriptions it is necessary to know the masses of the stars involved. This is not known directly, except perhaps for supernova remnants for which mass estimates are controversial.

By identifying observed stars with theoretical evolutionary stages (in practice the main sequence), the rate of evolution (and hence of death) can be inferred. Ostriker, Richstone and Thuan (1974) give the estimates for the solar neighborhood. If, for illustration, we assume all stars of mass $M > M_u$ die to give the event in question, then the event rate (per unit area) in the Galactic Disk near the sun is given in Table 4. Recent work by Scalo (1978) suggests that these rates are too low; the numbers may be good to factors of three (?). For purely disk phenomena, the Galactic rate can be found by multiplying by the effective area of the disk.

Table 4. Stellar Death Rates in the Galactic Disk

$M_u/M_\odot$	$\dot{\sigma}/10^{-11}$ pc^{-2} yr^{-1}
12	0.22
8	0.66
5	1.7
3.2	3.8
2.4	10.0

The existence of new white dwarfs in young galactic clusters implies that at least some fairly massive stars, $M \sim M_L$, become white dwarfs rather than collapsing. From the observations it appears that $M_L \gtrsim 4$ to $6\ M_\odot$. This is roughly consistent with the supernovae rate (at least in a statistical sense).

Finally it should be mentioned that other quantities besides mass (such as rotational state, binary companion type, etc.) might be important in determining the way stars die.

4.8 Event Rate as a Function of Distance

A moderately good ($\sim$ factor of 2) estimate of the number of collapse events, in a given volume centered at the Earth, can be made as follows. The luminosity of the Galaxy is

$$L_{Gal} \simeq 10^{10.5}\ L_\odot$$

and the event (SN, SNR, PSR) is about

$$R_{Gal} \simeq (30\ yr)^{-1}.$$

If the Galaxy is composed of a typical mix of objects in extragalactic space (which is thought to be the case), then for a volume luminosity L, the event rate is

$$R \simeq (L/L_{Gal})R_{Gal},$$
$$\simeq (L/10^{12}L_{\odot})yr^{-1}.$$

If the "cosmic emissivity" is $\varepsilon = 10^{8 \cdot 2}L_{\odot}Mpc^{-3}$ (see Ostriker, this volume), then $L \simeq (4\pi/3)D^3\varepsilon$, so

$$R \simeq 10^{-3 \cdot 2}\ D^3\ Mpc^{-3}\ yr^{-1}.$$

This estimate is too low for the Local Group of galaxies because of a local density enhancement. For the Virgo cluster of galaxies, the distance is $D \simeq 10$ Mpc so $R \simeq 10^{-1}\ yr^{-1}$. If for the Local Group, $L \sim 3\ L_{Gal}$ so $R \simeq 3\ R_{Gal} \simeq 0.1\ yr^{-1}$.

Note that the actual discovery rate of supernovae (all extragalactic) is only of the order of several/year. If the typical apparent magnitude is 14 and the absolute magnitude is -18 near maximum, the typical distance is 25 Mpc. The estimate above gives $R \simeq 10^{-3 \cdot 2}\ (25)^3\ yr^{-1} \simeq 10\ yr^{-1}$; our searches may be incomplete or our estimated rate a bit high.

REFERENCES

Arnett, W.D. (1969). Exploding star models and supernovae. In Supernovae and their Remnants, ed. P. Brancazio & A.G.W. Cameron, pp. 89-109. Gordon and Breach, New York.

Arnett, W.D. (1972). Hydrostatic oxygen burning in stars. I. Oxygen stars. Astrophys. J. 173, pp. 393-400.

Arnett, W.D. (1974). Nuclear reactions and neutrinos in stellar evolution. In Late Stages of Stellar Evolution, ed. R. Taylor, pp. 1-14. D. Reidel, Dordrecht, Netherlands.

Arnett, W.D. (1977a, denoted "A77"). Neutrino trapping during gravitational collapse of stars. Astrophys. J. 218, pp. 815-833.

Arnett, W.D. (1977b). Advanced evolution of massive stars. VII. Silicon burning. Astrophys. J. Suppl. 35, pp. 145-159.

Bethe, H.A., Brown, G.E., Applegate, J. & Lattimer, J.M. (1978). Equation of state in the gravitational collapse of stars. Preprint.

Bodansky, D., Clayton, D.D. & Fowler, W.A. (1968). Nuclear quasi-equilibrium during silicon burning. Astrophys. J. Suppl. 16, pp. 299-371.

Bruenn, S.W. (1975). Neutrino interactions and supernovae. Ann. N.Y. Acad. Sci. 262, pp. 80-94.

Bruenn, S.W., Buchler, F.R. & Yueh, W.R. (1978). Shock structure and neutrino radiation in stellar collapse. Astrophys. J. (Lett.) 221, pp. L83-86.

Clark, D. (1978). Supernova remnants. In Supernovae: Proceedings of Astronomy Summer School 'E. Majorana,' Erice, Sicily, ed. J. Danziger & A. Renzini, in press.

Colgate, S.A. & White, R.H. (1966). The hydrodynamic behavior of supernovae explosions. Astrophys. J. 143, pp. 626-681.

Epstein, R.I. (1977). Private communication.

Epstein, R.I. & Arnett, W.D. (1975). Neutronization and thermal disintegration of dense stellar matter. Astrophys. J. 201, pp. 202-211.

Freedman, D.Z., Schramm, D.N. & Tubbs, D. (1978). The weak neutral current and its effects in stellar collapse. Ann. Rev. Nuclear Sci. 27, pp. 167-207.

Lattimer, J.M., Lamb, D.Q., Pethick, C. & Ravenhall, D.G. (1978). In preparation.

Mazurek, T.J. (1975). Chemical potential effects on neutrino diffusion in supernovae. Astrophys. Space Sci. 35, pp. 117-135.

Mazurek, T.J. (1978). Talk given at Aspen workshop on "Neutrino Physics and Gravitational Collapse."

Ostriker, J.P., Richstone, D.O. & Thuan, T.X. (1974). On the numbers, birthrates, and final states of moderate- and high-mass stars. Astrophys. J. (Lett.) 188, pp. L87-90.

Scalo, J. (1978). In preparation.

Talbot, R.J., Jr. & Arnett, W.D. (1975). The evolution of galaxies. IV. Highly flattened disks. Astrophys. J. 197, pp. 551-570.

Tammann, G. (1978). Supernova rates. In Supernovae (Erice), op. cit.

Taylor, J.H. & Manchester, R.N. (1977). Galactic distribution and evolution of pulsars. Astrophys. J. 215, pp. 885-896.

van den Bergh, S. (1978). Comment on supernova rates. In Supernovae (Erice), op. cit.

Van Riper, K.A. (1978). The hydrodynamics of stellar collapse. Astrophys. J. 221, pp. 304-319.

Van Riper, K.A. (1979). General relativistic hydrodynamics and the adiabatic collapse of stellar cores. Submitted for publication.

Van Riper, K.A. & Arnett, W.D. (1978). Stellar collapse and explosion: Hydrodynamics of the core. Astrophys. J. (Lett.), in press.

Weaver, T.A., Zimmerman, G.B. & Woosley, S.E. (1978). Presupernova evolution of massive stars. Astrophys. J., in press.

Wilson, J.R. (1971). A numerical study of gravitational stellar collapse. Astrophys. J. 163, pp. 209-219.

Wilson, J.R. (1979). In this volume, and private communication.

Woosley, S.E., Arnett, W.D. & Clayton, D.D. (1973). The explosive burning of oxygen and silicon. Astrophys. J. Suppl. 26, pp. 231-312.

STELLAR COLLAPSE AND SUPERNOVAE

James R. Wilson

Lawrence Livermore Laboratory

INTRODUCTION

The present work is a revision of the calculations of Wilson (1978), prompted by suggestions made at the 1978 Copehagen meeting on Supernovae (Bethe et al 1978). The principal change is in the equation of state which the statistical weight, is now taken from Fowler, Engelbrecht and Woosley (1978). In the older calculations the heavy nuclei decomposed at a density of about 10^{13}gm/cc and bounce ensued at a density of 5×10^{13}gm/cc. The new statistical weight enables the nuclei to hold together longer, and the star bounce is at nuclear density. As will be seen later, the bounce being at higher density drastically alters the outward shock which now is neutrino radiation dominated. In section II the equation of state and electron capture model will be given and then in section III the results of a few preliminary calculations are given. For a more general discussion of the physics involved see Arnett in this volume.

MODEL

The stellar material is considered to be made up of heavy nuclei (Fe and higher A), He, protons, neutrons, and electrons. The free energy $\tilde{F}$ for the nuclear particles is taken as

$$\tilde{F} = \Sigma X_i \left[\varepsilon_i + T \log (\alpha\rho\, X_i/h_i A_i^{5/2} T^{3/2})/A_i\right], \qquad (1)$$

$$i = \text{Fe, He, b}$$

where X_i is mass fraction of heavies, He, and free baryons b, ε_i is the binding energy, A_i is atomic number, ρ is density, T is temperature, h_i is statistical weight, and α is a numerical constant. From Baym, Bethe, Pethick (1971) we take the difference of chemical potentials of neutron and proton as

$$\mu_n - \mu_p = 250(0.464 - Z) \text{ Mev} \qquad (2)$$

where Z is the average charge on the nucleus. See also Lattimer and Ravenhall (1977). We use c.g.s. units, but temperature and particle energies are in Mev. The change of binding energy of the nucleus from its normal value due to the change of Z is then

$$\Delta\varepsilon = 125(464 - Z)^2 \tag{3}$$

We let $\varepsilon_{Fe} = \Delta\varepsilon$, $\varepsilon_{He} = 8.9 - 0.8(8.9 - \Delta\varepsilon)$ and $\varepsilon_b = 8.9$

The statistical weights are taken as

$$h_b = 2, \quad h_{He} = 1, \text{ and} \tag{4}$$

$$h_{Fe} = 3 \{1 + \exp \ (- 100.8 - 5665T + 11410T^2)/(T + 1091T^2) \}$$

The latter is three times a fit to Fowler et al (1978) supplied privately by Woosley. The three is a guess of the number of nuclear species present. Statistical weights should be of the form exp (Af(T)); however, the 1/A term in the free energy expression almost cancels the A in the statistical weight. Hence the effect of a change of A is ignored in the equation of state for the nuclear matter. The restriction of the charge on the helium is ignored in solving (1) for the mass fractions X_i. The charge on the heavy nucleii is found by the electron capture model given below. The charge on the free baryons is determined by the equations given in Wilson (1978). The charge in the He is then determinded so that overall charge is conserved. The helium charge is usually close to .5 but does fall to about .4 in the center at bounce. The pressure and energy are found from F by the usual thermodynamic formulas, except that the pressure of the free baryons has degeneracy pressure added in as follows

$$P_o = 3.1 \times 10^9 \, \rho_b^{5/3}[Z^{5/3} + (1 - Z)^{5/3}]$$

$$P_b = \sqrt{P_o^2 + (n_b \, kT)^2} \tag{5}$$

The pressure due to compression of nuclear matter is taken from Baym, Bethe, Pethrick (1971) as:

$$P_n = .0015 \, \rho^{5/2}, \; E_n = .001 \, \rho^{3/2} \text{ cgs units} \tag{6}$$

This pressure is simply added to the pressure from the other terms. For the electron pressure we start with the exact Fermi zero temperature pressure P_{eo} and from this subtract a Coulomb pressure

$$P_c = 5.64 \times 10^{12} \; n_e^{1/3} \, (AZ_{Fe})^{2/3} \, X_{Fe} \tag{7}$$

Then the total electron pressure is taken as

$$P_e = \sqrt{(P_{eo} - P_c)^2 + (n_e \ kT)^2} \tag{8}$$

Radiation and pair pressure are the same as used in Wilson (1978).

The atomic weight A is given by

$$A = X/(1 + 2 \times 10^{-7} X^2) + 1 \tag{9}$$

where

$$X = \exp\left[8 \times 10^{-5}(\rho^{1/3} - 400)/(1. + 5 \times 10^7 (T/(\rho^{1/3} - 400)^2))\right] \tag{10}$$

This is a fit to data given at the Copenhagen meeting by Lattimer and Ravenhall. The atomic weight, A, is used only in the Coulomb energy and neutrino elastic scattering.

The electron captures by Fe is based on a model given by Bethe at the Copenhagen meeting. The energy of the captured electron is taken as

$$\varepsilon_e = \mu_n - \mu_p + \varepsilon_\nu + \varepsilon^* \tag{11}$$

where $\mu_n - \mu_p$ is given by (2), ε_ν is the energy of the emitted neutrino, and $\varepsilon^* = 3\text{MeV}$ is the nuclear excitation produced by the capture ($\varepsilon^* = 0$ if $T > 3\text{MeV}$). The capture cross section is

$$\sigma = 1.7 \times 10^{-44} (\varepsilon_\nu / m_e)^2 \equiv \sigma_o \varepsilon_\nu^{\ 2}$$

The rate of production of neutrino energy density is

$$\frac{\partial F(\varepsilon_\nu)}{\partial t} = \sigma_o \rho N_o \ X_{Fe} \ |Z - Z_{eq}| \varepsilon_\nu^{\ 3} (f(\varepsilon_e) - F/k \ \varepsilon_\nu^{\ 3}) \varepsilon_e^{\ 2} \tag{12}$$

N_o is Avagodros number, $f(\varepsilon_e)$ is the Fermi distribution function for electrons of energy ε_e, and $k \ \varepsilon_\nu^3$ is the energy phase space volume for neutrinos of energy ε_ν.

The diffusion coefficient for neutrino flow has been altered as recommended by Lamb (1976) to include the effects of neutrino degeneracy. Otherwise the present calculation treats neutrinos the same as Wilson (1978). To achieve a better estimate of bounce density the relativistic formula for gravitational acceleration is used in the calculations discussed below. It raises the bounce density by about a factor 1.7.

CALCULATIONAL RESULTS

We use as an initial configuration the inner 1.8M⊙ core from the stellar evolution calculation of a 15M⊙ star supplied by Weaver and Woosley. The central state variables at the start of our calculation are $\rho = 6 \times 10^9$gm/cc, T = .72MeV, Y_e = .440 (Ye is the average charge per nucleon). The star has just become unstable. The neutrino cross sections are taken from Schramm and Tubbs (1975) using a Weinberg angle given by $\sin^2\theta = .35$. Mu and tau neutrinos are included. For elastic scattering two runs were made, the first with

$$\sigma_s = .082\ \sigma_{os}\ A^2\varepsilon^2 \tag{13}$$

and the second with

$$\sigma_s = 1.0\ \sigma_{os}\ A^2\ \varepsilon_\nu^{\ 2}\ . \qquad \sigma_{os} = 1.7 \times 10^{-44}\ x(.67/(M^2_{\ e})) \tag{14}$$

The latter calculation was performed because with the low scattering cross section in the first calculation the outgoing shock was damped by the radiation to the point where no explosion resulted. In Figs. 1 and 2 the radius vs time for several zones is given for the two

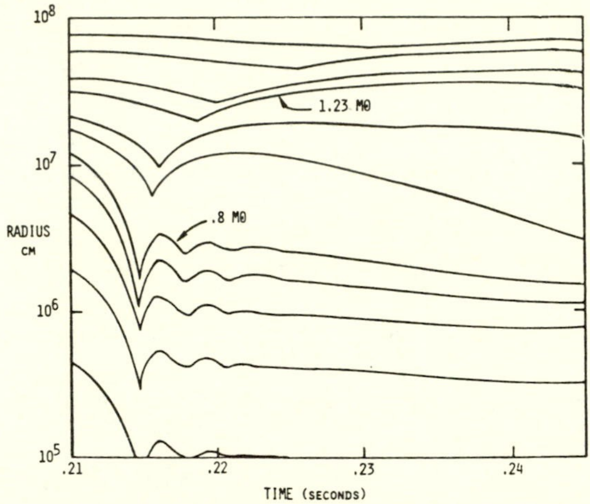

Figure 1 The radius time trajectories of selected mass points for the example with equation (13).

calculations. For the calculation using the small nuclear scattering cross section, the shock becomes very weak towards the end of the calculation and the velocities of the outward moving matter are less than half the escape velocity. In Fig. 2 we can see substantial velocities at the end of the calculation. About .2M⊙ of

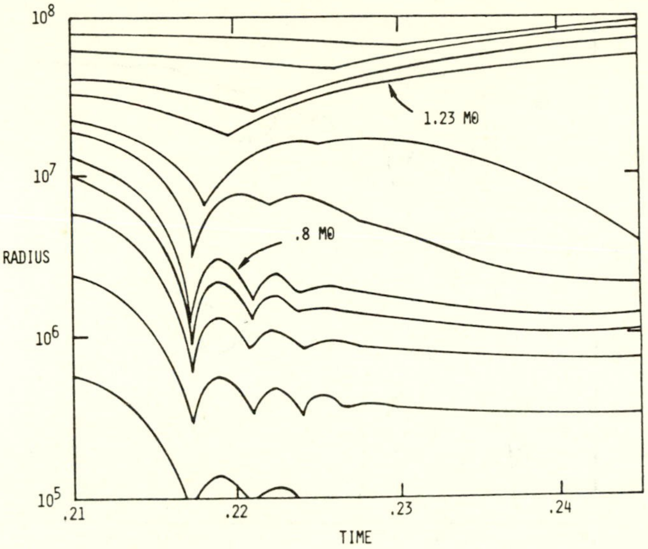

Figure 2 Radius-time trajectories of selected mass point for the example with equation (14).

material is above escape velocity and has a total energy (internal + kinetic) greater than its gravitational binding energy of about 1.0×10^{51} ergs. The excess of kinetic over gravitational energy of this material is only 2×10^{49} ergs. The present calculations take too much computer time to run far enough to clearly determine the explosion energies. Hence the final energy in the explosion is hard to estimate. The mass of the core inside the inner zone with supra-escape velocity is $1.4M_\odot$. More material will probably fall back in and the final neutron star might have a rest mass considerably above $1.4M_\odot$.

Immediately after bounce the star can be though of as split into two parts, an inner core of $.8M_\odot$ which has bounced adiabatically and an outer region which is shocked by the elastic bouncing of the inner core. The decay of the oscillations of the inner core is crucial to the gravitational radiation production senario discussed by Shapiro (1979). The computer calculation has some artificial numerical damping so that the true damping should be somewhat less than that indicated by Fig. 3.

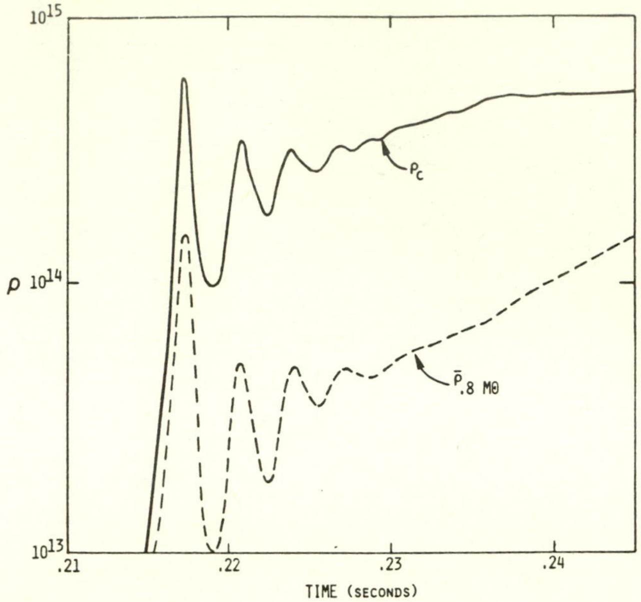

Figure 3 The solid curve gives the density as a function of time for the central zone and the dashed curve gives the mean density in the inner .8M⊙ as a function of time. These curves were made for the example with equation (14).

The fact that only .8M⊙ or about 2/3 of the collapsing core is oscillating in a simple adiabatic mode means damping will be strong by the outer core material. The entropy of the inner .8M⊙ starts at about .8k per nucleon and rises to about 1.0k per nucleon at bounce and remains there even after see (Fig. 4). When the central density

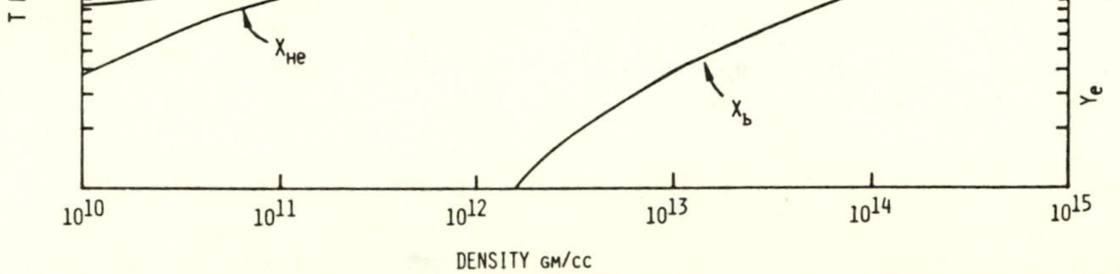

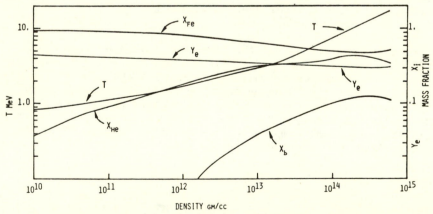

Figure 4 Temperature, mass fractions of Fe, He, free baryons and average charge versus density in the central zone for the calculation with equation (14).

has risen to 1.4 x 10^{12}gm/cc the total change of central lepton number is only .04. The change in entropy due to these escaping neutrinos (see Arnett section 2.7 this volume) is only a few tenths. The mean charge, Ye, at the center has only dropped to .29 by the end of the calculation. At the time of first bounce the central density is about 5.5 x 10^{14}gm/cc, the temperature is 17 MeV, the average electron number, Ye, is .31, and the average neutrino mean free path is 10 cm. The peak infall kinetic energy is 8 x 10^{51} ergs, most of which is dissipated in thermal energy. When the center of the star reaches nuclear density the temperature is 10.5 Mev and material is 50% Fe, 40% He, and 10% free baryons. See Fig. 5 for several central quantities versus density.

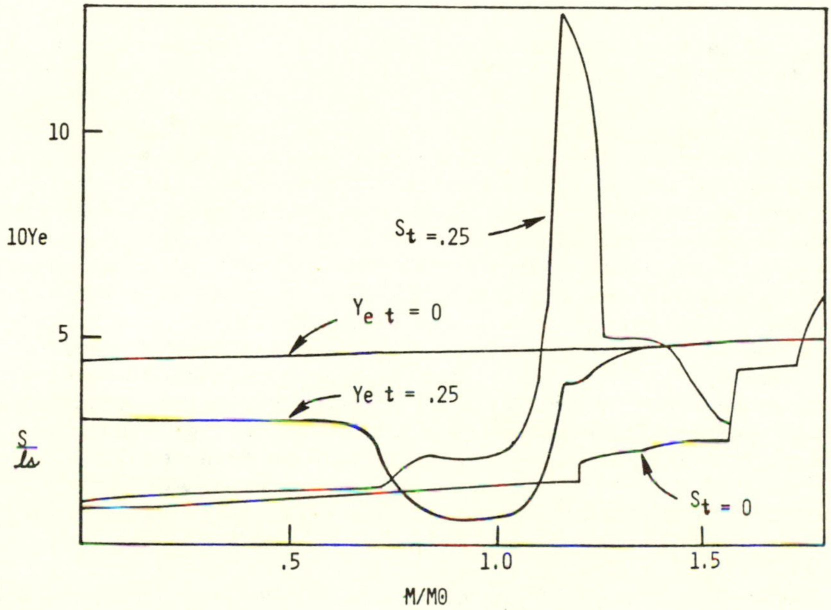

Figure 5 Entropy per nucleon in units of k and electron number per nucleon at the beginning and end of the calculation using equation (14).

In the calculation with the low elastic nuclear scattering cross-section when the shock reaches a density between 10^{10} and 10^{11} cooling becomes very important. The velocity jump across the shock is about 5 x 10^9 cm/sec. This raises the temperature to nearly 7 Mev. Pair production of mu and tau neutrinos is very high at this temperature and density, and the temperature quickly falls to about 4 Mev. With the higher scattering cross section the temperature only falls to about 6Mev behind the shock. Besides sustaining the shock later on, the high scattering cross section is very important in giving a substantial radiation pressure on the outer layers of iron. The radiation pressure rises to about half the Eddington limit for about .02 seconds, and is crucial to the explosion. The mu and tau neutrinos are the most important energy and momentum carriers after bounce. They carry out .7 to .8 of total neutrino energy flux from the shock. If only electron neutrinos are included in the calculation the cooling of the shock is reduced sufficiently that the calculation comes very close to an explosion.

We have seen from these calculations a high central density is reached with a fairly good bouncing core which is good so far as gravitational radiation is concerned. However, the question of the relation of these calculations to real supernova is still murky. We have had to use an unpopular elastic cross section in order to obtain an explosion.

Of the changes in the model, the change in the equation of state and in particular the statistical weight of the Fe are the most important changes. The results of the calculations are relatively insensitive to the electron capture rate. The electron capture cross section was reduced by a factor ten and the collapse was recalculated until bounce time. It took .025 seconds longer to collapse, but the central density at bounce was only 10% higher The peak infall kinetic energy was about 5% higher. The entropy change due to electron captures comes about in a rather indirect way numerically. In capture, number of leptons and total energy are conserved. The representation of the neutrino spectrum by all a finite number of energy groups implants an entropy error. We are currently working with D. Tubbs to compare neutrino equilibration as calculated with Monte Carlo methods with the Fokker-Planck method used in these calculations. The preliminary results show only small differences.

The author is indebted to R. Bowers for helping implement the new equation of state and electron capture model. This work was performed under the auspices of the US Department of Energy under contract W-7405-Eng-48.

REFERENCES

Baym, G., Bethe, H.A., and Pethick, C.J. (1971). Neutron Star Matter. Nucl. Phys., A175, pp. 225-271.

Bethe, H.A., Brown, G.E. Applegate, J., and Lattimer, J.M. (1978) Equation of State in the Gravitational Collapse of Stars. Nordita Preprint.

Fowler, W.A., Engelbrecht, C.A., and Woosley, S.E., (1978). Nuclear Partition Functions, Orange Aid Preprint 520, April 1978.

Lamb, D.Q., and Pethic, C.J., (1976). Effects of Neutrino Degeneracy in Supernova Models. Ap. J., 209, pp. L77-81.

Lattimer, J.M., Ravenhall, D.G., (1977) Neutron Star Matter at High Temperatures and Densities. Illinois Preprint, Oct. 1977.

Shapiro, S. (1979). This volume.

Tubbs, D.L. and Schramm, D.N. (1975). Neutrino Opacities, Ap. J., 201, pp. 467-488.

Wilson, J.R. (1978). Neutrino Flow and Gravitational Collapse, Proc. of Int. Sch. of Phy. Fermi Course LXV, North Holland, Amsterdam.

NEUTRINO COMPETITION WITH GRAVITATIONAL RADIATION DURING COLLAPSE

Demosthenes Kazanas and David N. Schramm*

Department of Physics and The Enrico Fermi Institute

The University of Chicago

*Also Department of Astronomy and Astrophysics

INTRODUCTION

The purpose of this paper will be to review the situation with regard to neutrino emission during the gravitational collapse of the cores of massive stars and to compare the effectiveness of neutrino emission with gravitational wave emission as mechanisms for removing energy from the collapsing object. This paper will tend to concentrate on the general character of the role of neutrinos rather than the hydrodynamic details of any specific collapse calculations. (See, e.g., Wilson, this volume.)

To provide a basic framework for discussion, let us briefly review the relevant physics of the collapse situation. In particular let us remember that it is generally agreed that single, non-rotating massive stars ($M \gtrsim 8\ M_\odot$) eventually evolve to a configuration with a $\sim 1.4\ M_\odot$ Fe-Ni core surrounded by concentric shells of Si, O, Ne, C, He, and H. (Rotation and/or magnetic fields may increase the central core mass but should not change the general configuration.) Since Fe-Ni has the maximum binding energy per nucleon, it is clear that unlike the star's previous core collapse stages the subsequent collapse of the Fe-Ni core will not be stopped by thermonuclear burning. Stellar evolution calculations (c.f. Arnett 1973; Weaver, Woosley & Zimmerman 1978) indicate that the temperature and density of the core at the beginning of this final collapse are near 10^{10} K and 10^{10} g cm^{-3}. As the collapse continues and the density and temperature rise, it is obvious from an examination of the microscopic processes (see review by Freedman, Schramm & Tubbs 1977 and references therein) that neutrinos will be copiously emitted.

In particular the high density and thus high electron Fermi level will drive electron capture reactions, releasing ν_e's; the high temperature will in turn drive various $\nu\bar{\nu}$ pair-creating processes such as $e^+e^- \rightarrow \nu\bar{\nu}$. These pair processes will not only produce ν_e's but also ν_μ's and presumably ν_τ's (or any other massless ν species) via neutral currents. The detailed time structure of the flux of these escaping neutrinos depends on the details of the poorly known equation of state of the collapsing matter. However the basic fact that no other mechanism radiates as efficiently at these densities implies that the neutrinos carry away the bulk of the binding energy of the collapsing material. This statement is independent of the details of the equation of state and is a general feature of a collapse going to high densities and temperatures. Since the binding energy of a neutron star is $\sim 10^{53}$ ergs, then $\gtrsim 10^{53}$ ergs of neutrinos must eventually be emitted if the collapse is to lead to a neutron star or black hole.

We will return later and look at this basic physics in more detail. However, at this point it is useful to mention and dismiss a "red herring" which tends to obfuscate in some people's minds the basic fact that $\sim 10^{53}$ ergs of neutrinos are emitted. The point of obfuscation is whether or not these neutrinos provide the explicit mechanism whereby the collapsing core is able to eject the outer part of the star in a supernova outburst. Over the past decade there has been much speculation and debate on the possibility that the escaping neutrinos might cause mass ejection by either energy or momentum deposition (cf. Bruenn, Arnett & Schramm 1977; review by Schramm 1977 and references therein). Unfortunately the final resolution of the role of neutrinos in the mass ejection is tied up in the details of the interplay between the hydrodynamics and the equation of state and the results are extremely model dependent. However, the key point to remember here is that regardless of the role of the neutrinos in the ejection, the neutrinos do carry away the bulk of the binding energy. If $\lesssim 1\%$ of this energy can be tapped for an ejection mechanism that would be fine, but such "tapping" is irrelevant to the basic energetics of the neutrino emission. There may even be events which form black holes directly with no mass ejection but they would still release neutrinos.

Clearly in the idealized case of spherically symmetric collapse, gravitational radiation will not occur and neutrinos are the only means of removing the binding energy. Thus, to the degree that the collapse is spherical, gravitational wave detectors would be useless with regard to detecting gravitational collapse. However, in the case of non-spherical collapse, the emission of some gravitational radiation is possible. We will see that for any realistic case such emission is still small compared to the neutrino emission. To understand this let us examine the physical nature of the collapse in more detail so that we can assess the relative roles of neutrino and gravitational radiation in non-spherical collapse. We will then show that even non-spherical perturbations radiate more rapidly via neutrinos and thus neutrinos dominate as a damping mechanism. We will note that although this damping is faster than that due to gravitational radiation (hereafter referred to as GR), it is not sufficiently rapid to totally exclude some GR from being emitted. At the end of the paper we will briefly review the observational evidence relating to the degree of rotation of the collapsing core. We will also briefly mention the observational prospects for detecting the copiously emitted neutrinos from gravitational collapse and show that a complete observational understanding of collapse will require observations of both neutrinos and gravity waves.

GENERAL COLLAPSE SCENARIO

The Fe-Ni core collapses due to the simultaneous occurence of: 1) $\bar{\nu}$-emission removing energy; 2) electron captures removing the supporting electrons and emitting ν_e's; 3) the surrounding Si shell increasing the core mass above the Chandrasekhar mass; 4) energy going into excited states of complex nuclei. Based on simple virial theorem arguments it is easy to see that as the core collapses it heats up. The neutrino emission is not sufficiently rapid to cool the core; it is only rapid enough to carry away the binding energy which must be radiated for the collapse to continue. In fact, for densities greater $\sim 10^{11}$ g cm^{-3} the neutrinos no longer stream out but diffuse out due to the mean free path becoming significantly less than the size of the core. The diffusion times are $\sim$ 0.01 to 1 sec depending on the density-temperature regime. In fact, in the central core the diffusion times become so long that the neutrinos fill phase space. Even though these times are long compared to free fall ($\sim 10^{-3}$ sec) they are still short compared to the emission rate of any other particle or radiation and this is the reason why these are the dominant mode for removing the binding energy. Because the collapsing material is hot and because it takes a while for the neutrinos to get out, it is obvious that the collapse will be far slower than free fall.

From the point of view of neutrino astronomy it is useful to note that the neutronization of the collapsing core accounts for $\lesssim 10^{52}$ ergs of $\sim$ 10 MeV ν_e's, (the number of ν_e's = number of protons in core $\approx 10^{57}$), whereas the bulk of the 10^{53} ergs comes out in $\nu\bar{\nu}$ pairs. The neutronization neutrinos do dominate the early part of the collapse and the time resolution of these relative to rest may be quite interesting (see Nadyozhin in these proceedings). It is also interesting to note that because of the location of the neutrino photosphere, the energies of the escaping ν's will probably be $\sim$ 10 MeV.

Lattimer, Lamb, Pethick and Ravenhall (1978) and Bethe and Brown (1978) point out that although the collapse is not hot enough to cause significant photodisintegration of the nuclei, it is sufficiently hot to put the nuclei into highly excited states. Because the nucleons remain bound, the pressure of collapsing gas will be primarily from the electrons until nuclear densities are reached. It had once been thought (Wilson et al. 1976) that the transition to a thermal neutron gas would occur at $\sim 10^{13}$ g cm^3 and thus cause a stiffening at that point of the equation of state from a $\gamma = 4/3$ relativistic electron gas to a non-relativistic $\gamma = 5/3$ neutron gas. However the current situation (Lattimer *et al.* 1978; Bethe & Brown 1978) seems to indicate that a stiffening will not occur until nuclear densities and thus the hydrodynamic bounce caused by this transition will not occur until then. (Van Riper 1978 also found a tendency toward high density bounces when general relativity was included in the hydrodynamics.) This point is useful for GR because the higher the density of the bounce the more relativistic the object is and the greater the chance for significant GR emission. Thus although the collapse is slow, which hinders GR, the fact that it seems to go to high densities helps. Following the core bounce it is expected that the shock wave and/or the outflow of neutrinos will eject the outer part of the star and leave the central core to collapse down to make a neutron star or black hole. It is during this bounce and subsequent collapse that the bulk of the neutrinos radiate away and that the possibility of GR occurs.

The fact that the bouncing homologous core ($M > 1.1\ M_{\odot}$) extends out to a density of $\sim 10^{11}$ g cm^3 and that is about the neutrino photosphere is one of the reasons why some feel that neutrino transport may have some influence on the ejection of matter even if it is only via its effect on the equation of state at that crucial boundary.

Currently there is much disagreement (e.g., see papers in these proceedings by Arnett, Shapiro or Nadyozhin; or Wilson 1978) from model to model on the details of the bounce and the strength of the generated shock and the possible role of neutrinos in this shock. Thus the detailed time structure of the neutrino emission and the possible emission of GR is clearly unresolved and not likely to be resolved in the near future without some observations. However, the fact that neutrinos dominate is not in question.

DAMPING OF NON-SPHERICAL PERTURBATIONS

Shapiro, in his paper, will discuss the details of GR versus ν's for non-spherical collapse. Here, following Kazanas and Schramm (1976, 1977), we will examine the damping times, due to neutrino losses, of small non-axisymmetric perturbations of a collapsing core model, about its equilibrium, and these will be compared against those due to GR; since other than acoustic damping the latter is the only alternative means for damping out such oscillations, this will be indicative of the most efficient mechanism in radiating away the energy associated with such oscillations.

For this reason we employ a particularly simple model for a quasistatically collapsing, non-rotating stellar core, or a young star, undergoing small oscillations about its equilibrium position, very similar to the model cores used for estimates of GR in stellar collapse. The model under consideration consists of a spherical (in equilibrium) mass of fluid, homogeneous and obeying an ideal gas law. The assumption of homogeneity, crucial in our case, serves in simplifying considerably the calculation, to allow analytic estimates to be obtained. Then starting with the equation of motion for the perturbations of the fluid

$$\frac{\partial}{\partial t} \underset{\sim}{U} = -\underset{\sim}{\nabla}(\delta\Phi) + \frac{\delta\rho}{\rho_0^2} \underset{\sim}{\nabla} p_0 - \frac{1}{\rho_0} \underset{\sim}{\nabla} (\delta p) \tag{1}$$

(δ denotes the Eulerian change of the corresponding quantity, while the subscript o denotes the unperturbed quantities; Φ is the gravitational potential, p the pressure, ρ the density, and $\underset{\sim}{U}$ the velocity of the fluid) and using the first law of thermodynamics including the neutrino losses, which in this case takes the form

$$\frac{\partial}{\partial t} (\delta p) + \underset{\sim}{U} \cdot \underset{\sim}{\nabla} p_0 - \frac{\Gamma p_0}{\rho_0} \frac{\partial}{\partial t} (\delta\rho) = -(\Gamma - 1) L_\nu \tag{2}$$

(Γ is the adiabatic index and L_ν is the neutrino emissivity) one can obtain the energy conservation equation for the perturbations:

$$\frac{dE}{dt} = -(\Gamma - 1)\int\{T \cdot \frac{\partial L_\nu}{\partial T} (\Gamma - 1) + \rho \frac{\partial L_\nu}{\partial \rho}\}\left(\frac{\delta\rho}{\rho_o}\right)_a^2 dV. \tag{3}$$

The subscript a denotes quantities corresponding to the solution of the adiabatic problem, i.e., the one with $L_\nu = 0$, which is well known (Pekeris 1939). The damping time can then be defined as the inverse rate of change of the pulsational energy due to neutrino losses

$$\frac{1}{t} = \sigma' = -\frac{1}{2}\frac{(dE/dt)}{E} = -\frac{1}{2}\frac{(dE/dt)}{1/2\ \sigma_a^2 \int |\delta r|_a^2 \rho_0 dV} \tag{4}$$

The above expression can be evaluated once a particular neutrino producing mechanism is employed in the expression for dE/dt. However, since, for densities $\rho \gtrsim 10^{11}$ g cm^{-3}, where these estimates are going to be valid, neutrinos do not stream freely out of the core, but they diffuse, one has to account for that by reducing the neutrino luminosity by a factor equal to the ratio of diffusion to free streaming timescales. (It turns out that this ratio is equal to R/λ, the core radius over the neutrino mean free path.)

Using the dominant thermal neutrino producing process in these temperatures and densities (i.e., pair annihilation) as the only neutrino mechanism contributing to the energy losses, one can obtain an upper estimate for the neutrino damping times at various densities. These damping times can then be compared against those due to GR of a similar configuration of the corresponding density given by Detweiler (1975). Our conservative estimates (non-inclusion of the non-thermal neutrino producing reaction $\rho^- p \to \nu_e n$ which is dominant during neutronization) yield comparable damping times for $\rho \sim 10^{15}$ g cm^{-3} (0.1 sec) for neutrinos and GR and much shorter neutrino ones for $\rho = 10^{14}$ g cm^{-3} ($\sim$ 10 sec for GR and $\sim$ 0.2 sec for neutrinos). For lower densities the differences are even higher since the GR damping times are almost infinite while the neutrino ones are of the order of a second. Only at densities above nuclear does GR effectively compete (at $\rho = 10^{15}$ $\sim$ 0.1 for GR and 0.7 for neutrinos). Neutrinos therefore appear to contribute more significantly in the damping of such oscillations. It should be remembered that for most densities the hydrodynamic times are more rapid than either of these damping times, thus the neutrinos probably do not have sufficient time to damp out all perturbations and so some GR can still be generated. (Remember $\tau_{freefall} \sim 500\ \rho^{-1/2}$ sec and typical collapse times are $\sim$ 5 to 10 times freefall.) A definite answer to the problem cannot be obtained without a detailed examination of the collapse microphysics, mainly the equation of state and the neutrino transport. Neutrinos themselves can actually play an important role in determining the equation of state, since they appear not to have enough time to escape from a dynamically collapsing core (Arnett 1977) and hence they get trapped. Being fermions they become degenerate as the collapse proceeds further and increase their contribution to the pressure of the collapsing core. This can therefore affect the core's bounce density and consequently the GR emitted. If this is actually the case neutrinos will eventually escape out in diffusion timescales as the remnant settles down to a cold neutron star.

The trapped neutrinos except for causing the core to bounce at lower densities may also have some other interesting consequences. Because of their long mean free paths and high energy density (due to degeneracy) they render the core fluid viscous with a kinematic viscosity of the order of $\sim 10^{11}$ cm^2 s^{-1}. This neutrino viscosity can smooth out velocity gradients in viscous time scales ($\sim$ 0.1-1 sec; note that they are of the same order or shorter than diffusion timescales); it can also stabilize rotating stellar cores with $|T|W| > 0.14$ collapsing quasistatically, up to the point of onset of dynamical instability (Lindblom & Detweiler 1977; Kazanas 1978). Such quasistatically collapsing cores have been considered as the best candidates for sources of large amounts of GR since they would allow enough time for non-axisymmetric perturbations to grow and drive a deformed McLaurin spheroid to a nonrotating triaxial Dedekind ellipsoid (Miller 1974). As evolution proceeds further, neutrinos diffusing out reduce the collapsing core's angular momentum (Kazanas 1977; Epstein 1977; Mikaelian 1977) so that eventually the collapsed object is stable.

CONCLUSION

The above discussion and also the current detailed numerical collapse models and gravitational radiation estimates allow several conclusions to be drawn concerning the importance of neutrinos in gravitational collapse and their influence on the GR emitted. On one hand neutrinos appear to be responsible for carrying away most of the energy associated with the collapse, thus not allowing much to be radiated gravitationally. In addition, rapid damping by neutrinos would tend to hinder the GR expected to be emitted. Therefore, before strong claims can be made on the GR emission and the possibility of its detection, estimates should be made, incorporating as much as possible of the detailed physics involved in the collapse, and in particular realistic equations of state and neutrino transport.

Since neutrinos and GR are the prime methods of energy transport in gravitational collapse (which presumably results in a supernova explosion) they constitute unique means for probing into the "heart" of such astrophysical events.

As mentioned elsewhere in these proceedings, GR detectors are already being developed. Although these may still be a long way from detecting any GR from collapse (Turner & Wagoner 1977) they may provide some useful limits. Neutrinos may prove to be even more favorable for detection due to the higher energy luminosity associated with them. Neutrino astronomy is just in its beginning. Over the past few years a few neutrino telescopes intended to detect neutrinos from supernovae within our galaxy have become operational. These have been reviewed recently by Lande (1978) and involve water Cerenkov detectors in the U.S. (Homestake gold mine), in the Mount Blanc Tunnel, and in the Soviet Union (Caucasus). Unfortunately these water Cerenkov detectors are more sensitive to $\overline{\nu}_e$'s which are suppressed by factors of 10 to 100 relative to other neutrino species because of the degeneracy of $\overline{\nu}_e$'s in the core. Since different collapse models yield different neutrino pulse profiles (in energy and in time) it should eventually be possible to use an observed profile to select between models (see Nadyozhin in these proceedings). (The water Cerenkov neutrino detectors detect $\overline{\nu}_e$ via $p + \overline{\nu}_e \rightarrow n + e^+$ yielding a fast positron.) These Cerenkov detectors provide real time neutrino detection with time resolution of much better than 10^{-3} sec. This is contrasted with Davis's solar neutrino experiment (also in the Homestake gold mine) which integrates all counts over the ^{37}Ar lifetime ($\sim$ 34 days). Bahcall (1978) has pointed out that a gravitational collapse in our galaxy would probably be detectable by the Davis experiment which is sensitive to ν_e's. However, this total time integration technique will not enable neutrinos to give an early warning nor will it give fine-timing information on neutrino pulses.

One point which the Mount Blanc Tunnel team has recently mentioned which makes ν-detection particularly relevant for people interested in GR detection, is that for some ν-pulse profiles, it may be possible for the ν interactions in the GR detector to trigger the detector independent of GR. This could only be resolved if a ν-detector were simultaneously operating.

Unfortunately one severe limitation for the future of ν-detection is the size of the detector required to detect events outside of the galaxy. Since the supernova rate in our galaxy is ∿ 1/30 years, this may yield a long wait between events. Although a 100 to 1000 ton detector can detect collapse events within the galaxy, it would take a 10^7 to 10^8 ton detector fitted with a photomultiplier for every ton of water to detect ∿ 10 MeV neutrinos from any collapse in the local supercluster. (The expected rate of supernova would then be $\gtrsim$ 1/yr.) Even DUMAND (Roberts 1977), with its 10^9 ton detector, will not have a sufficient density of photomultipliers to detect 10 MeV neutrinos. Thus for cost-effectiveness DUMAND is being aimed at $\gtrsim$ 1 TeV neutrinos with a relatively sparse array rather than the extragalactic collapse neutrinos. Hopefully all can be solved if a supernova goes off in our galaxy in the near future.

REFERENCES

Arnett, W.D. (1973). Some quantitative calculations of final stages of stellar evolution. In _Explosive Nucleosynthesis_, ed. D.N. Schramm & W.D. Arnett, pp. 236-247.

Arnett, W.D. (1977). Neutrino trapping during gravitational collapse. _Astrophys. J_. 218, pp. 815-833.

Bahcall, J. (1978). Solar Neutrino Experiments. Preprint.

Bethe, H. & Brown, G. (1978). Talk at Copenhagen workshop.

Brown, G., Bethe, H. & Lattimer, J. (1978). NORDITA preprint.

Chia, T.T., Chau, W.Y. & Henriksen, R.N. (1977). Gravitational radiation from a rotating collapsing gaseous shape ellipsoid. _Astrophys. J_. 214, pp. 576-583.

Detweiler, S.L. (1975). Variational calculations of eigen frequencies. _Astrophys. J_. 197, pp. 203-215.

Epstein, R. (1978). Neutrino angular momentum loss in rotating stars. _Astrophys. J. (Lett.)_ 219, pp. L39-41.

Freedman, D.Z., Schramm, D.N. & Tubbs, D.L. (1977). The weak neutral current and its effect in stellar collapse. Ann. Rev. Nuc. Sci. 27, pp. 167-207.

Kazanas, D. (1977). Neutrino angular momentum losses in stellar collapse. Nature 267, pp. 501-502.

Kazanas, D. (1978). On neutrino viscosity in collapsing stellar cores. Astrophys. J. (Lett.) 222, pp. L109-L111.

Kazanas, D. & Schramm, D.N. (1976). Competition of neutrino and gravitational radiation. Nature 262, pp. 671-672.

Kazanas, D. & Schramm, D.N. (1977). Neutrino damping of non-radial pulsations. Astrophys. J. 214, pp. 814-825.

Lattimer, J., Lamb, D., Pethic, C. & Ravenhall, G. (1978). University of Illinois preprint.

Laudé, K. (1978). In Proceedings of Neutrino '78, West Lafayette, Indiana, in press.

Lindblom, L. & Detweiler, S.L. (1977). On the secular instabilities of the McLaurin spheroids. Astrophys. J. 211, pp. 565-567.

Mikaelian, K.O. (1977). New mechanism for slowing down the rotation of dense stars. Astrophys. J. (Lett.) 214, pp. L23-L24.

Miller, B.D. (1973). Secular stability of uniformly rotating fluid masses. Astrophys. J. 181, pp. 497-512.

Pekeris, C.L. (1939). Non-radial oscillations of stars. Astrophys. J. 88, pp. 189-199.

Roberts, A. (1977). Current status of DUMAND project. In Proceedings of DUMAND Workshop, pp. 5-15. Fermi National Accelerator Laboratory, Batavia, Illinois.

Schramm, D.N. (1977). In Proceedings of Ben Lee Memorial Conference. Fermi National Accelerator Laboratory, Batavia, Illinois.

Shapiro, S.L. (1977). Gravitational radiation from stellar collapse. Astrophys. J. 214, pp. 556-575.

Shapiro, S.L. & Saenz, R.A. (1978). Cornell University preprint.

Turner, M. & Wagoner, R.V. (1977). Gravitational radiation from slowly rotating supernovae. Stanford University preprint.

Weaver, T.A., Zimmerman, G.B. & Woosley, S.E. (1978). Presupernova evolution of massive stars. *Astrophys. J.*, in press.

Wilson, J., Couch, R., Cochran, G., Le Blanc, J. & Barkat, Z. (1976). *Ann. N.Y. Acad. Sci.* 262.

GRAVITATIONAL AND NEUTRINO RADIATION FROM STELLAR CORE COLLAPSE: ELLIPSOIDAL MODEL CALCULATIONS

Stuart L. Shapiro

Center for Radiophysics and Space Research
Cornell University

I. ELLIPSOIDAL MODELING: MOTIVATION AND LIMITATIONS

In this paper we summarize some of the key results which have emerged recently from several quasi-Newtonian ellipsoidal model calculations of gravitational and neutrino emission from stellar collapse. The motivation for these numerical calculations is obvious: the design and construction of gravitational wave detectors with increased sensitivity [see report by Weiss (1978), this volume] have intensified the need for reliable theoretical calculations of gravitational radiation from all possible energetic astrophysical sources, including collapsing stellar cores. The difficulties associated with such calculations - including the formal, mathematical difficulties encountered when attempting to solve the full, non-linear Einstein field equations and the complexities of developing an accurate, hydrodynamical-neutrino transport model of stellar collapse - are well known. These problems are aggravated whenever the effects of rotation, magnetic fields, deviations from spherical and axisymmetry, etc., must be considered, as in the case of wave generation during core collapse. Although progress in handling these numerical difficulties has been considerable in recent years [see, e.g., reports by Smarr (1978) & Wilson (1978), this volume], the advances of gravity wave detector technology demand immediate (albeit preliminary) estimates of the character and amount of the radiation emitted from diverse astrophysical events. Accordingly, rough estimates can be readily furnished in the case of stellar collapse by employing homogeneous, ellipsoidal configurations to model the collapse in those regimes in which the weak-field, slow-velocity limit of general relativity is applicable. The resulting calculations, which avoid many of the numerical complexities of more realistic treatments, usually apply only to the initial dynamical stages of (nonspherical)gravitational core collapse but they provide a good qualitative picture of the initial radiation pulse(s) emitted during this event.

Ellipsoidal modeling offers the following advantages for treating core collapse: it (1) provides a rapid (i.e. inexpensive) means of integrating Euler's equations, as all partial differential equations readily reduce to ordinary differential equations (see e.g., Chandrasekhar (1969); (2) facilitates the study of global asymmetries (e.g., flattening due to rotation or magnetic fields) and the computation of quadrupole moments; (3) readily reveals the qualitative dependence of final results on the adopted input physics (e.g., on the assumed values of the core mass M, angular momentum J, initial eccentricity e_i, equation of state $P(\rho,T)$, neutrino opacity $\kappa(\rho,T)$, etc.); (4) offers a potentially useful check and/or guide for more realistic hydrodynamical calculations; and (5) provides quick, quantitative estimates of gravitational and neutrino radiation losses. On the other hand, ellipsoidal models suffer from the following constraints: they (1) restrict the density profile to be homogeneous and prevent central mass concentration; (2) restrict the velocity profile, including vorticity, to be a linear function of the coordinates (e.g., any rotation must be uniform) and the pressure profile to be a quadratic function of the coordinates; (3) constrain the shape of all configurations to be perfectly ellipsoidal; and (4) employ Newtonian equations of motion, requiring slow velocities $v^2 \ll c^2$ and weak fields $\phi \ll c^2$ (although secular gravitational radiation-reaction forces can be included).

The numerical advantages of ellipsoidal modeling are obvious; the physical consequences of the implied constraints are not so apparent. For example, homogeneity ensures homologous dynamical behavior, which tends to accelerate the rate of core collapse somewhat above the rate found from more realistic hydrodynamical calculations. This effect typically results in a higher energy dissipation rate from gravitational radiation relative to neutrino emission, which must be recalled when interpreting the ellipsoidal results. Yet it is significant that the recent spherical, hydrodynamical collapse calculations of Van Riper (1978), Arnett (1977), and Van Riper & Arnett (1978) reveal that a substantial fraction of the core mass (the "inner core") may in fact participate in homologous motion after the first bounce, justifying the ellipsoidal approximation for these zones. Smarr (1977) has argued from his fully relativistic calculation of two colliding black holes that wave generation in the presence of strong fields may not be important energetically since the combined effects of gravitational redshift, time dilation and wave recapture serve to eliminate any expected enhancement above weak-field, slow-velocity estimates. The post-Newtonian calculations of nearly spherical dust collapse of Epstein (1976) and Wagoner (1977) confirm this behavior. Accordingly, once again the restrictive nature of quasi-Newtonian ellipsoidal modeling may ultimately prove to be no obstacle in obtaining reliable estimates of gravity wave generation for a very general class of core collapse scenarios.

II. SECULAR VS DYNAMICAL CALCULATIONS

Published calculations employing homogeneous ellipsoids to study the generation of gravitational radiation (GR) from stars fall into two general categories: (1) secular evolution calculations, which follow the slow evolution of equilibrium configurations on dissipative (GR) timescales, and (2) dynamical evolution calculations, which follow the evolution of nonequilibrium configurations on dynamical (e.g. collapse) timescales. Examples and highlights of both types of calculations are summarized below.

The secular instability of Maclaurin spheroids due either to viscosity or to gravitational radiation has been studied in detail by Chandrasekhar (1969; 1970 a,b). He has shown that the basic effect of such an instability is to excite nonaxisymmetric toroidal modes whenever the ratio of rotational kinetic energy T to gravitational potential energy satisfies $T/|W| \geq 0.138$ (eccentricity $e \geq 0.813$), where the equality denotes the point at which the Jacobi and Dedekind sequences bifurcate from the Maclaurin sequence. Chandrasekhar (1970c) and Miller (1974) have integrated the equations of motion for a Jacobi ellipsoid and other Riemann S-type ellipsoids for different initial configurations whose subsequent evolution is driven by radiation-reaction. Employing the gravitational radiation-reaction potential derived by Burke (1969) and Thorne (1969), they have shown that the initial configurations invariable evolve toward a nonradiating state. Specifically, Miller found that this state is a (1) Maclaurin spheroid, if the initial figure is a Jacobi-like (S-type) ellipsoid with angular velocity $|\Omega|$ > vorticity $|\Lambda|$, or a Dedekind-like ellipsoid, with $|\Lambda| > |\Omega|$, above the Jacobi-Dedekind sequence; and a (2) nonrotating, triaxial Dedekind ellipsoid for an initial Dedekind-like figure below the Jacobi-Dedekind sequence. Basically, this analysis proved that gravitational radiation can play a significant role in dissipating the energy and angular momentum of sufficiently eccentric equilibrium configurations (e.g. newly formed neutron stars) and in determining their ultimate (potentially nonaxisymmetric) equilibrium shape.

Lindblom & Detweiler (1977) have discussed the combined effects of gravitational radiation reaction and of viscosity on the stability of Maclaurin spheroids and showed that, when operating together, these instabilities tend to cancel each other. The sequence of stable Maclaurin spheroids reaches beyond the bifurcation point to a new point governed by the ratio of the strengths of viscous and radiative forces and in principle can be extended all the way to the point of the onset of the dynamical instability ($T/|W|=0.274$, $e=0.953$). Although the viscosity of an ideal, cold degenerate gas is orders of magnitude too small to stabilize a rotating white dwarf or neutron star beyond the bifurcation point, the estimated neutrino viscosity in a hot, newly formed neutron star may indeed be sufficient to extend its rotational stability domain somewhat (Kazanas 1978).

Kazanas & Schramm (1976, 1977) have modeled the collapse of a stellar core by employing a sequence of ellipsoids of increasing density and temperature in spherical equilibrium. Linearly perturbing this model, they conclude that neutrino radiation is typically more efficient than gravitational radiation in damping out nonradial pulsations during gravitational collapse. Although their use of equilibrium models to analyze dynamical collapse is questionable, their results may certainly be applied to the final settling of a rapidly cooling neutron star and may in fact be true in general.

Since in this paper we shall deal principally with the dynamical stages of core collapse, where most of the gravitational radiation is produced, we will not discuss further the results of other secular evolution model calculations. These calculations clearly are relevent to the evolution of (equilibrium) degnerate cores on the verge of collapse and to newly formed neutron stars settling into their final (equilibrium) state.

In their pioneering application of homogeneous ellipsoids to estimate GR losses from dynamical collapse, Thuan & Ostriker (1974) calculated the "free-fall" radiation generated by a cold uniformly rotating spheroid collapsing to a thin pancake. Epstein & Wagoner (1975), Epstein (1976) and Wagoner (1977) provided a post-Newtonian treatment of the above problem. Novikov (1975) estimated the enhancement to the emitted radiation generated near maximum compression due to the rapid deceleration of the highly oblate core by internal pressure. Shapiro (1977) and Saenz & Shapiro (1978) calculated the gravitational radiation generated during the first bounce of a collapsing ellipsoidal core with internal (polytropic) pressure with and without rotation. Recently, Saenz & Shapiro (1979) extended their calculations by considering a more realistic equation of state, allowing for neutrino dissipation and following the evolution of the core beyond the first bounce. The key results of the above dynamical analyses will be explored in the following sections.

III. ELLIPSOIDAL EQUATIONS OF MOTION

Nothing better dramatizes the convenience of ellipsoidal modeling than a glance at the ellipsoidal equations of motion for a core with a polytropic (central) equation of state and angular velocity and vorticity both along the semi-minor (c) axis in the $\hat{z}$ direction [see Box 1, from, e.g., Miller 1974; Saenz & Shapiro 1978]. The equations include a 2 1/2 post-Newtonian radiation-reaction potential correctly describing the lowest order dissipative losses due to GR, and are appropriate whenever the (conservative) first and second post-Newtonian terms are small. For an arbitrary equation of state and dissipation by neutrino emission $P_c=P_c(\rho, T_c)$ and an additional entropy evolution equation must be solved simultaneously for the central temperature, T_c (see §V.b). In the absence of GR, the circulation C, angular momentum J, and total energy E of the configuration are conserved [see Box 2]. In the presence of wave emission, C is still conserved (Miller 1974), J is conserved only if the collapse is axisymmetric or if (in the case of ellipsoidal symmetry) $J\equiv 0$, and E is conserved only if the collapse is spherical. The GR dissipation rates (averaged over many wavelengths) and wave amplitudes [see Box 3 from Saenz & Shapiro 1978; geometrized units are employed with $G\equiv c\equiv 1$] take on simple forms in the case of ellipsoidal motion. The amplitudes represent the nonvanishing components of the metric perturbation $h^{TT}_{\mu\nu}(\underset{\sim}{x},t)$ expressed in the TT gauge and measured along orthonormal basis vectors by a stationary observer in the radiation zone at (t,r,θ_0,ϕ_0). The right hand sides of the expressions in Box 3 are to be evaluated at retarded time $t'=t-r$. Note that for axisymmetric collapse, one of the two possible polarization states of the emitted wave always vanishes.

BOX 1. ELLIPSOIDAL EQUATIONS OF MOTION (WITH RADIATION REACTION)

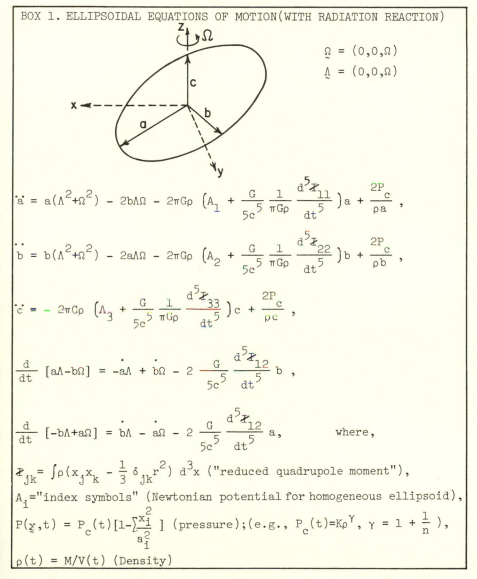

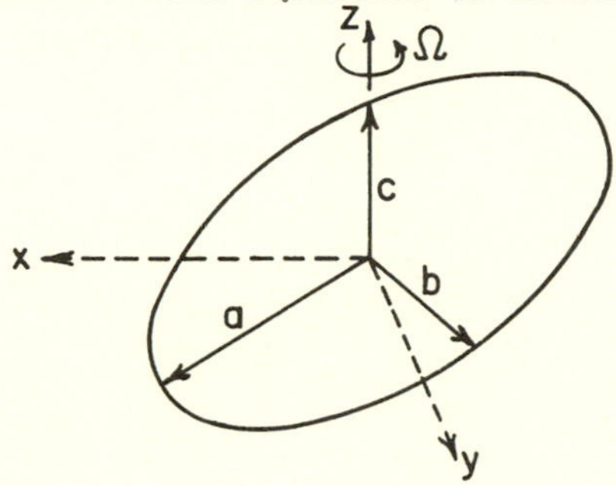

$\underset{\sim}{\Omega} = (0,0,\Omega)$

$\underset{\sim}{\Lambda} = (0,0,\Omega)$

$$\ddot{a} = a(\Lambda^2+\Omega^2) - 2b\Lambda\Omega - 2\pi G\rho\left(A_1 + \frac{G}{5c^5}\frac{1}{\pi G\rho}\frac{d^5 I\!\!\!-_{11}}{dt^5}\right)a + \frac{2P_c}{\rho a},$$

$$\ddot{b} = b(\Lambda^2+\Omega^2) - 2a\Lambda\Omega - 2\pi G\rho\left(A_2 + \frac{G}{5c^5}\frac{1}{\pi G\rho}\frac{d^5 I\!\!\!-_{22}}{dt^5}\right)b + \frac{2P_c}{\rho b},$$

$$\ddot{c} = -2\pi G\rho\left(A_3 + \frac{G}{5c^5}\frac{1}{\pi G\rho}\frac{d^5 I\!\!\!-_{33}}{dt^5}\right)c + \frac{2P_c}{\rho c},$$

$$\frac{d}{dt}[a\Lambda - b\Omega] = -\dot{a}\Lambda + \dot{b}\Omega - 2\frac{G}{5c^5}\frac{d^5 I\!\!\!-_{12}}{dt^5}b,$$

$$\frac{d}{dt}[-b\Lambda + a\Omega] = \dot{b}\Lambda - \dot{a}\Omega - 2\frac{G}{5c^5}\frac{d^5 I\!\!\!-_{12}}{dt^5}a, \qquad \text{where,}$$

$I\!\!\!-_{jk} = \int\rho(x_j x_k - \frac{1}{3}\delta_{jk}r^2)\, d^3x$ ("reduced quadrupole moment"),

A_i="index symbols" (Newtonian potential for homogeneous ellipsoid),

$P(\underset{\sim}{x},t) = P_c(t)[1-\sum \frac{x_i^2}{a_i^2}]$ (pressure); (e.g., $P_c(t)=K\rho^\gamma$, $\gamma = 1 + \frac{1}{n}$),

$\rho(t) = M/V(t)$ (Density)

BOX 2. DISSIPATION DUE TO GRAVITATIONAL RADIATION

A. Newtonian Conserved Quantities

Circulation: $\frac{C}{\pi} = 2ab\Omega - (a^2 + b^2)\,\Lambda$ (Conserved)

Angular Momentum: $\frac{J}{(M/5)} = (a^2 + b^2)\,\Omega - 2ab\Lambda$ (Conserved if $a \equiv b$ or if $J \equiv 0$)

Energy: $$\frac{E}{(M/5)} = \frac{1}{2}\sum_i \dot{a}_i^2 + \frac{1}{2}(a^2 + b^2)(\Omega^2 + \Lambda^2) - 2ab\Omega\Lambda - 2\pi G\rho \sum_i a_i^2 A_i + 2nP_c/\rho$$ (Conserved if $a \equiv b \equiv c$)

B. Dissipation Rates

Power Radiated: $\frac{dE}{dt} = \frac{1}{5}\left\langle \dddot{I\!\!\!-}_{jk}\, \dddot{I\!\!\!-}_{jk} \right\rangle$

[e.g., $= \frac{2}{375} M^2 (\dddot{a^2} - \dddot{c^2})^2$ when $a \equiv b$]

Power Spectrum: $\frac{dE}{d\nu} = \frac{2}{5}\sum_{jk} \left| \int_{-\infty}^{\infty} \dddot{I\!\!\!-}_{jk}\, e^{i\omega t}\, dt \right|^2$

[e.g., $= \frac{4}{375} M^2 \left| \int^{\infty} (\dddot{a^2} - \dddot{c^2})\, e^{i\omega t}\, dt \right|^2$

when $a \equiv b$]

Angular Momentum Radiated: $\frac{dJ_j}{dt} = \frac{2}{5}\,\varepsilon^{jkl} \left\langle \dddot{I\!\!\!-}_{ka}\, \dddot{I\!\!\!-}_{al} \right\rangle$

[e.g., $= 0$ when $a = b$ $\forall J$

$= 0$ when $J = 0$ $\forall a,b,c$]

BOX 3. GENERATION OF GRAVITATIONAL WAVES

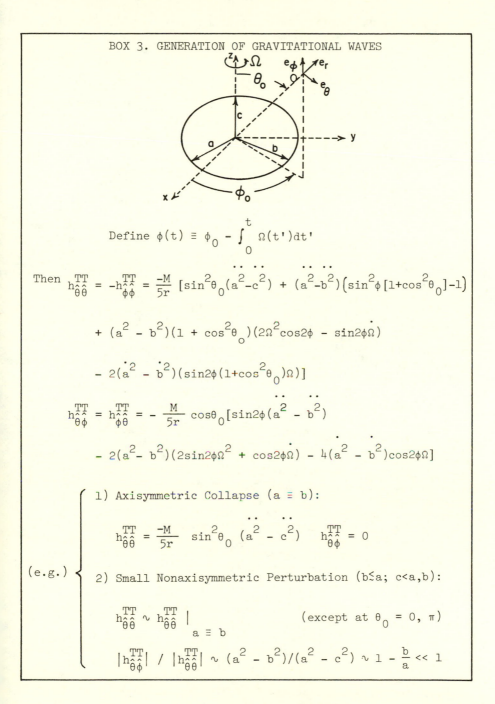

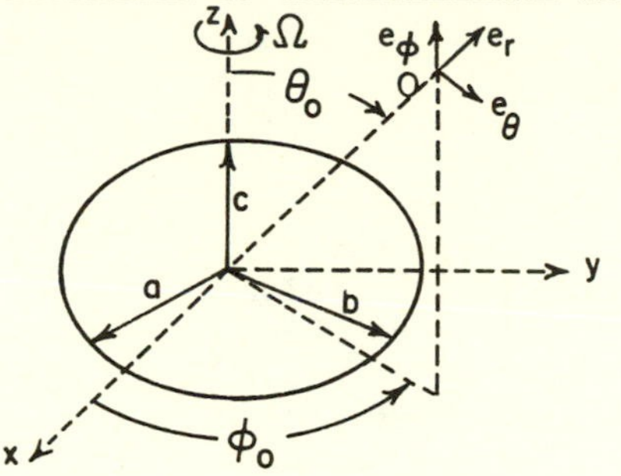

Define $\phi(t) \equiv \phi_0 - \int_0^t \Omega(t')dt'$

Then
$$h^{TT}_{\hat\theta\hat\theta} = -h^{TT}_{\hat\phi\hat\phi} = \frac{-M}{5r}\Big[\sin^2\theta_0(\ddot{a^2}-\ddot{c^2}) + (\ddot{a^2}-\ddot{b^2})\big(\sin^2\phi[1+\cos^2\theta_0]-1\big)$$
$$+ (a^2 - b^2)(1 + \cos^2\theta_0)(2\Omega^2\cos 2\phi - \sin 2\phi\dot\Omega)$$
$$- 2(\dot{a^2} - \dot{b^2})(\sin 2\phi(1+\cos^2\theta_0)\Omega)\Big]$$

$$h^{TT}_{\hat\theta\hat\phi} = h^{TT}_{\hat\phi\hat\theta} = -\frac{M}{5r}\cos\theta_0\Big[\sin 2\phi(\ddot{a^2} - \ddot{b^2})$$
$$- 2(a^2 - b^2)(2\sin 2\phi\,\Omega^2 + \cos 2\phi\dot\Omega) - 4(\dot{a^2} - \dot{b^2})\cos 2\phi\,\Omega\Big]$$

(e.g.)

1) Axisymmetric Collapse ($a \equiv b$):

$$h^{TT}_{\hat\theta\hat\theta} = \frac{-M}{5r}\sin^2\theta_0\,(\ddot{a^2} - \ddot{c^2}) \qquad h^{TT}_{\hat\theta\hat\phi} = 0$$

2) Small Nonaxisymmetric Perturbation ($b \lesssim a$; $c < a,b$):

$$h^{TT}_{\hat\theta\hat\theta} \sim h^{TT}_{\hat\theta\hat\theta}\Big|_{a\,\equiv\,b} \qquad \text{(except at } \theta_0 = 0, \pi)$$

$$|h^{TT}_{\hat\theta\hat\phi}| \,/\, |h^{TT}_{\hat\theta\hat\theta}| \sim (a^2 - b^2)/(a^2 - c^2) \sim 1 - \frac{b}{a} \ll 1$$

IV. GRAVITATIONAL RADIATION FROM STELLAR CORE COLLAPSE: SUMMARY OF KEY INITIAL RESULTS

A) Qualitative Overview

Thuan & Ostriker (1974) calculated the total energy and frequency spectrum of the radiation generated by a cold, rotating, collapsing spheroid, initially at rest and asymptotically spherical (though infinitesimally perturbed) at infinity. By following the collapse up to the arbitrary moment at which the spheroid flattened to a pancake of zero thickness they found that the quadrupole "free-fall" emission scaled as $\Delta E_{ff}/Mc^2=0.0370(2M/a_m)^{7/2}$ when $T/|W|=0(J\equiv 0)$ at maximum compression and increased to $\Delta E_{ff}/Mc^2=0.110(2M/a_m)^{7/2}$ for $T/|W|=0.5(J\equiv\infty)$. Here a_m is the semi-major axis of the flattened spheroid ($c_m=0$). For the same value of a_m, rotation thus increases the radiation efficiency of stellar collapse. The corresponding frequency spectrum reveals a broadband profile with a typical frequency cut-off $\nu_{ff}\sim(M/a_m)^{3/2}$ associated with the free-fall timescale of the spheroid near maximum compression. Post-Newtonian corrections to these calculations by Epstein & Wagoner (1975) and Epstein (1976) indicated that the principal post-Newtonian effect is to prolong the collapse as viewed from infinity. Consequently, the observed luminosity and total energy radiated during the collapse, which are strongly inversely dependent on the collapse timescale, are reduced below the values estimated from the results of Newtonian calculations. This effect is also observed in the calculated wave amplitudes recently calculated for slightly perturbed spherical dust collapse by Wagoner (1977).

Novikov (1975) argued qualitatively that the inclusion of pressure leads to a considerable enhancement in the total amount and characteristic frequency of the observed emission. This increase results from the sudden, pressure-induced deceleration of the semi-minor axis of the spheroid near maximum compression and bounce, which enhances the wave generation rate again due to the strong inverse dependence of the emission rate on the collapse timescale [see Box 2]. For a polytropic law $P_c=K\rho^{1+1/n}$ (K,n constant), Novikov's relations imply a "bounce" emission $\Delta E_B\sim\Delta E_{ff}(a_m/c_m)/n^2\gg\Delta E_{ff}$ and associated "bounce" frequency $\nu_B\sim\nu_{ff}(a_m/c_m)/\sqrt{n}\gg\nu_{ff}$ whenever $c_m\ll a_m$. Thus, the emission and characteristic frequency increases for harder equations of state (i.e. smaller n).

Shapiro (1977) has investigated quantitatively the combined effects of internal pressure and rotation on spheroidal collapse. The special case of zero energy, time-symmetric bounce and rebound, where $\dot{a}\equiv 0\equiv\dot{c}$ at maximum compression was considered for several polytropic-equations of state ($n=3,3/2$, and 1). The results were employed to analyze the initial energy burst from isentropic stellar collapse to a neutron star and from isentropic collapse of an entropy-deficient supermassive star. It was shown that, in general, for a configuration of given mass M and specific entropy S, there exists a critical angular momentum J_c ($T/|W|_c \lesssim 0.3$) for which the GR radiation loss achieves a maximum on the first bounce. For $J \ll J_c$ the spheroid remains nearly spherically symmetric even at maximum compression while for $J \gg J_c$ the configuration never collapses to high densities (i.e. short collapse timescales) due to the large equatorial centrifugal forces. Thus the emitted energy decreased as J departs from J_c in either direction and there is no substantial pressure-induced enhancement of the gravitational radiation efficiency. The maximum efficiency of GR energy release was found to be $\Delta E/Mc^2 \simeq 2\times10^{-3}(M/1.4M_\odot)^{14/3}$ for $J_c \simeq 2\times10^{49}(M/1.4M_\odot)^{4/3}$g erg-s from the collapse of cold, nonrelativistic degenerate neutrons ("cold collapse") compared to $\Delta E/Mc^2 \simeq 4\times10^{-6}(M/1.4M_\odot)^{14/3} \exp[-7m_n(S-S_*)/3k]$ for $J_c \simeq 4\times10^{49}(M/1.4M_\odot)^{4/3} \exp[m_n(S-S_*)/3k]$ from the adiabatic collapse of hot, nondegenerate neutrons ("hot collapse"). Here S_* is the specific entropy of an ideal neutron gas with $\rho_*=3\times10^{13}$gm cm^{-3} and $T_*=3.5\times10^{11}$K (typical central values at maximum compression; Wilson 1974), k is Boltzmann's constant and m_n is the neutron rest mass. The corresponding frequency spectrum is "double-humped", peaking at $\nu \sim \nu_{ff}$ and $\nu \sim \nu_B$.

Chia, Chau & Henriksen (1977) have attempted to model "thermal" photon and neutrino losses in their analysis of the collapse of a rotating spheroid with internal thermal gas pressure. They draw the reasonable conclusion that whenever the "thermal" cooling rate (parametrized by a single term $\propto \kappa \cdot T_c$, where κ is an adjustable constant) is large, thermal emission dominates gravitational radiation as a dissipative agent. Unfortunately, their results are suspect since they handle GR dissipation by inserting a GR "cooling" term into the entropy equation for $\dot{T}_c$ (equation (11) of their paper), rather than by properly including a radiation reaction potential in the equations of motion (see Box 1). The effect of their equations is to (erroneously) extract the energy dissipated by wave emission from the internal energy of the spheroid, rather than from the bulk kinetic energy of motion.

B) A Numerical Illustration

Saenz & Shapiro (1978) extended the analysis of Shapiro (1977) by treating the collapse from rest of homogeneous ellipsoids which described more general and realistic initial configurations to model degenerate $1.4M_\odot$ cores on the verge of gravitational collapse at a density $\rho_i=10^{10}$g cm^{-3}. They systematically considered initially oblate and prolate spheroidal configurations with and without rotation; the nonspherical, nonrotating cores were assumed to be deformed by internal magnetic stresses. They also investigated ellipsoidal cores characterized by both large and small initial nonaxisymmetric deformations. The microphysics was modeled by following Shapiro (1977) and considering both hot and cold equations of state for a pure neutron gas: $P_c=K\rho^{5/3}$ where

$$K = \frac{1}{5}\left(\frac{3\pi^2 h^3}{m_n^4}\right)^{2/3} = 5.38 \times 10^9 \text{ cgs} \qquad \text{(IVB.1a)}$$

for cold, degenerate (NR) collapse and

$$K = 2.99 \times 10^{10} \text{ cgs} \qquad \text{(IVB.1b)}$$

for hot, isentropic collapse ($S=S_*$). The "cold" collapse scenario presumably applies if during neutronization the central regions of the collapsing core cool quickly via neutrino dissipation as in Colgate & White (1966), or to the late stages of collapse after the initially hot "inner core" has cooled via neutrino and shock dissipation and is undergoing large scale, homologous oscillations at high density (Van Riper 1978; Saenz & Shapiro 1979). The "hot" collapse scenario applies if, during the initial infall, the emitted neutrinos in the core are substantially reabsorbed, allowing the collapse to procede adiabatically (Arnett 1977). The results of their study, which included computations of the total emitted gravitational radiation, power spectrum, radiated angular momentum and waveforms, and emphasized the initial bounce and rebound, are summarized in the following paragraphs.

Briefly, it was found that for a wide range of initial configurations the GR efficiency during the first bounce and rebound never exceeds $\Delta E/Mc^2 \simeq 10^{-3}$. Since the adopted equations of state ensured that the collapse proceeded on rapid, $\sim$ free-fall (in contrast to secular, neutrino-dissipation) timescales prior to maximum compression, the quoted efficiencies probably represent upper limits to the total gravitational energy radiated during this phase. For initially oblate collapse with rotation, the efficiency reaches a maximum for a finite $J=J_c$, corresponding to an initial eccentricity $e_i \simeq 0.3$ and decreased as J and e_i departed from these values. For oblate and prolate collapse without rotation, the efficiency increases with e_i for $e_i \ll 1$ and remains $\sim$ constant for $e_i \gtrsim 0.3$. Small initial nonaxisymmetric perturbations do not have sufficient time to grow substantially during the initial (dynamical) collapse and bounce phases; large initial nonaxisymmetric deformations do not significantly enhance the radiative efficiencies. The calculated waveforms carry detailed information on the nature of the initial collapse and subsequent oscillations and can in principle, serve as probes regarding the shape and dynamical behavior of the core near maximum compression. A rough estimate of the total neutrino loss during this initial dynamical phase is $\Delta E_\nu/Mc^2 \lesssim 10^{-2}$, typically, originating principally from electron capture onto protons well before maximum compression.

It is interesting to examine and compare some specific findings of the above analysis:

(1) Oblate Spheroids. The radiation efficiency $\Delta E/Mc^2$ during the first bounce is shown in Figure (1a) as a function of J, e_i and the initial rotation period P for the hot and cold rotating oblate se-

quences. Here $J=J_M(e_i)$ is chosen to be the value of angular momentum found in an (equilibrium) Maclaurin spheroid with density $\rho_i=10^{10}$ g cm^{-3}, mass $M=1.4M_\odot$ and eccentricity e_i. The results are similar to those found earlier by Smarr (1977) and Shapiro (1977) and show that (1) the efficiency achieves a maximum for $J_c\sim10^{49}$erg-s, corresponding to $e_i\sim0.1$-0.3 and (2) the cold, degenerate configurations which achieve higher densities (and shorter collapse timescales) near maximum compression radiate more efficiently than the hot configurations. In the former case, the cores actually undergo complete gravitational collapse when $e_i\lesssim0.1$, according to the "hoop" conjecture of Thorne (1972), since $R_{max}=\max(a,b,c)<2M$ at some moment during the collapse; all numerical integrations were abrubtly terminated at this instant. For nearly spherical collapse with $e_i<<1$ the efficiency satisfies $\Delta E/Mc^2\approx\alpha e_i{}^4\simeq\alpha(J/6\times10^{48}\text{erg-s})^4$, where $\alpha\simeq5$ for cold collapse and $\alpha\simeq0.1$ for hot collapse. The efficiency is shown in Figure (1b) as a function of e_i induced by an assumed initial poloidal magnetic field B_p estimated from the relation

$$e_i^2 = \frac{21}{2}\frac{\mathcal{M}_p}{|W|}\,,\ \mathcal{M}_p = \frac{B_p^2}{8\pi}\cdot\frac{M}{\rho_i} \qquad \text{(IVB.2)}$$

(Wentzel 1961). The efficiency again varies as e_i^4 ($\propto B_p^4$) for $e_i<<1$ and is larger for the colder equation of state. However, the emissions obtains a limiting value nearly independent of e_i for large $e_i\gtrsim0.1$ ($B_p\gtrsim10^{13}$gauss), since, with $J\equiv0$, there are no large centrifugal forces capable of supporting the configuration at large radii and low densities. [The oscillatory nature of ΔE with increasing e_i above ~ 0.5 results from the large number of core bounces near maximum compression along the semi-minor (c) axis during one complete collapse cycle as defined by Saenz & Shapiro, so it is, in part, an artifact of the integration termination point]. Such high magnetic field strengths in degenerate cores cannot be ruled out by observations (Ostriker & Hartwick 1968) and are in fact the logical outcome of the collapse and fragmentation into stars of "standard" interstellar clouds threaded by frozen-in magnetic fields with $B\sim3\times10^{-6}$ gauss (Spitzer 1968).

The power spectrum of the observed radiation from hot, oblate collapse is plotted in Figure (2) for $e_i=10^{-3}$, 10^{-2}, 0.5, and 0.9. The rotating sequence exhibits the usual double-humped spectrum described above, except at very low e_i where the two peaks merge (since $\nu_B\sim\nu_{ff}$). For this sequence the peak frequency steadily decreases with increasing e_i since the density and associated bounce timescale near maximum compression steadily decrease. For the nonrotating sequence the maximum density and corresponding peak frequency do not significantly change with e_i. Typically $\nu_{max}\sim10^3$Hz; the power spectra for the cold collapse sequence is qualitatively similar. The zero frequency limit of the distribution satisfies $dE/d\nu\sim\text{const}\neq0(\nu\to0)$, which is apparent from the (quadrupole) expression for the power spectrum given in Box 2: as $\nu\to0$, $dE/d\nu$ is given by an integral of a total derivative,

$$\frac{dE}{d\nu} \rightarrow \sum_{j,k} |[\ddot{I\!\!\!-}_{jk}(t)]_{t_1}^{t_2}|^2 \rightarrow \text{constant} \neq 0 \ (\nu \rightarrow 0) \ , \tag{IVB.3}$$

since $\ddot{I\!\!\!-}_{jk}(t_2) \neq \ddot{I\!\!\!-}_{jk}(t_1)$ during the integration (bounce) cycle.

Representative waveforms for the hot, oblate sequence are shown in Figure (4) for $e_i=10^{-3}$, 0.5, 0.9. Wave amplitudes convey detailed dynamical information regarding the symmetries, rotation properties and timescales associated with collapse. They invariably attain their peak values near maximum compression. For $e_i \gtrsim 0.3$ rotation serves to reduce the peak value below and broaden the wave about the peak in comparison to the corresponding J=0 spheroid with the same value of e_i (thus reflecting the manner in which centrifugal forces slow down the collapse in the equatorial plane). By comparing Figure (3) with a plot showing the time-dependent evolution of a and c (with and without rotation), one finds that the "glitches" in the waveforms which appear superposed on the overall envelopes are associated with a bounce of one of axes at maximum compression. Moreover, the <u>sign</u> of the glitch invariably reveals which axis has bounced: for oblate collapse, positive (negative) glitches occur when $\dot{c}=0(\dot{a}=0)$. The core may oscillate along the c-axis several times near maximum compression before rebounding back out to low density. The corresponding waveforms for cold collapse are qualitatively similar (though of larger amplitude) except for low e_i, where total gravitational collapse occurs. Since these models collapse inside event horizons, they never exhibit an observable sharp increase in amplitude near maximum compression.

(2) <u>Prolate Spheroids</u>. A toroidal magnetic field, B_T can induce a prolate eccentricity e_i approximately by

$$e_i^2 \simeq 8 \, \frac{\mathcal{M}_T/|W|}{1-1.5 \, \mathcal{M}_T/|W|} \ , \quad \mathcal{M}_T = \frac{B_T^2}{8\pi} \, \frac{M}{\rho_i} \tag{IVB.4}$$

(Wentzel 1961). In fact the nonrotating compressible-fluid equilibrium models of Wentzel (1961) and the rotating magnetic white dwarf models of Ostriker & Hartwick (1968) are dominated by toroidal fields and are prolate. The efficiency of GR dissipation is shown in Figure (3) for the hot collapse of initially prolate spheroids (rotating and nonrotating) as a function of B_T and e_i. For J=0, the results are comparable to the oblate sequence. For J$\neq$0, the behavior is quite different for small and large values of e_i, due to the fact that the ratio $T/|W|$, which measures the relative importance of centrifugal to gravitational forces, grows as r^{-1} during the collapse. For large e_i, this growth is not sufficient to influence the motion appreciably and the collapse (and corresponding efficiencies) closely resembles J=0 prolate collapse for the same value of e_i. For very small values of e_i, rotation eventually dominates and drives the spheroid oblate near maximum compression. Consequently, the low e_i, J$\neq$0 regime closely resembles the initially oblate,

$J=J_M(e_i)$ sequence, yielding the same efficiencies and erasing all imprints of prolate behavior near maximum compression. The "anomalously" low efficiencies which occur at intermediate values of e_i result from the competition and mutual cancellation of rotation and initially prolate geometry, which produce nearly spherical configurations near bounce. The rotating, prolate models may actually represent the canonical cases found in Nature, since degenerate stellar cores are likely to be endowed with both rotation and (toroidal) magnetic fields.

Waveforms for the J=0, prolate models are, typically, inverted from their oblate counterparts [$h^{\phi}{}_{\phi}(P)=-h^{\phi}{}_{\phi}(O)$], due to the role-reversal of the a and c axes (see Box 3). For $J\neq0$ and low e_i, the forms are indistinguishable from their oblate counterparts as expected; for high e_i, rotation is not important and the forms appear as slightly broadened counterparts of the corresponding J=0 prolate forms.

(3) Nonaxisymmetric Ellipsoids. Nonaxisymmetric deformations - even if large initially - do not result in very significant changes in the integrated energy output during the initial cataclysmic dynamical (∿free-fall) stages of core collapse. Moreover, when the hot oblate sequence is subjected to initially small nonaxisymmetric distortions, the distortions remain small. Since the collapse timescale is roughly free-fall in these idealized, low-entropy models, substantial exponentiation of unstable, nonaxisymmetric modes does not occur during the first bounce, even when $T/|W|$ exceeds the dynamical instability limit of 0.27. [This is in striking contrast to quasi-steady collapse on dissipative timescales which can lead to the substantial growth of nonaxisymmetric perturbations whenever $T/|W|\gtrsim 0.14$ (Miller 1974)]. Although J is no longer conserved $\Delta J/J$ never exceeds $\sim 10^{-4}$ typically.

The diagonal component of the observed metric perturbation $h^{TT}_{\hat\phi\hat\phi}$ varies with the azimuthal position of the observer, ϕ_o, but is characterized by the same amplitude, roughly, and time behavior as its axisymmetric counterpart. The amplitude of the off-diagonal component $h^{TT}_{\hat\phi\hat\theta}$ is much smaller; typically, for small nonaxisymmetric deformations, the amplitude satisfies (see Box 3, $\theta_o\neq0,\pi$)

$$|h^{TT}_{\hat\phi\hat\theta}|/|h^{TT}_{\hat\phi\hat\phi}| \sim \frac{a^2-b^2}{a^2-c^2} \sim 1 - \frac{b}{a} \lesssim 1. \qquad \text{(IVB.5)}$$

Clearly, the relative amplitude of the two polarization states provides a direct measure of deviations from axisymmetry during core collapse.

V. GRAVITATIONAL RADIATION AND NEUTRINO DISSIPATION FROM STELLAR COLLAPSE: ELLIPSOIDAL CALCULATIONS WITH A MORE REALISTIC EQUATION OF STATE

A. Motivation

Recently, Saenz & Shapiro (1979) have attempted to improve on and extend existing ellipsoidal collapse calculations by modeling the core microphysics more realistically. They were specifically motivated by a desire to (1) include entropy loss via neutrino (ν) and anti-neutrino ($\bar{\nu}$) dissipation in the dynamical equations, (2) incorporate the effects of inverse β-decay and a "hard core" nuclear potential on the composition and equation of state of the collapsing material, (3) reliably estimate neutrino luminosities ($L_\nu, L_{\bar{\nu}}$) as well as gravitational radiation luminosities (L_{GR}) as a function of time and to compare their integrated efficiencies, and (4) follow core collapse beyond the initial infall and rebound and track the approach to a final, equilibrium state. Details of their calculations will appear elsewhere but some of the key features and preliminary results are summarized below.

B. The Basic Model

The "canonical" sequence considered by Saenz & Shapiro (1979) is the sequence of rotating, oblate (Maclaurin) spheroids with mass $M=1.4M_\odot$, initially in hydrostatic equilibrium at a density $\rho_i=4\times10^9$g cm^{-3} (cf. Arnett 1977) with eccentricity e_i and $J=J_M(e_i)$. Variations in initial global core parameters, including the values of ρ_i and M, some initial pressure deficiency, and nonaxisymmetric deformations were also explored. The core was assumed to consist entirely of free neutrons (n), protons (p), and electrons (e^-), in beta equilibrium. Collapse in the canonical case is triggered and controlled by ν and $\bar{\nu}$ dissipation, in contrast to the constant entropy models described in §IV, where collapse always occurs because of initial pressure deficiency.

The composition and equation of state of a gas consisting of n, p, e- and ν in β-equilibrium is determined as a function of baryon density n, T_c and the neutrino chemical potential $\mu_\nu \equiv \eta_\nu kT_c$ by employing a modified version of the numerical scheme of Van Riper & Bludman (1977). The validity of the assumption of β-equilibrium and the applicability of the free-nucleon approximation following thermal dissociation ($T \gtrsim 10^{10}$K) is discussed in this reference; the usefulness of adopting a pure n, p, e- composition has already been demonstrated for spherical hydrodynamical calculations by Schramm (1976). The adopted scheme is most suitable when the core remains in a regime in which the electrons are extremely degenerate (ED) and relativistic (ER) and the nucleons are nonrelativistic (NR). To model the effects of hard core nuclear interactions at high density ($\rho \gtrsim 2\times10^{14}$g cm^{-3}) Saenz & Shapiro multiplied the ED, NR contribution to the neutron pressure (and energy density) by a "correction" factor $\gtrsim 1$ whenever $\rho \gtrsim 2\times10^{14}$g cm^{-3}:

$$P_n^*(ED,NR) = P_n(ED,NR)\ [9.4\times10^{-6}(\rho-1\times10^{14})^{0.36}]$$

Accordingly, the nucleon contribution to the equation of state approaches that of a thermal ideal gas when the neutrons are hot and it approximates the (hard) Bethe-Johnson (1974) equation of state when they are cold and $\rho \gtrsim 2\times10^{14}$g cm^{-3}, (cf. Van Riper & Arnett 1978). The normalized neutrino chemical potential, η_ν, required to solve the equilibrium equation, is determined at each moment self-consistently from the neutrino emissivity and absorption coefficients for either ν-transparent or opaque conditions, whichever apply (see §VC).

The (electron) neutrino emission processes considered were non-thermal electron capture ($e^-+p\to n+\nu$) and the main thermal pair processes included pair annihilation ($e^++e^-\to\nu+\bar{\nu}$), plasmon decay ($\text{plasmon}\to\nu+\bar{\nu}$), and bremsstrahlung ($e+p\to e+p+\nu+\bar{\nu}$). [For $Y_e\rho \gtrsim 10^5$ and $T_9 \gtrsim 0.7$ thermal photoneutrino emission is not important, and for the composition considered here, capture dominates bremsstrahlung]. For anti-neutrino emission, positron capture ($e^++n\to p+\bar{\nu}_e$) was also considered in addition to the above processes. Rates for e^- and e^+ capture, bremsstrahlung and the remaining thermal pair processes were taken from Bludman & Van Riper (1977), Dicus et al. (1976), and Hansen (1968), respectively. [Neutral current corrections to the pair production rates computed by Hansen (1968) are small (Dicus 1972)]. Only electron neutrinos were considered.

The principle neutrino opacity sources treated were absorption ($\nu+n\to e+p$), conservative nucleon scattering ($\nu+n,p\to\nu+n,p$) and nonconservative electron scattering ($\nu+e\to\nu+e$). Corresponding reactions were considered for the anti-neutrinos. The required cross sections were taken from Schramm & Arnett (1975) and Tubbs & Schramm (1975).

At each timestep during the collapse the local central emissivities for each process i, Λ_i(erg s^{-1}cm^{-3}) are "corrected" for neutrino degeneracy ("stimulated absorption") and absorption to yield a <u>net</u> total emission rate from the stellar surface, $L_{\nu+\bar{\nu}}$ in erg s^{-1} (see §VC) and a corresponding <u>net</u> central emissivity $\Lambda_{\nu+\bar{\nu}}$[$=2.5\ L_{\nu+\bar{\nu}}/(4\pi abc/3)$, where the first factor results from ellipsoidal averaging]. In this way a reasonable entropy equation can be written for T_c:

$$\dot{T}_c = \{[\frac{U_c+P_c}{n} - (\frac{\partial U_c}{\partial n})_{T_c}]\ \dot{n} - \Lambda_{\nu+\bar{\nu}}\}/(\frac{\partial U_c}{\partial T_c})_n \qquad \text{(VB.1)}$$

where U_c is the central energy density (including rest mass).

C. Ellipsoidal Treatment Of Neutrino Transport

Neutrino transport in the collapsing core is handled at much the same level of approximation that radiative transport has been treated in recent studies of x-ray emission from accretion disks around black holes (see, e.g., Novikov & Thorne (1973) or Lightman, Rees & Shapiro (1978), for a discussion and references). Saenz & Shapiro (1979) construct, in effect, a "one-zone" model atmosphere about the core characterized by a single density ρ and temperature T_c. They evaluate all cross sections by dividing the emitted neutrinos and anti-neutrinos into two characteristic energy "bins": (1) $E_c \sim 2/3 u_e$ for all e^--capture neutrinos (Schramm & Arnett (1975)), where u_e is the e^- chemical potential, and (2) $E_{th} \sim 3kT_c$ for all thermal neutrinos and anti-neutrinos. For each bin total absorption and scattering depths (τ_a and τ_s, resp.)are computed for the core by multiplying the corresponding opacities by $R_{min}=\min\ [a,b,c]$, thereby enforcing the preference for neutrinos to escape via the shortest route. Defining an effective absorption depth for each bin, $\tau^* \sim \max[(\tau_a\tau_s)^{1/2},\tau_a]$, which accounts for the scattering-induced random-walk of neutrinos in the core prior to absorption whenever $\tau_s > \tau_a$ (see e.g., Felten & Rees (1972), Novikov & Thorne (1973) for the photon analogue), there are two distinct regimes: the transparent regime in which $\tau^* \lesssim 1$ and net luminosities for each mechanism i are related to emissitivies by

$$L^i_\nu \sim \Lambda^i_\nu \cdot (4\pi abc/3),\ \tau^* < 1, \qquad \text{(VC.1)}$$

and the opaque (diffusion) regime, where

$$L^i_\nu \sim 4\pi R^2_{max}\ \nabla_{\tau_m}\ U^i_\nu \sim 4\pi \cdot \max\ (a_i a_j)\ \frac{U^i_\nu}{\tau_m}\ . \qquad \text{(VC.2)}$$

In eqn. (VC.2), $\tau_m(E_\nu)=\tau_a+\tau_s$ is the total optical depth and U^i_ν is the central ν energy density, calculated from

$$U^i_\nu \sim \Lambda^i_\nu\ \frac{R_{min}}{c} \times \begin{cases} \tau_a^{-1} & ,\ \tau^* > 1 \\ \tau_s & ,\ \tau^* < 1,\ \tau_s > 1 \\ 1 & ,\ \tau^* < 1,\ \tau_s < 1 \end{cases} \qquad \text{(VC.3)}$$

The above equations are appropriate only for conservative scattering but nevertheless apply during core collapse since ν degeneracy ensures, typically, that, elastic nucleon scattering dominates inelastic e^--scattering (Lamb & Pethick (1976)). To account for ν degeneracy, blocking factors, $B^i_\nu \lesssim 1$, must be introduced to reduce the emissivitives appearing in eqns. (VC.1) and (VC.2): $\Lambda^i_\nu \rightarrow \Lambda^i_\nu \cdot B^i_\nu$. Following Arnett (1977), these factors are defined for each energy bin E_ν according to

$$B^i_\nu(E_\nu) \sim 1 - \frac{U^i_\nu(E_\nu)}{U_{\nu,max}} \sim 1 - \frac{U^i_\nu(E_\nu)}{(\pi E^4_\nu/h^3c^3)} \tag{VC.4}$$

The potential η_ν is then determined from the total energy density $U_\nu = \sum_i U^i_\nu$ according to the (ED) relation

$$\eta_\nu \sim \left(\frac{U_\nu h^3 c^3}{\pi}\right)^{1/4} / T_c \tag{VC.5}$$

D. Ellipsoidal Treatment Of Reflection Shocks

As the parametrized, adiabatic hydrodynamic calculations of Van Riper (1978) dramatically illustrate, a reflective or bounce shock may form at the boundary of the rebounding "inner core" as the infalling mantle ("outer core") undergoes deceleration and velocity reversal. This outward propagating shock removes kinetic energy from the inner core, which thereafter bounces homologously. The size of the oscillating inner core of a 1.4 $M_\odot$ configuration depends on many factors (initial conditions, the adopted equation of state, etc.), varying between $0.3 \lesssim M_s/M_\odot \lesssim 1.1$ in the examples considered by Van Riper (1978), with the larger value associated with a harder equation of state.

To mimic the effects of a reflection shock in their ellipsoidal models, Saenz & Shapiro (1979) adopt the following (naive) proceedure: at the moment of maximum core expansion velocity during the initial rebound, (1) the outward velocity of the ellipsoid is abruptly set equal to zero in all directions and (2) the mass and angular momentum are reduced so that $M \to M_s < 1.4 M_\odot$ and $J \to J_s = J$ (interior to M_s), implying $\Omega \to \Omega_s = \Omega$. The evolution is continued for the "inner core" using the current values of ρ, T_c and eccentricity; the lost kinetic energy is envisioned to be deposited in the overlying mantle.

E. Some Numerical Results

To judge the validity and limitations of the above scheme, spherically symmetric ellipsoidal collapse with J=0 can be compared with the hydrodynamical collapse calculations of Arnett (1977) for a $1.4 M_\odot$ core initially in hydrostatic equilibrium with $\rho_c = 4\times10^9$ g cm^{-3}. At maximum compression, the "canonical" ellipsoidal model gives for first (third) bounce $\rho = 0.8(9)\times10^{14}$ g cm^{-3}, $T_c = 4(6)\times10^{11}$K, $\dot{a}_{max} = 3(8)\times10^9$ cm s^{-1}, $P_n/P = 0.4(0.9)$, $Y_e = 0.3(0.3)$ (e^- per nucleon), $\eta_e = 4(7)$, $\eta_\nu \sim 2/3\eta_e$, $\eta_n = -1(1.7)$, $\tau^*(E_c) = 2(25)\times10^4$, $B_\nu(E_c) \sim 0.01(0.02)$, $L_{\nu,max} \sim 5(2)\times10^{52}$ erg s^{-1}, $L_{\bar\nu,max} \sim 7(.4)\times10^{49}$ erg s^{-1}, $\Delta E_{\nu+\bar\nu}/Mc^2 \sim 2(3)\times10^{-3}$, and $\Delta t \sim 0.1(0.1)$s.

Arnett finds, near the center at maximum compression, $\rho_c = 0.2\times10^{14}$, $T_c \sim 0.8\times10^{11}$K, $\dot{a}_{max} = 2\times10^9$ cm s^{-1}, $Y_e = 0.2$, $\eta_e \sim 9$, $\eta_\nu \sim 2/3\eta_e$, $\eta_n \sim 1.3$, $\tau_s(E_\nu \sim E_c) \sim 10^4$, $L_{\nu,max} \sim 8\times10^{52}$ erg s^{-1}, $\Delta E_\nu/Mc^2 \sim 2\times10^{-3}$, and $\Delta t \sim 1$s. His calculations do not continue beyond the first bounce.

On the first bounce, the ellipsoidal core apparently collapses to somewhat higher density and temperature (due to adiabatic compression) than the inhomogeneous core. This results from the greater fraction of mass participating in homologous motion in the homogeneous star, which gives higher infall velocities, greater overshoot and shorter collapse timescales. Nevertheless, the central core parameters and maximum ν-luminosities are certainly comparable to their inhomogenous counterparts, which is remarkable considering the crude nature of the ellipsoidal approximations. The ellipsoidal results for $\rho_i=1\times10^9$g cm^{-3} are in better agreement with the inhomogeneous core parameters, presumably due to the closer agreement of initial mean densities and collapse timescales in this case. More significant than any precise numerical identity of parameters near maximum compression, which may depend on initial conditions, etc., and has yet to be achieved by different hydrodynamicists, is the qualitative agreement found on the following physical points: (1) the collapse is very nearly adiabatic, due to the inhibition of neutrino cooling, (2) many of the emitted neutrinos are trapped due to the high optical depths, become degenerate and keep the lepton number (Y_e) high, (3) conservative neutron scattering dominates inelastic e^-scattering and absorption [but coherent scattering off heavy nuclei dominates all processes in Arnett (1977)], (4) e^--capture is the dominant emission mechanism, (5) the maximum luminosity remains well below the critical Eddington luminosity required to blow off the outer mantle, L_ν(Edd)$\sim10^{54}$erg s^{-1} (Schramm 1976). [Similar results have been reported by Wilson (1978), this volume].

Luminosities are illustrated in Figure (5) as a function of time for a typical, slowly rotating ellipsoidal core with $e_i=10^{-2}$ ($J=7.2\times10^{47}$erg-s, initial period = 30s). The values for ν and $\bar{\nu}$ are comparable to their spherical counterparts and decay slowly with time due to ν and $\bar{\nu}$ trapping. The quadrupole GR luminosity steadily increases on successive bounces due to (1) the steady growth of the eccentricity and (2) the increase (decrease) in density (collapse time) near maximum compression, as ν-dissipation steadily drives the core closer to its final (cold) equilibrium state. The development of an homologously contracting "inner core" near bounce of mass$\sim$1/2 M in the spherical calculation of Arnett (1977) motivates the choice of a post-shock ellipsoidal mass $M_s=0.7M_\odot$, which is also illustrated in the figure. Post-shock GR luminosities are smaller than no-shock values, since the dissipation of kinetic energy ($\Delta E_{kin}/Mc^2\sim6\times10^{-3}$) prevents the shocked core from reaching the high densities attained in the absence of a shock. Waveforms for this case are plotted in Figure (6).

The dissipative efficiencies of ν, $\bar{\nu}$, and GR emission through the first (third and fifth) bounce are plotted in Figure (7a,b) as a function of e_i and J. Neutrino emission dominates in all cases but decreases slowly with time. The GR plot for the first bounce has a familiar shape (cf. Fig.(1)), but the peak value is well below the idealized low entropy-values described in §IV. As the core continues to bounce homologously, the GR efficiency climbs so that, by the third bounce, $\Delta E_{GR}/Mc^2 \lesssim 10^{-4}$ for all $e_i \lesssim 0.5$; GR emission continues to increase as the core oscillates, cools and contracts. Apparently, the integrated GR efficiency after a few bounces is not very sensitive to J for $10^{-3}J_c \ll J \ll J_c \sim 4\times10^{49}$ erg-s since all low-J configurations eventually reach the same high densities ($\rho \gtrsim 5\times10^{14}$ g cm^{-3}) and become highly eccentric after a few oscillations. The presence of a reflection shock upon rebound does not alter the results dramatically, although significant acoustic and/or neutrino damping of successive bounces (Van Riper 1978) would certainly lower the GR efficiencies. However, the choice of $M_s \simeq M/2$, is quite crucial in this regard since the adopted ellipsoidal prescription for handling reflection shocks would give no core motion following the onset of the shock if $M_s \equiv M$.

VI. CONCLUSIONS

The most recent ellipsoidal model calculations suggest that significant GR dissipation may be expected from the homologous, oscillations of the "inner" regions of a $1.4M_\odot$ collapsing, rotating core for all $10^{-3}J_c \lesssim J \lesssim J_c \sim 10^{49}$ erg-s. Although neutrino emission dominates the dissipation, dissipation is slow due to ν-trapping, hence the oscillations are large amplitude and quasi-adiabatic (cf. Arnett 1977; Van Riper 1978). Core eccentricities and maximum densities grow on successive bounces, causing the GR efficiencies to increase with time for all low-J configurations. The ellipsoidal calculations are terminated after a few bounces, [they eventually become invalid due to the neglect of accretion of the outer layers], at which point they give efficiencies close to $\Delta E_{GR}/Mc^2 \sim 10^{-4}$ for $J \ll J_c$. However, if the total net ν-dissipation rate remains small and the oscillations and eccentricities are not severely damped, the GR efficiencies may continue to rise as the core cools. The ellipsoidal models already indicate efficiencies as high as $\sim 10^{-2}$ and wave amplitudes $|h| \sim 0.1(M/r)$ by the fifth bounce; cf. Figs.(6,7b). Consequently, special initial conditions - i.e., large $J \sim J_c$ or strong B fields - may not be required to generate high GR dissipative efficiencies during the later phases of core collapse. This statement only applies if the core oscillates and does not catastrophically collapse to a black hole on the first infall. Collapse to a black hole may not be as efficient a source of GR as collapse to a neutron star. A fully relativistic, nonspherical hydrodynamical calculation with neutrino transport, which follows the core for many bounces, is required to verify this conclusion. If true, then core collapse may invariably result in GR dissipative efficiencies comparable to those estimated from neutron star binary systems (cf. Clark & Eardley 1977) and may alter conclusions regarding the relative significance for GR detection of the two astrophysical events (cf. Clark, van den Heuvel, & Sutantyo 1978, in press).

This work was supported by NSF Grant AST 75-21153 at Cornell. Acknowledgement is made to the National Center for Atmospheric Research which is sponsored by the National Science Foundation, for computer time uses in this research.

REFERENCES

Arnett, W.D. (1977). Neutrino Trapping During Gravitational Collapse of Stars. Ap.J., 218, 815.

Bethe, H.A. & Johnson, M.B. (1974). Dense Baryon Matter Calculation with Realistic Potentials. Nucl.Phys., A230, 1.

Bludman, S. & Van Riper, K. (1977). Equation of State of an Ideal Fermi Gas. Ap.J., 212, 859.

Burke, W. (1969). Unpublished Ph.D. Thesis, California Institute of Technology.

Chandrasekhar, S. (1969). Ellipsoidal Figures of Equilibrium. Yale University Press, New Haven.

Chandrasekhar, S. (1970a). Solutions of Two Problems in the Theory of Gravitational Radiation. Phys.Rev.Letters, 24, 611.

Chandrasekhar, S. (1970b). The Effect of Gravitational Radiation on the Secular Stability of the Maclaurin Spheroid. Ap.J., 161, 561.

Chandrasekhar, S. (1970c). The Evolution of the Jacobi Ellipsoid by Gravitational Radiation. Ibid, p.571.

Chia, T.T., Chau, W.Y. & Henriksen, R.N. (1977). Gravitational Radiation from a Rotating Collapsing Gaseous Ellipsoid. Ap.J., 214, 576.

Clark, J.P.A. & Eardley, D.M. (1977). Evolution of Close Neutron Star Binaries. Ap.J., 215, 311.

Clark, J.P.A., van den Heuvel, E.P.J. & Sutantyo, W. (1978). Formation of Neutron Star Binaries and Their Importance for Gravitational Radiation. Astron.Ap., in press.

Colgate, S.A. & White, R.H. (1966). The Hydrodynamic Behavior of Supernovae Explosions. Ap.J., 143, 626.

Dicus, D. (1972). Stellar Energy-Loss Rates in a Convergent Theory of Weak and Electromagnetic Interactions. Phys.Rev.D., 66, 941.

Dicus, D., Kolb, F., Schramm, D.N. & Tubbs, D. (1976). Neutrino Pair Bremsstrahlung Including Neutral Current Effects. Ap.J., 210, 481.

Epstein, R. (1976). Unpublished Ph.D. Thesis, Stanford University.

GRAVITATIONAL AND NEUTRINO RADIATION

Epstein, R. & Wagoner, R.V. (1975). Post-Newtonian Generation of Gravitational Waves. Ap.J., 197, 717.

Felten, J.E. & Rees, M.J. (1972). Continuum Radiative Transfer in a Hot Plasma, with Application to Scorpius X-1. Astron.Ap., 17, 226.

Hansen, C.J. (1968). Some Weak Interaction Processes in Highly Evolved Stars. Ap.Space Sci., 1, 499.

Kazanas, D. (1978). On Neutrino Viscosity in Collapsing Stellar Cores. Ap.J.Letters, 222, L109.

Kazanas, D. & Schramm, D.N. (1976). Competition of Neutrino and Gravitational Radiation in Neutron Star Formation. Nature, 262, 671.

Kazanas, D. & Schramm, D.N. (1977). Neutrino Damping of Nonradial Pulsations in Gravitational Collapse. Ap.J., 214, 819.

Lamb, D.Q. & Pethick, C. (1976). Effects of Neutrino Degeneracy in Supernova Models. Ap.J., 209, L77.

Lightman, A.P., Rees, M.J., & Shapiro, S.L. (1978). Accretion Onto Compact Objects. In Varenna Lectures, 1976, in press.

Lindblom, L. & Detweiler, S.L. (1977). On the Secular Instabilities of the Maclaurin Spheroids. Ap.J., 211, 565.

Miller, B.D. (1974). The Effect of Gravitational Radiation-Reaction on the Evolution of the Riemann S-Type Ellipsoids. Ap.J., 187, 609.

Novikov, I.D. (1975). Gravitational Radiation from a Star Collapsing into a Disk. Soviet Astr.-AJ., 19, 398.

Novikov, I.D. & Thorne, K.S. (1973). Astrophysics of Black Holes. In Black Holes, eds. C. DeWitt and B. DeWitt, pp.343-450. Gordon and Breach, New York.

Ostriker, J.P. & Hartwick, F.D.A. (1968). Rapidly Rotating Stars. IV. Magnetic White Dwarfs. Ap.J., 153, 797.

Saenz, R.A. & Shapiro, S.L. (1978). Gravitational Radiation from Stellar Collapse: Ellipsoidal Models. Ap.J., 221, 286.

Saenz, R.A. & Shapiro, S.L. (1979). Gravitational and Neutrino Radiation from Stellar Core Collapse: Improved Ellipsoidal Model Calculations. In preparation.

Schramm, D.N. (1976). Talk delivered at the Deep Underwater Muon Neutrino Detector Workshop (DUMAND), Honolulu, Hawaii, 1976.

Schramm, D.N. & Arnett, W.D. (1975). The Weak Interaction and Gravitational Collapse. Ap.J., 198, 629.

Shapiro, S.L. (1977). Gravitational Radiation from Stellar Collapse: The Initial Burst. Ap.J., 214, 566.

Smarr, L. (1977). Spacetimes Generated by Computers: Black Holes with Gravitational Radiation. Ann. N.Y. Acad. Sci., in press.

Spitzer, L., Jr. (1968). Diffuse Matter In Space, Wiley, New York.

Thorne, K.S. (1969). Nonradial Pulsation of General-Relativistic Stellar Models. III. Analytic and Numerical Results for Neutron Stars. Ap.J., 158, 1.

Thorne, K.S. (1972). In Magic Without Magic: John Archibald Wheeler, ed. J. Klauder, p.231. Freeman, San Francisco.

Thuan, T.X. & Ostriker, J.P. (1974). Gravitational Radiation from Stellar Collapse. Ap.J.Letters, 191, L105.

Tubbs, D. & Schramm, D.N. (1975). Neutrino Opacities at High Temperature and Densities. Ap.J., 201, 467.

Van Riper, K. (1978). The Hydrodynamics of Stellar Collapse. Ap.J., 221, 304.

Van Riper, K. & Arnett, W.D. (1978). Stellar Collapse and Explosion: Hydrodynamics of the Core. Preprint.

Van Riper, K. & Bludman, S.A. (1977). Composition and Equation of State of Thermally Dissociated Matter. Ap.J., 213, 239.

Wagoner, R.V. (1977). Gravitational Radiation from Slowly-Rotating Collapse: Post-Newtonian Results. In Proceedings of the Academia Nazionale dei Lincei International Symposium on Experimental Gravitation. Pavia, Italy, September 17-20, 1976.

Wentzel, D. (1961). On the Shape of Magnetic Stars. Ap.J., 133, 17(

Wilson, J.R. (1974). Coherent Neutrino Scattering and Stellar Collapse. Phys.Rev.Letters, 32, 849.

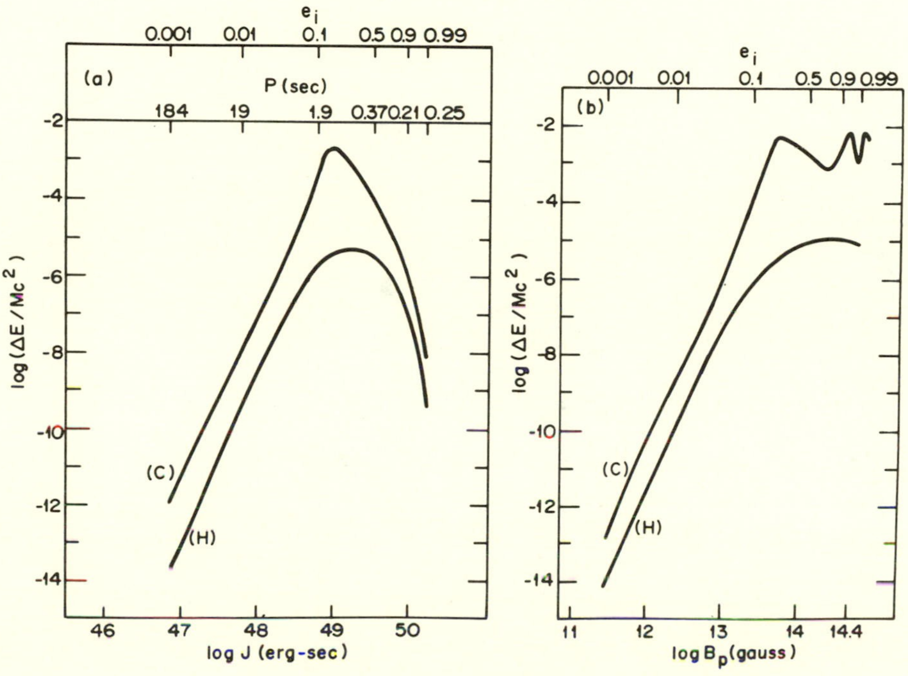

Fig.1 - The efficiency of energy loss via gravitational radiation (a) as a function of angular momentum J for rotating configurations and (b) as a function of poloidal magnetic field B_p for nonrotating configurations. These spheroidal configurations represent stellar cores, initially oblate and at rest on the verge of collapse, all with mass $M=1.4M_\odot$ and density $\rho_i=10^{10}$g cm^{-3}. The collapse is characterized by either a hot (H) or cold (C) equation of state (see text); e_i is the initial eccentricity and P the corresponding rotation period. From Saenz & Shapiro (1978).

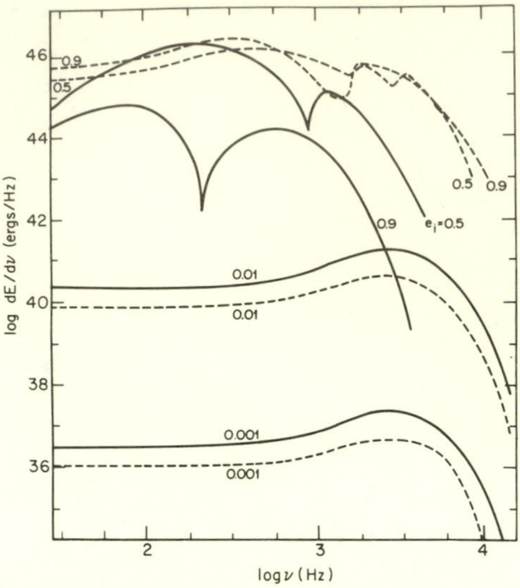

Fig.2 - The frequency spectra of the gravitational radiation for spheroidal (oblate)"hot" collapse cases with J=0 (dashed line) and J≠0 (solid line) are shown for several different values of initial eccentricity, e_i. From Saenz & Shapiro (1978).

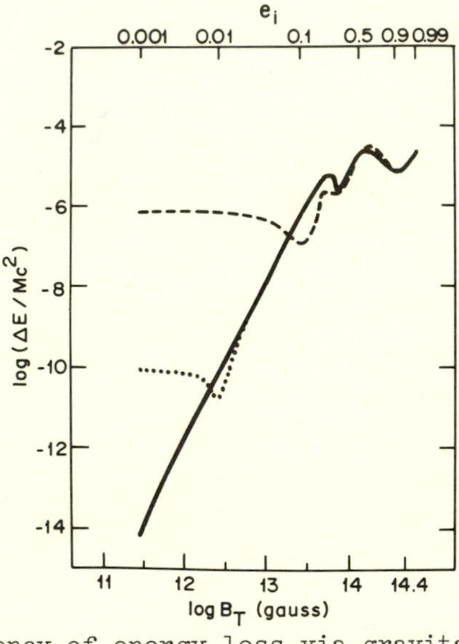

Fig.3 - The efficiency of energy loss via gravitational radiation as a function of the initial eccentricity e_i induced by a toroidal magnetic field, B_T. Results are for configurations with $J_M(e_i)$=0 (solid line), J=J_M(0.01)(dotted line), and J=J_M(0.1)(dashed line). These spheroidal configurations represent stellar cores, initially prolate and at rest on the verge of collapse, all with mass $M=1.4M_\odot$ and density $\rho_i=10^{10}$g cm^{-3}. The collapse is governed by a "hot" equation of state (see text). From Saenz & Shapiro (1978).

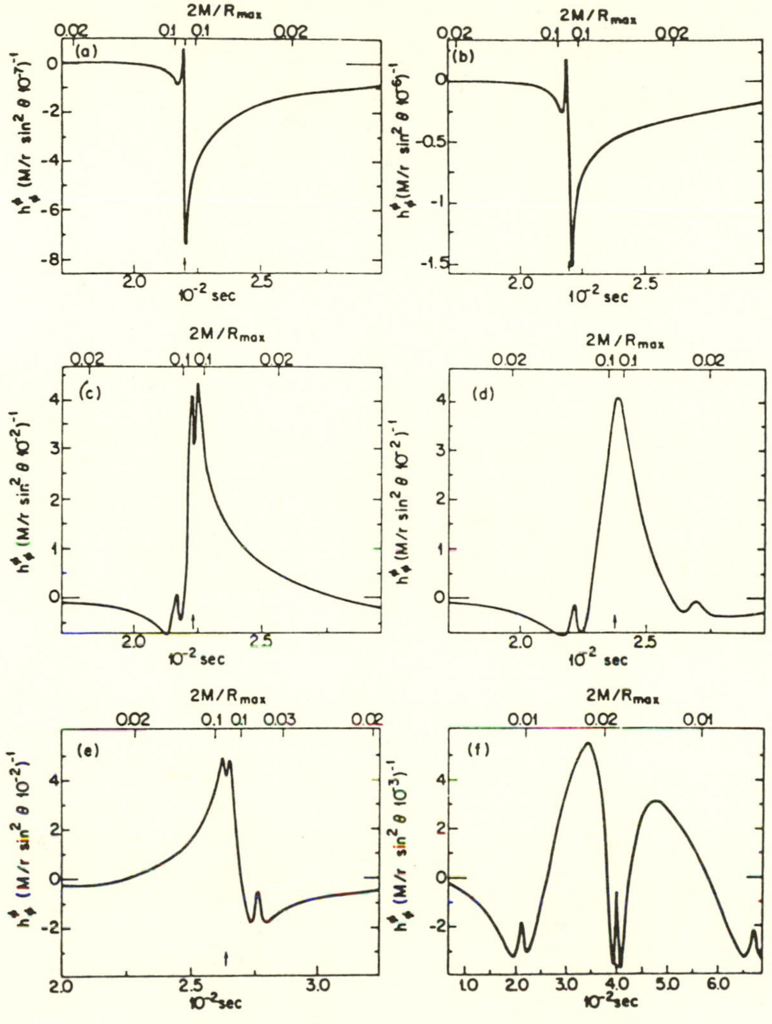

Fig.4 - Catalog of waveforms for "hot" spheroidal (oblate) collapse with initial eccentricity e_i=0.001 for (a) and (b); e_i=0.5 for (c) and (d); and e_i=0.9 for (e) and (f). Wave amplitudes are plotted as a function of (retarded) time from the onset of collapse; arrows denote the moment of maximum compression. Plots in the left-hand column are characterized by $J_M(e_i)$=0 and the right-hand column by $J=J_M(e_i)$, the value for an equilibrium Maclaurin spheroid with the same e_i, density and mass. All waveforms are given with respect to orthonormal basis vectors. From Saenz & Shapiro (1978).

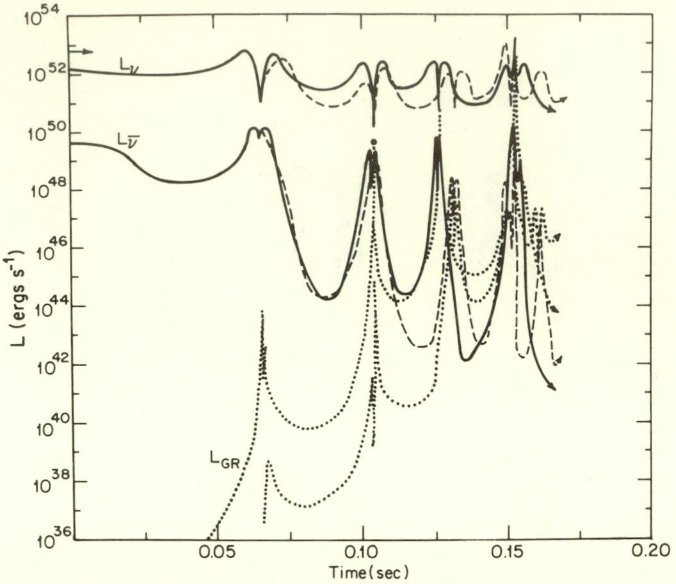

Fig.5 - Luminosity versus time for a collapsing core of mass $M=1.4M_\odot$ with initial density $\rho_i=4\times10^9$g cm^{-3}, eccentricity $e_i=10^{-2}$, and angular momentum $J=J_M(e_i)$. The effects of a reflection shock are indicated by dashed lines for neutrino dissipation and the lower dotted line for GR.

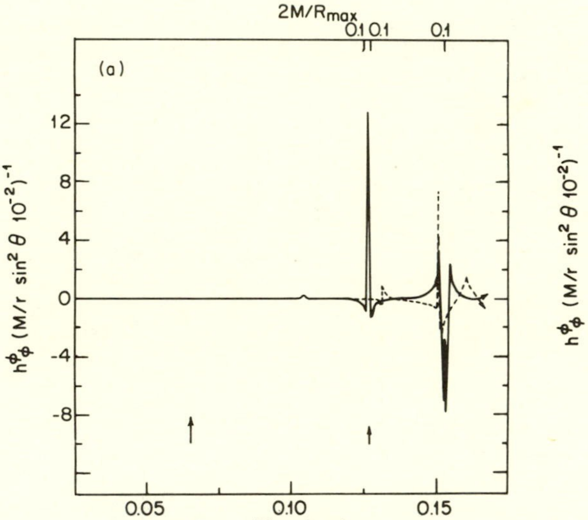

Fig.6 - Waveforms for the case illustrated in Fig.5. The effect of a reflection shock is indicated by the dashed line. Arrows locate core bounces. (Shocked waveforms have been multiplied by two).

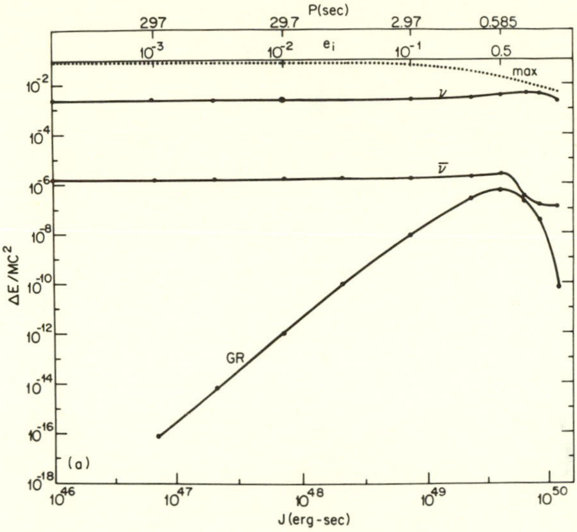

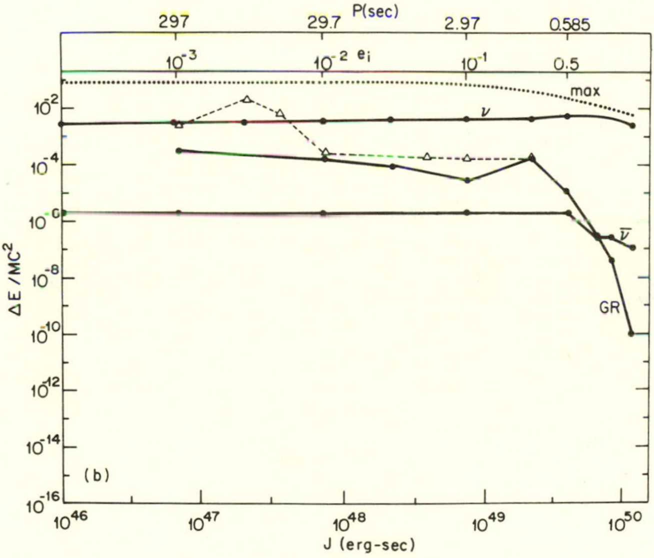

Fig.7 - Dissipative efficiencies (a) after one bounce and (b) after three and (dashed line) five bounces of a rotating, collapsing core plotted as a function of $J=J_M(e_i)$. The maximum allowable dissipation prior to settling a final equilibrium state is denoted by the dotted line ("max").

GRAVITATIONAL RADIATION FROM SLOWLY-ROTATING "SUPERNOVAE": PRELIMINARY RESULTS

Michael S. Turner and Robert V. Wagoner

Institute of Theoretical Physics, Department of Physics
Stanford University, Stanford, California 94305

ABSTRACT

We consider the gravitational radiation produced by rotationally-induced perturbations of spherically-symmetric stellar collapse, within a post-Newtonian multipole formalism developed by Wagoner. If the angular velocity Ω of the collapsing star is non-uniform, octupole (post-Newtonian) radiation is generated at order Ω. If the angular velocity is uniform (or non-uniform), quadrupole (Newtonian) radiation is generated at order Ω^2. We apply this perturbation method at order Ω to three "input" non-rotating models: (1) Analytically to an isentropic, non-relativistic perfect gas ($p \propto \rho^{5/3}$), (2) Numerically to an unpublished supernova calculation of Wilson, and (3) Numerically to a detailed supernova study of Van Riper. It appears that most of the radiation is released during a very short time interval at the bounce. Detailed knowledge of the behavior of the matter at that time and of the distribution of angular momentum are necessary to make reliable estimates of the gravitational-wave flux.

1 INTRODUCTION

Among all potential sources of gravitational radiation which have been <u>observed</u>, supernovae have been thought to be the most powerful emitters, especially in the kilohertz frequency range probed by laboratory detectors. However, attempts at estimating the energy radiated in gravitational waves have been hampered by a lack of knowledge of the behavior of the matter in the collapsing core of the presupernova star. The maximum energy which could be emitted is $\sim 10^{53}$ ergs, the binding energy of a typical neutron star (Press & Thorne, 1972; Rees, Ruffini & Wheeler, 1974; Misner, 1974; Thorne, 1978). However, various factors, the most important of which are probably slow rotation and high entropy, are likely to reduce the energy generated by a typical supernova well below this upper limit, as we shall see. On the other hand, the observed supernova rate has recently been revised upward by Tammann (1977), who quotes an average time between supernovae in giant spiral galaxies of $\sim 20 \pm 10$ years. For a recent summary of the properties of supernovae, see Schramm (1977).

As detectors of gravitational radiation increase in number and sensitivity (DeWitt-Morette, 1974; De Sabbata & Weber, 1977; Bertotti, 1977), the need for reliable estimates of the expected flux becomes more acute. When detection is finally achieved, the analysis of the gravitational-wave signals will provide a unique probe into the heart of such spectacular astrophysical events. Because of their coherence and negligible absorption by the supernova, gravitons can in principle provide more information than even neutrinos.

Various idealized models of stellar collapse leading to the production of gravitational radiation have been investigated. Thuan & Ostriker (1974) considered the collapse of a uniform density, uniformly-rotating, zero-pressure spheroid in the Newtonian limit, while Epstein (1976) added post-Newtonian corrections. Novikov (1975) found that the inclusion of pressure, giving rise to a bounce, could significantly affect the generation of waves. Shapiro (1977) included pressure in a more complete way, using various equations of state. This calculation was extended by Saenz & Shapiro (1978) to more general initial configurations (oblate, prolate, and nonaxisymmetric). Chia, Chau & Henriksen (1977) have also included other forms of energy loss.

In contrast, we shall choose an approach which can accommodate "realistic" models of non-rotating, spherically-symmetric supernovae (supplied to us by the supernova model-building industry). We will show how rotation can be added to such models, and the resulting gravitational radiation computed. Of course, we must pay a price for such "realism" - the assumption that the body remains slowly rotating. For a star with mass M, radius a, and equatorial angular velocity Ω, we thus require that $\Omega_*^2 \equiv \Omega^2/(GM/a^3) \sim$ (centrifugal force/gravitational force) $\ll 1$.

What is the evidence that slow rotation is a valid assumption for most supernovae? The most direct evidence comes from observations of pulsars. For the most rapidly-rotating pulsar (the Crab), $\Omega_*^2 \sim 10^{-3}$ (assuming it is a typical neutron star), but for most pulsars, $10^{-7} \lesssim \Omega_*^2 \lesssim 10^{-5}$. Ruderman (1972) has presented arguments that the Crab never rotated significantly faster than at present, but these arguments cannot be applied to the slower pulsars. Other evidence comes from observations of white dwarfs, which may be relevant because the stellar core whose collapse is believed to initiate a supernova closely resembles a white dwarf (Arnett, 1973). No white dwarf has ever been observed to have a rotation rate which would lead to a value of $\Omega_*^2 \gtrsim 10^{-3}$ under angular-momentum-conserving collapse to a neutron-star radius (Greenstein & Peterson, 1973). This indicates that such cores could have transferred (magnetically) most of their angular momentum to the outer layers of the star, probably during the red-giant phase of evolution (Hardorp, 1974; Ruderman, 1972). In summary, we have found no evidence that contradicts our assumption that most supernovae are slowly rotating. (See, however, Ostriker, this volume).

In this calculation, we employ the post-Newtonian approximation scheme of Epstein and Wagoner (1975), as modified by Wagoner (1977). This involves an analysis of the radiation emitted at each order in a relativistic expansion parameter $\varepsilon \sim v/c \sim (GM/ac^2)^{1/2}$. At each order in Ω_* which we shall consider (Ω_* and Ω_*^2), only the lowest-order effects in ε will be included. We do this because the errors made in such an approximation ($\varepsilon^2 \lesssim 0.1$) are much smaller than the present uncertainties in the non-rotating "input" supernova models. An important result emerges: To lowest order in Ω_*, the radiation is not given by the usual "Newtonian quadrupole formula", but is actually of post-Newtonian order if the star is rotating non-uniformly. This radiation is generated not by a non-spherical distribution of matter (occurring at order Ω^2), but by internal motions ($v^r v^\phi$) occurring at order Ω.

2 FORMALISM

In what follows we shall often employ the notation, conventions (G = c = 1, etc.), and results of Wagoner (1977), which should be read as background to this investigation. Here, somewhat different assumptions will be made:

a) The matter can be described as a perfect fluid insofar as it obeys the Newtonian equations of motion

$$\partial\rho/\partial t + \nabla\cdot(\rho \mathbf{v}) = 0 \qquad (1)$$

$$\partial \mathbf{v}/\partial t + (\mathbf{v}\cdot\nabla)\mathbf{v} + \nabla p/\rho - \nabla U = 0. \qquad (2)$$

Here ρ is the mass density, p the pressure, v the velocity of the fluid, and U the Newtonian potential. Although we shall neglect effects such as the production of gravitational radiation by any emitted neutrinos (see, however, Epstein, 1978, or Turner, 1978), we do allow for their effects on the equation of state. We make no assumptions about the form of the equation of state in this study.

b) The body possesses axial symmetry (no dependence on the angular coordinate ϕ) and reflection symmetry about the equatorial plane. This assumption is likely to be realized to a sufficient degree for the following reason. In the pre-collapse equilibrium state, the major deviation from spherical symmetry will be the axisymmetric one caused by centrifugal force. Since all perturbations grow with roughly the same time-scale (the collapse time), any other perturbation which develops will remain smaller than the rotationally-induced distortion, at least until after the bounce. In addition, damping of non-axisymmetric perturbations by neutrino emission (Kazanas & Schramm, 1977) will be more efficient than the

damping of the axisymmetric rotational perburbation through the angular momentum carried off by the neutrinos.
c) At all times during the evolution of the body, $\Omega_* \ll 1$, which implies that the collapsing star remains nearly spherically-symmetric.

Although these assumptions preclude any transfer of angular momentum during collapse, this is not a serious omission. The reason is that most of the gravitational radiation is produced during a short time interval at the bounce, so what really matters is the assumed angular momentum distribution at that time.

In order to employ most easily our (assumed) knowledge of the behavior of the non-rotating star, we transform to coordinates slightly different than the usual orthonormal spherical set t, r, θ, ϕ. Generalizing a method used by Hartle (1967) to analyze stationary, slowly-rotating stars, we label a surface of constant density in the slowly-rotating body by R, the radius of that surface in the non-rotating star which has the same density. The relation between the two coordinate systems is shown in Figure 1,

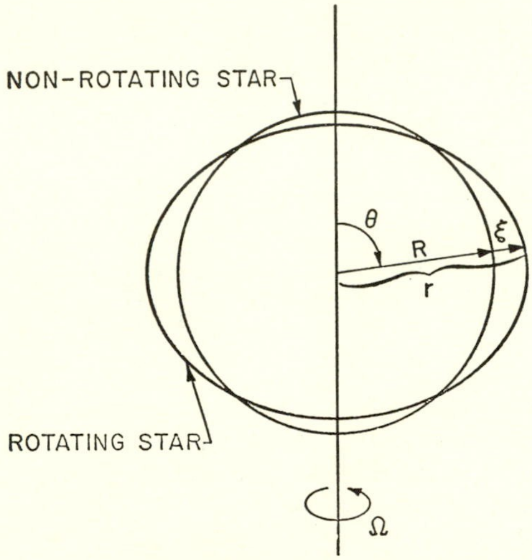

Fig. 1. A surface of the same constant density is shown for a rotating and non-rotating configuration having the same central density. The surface in the rotating body is labeled by R, the radial coordinate of the surface of equal density in the non-rotating body. The radial coordinate of the surface in the rotating body is $r = R + \xi(R,\theta,T)$.

and is expressed by the equations

$$\rho(r,\theta,t) = \rho_{(0)}(R,T) \tag{3}$$

$$r = R + \xi(R,\theta,T) \tag{4}$$

$$t = T + \tau(T) \ , \tag{5}$$

where $\rho_{(0)}(R,T)$ refers to the non-rotating body, and $\xi\ (\sim\Omega_*^2 R)$ measures the distortion of the rotating body. In order to compare rotating and non-rotating bodies of the same central density we demand that $\xi\,(0,\theta,T) = 0$. Since the time dependence of the central density is different in the two cases, it is then necessary to adjust the time coordinate by an amount $\tau \sim \Omega_*^2 T$. The coordinates θ and ϕ remain unchanged. The orthonormal components of the gradient then become

$$\nabla_r = (1 - \frac{\partial\xi}{\partial R})\frac{\partial}{\partial R} + \mathcal{O}(\Omega_*^4) \tag{6}$$

$$\nabla_\theta = \frac{1}{R}(\frac{\partial}{\partial\theta} - \frac{\partial\xi}{\partial\theta}\frac{\partial}{\partial R}) + \mathcal{O}(\Omega_*^4) \tag{7}$$

$$\nabla_t = (1 - \frac{d\tau}{dT})\frac{\partial}{\partial T} - \frac{\partial\xi}{\partial T}\frac{\partial}{\partial R} + \mathcal{O}(\Omega_*^4)\ . \tag{8}$$

Denoting a quantity of order Ω_*^n by $q_{(n)}$, the other variables of interest are

$$v^r(r,\theta,t) = v^r_{(0)}(R,T) + v^r_{(2)}(R,\theta,T) \tag{9}$$

$$v^\theta(r,\theta,t) = v^\theta_{(2)}(R,\theta,T) \tag{10}$$

$$v^\phi(r,\theta,t) = v^\phi_{(1)}(R,\theta,T) \tag{11}$$

$$U(r,\theta,t) = U_{(0)}(R,T) + U_{(2)}(R,\theta,T) \tag{12}$$

$$p(r,\theta,t) = p_{(0)}(R,T) + p_{(2)}(R,\theta,T) \tag{13}$$

Employing the angular decompositions

$$\xi = \sum_{\ell\ \mathrm{even}} \xi_\ell(R,T)\,P_\ell(\cos\theta) \tag{14}$$

$$U_{(2)} = \sum_{\ell\ \mathrm{even}} U_\ell(R,T)\,P_\ell(\cos\theta) \tag{15}$$

$$p_{(2)} = \sum_{\ell\ \mathrm{even}} p_\ell(R,T)\,P_\ell(\cos\theta) \tag{16}$$

$$v^r_{(2)} = \sum_{\ell\ \mathrm{even}} v^r_\ell(R,T)\,P_\ell(\cos\theta) \tag{17}$$

$$v^\theta_{(2)} = \sum_{\ell\ \mathrm{odd}} v^\theta_\ell(R,T)\,dP_\ell/d\theta \tag{18}$$

$$v^\phi_{(1)} = \sum_{\ell\ \mathrm{odd}} -v^\phi_\ell(R,T)\,dP_\ell/d\theta \tag{19}$$

$$[v^\phi_{(1)}]^2 = \sum_{\ell\ \mathrm{even}} S_\ell(R,T)\,P_\ell(\theta) = \sum_{\ell\ \mathrm{even}} \bar{S}_\ell(R,T)\tan\theta\,dP_\ell/d\theta\ , \tag{20}$$

the equations of motion (1) and (2) become, at $\mathcal{O}(\Omega^0)$ (spherical collapse):

$$\frac{\partial \rho_{(0)}}{\partial T} + \frac{1}{R^2}\frac{\partial}{\partial R}(R^2 \rho_{(0)} v^r_{(0)}) = 0 , \tag{21}$$

$$\frac{\partial v^r_{(0)}}{\partial T} + v^r_{(0)} \frac{\partial v^r_{(0)}}{\partial R} + \frac{1}{\rho_{(0)}}\frac{\partial p_{(0)}}{\partial R} - \frac{\partial U_{(0)}}{\partial R} = 0; \tag{22}$$

at $\mathcal{O}(\Omega)$, for $\ell \geqslant 1$:

$$\frac{\partial v^\phi_\ell}{\partial T} + v^r_{(0)} \frac{\partial v^\phi_\ell}{\partial R} + \frac{v^r_{(0)} v^\phi_\ell}{R} = 0; \tag{23}$$

and at $\mathcal{O}(\Omega^2)$, for $\ell \geqslant 1$:

$$\frac{\partial \rho_{(0)}}{\partial R}\frac{\partial \xi_\ell}{\partial T} - \left(\frac{2\xi_\ell}{R} + \frac{\partial \xi_\ell}{\partial R}\right)\frac{\partial \rho_{(0)}}{\partial T}$$
$$-\frac{1}{R^2}\frac{\partial}{\partial R}[\rho_{(0)} R^2 (\frac{2v^r_{(0)}}{R}\xi_\ell + v^r_\ell)] + \ell(\ell+1)\frac{\rho_{(0)}}{R} v^\theta_\ell = 0, \tag{24}$$

$$\frac{\partial v^r_\ell}{\partial T} + v^r_{(0)} \frac{\partial v^r_\ell}{\partial R} + \frac{\partial v^r_{(0)}}{\partial R} v^r_\ell + \frac{\partial v^r_{(0)}}{\partial T}\frac{\partial \xi_\ell}{\partial R} - \frac{\partial v^r_{(0)}}{\partial R}\frac{\partial \xi_\ell}{\partial T}$$
$$+ \frac{1}{\rho_{(0)}}\frac{\partial p_\ell}{\partial R} - \frac{\partial U_\ell}{\partial R} - \frac{S_\ell}{R} = 0 , \tag{25}$$

$$\frac{\partial v^\theta_\ell}{\partial T} + v^r_{(0)} \frac{\partial v^\theta_\ell}{\partial R} + \frac{v^r_{(0)}}{R} v^\theta_\ell + (\frac{\partial v^r_{(0)}}{\partial T} + v^r_{(0)} \frac{\partial v^r_{(0)}}{\partial R})\frac{\xi_\ell}{R}$$
$$+ \frac{p_\ell}{R\rho_{(0)}} - \frac{U_\ell}{R} - \frac{\overline{S}_\ell}{R} = 0. \tag{26}$$

The potentials $U_{(0)}$ and $U_{(2)}$ are defined by the Newtonian field equation $\nabla^2(U_{(0)} + U_{(2)}) = -4\pi\rho_{(0)}$, with solutions

$$\frac{\partial U_{(0)}}{\partial R} = -\frac{4\pi}{R^2}\int_0^R \rho_{(0)} R_0^{\,2}\, dR_0 , \tag{27}$$

$$U_\ell = \frac{\partial U_{(0)}}{\partial R}\xi_\ell - \frac{4\pi}{2\ell+1}\Big[\frac{1}{R^{\ell+1}}\int_0^R \xi_\ell \frac{\partial \rho_{(0)}}{\partial R_0} R_0^{2+\ell} dR_0$$
$$+R^\ell \int_R^a \xi_\ell \frac{\partial \rho_{(0)}}{\partial R_0} R_0^{1-\ell}\, dR_0\Big]. \tag{28}$$

The equations of motion can be greatly simplified by the introduction of the comoving coordinates $m = 4\pi \int_0^R \rho_{(0)}(R_0,T)R_0^2 dR_0$, $T_c = T$. Note that these coordinates are only comoving with respect to the non-rotating star. The comoving derivatives are then given by

$$\frac{\partial}{\partial R} = 4\pi\ \rho_{(0)}\ R^2\ \frac{\partial}{\partial m} \tag{29}$$

$$\frac{\partial}{\partial T} = \frac{\partial}{\partial T_c} - 4\pi\ \rho_{(0)}\ v^r_{(0)}\ R^2\ \frac{\partial}{\partial m}\ . \tag{30}$$

All quantities, such as R, $\rho_{(0)}$, ξ_ℓ, ..., are now considered functions of m and T_c. From equations (29), (21), and (22), we obtain the $\mathcal{O}(\Omega^0)$ relations

$$\rho_{(0)} = (4\pi R^2 R')^{-1} \tag{31}$$

$$v^r_{(0)} = \dot{R} \tag{32}$$

$$\ddot{R} + 4\pi R^2\ p'_{(0)} + mR^{-2} = 0, \tag{33}$$

where we have let $' = \partial/\partial m$ and $\cdot = \partial/\partial T_c$. At $\mathcal{O}(\Omega)$, equation (23) then gives

$$v^\phi_\ell = h_\ell(m)/R, \tag{34}$$

representing conservation of angular momentum per unit mass. At $\mathcal{O}(\Omega^2)$, equations (24), (25), and (26) become

$$[2R + (R/R')^2 R'']\,\dot{\xi}_\ell - (R/R')^2\ \dot{R}'\,\xi'_\ell - (R/R')^2\ R'' v^r_\ell - 2\dot{R}\xi_\ell$$

$$+ (R^2/R')v^{r\prime}_\ell - \ell(\ell+1)\ R\ v^\theta_\ell = 0\ , \tag{35}$$

$$\partial(R'v^r_\ell)/\partial T_c + \ddot{R}\,\xi'_\ell + 4\pi R^2 R' p'_\ell - \dot{R}'\,\dot{\xi}_\ell - U'_\ell - S_\ell R^{-1} R' = 0\ , \tag{36}$$

$$\partial(R v^\theta_\ell)/\partial T_c + \ddot{R}\,\xi_\ell + 4\pi R^2 R' p_\ell - U_\ell - \overline{S}_\ell = 0\ , \tag{37}$$

while the potential given by equation (28) assumes the form

$$U_\ell = -\frac{m}{R^2}\ \xi_\ell - \frac{1}{2\ell+1}\Big[\ \frac{1}{R^{\ell+1}}\int_0^m \frac{\rho'_{(0)}}{R_0{}'\rho_{(0)}}\ \xi_\ell\ R_0^\ell\ dm$$

$$+ R^\ell \int_m^M \frac{\rho'_{(0)}}{R_0{}'\rho_{(0)}}\ \xi_\ell\ R_0^{-(\ell+1)}\ dm\Big]\ . \tag{38}$$

Once the pressure perturbation $p_\ell(R,T)$ has been determined from the equation of energy conservation and the equation of state provided, equations (35)-(38) can be solved to find ξ_ℓ, v^r_ℓ, v^θ_ℓ, and U_ℓ. Note that if the equation of state is of the simple form $p = p(\rho)$, $p_\ell = 0$.

The most convenient way to describe the emitted gravitational radiation is by a multipole expansion of the transverse-traceless (TT) part of the metric perturbation $h_{ij} = g_{ij} - \delta_{ij}$ $(i,j\ldots=1\text{-}3)$

(Wagoner, 1977). This has the form

$$\underset{\sim}{h}_{TT} = r^{-1} \sum_{L\geq 2} \sum_{M=-L}^{L} [A^{E2}_{LM}(t-r)\, \underset{\sim}{T}^{E2}_{LM}(\theta,\phi) + A^{B2}_{LM}(t-r)\, \underset{\sim}{T}^{B2}_{LM}(\theta,\phi)] \ , \tag{39}$$

at distances r from the source much greater than the characteristic wavelength of the radiation. We will adopt the "electric" and "magnetic" tensor spherical harmonics $\underset{\sim}{T}^{E2}_{LM}$, $\underset{\sim}{T}^{B2}_{LM}$ advocated by Thorne (1978b), which have opposite parity to those used by Wagoner (1977). All the information carried by a wave is contained in the amplitudes A^{E2}_{LM}, A^{B2}_{LM}. For instance, the total energy radiated is given by

$$E = (32\pi)^{-1} \sum_{L,M} \int (|\partial A^{E2}_{LM}/\partial t|^2 + |\partial A^{B2}_{LM}/\partial t|^2) dt. \tag{40}$$

It can also be shown that a post-Newtonian expansion (in powers of ε) generates such a multipole expansion (Wagoner, 1977). In fact, to lowest order in ε, only one amplitude is present at order Ω and at order Ω^2. They are

$$A^{B2}_{30} = -\frac{16}{7}\left(\frac{\pi}{105}\right)^{1/2} \frac{d^3}{dt^3} \int_0^M h_3(m)\, R^2 dm \qquad (\propto \Omega) \ , \tag{41}$$

$$A^{E2}_{20} = \frac{8}{5}\left(\frac{\pi}{15}\right)^{1/2} \frac{d^2}{dt^2} \int_0^M \left(\frac{4\,\xi_2}{R} + \frac{\xi_2'}{R'}\right) R^2 dm \qquad (\propto \Omega^2) \ . \tag{42}$$

Note that since $A^{E2}_{20} \sim M\varepsilon^2\Omega_*^2$ (Newtonian order), while $A^{B2}_{30} \sim M\varepsilon^4\Omega_*$ (post-Newtonian order), the post-Newtonian amplitude will actually dominate for sufficiently slow rotation ($\Omega_* \lesssim \varepsilon^2$). If the collapse becomes sufficiently relativistic, this amplitude can still represent a significant amount of radiation. However, for purely uniform rotation, only $h_1(m)$ is non-zero, so $A^{B2}_{30} = 0$.

In order to compute the angular distribution of the radiation, one must employ the relevant tensor spherical harmonics

$$\underset{\sim}{T}^{B2}_{30} = -(105/64\pi)^{1/2} \sin^2\theta \cos\theta\, (\underset{\sim}{e}_\theta \otimes \underset{\sim}{e}_\phi + \underset{\sim}{e}_\phi \otimes \underset{\sim}{e}_\theta) \tag{43}$$

$$\underset{\sim}{T}^{E2}_{20} = (15/64\pi)^{1/2} \sin^2\theta (\underset{\sim}{e}_\theta \otimes \underset{\sim}{e}_\theta - \underset{\sim}{e}_\phi \otimes \underset{\sim}{e}_\phi) \ . \tag{44}$$

It is also often convenient to express the metric perturbation in terms of its "+" and "x" polarization components (Misner, Thorne & Wheeler, 1973), defined by

$$\underset{\sim}{h}_{TT} = h_+ (\underset{\sim}{e}_\theta \otimes \underset{\sim}{e}_\theta - \underset{\sim}{e}_\phi \otimes \underset{\sim}{e}_\phi) + h_x (\underset{\sim}{e}_\theta \otimes \underset{\sim}{e}_\phi + \underset{\sim}{e}_\phi \otimes \underset{\sim}{e}_\theta) \ . \tag{45}$$

These polarization states are simply related to the multipole amplitudes by

$$h_\times = -(105/64\pi)^{1/2}\, r^{-1} \sin^2\theta \cos\theta\, A^{B2}_{30}(t-r) \tag{46}$$

$$h_+ = (15/64\pi)^{1/2}\, r^{-1} \sin^2\theta\, A^{E2}_{20}(t-r)\ . \tag{47}$$

The angular distribution of the total energy radiated is given by

$$dE/d\Omega = (32\pi)^{-1} \Big\{ |\underset{\sim}{T}^{B2}_{30}|^2 \int |\partial A^{B2}_{30}/\partial t|^2 dt$$

$$+ |\underset{\sim}{T}^{E2}_{20}|^2 \int |\partial A^{E2}_{20}/\partial t|^2 dt \Big\}. \tag{48}$$

Because the tensor spherical harmonics are orthogonal (under contraction over both indices), there are no interference terms, leading to a radiation pattern represented by an incoherent sum of the patterns generated by each multipole amplitude.

In summary, in order to calculate the gravitational radiation emitted by a slowly-rotating, non-relativistic star, one proceeds as follows:

1) Choose a non-rotating model of collapse, described by the function $R(m,T_c)$.

2) Specify the amount of non-uniform rotation by choosing the function $h_3(m)$, related to the rotational velocity by $v^\phi{}_{(1)} = -\sum_\ell h_\ell(m)\, R^{-1}\, dP_\ell/d\theta$. Compute the gravitational radiation emitted at order Ω from equation (41).

3) Determine the distortion parameter $\xi_2\,(m,T_c)$ by solving equations (35)-(38) after specifying the appropriate initial conditions and the function $p_\ell(R,T)$. Compute the gravitational radiation emitted at order Ω^2 from equation (42).

In the next section we will apply steps (1) and (2) of this procedure to three supernova models, one analytic and two numerical. We reserve the application of step (3) to a future paper for two reasons: The order Ω^2 radiation will not dominate unless $\Omega_* \gtrsim \varepsilon^2$ or the rotation is nearly uniform; and the computational effort is much greater in step (3).

3 APPLICATION TO THREE MODELS

3.1 Analytic Model

Consider a supernova core which, at least in some region Δm, satisfies the following two conditions:

i) The density distribution is homologous; that is, $\partial R/\partial m \equiv R' = R/f(m)$, which implies, by equation (31), that

$$\rho_{(0)} = f(m)/4\pi R^3\ . \tag{49}$$

ii) The gas (whether degenerate or non-degenerate) has uniform entropy per baryon s. If the pressure is dominated by a single, non-interacting, non-relativistic component, it then follows that the equation of state has the form

$$p = K\rho^{5/3} . \tag{50}$$

It is expected that it will be the pressure of the neutrons (whether cold or hot) that will be responsible for any "bounce".

With these assumptions, the non-rotating equation of motion (33) can be integrated to give (within Δm)

$$\dot{R}^2 + g(m)R^{-2} - 2mR^{-1} = E(m) , \tag{51}$$

where

$$g = 5K\,(1 - f'/3)(f/4\pi)^{2/3} . \tag{52}$$

For those supernovae of interest (those which produce a significant amount of gravitational radiation), collapse will begin at a radius R much greater than the minimum radius R_0 reached at "bounce". Therefore it is a good approximation to take the energy $E(m) = 0$. In this case another integration gives implicitly the required function $R(m,T)$:

$$\pm[T - T_0(m)] = (2/9m)^{1/2}\,[(R-R_0)^{3/2} + 3R_0(R-R_0)^{1/2}] , \tag{53}$$

where the minimum radius is

$$R_0(m) = g(m)/2m . \tag{54}$$

That is, $R = R_0(m)$ and $T = T_0(m)$ when $\dot{R}(m,T) = 0$.

Let us focus on the gravitational radiation emitted by a comoving spherical shell of matter of thickness Δm and interior mass m. Using equation (41) and the above results, we then obtain the multipole amplitude

$$\frac{\Delta A^{B2}_{30}}{\Delta m} = \pm\,\frac{32}{7}\,(\pi/105)^{1/2}\,\frac{m h_3(m)}{R^2}\,[\frac{2m}{R}\,(1 - \frac{R_0}{R})]^{1/2} . \tag{55}$$

Let us introduce the dimensionless parameters

$$\varepsilon_0{}^2 = m/R_0 , \tag{56}$$

$$\Omega_0{}^2 = h_3{}^2/mR_0 . \tag{57}$$

The constant ε_0 is a measure of how relativistic the star becomes at bounce. The constant Ω_0 is approximately equal to the largest value of Ω_* (non-uniform part) achieved during the collapse, which of course also occurs at the "bounce". Defining $x \equiv (R/R_0) - 1$, and employing equations (45) and (46), we obtain the metric perturbation as a function of time from the equations

$$\Delta h_{TT}^{\theta\phi} = \mp\Omega_0 \varepsilon_0{}^4 \Delta m\, r^{-1} \sin^2\theta \cos\theta\, F_1(x) \tag{58}$$

$$T - T_0 = \pm m\, \varepsilon_0{}^{-3} F_2(x)\ . \tag{59}$$

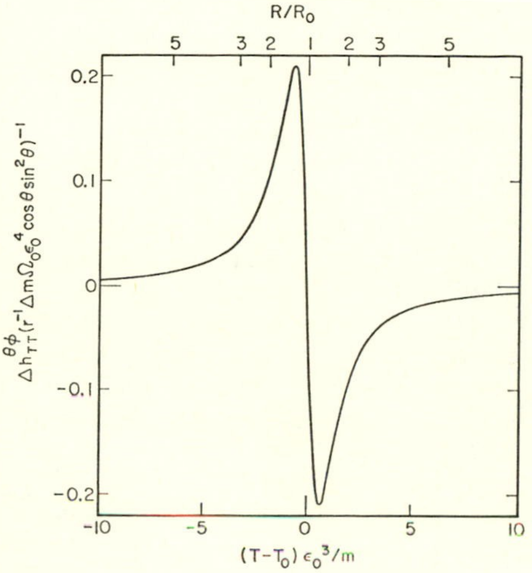

Figure 2. The time dependence of the normalized radiative metric perturbation produced by our analytic model of a supernova bounce. The metric perturbation is proportional to the multipole amplitude $A_{30}^{B2}(t-r)$.

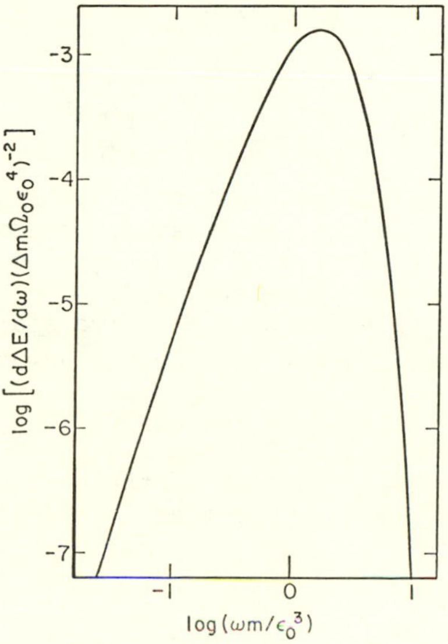

Figure 3. The energy spectrum of the radiation produced by the shell of mass Δm which generated the amplitude shown in Figure 2.

In Figure 2 we have plotted F_1 vs. F_2, where

$$F_1 = (4\sqrt{2}/7)\ x^{1/2}\,(1 + x)^{-3}\ , \tag{60}$$

$$F_2 = (\sqrt{2}/3)\ x^{1/2}\,(3 + x)\ . \tag{61}$$

Recall that the change in separation of two free test masses is proportional to this metric perturbation (Press & Thorne, 1972; Misner, Thorne & Wheeler, 1973).

If we neglect the coherent contribution from other shells, we obtain the energy ΔE radiated by Δm by using equation (40).

Fourier-analyzing the amplitude, we obtain the energy spectrum (shown in Figure 3)

$$\frac{d\Delta E}{d\omega} = \frac{36\Omega_0^{\,2}\varepsilon_0^{\,8}(\Delta m)^2\omega_*^{\,4}}{35\cdot 49\pi}\, K^2_{1/3}(\omega_*) \ , \tag{62}$$

where $K_{1/3}$ is a modified Bessel function (Abramowitz & Stegun, 1965), and the dimensionless frequency is

$$\omega_* = (2\sqrt{2}/3)\, m\varepsilon_0^{\,-3}\omega \ . \tag{63}$$

The integral of this spectrum gives the total energy

$$\Delta E = \frac{11\pi\Omega_0^{\,2}\varepsilon_0^{\,11}}{64\cdot 98\sqrt{2}}\cdot\frac{(\Delta m)^2}{m} = 3.90 \times 10^{-3}\,\Omega_0^{\,2}\,\varepsilon_0^{\,11}(\Delta m)^2/m \ . \tag{64}$$

This result illustrates the general dependence of the energy radiated on the rotation parameter Ω_0 and the relativity parameter ε_0. Note that for Newtonian quadrupole radiation, $\Delta E \sim \Omega_0^{\,4}\varepsilon_0^{\,7}(\Delta m)^2/m$. (We note that if our assumption of homology was extended to the entire star, the density $\rho_{(0)}$ and the bounce time T_0 would have to be independent of m.)

It is useful in estimating the energy radiated to recall that the constant K appearing in the equation of state (50) is an increasing function of the assumed entropy. From equations (52) and (54) we also see that

$$\varepsilon_0^{\,2} \sim m^{4/3}/K \ . \tag{65}$$

Thus the amount of gravitational radiation produced depends strongly on the entropy of the matter, as well as on the amount of non-uniform rotation.

3.2 Wilson's Numerical Model

James R. Wilson (1977) has very kindly supplied us with the computer results from his standard nonrelativistic collapse model, which includes the effects of neutrino transport. In this model the radially collapsing 1.4 solar mass core is "zoned" into mass shells, whose evolution is followed numerically. Trajectories R(m,T) of some of these shells are shown in Figure 4.

The key feature of most of these trajectories is the very sharp, cusp-like initial bounce. Note that in the central region of the core, where the bounce is strongest, the collapse is approximately homologous, with the bounce time T_0 almost independent of m. The bounce creates a shock wave which propagates through the outer regions, and in this model has sufficient strength to eject the outer layers of matter. The ability of a model to eject matter depends sensitively on a number of factors (Wilson 1971, 1974; Wilson et al., 1975; Bruenn, 1975; Schramm & Arnett, 1975; Mazurek, 1975; Bruenn, Arnett & Schramm, 1977; Chechetkin,

Imshennik, Ivanova & Nadyozhin, 1977; Epstein, 1977). It is beyond the scope of this paper to consider such details. However, even if no matter is ejected, the central region still "bounces", so that gravitational radiation can be produced without a supernova.

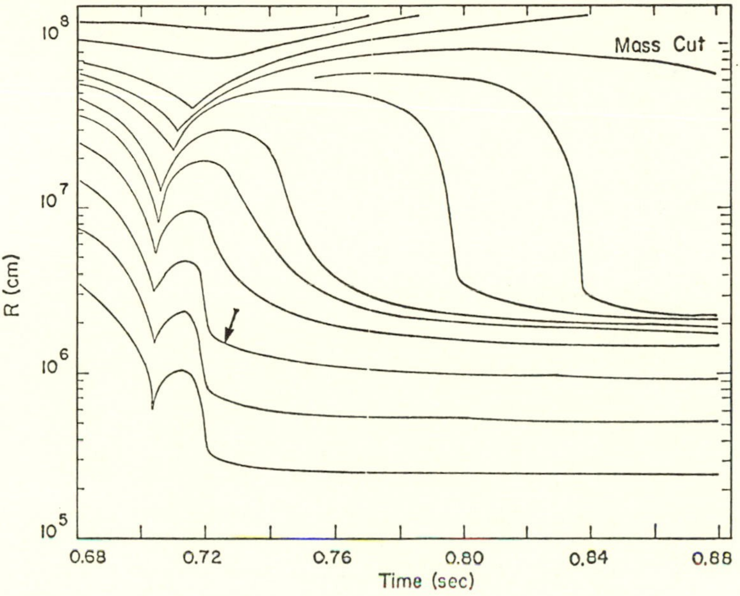

Figure 4. The trajectories R(m,T) of various mass shells, from the non-rotating supernova computations of James R. Wilson. In this particular model, the shells above the mass cut are expelled. The mass shell used in Figure 5 is indicated by the arrow.

The timescale of the bounce is comparable to Wilson's printout interval ($\sim 1 \times 10^{-3}$ sec), but presumably longer than the differential interval ($\sim 3 \times 10^{-5}$ sec) he used in integrating the equations of motion. As in other models, the bounce is precipitated by the rapid increase in the neutron pressure relative to the gravitational force, as indicated in the analytic model discussed above.

We have computed the contribution of the third mass shell indicated in Figure 4 to the amplitude A^{B2}_{30}, using equation (41). The mass of this shell is $\Delta m = 0.08\ M_{\odot}$. The results of this calculation are shown in Figure 5. Because of the coarseness of the printout interval in Wilson's results for R(m,T), Figure 5 is accurate to only a factor of about two. Despite this uncertainty, some general features are evident. The bulk of the gravitational radiation is emitted during a short interval (a few milliseconds) centered at the initial bounce. When the contributions from the

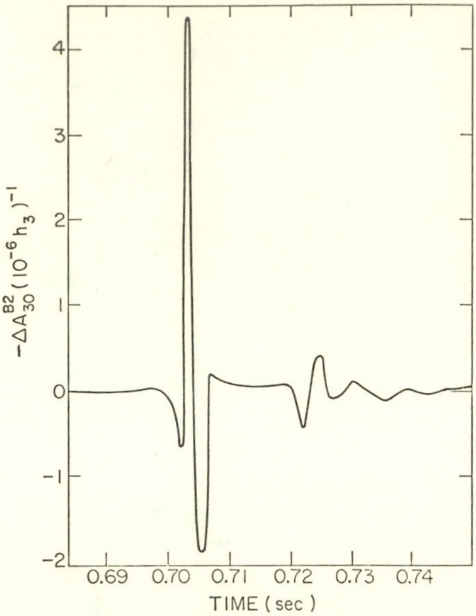

Figure 5. The radiation amplitude, ΔA^{B2}_{30}, produced by the mass shell indicated in Figure 4, in terms of the parameter h_3 which characterizes the amount of nonuniform rotation.

other mass shells are added, the burst will broaden somewhat because the bounce times are not exactly simultaneous, but the broadening should leave the timescale of the burst in the millisecond range.

It is interesting to compare the amplitude shown in Figure 5 with the corresponding result for the analytic model (Figure 2). The pulse shown in Figure 5 is not symmetrical, indicating that the bounce was not perfectly "elastic", due in part to loss of energy by neutrino emission. However, the qualitative nature of the waveforms are similar, indicating that the assumptions made in section 3.1 are not very unrealistic.

3.3 Van Riper's Numerical Model

Kenneth Van Riper (1978) has kindly supplied us with his numerical supernova model results in the bounce region in great detail. Van Riper's calculations follow the collapse of a nonrotating 1.4 solar mass supernova core which eventually ejects 0.1 solar mass. The core is divided into 60 mass shells; each of the first forty contains 2 percent of the total mass. In the final 20 shells, the mass contained in each shell decreases linearly, with the outermost shell containing 0.25 per cent of the total mass. The evolution of his model is governed by nonrelativistic, adiabatic hydrodynamics. The effective adiabatic index γ is a function of density only. The dependence on density is chosen to simulate more sophisticated models of collapse which include the effects of energy transport (Arnett, 1977). The simplified model has the same general behavior, but requires significantly less machine time to generate. In the model we have used (referred to as the "hard" model by Van Riper, 1978) the inner 40 shells evolve almost homologously near the hydrodynamic bounce.

Van Riper provided R(n, T) and dR(n,T)/dT in the interval (T_0 - 5 msec, T_0 + 5 msec) in time steps of 0.02 msec; where n refers to mass shell number and T_0 is the bounce time of the inner 40 shells. The trajectories R(n,T) are shown in Figure 6. Be-

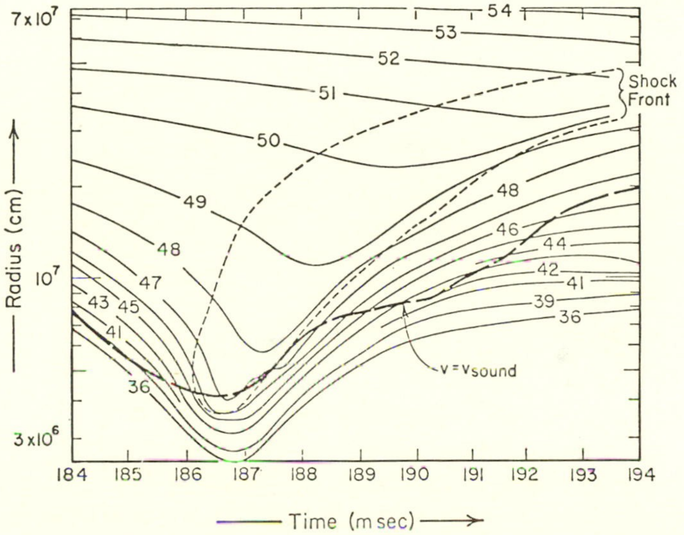

Figure 6. The trajectories R(n,T) from Van Riper's "hard" model for mass shells 36 through 54. The first forty shells evolve almost homologously. About 0.1 $M_\odot$ of the initial 1.4 $M_\odot$ core is hydrodynamically ejected by a shock wave (taken from Van Riper, 1978).

cause we have been given both R and dR/dT only two time derivatives are needed to compute the amplitude A^{B2}_{30} given by equation (41). These derivatives are calculated by two separate methods. In the first method the two derivatives are computed from the raw data by a finite difference scheme. Because the quantity of interest, R^2, has large high order derivatives, the accuracy of this method is not expected to be good. In the second method, a spline is fit to 2R dR/dT over the 10 msec interval of interest. The spline used consisted of a set of 83 cubic polynomials each least-squares fit to an interval of ~0.12 msec, and was constructed with continuous first and second derivatives at the interval boundaries. Such a spline by its nature cannot fit structure with timescales of $\lesssim 0.24$ msec. Therefore it acts to filter out any detail of frequency $\gtrsim 4$ kHz.

The raw data (2R dR/dT) for a typical shell in the homologous core is shown in Figure 7. Two things should be noted: (i) the gross shape of 2R dR/dT will lead to a waveform with low frequency behavior similar to that of Figure 2, and (ii) there is an abundance of high frequency structure which actually dominates the contribution of this shell to the waveform A^{B2}_{30}.

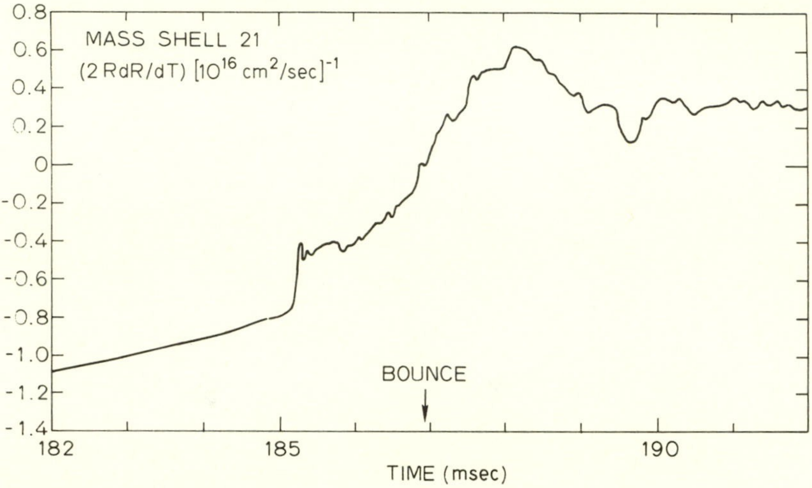

Figure 7. The raw data (2R dR/dT) for mass shell 21 in Van Riper's model plotted at 0.02 msec intervals. Note the abundance of high frequency structure.

In order to obtain the waveform from the entire supernova core, a model for the initial rotation must be chosen and $h_3(m)$ must be calculated. There is no theoretical consensus on the distribution of angular momentum during the final stages of stellar evolution (LeBlanc & Wilson, 1970; Endal & Sofia, 1977; Wiita, 1978). For our exploratory purposes a simple model is considered. In this model, the initial rotation is given by

$$\Omega = \Omega_b [1 - (R/b)^2 \sin^2\Theta] , \qquad (66)$$

where b is greater than the surface radius of the supernova core. This model is physically reasonable and fits the inner region of any stable, stationary presupernova model. In this case $h_3(m)$ is given by

$$h_3(m) = 2\Omega_b R_o^4(m)/15b^2 , \qquad (67)$$

where $R_o(m)$ refers to the initial (presupernova) value of R(m,T). In Van Riper's model $R_o(m)$ can be fit by the formula

$$R_o(n) = (1.041)^n \, 1.7 \times 10^{7} \mathrm{cm} , \tag{68}$$

where n refers to the shell number. This relation is actually only valid for $12 \leqslant n \leqslant 58$, but shells 59 and 60 contribute insignificantly to A^{B2}_{30}, and this form of $h_3(m)$ weights the outer shells so heavily that the inner shells contribute very little to the waveform.

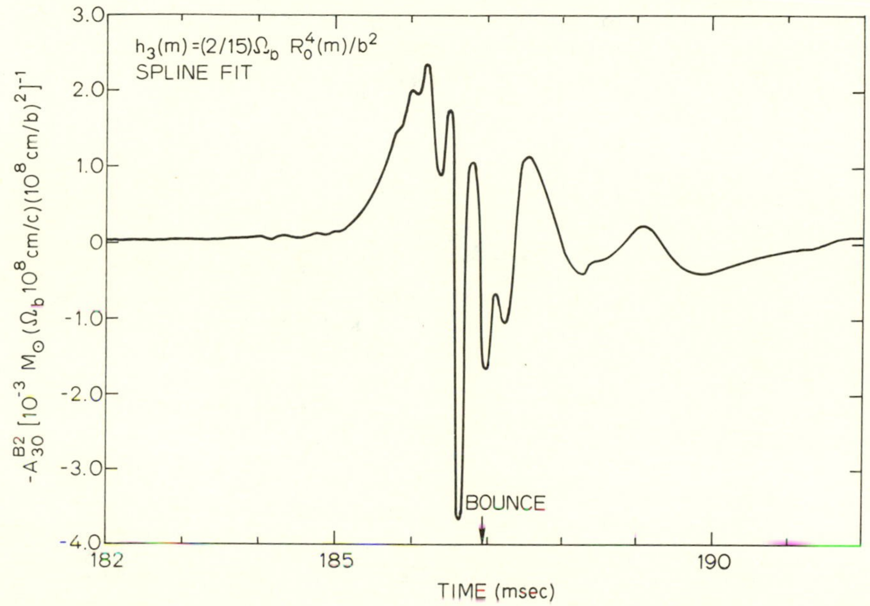

Figure 8. The multipole amplitude $A^{B2}_{30}(t)$ from Van Riper's results for the $h_3(m) = (2/15)\,\Omega_b R_o{}^4/b^2$ model of rotation. The curve shown is derived from the spline fit to the raw data. The waveform is similar to that in Fig. 2 in gross shape with additional high frequency structure, although the use of the spline effectively "filters out" all structure on a timescale of less than ~0.24 msec. The major contribution to the waveform is from mass shells 36 through 50.

The waveform from the spline fit to the data for this model of rotation is shown in Figure 8. The general shape seen in Figure 2 is apparent, but there is a lot of high frequency structure. Shells 51 through 60 contain little mass and undergo relatively gentle accelerations so they contribute little to the waveform. The major contribution to the waveform is from shells 36 through 50.

Energy spectra were obtained by performing a fast Fourier transform on both the raw data and the spline fits. The two methods are very consistent up to ~ 3 kHz; at this point the spline results begin to fall well below those from the raw data. The energy spectrum for the spline fit is shown in Figure 9. There are three surprising features in the spectrum: first, the dip in $d\Delta E/d\omega$ at ~ 1 kHz, second, the amount of energy radiated at high frequencies, and third, the amount of spectral structure. The physical reality of any of these effects is not clear. A supernova model [R(m,T)] is obtained by integrating a differential equation; the integration tends to smooth out most of the jiggles that may be present, real or otherwise. However, the waveform $A^{B2}_{30}(t)$, which depends upon the third time derivative of $R(m,T)^2$, is very sensitive to all the jiggles in the supernova model, real or otherwise.

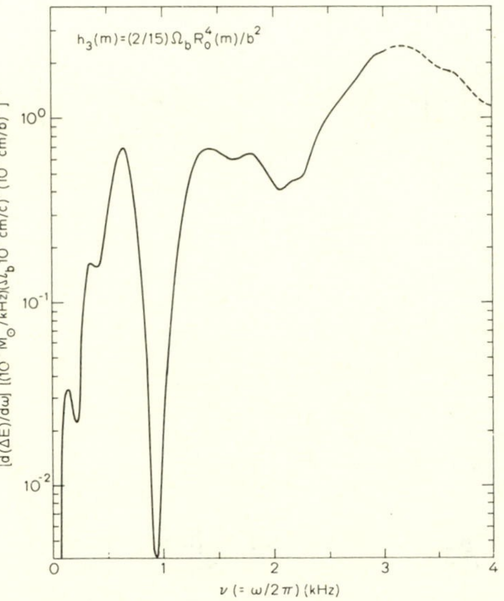

Fig. 9. The energy spectrum $d(\Delta E)/d\omega$ obtained from Fig.8. The results shown are derived from the fast Fourier transform of the spline. Up to ~ 3 kHz these results are consistent with those derived from the fast Fourier transform of the raw data. Above ~ 3 kHz the spline-derived results (broken line) fall below those derived from the raw data (Fig. 10). The amount of energy at high frequencies is surprising. Presumably, the results are only believeable up to some cutoff frequency (as of yet unknown) determined by the "input" model.

In the 10 msec interval around the bounce the integration time step used by Van Riper varied from 1 μsec to 20 μsec. Because of this a bump in the energy spectrum is expected in the range 50 kHz to 1 MHz. This artifact probably has a low frequency tail. We examined the energy spectrum (of the raw data) from 5 kHz to 25 kHz and found that the spectrum is spiky (the fast Fourier transform is near its limit of resolution as 1/frequency ~ time between data points) and grows roughly as ω^4, as shown in Figure 10. If we interpret this as a "numerical background," it implies that the real supernova physics (of Van Riper's model) stops at ~4 kHz. The spectrum in the 1-4 kHz range could be due to the mass zoning, since (speed of sound)/(thickness of zone) ~ 10^4 Hz at the bounce. This simple analysis by no means answers the question: Up to what frequency are the spectra presented here believeable? That question still invites input from supernova model builders.

Finally, the total energy radiated in gravitational waves can be estimated. Choosing b to be equal to the initial radius of the presupernova core, and integrating the energy spectrum shown in Figure 9 from 0 to 3 kHz we find that

$$\Delta E \sim 5 \times 10^{-13} M_{\odot} \Omega_0^2. \qquad (69)$$

The quantity Ω_0^2 is the maximum value of the ratio of centrifugal to gravitational forces, which of course occurs at the bounce. This result is consistent with the estimate one gets from equation (64) for this model.

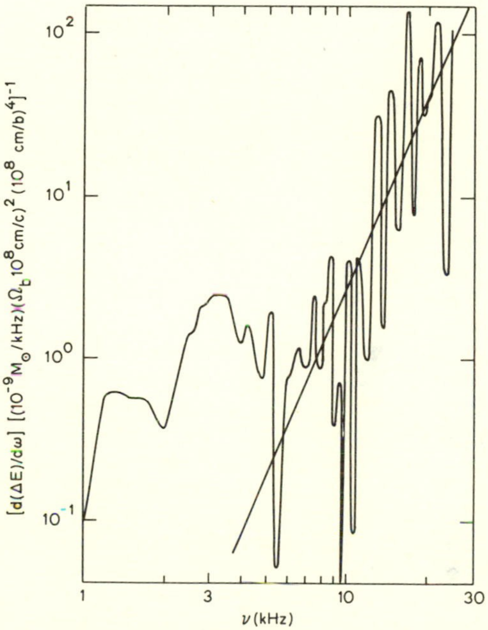

Fig. 10. Energy spectrum obtained from the fast Fourier transform of the raw data [Van Riper's "hard" model with rotation law given by (66).] The low frequency results ($\lesssim$ 3 kHz) are identical to those derived from the fast Fourier transform of the spline fit (Fig. 9). Note the logarithmic frequency scale.

There are basically two reasons why this energy is so much less than the binding energy of a neutron star, $\sim 10^{-1} M_{\odot}$. Most importantly, the entropy is high enough to produce a bounce at relatively small values of ε_0. In addition, the amount of mass in those zones which have a significant amount of differential rotation is small.

However, two factors which will be included in future calculations should produce deeper bounces, increasing ε_0 and therefore to a much greater extent, the energy radiated. The first is the inclusion of general-relativistic effects in the equation of motion, which can be of importance because the adiabatic index is close to the critical value of 4/3 during collapse (Van Riper & Arnett, 1978). The other factor is the recent discovery that nuclear excitation may absorb a large amount of the entropy (Schramm, 1978).

4 CONCLUSION

This paper represents only an exploratory investigation of the problem of calculating the amount of gravitational radiation produced by slowly-rotating supernovae (and other stars that "bounce" before becoming extremely relativistic). Our purpose has been to obtain an understanding of the factors which determine the nature of the radiation.

The strongest assumption that we have made is that the rotationally-induced distortion of spherically-symmetric collapse is small. Given a definite model [R(m,T)] of spherical collapse and a specification of the initial amount of rotation, the method we have presented provides a reliable way to calculate the radiation produced. In this paper we have only presented explicit results for the radiation at lowest order in angular velocity. This radiation, present whenever the initial rotation was non-uniform, is completely specified by the octupole amplitude $A^{B2}_{30}(t-r)$. We have, however, indicated the procedure involved in calculating the radiation emitted at second order in angular velocity, specified by the quadrupole amplitude $A^{E2}_{20}(t-r)$.

From these preliminary results the gross character of the radiation is already clear: it is emitted in a short burst at the initial bounce of the collapsing core. At this point, two pressing questions arise: (i) Down to what timescale are "input," non-rotating collapse models realistic? and (ii) What is the distribution of the angular momentum [specifically $h_3(m)$] in the supernova core? Investigation of other "input" models, extraction of time derivatives directly by using the equations of motion, and thoughtful comment from supernova model builders may help clarify the first question. We hope that both observation and stellar evolution theory can illuminate the subject of initial rotation.

Let us recall that it is the bounce, and not the ejection of matter, which gives rise to most of the gravitational radiation. The total number of stars that bounce from neutron-star configurations could be up to about ten times the observed number of supernovae.

Finally, we can place these results within the context of all sources of gravitational radiation by considering Figure 11, which is a schematic diagram illustrating the importance of angular momentum. Consider a system (star, binary, etc.) of size R and angular momentum J. The relativity parameter $\varepsilon^2 = GM/Rc^2 \sim (v/c)^2$ increases in the vertical direction, and the dimensionless measure of angular momentum $J_* = cJ/GM^2$ increases in the horizontal direction. The range of values of J_* for observed binaries, pulsars, and the sun is indicated. Curves of constant metric perturbation h are also included, for sources of fixed mass M and distance r. If differential rotation is significant, the radiation will be dominated by the post-Newtonian octupole contribution in the region

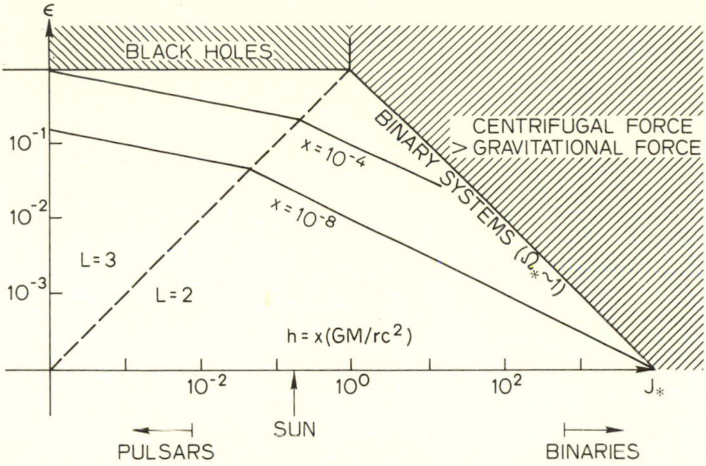

Fig. 11. The role of angular mementum in gravitational radiation. All symbols are defined in the text.

marked L = 3, and by the Newtonian quadrupole contribution in the remainder (L = 2), as long as relativistic effects are not very strong ($\varepsilon < 0.4$).

Consider a star more massive than the sun, for which $J_* > 1$ while on the main sequence. If it were to collapse while conser-ing angular momentum, it would most likely fission into a binary system when Ω_* reached a value of order unity. Thereafter gravitational radiation energy loss would cause it to slowly evolve upward along the line $\Omega_* \approx 1$, where any binary must lie. (For $J_* \gtrsim 10^3$ and $M \sim M_\odot$, this evolution time is greater than the age of the universe.)

On the other hand, a star which collapses with $J_* < 1$ will always be slowly rotating, and the considerations of this paper apply.

An intrinsically strong source of gravitational radiation must have both $\varepsilon \sim 1$ and $J_* \sim 1$. The number of systems in our universe which can reach such a state should be very small.

We are grateful to James Wilson for supplying us with the numerical results of his supernova calculations. We give special thanks to Ken Van Riper for rerunning his collapse codes to generate results in a format more suitable for us. We also acknowledge helpful conversations with J. Wilson, K. Van Riper, and W. D. Arnett.

This work was supported in part by National Science Foundation grant PHY 76-21454 A01. Michael Turner's present address is: Enrico Fermi Institute, University of Chicago, Chicago, Illinois 60637.

REFERENCES

Abramowitz, M. & Stegun, I.A. (1965). Editors, Handbook of Mathematical Functions. U.S. Government Printing Office, Washington, D.C.

Arnett, W.D. (1973). Some quantitative calculations of final stages of stellar evolution. In Explosive Nucleosynthesis, ed. D.N. Schramm & W.D. Arnett, pp. 236-247. University of Texas Press, Austin.

Arnett, W.D. (1977). Neutrino trapping during gravitational collapse of stars. Astrophys. J., 218, pp. 815-833.

Bertotti, B. (1977). Editor, International Meeting on Experimental Gravitation (Pavia, Italy). Academia Nazionale dei Lincei, Rome.

Bruenn, S.W. (1975). Neutrino interactions and supernovae. Annals N.Y. Acad. Sci., 262, pp. 80-94.

Bruenn, S.W., Arnett, W.D. & Schramm, D.N. (1977). Some criteria for mass ejection by stars undergoing gravitational collapse. Astrophys. J., 213, pp. 213-224.

Chechetkin, V.M., Imshennik, V.S., Ivanova, L.N. & Nadyozhin, D.K. (1977). Gravitational collapse, weak interactions, and supernova outbursts. In Supernovae, ed. D.N. Schramm, pp. 159-182. D. Reidel, Dordrecht.

Chia, T.T., Chau, W.Y. & Henriksen, R.N. (1977). Gravitational radiation from a rotating collapsing gaseous ellipsoid. Astrophys. J., 214, pp. 576-583.

DeSabbata, V. & Weber, J. (1977). Editors, Topics in Theoretical and Experimental Gravitation Physics. Plenum Press, London.

DeWitt-Morette, C. (1974). Editor, Gravitational Radiation and Gravitational Collapse. D. Reidel, Dordrecht.

Endal, A.S. & Sofia, S. (1977). Detectable gravitational radiation from stellar collapse. Phys. Rev. Letters, 39, pp. 1429-1432.

Epstein, R. (1976). The Post-Newtonian Theory of the Generation of Gravitational Radiation and Its Application to Stellar Collapse. Ph.D. Thesis, Stanford University.

Epstein, R. (1978). The generation of gravitational radiation by escaping supernova neutrinos. Astrophys. J., 223, pp. 1037-1045.

Epstein, R. & Wagoner, R.V. (1975). Post-Newtonian generation of gravitaional waves. Astrophys. J., 197, pp. 717-723.

Epstein, R.I. (1977). Mechanisms for supernova explosions. In Supernovae, ed. D.N. Schramm, pp. 183-192. D. Reidel, Dordrecht.

Greenstein, J.L. & Peterson, D.M. (1973). Line profiles and rotation in white dwarfs. Astron. & Astrophys., 25, pp. 29-34.

Hardorp, J. (1974). Rotation in late stages of stellar evolution. Astron. & Astrophys., 32, pp. 133-136.

Hartle, J.B. (1967). Slowly rotating relativistic stars I. Equations of structure. Astrophys. J., 150, pp. 1005-1029.

Kazanas, D. & Schramm, D.N. (1977). Neutrino damping of nonradical pulsations in gravitational collapse. Astrophys. J., 214, pp. 819-825.

LeBlanc, J.M. & Wilson, J.R. (1970). A numerical example of the collapse of a rotating magnetized star. Astrophys. J., 161, pp. 541-551.

Mazurek, T.J. (1975). Chemical potential effects on neutrino diffusion in supernovae. Astrophys. & Space Sci., 35, pp. 117-135.

Misner, C.W. (1974). Mechanics for the emission and absorption of gravitational radiation. In Gravitational Radiation and Gravitational Collapse, ed. C. DeWitt-Morette, pp. 3-15.

Misner, C.W., Thorne , K.S. & Wheeler, J.A. (1973). Gravitation. Freeman, San Francisco.

Novikov, I.D. (1975). Gravitational radiation from a star collapsing into a disk. Soviet Astronomy, 19, pp. 398-399.

Press, W.H. & Thorne, K.S. (1972). Gravitational-wave astronomy. Ann. Rev. Astron. & Astrophys, 10, pp. 335-374.

Rees, M., Ruffini, R. & Wheeler, J.A. (1974). Black Holes, Gravitational Waves and Cosmology: An Introduction to Current Research. Gordon & Breach, New York.

Ruderman, M. (1972). Pulsars: Structure and dynamics. Ann. Rev. Astron. & Astrophys., 10, pp. 427-476.

Saenz, R.A. & Shapiro, S.L. (1978). Gravitational radiation from stellar collapse: Ellipsoidal models. Astrophys. J., 221, pp. 286-303.

Schramm, D.N. (1977). Editor, Supernovae. D. Reidel, Dordrecht.

Schramm, D.N. (1978). Private communication.

Schramm, D.N. & Arnett, W.D. (1975). The weak interaction and gravitational collapse. Astrophys. J., 198, pp. 629-639.

Shapiro, S.L. (1977). Gravitational radiation from stellar collapse: The initial burst. Astrophys. J., 214, pp. 566-575.

Tammann, G.A. (1977). A progress report on supernovae statistics. In Supernovae, ed. D.N. Schramm, pp. 95-116. D. Reidel, Dordrecht.

Thorne , K.S. (1978a). General-relativistic astrophysics. In Theoretical Principles in Astrophysics and Relativity, ed. N.R. Lebovitz, W.H. Reid & P.O. Vandervoort, pp. 149-216. University of Chicago Press.

Thorne, K.S. (1978b). The generation of gravitational waves V. Multipole-moment formalisms. Rev. Mod. Phys., in press.

Thuan, T.X. & Ostriker, J.P. (1974). Gravitational radiation from stellar collapse. Astrophys. J. (Letters), 191, pp. L105-L107.

Turner, M.S. (1978). Gravitational radiation from supernova neutrino bursts. Nature, 274, pp. 565-566.

Van Riper, K.A. (1978). The hydrodynamics of stellar collapse. Astrophys. J., 221, pp. 304-319.

Van Riper, K.A. & Arnett, W.D. (1978). Stellar collapse and explosion: Hydrodynamics of the core. Astrophys. J., submitted for publication.

Wagoner, R.V. (1977). Gravitational radiation from slowly rotating collapse: Post-Newtonian results. In International Meeting on Experimental Gravitation (Pavia, Italy), ed. B. Bertotti, pp. 117-135.

Wiita, P.J. (1978). On rotation in the later stages of stellar evolution. Astrophys. J., submitted for publication.

Wilson, J.R. (1971). A numerical study of gravitational stellar collapse. Astrophys. J., 163, pp. 209-219.

Wilson, J.R. (1974). Coherent neutrino scattering and stellar collapse. Phys. Rev. Letters, 32, pp. 849-852.

Wilson, J.R. (1977). Unpublished calculations.

SLOWLY-ROTATING "SUPERNOVAE"

Wilson, J.R., Couch, R., Cochran, S., LeBlanc, J. & Barkat, Z. (1975). Neutrino Flow and the collapse of stellar cores. Annals N.Y. Acad. Sci., 262, pp. 54-64.

GRAVITATIONAL WAVES AND GRAVITATIONAL COLLAPSE IN CYLINDRICAL SYSTEMS

Tsvi Piran

Center for Relativity
The University of Texas at Austin

INTRODUCTION

Cylindrical systems are the simplest systems in which gravitational waves can be generated. Spherical configurations, though simpler for computation and of more direct astrophysical importance, cannot include gravitational radiation. On the other hand, realistic asymmetric configurations are extremely complicated and only recently have attempts been made to explore them numerically (see lectures by Smarr and Wilson in this volume). The simplicity of cylindrical systems, combined with the existence of exact solutions, makes them the best arena for developing and testing numerical methods for solutions of general relativistic problems. Moreover, a better understanding of cylindrical general relativistic systems will teach us a great deal about the basic qualitative features of gravitational wave generation in realistic configurations. Cylindrical configurations can also shed some light on the problems of Cosmic Censorship and general asymmetric collapse.

In this lecture we review the analytic work on cylindrical systems (for a very extensive review of this subject see Thorne, 1965a), and compare it with a numerical study of cylindrical collapse. The first section includes a discussion of the Newtonian approximation, static vacuum solutions (Levi-Civita metric), dynamic vacuum solutions (Einstein-Rosen and Kompaneets-Ehlers-Jordan waves), static matter solutions, and dynamically collapsing cylinders of matter coupled to gravitational radiation. Numerical calculations of cylindrical collapse are described in the second section. The numerical results are compared with the analytic solutions, emphasizing the more general features of gravitational wave generation which are likely to hold in any realistic collapse.

THE NEWTONIAN APPROXIMATION

An infinitely long cylinder obeying Newtonian mechanics resembles in many aspects a Newtonian sphere. The only difference is the appearance of different powers of r, the radial distance, in the equations. Chandrasekhar & Fermi (1953) derived a virial theorem for infinite cylinders in equilibrium, and Ostriker (1964) developed the theory of equilibrium of infinite polytropic (p = constant $\rho^{1+1/n}$) and isothermal cylinders. It is known that all equilibrium Newtonian cylinders display a sausage instability, i.e. they break up into many disconnected links along the axis of symmetry (Lamb, 1924).

The gravitational potential around an infinite cylinder of finite radius is $2GM\ln r$ + constant, where M is the mass per unit length and G the gravitational constant. This potential is remarkably different from the spherical one, as it diverges at large radii. Therefore given spherical symmetry the Newtonian approximation always holds at large distances, and sometimes breaks down at small distances, but given cylindrical symmetry the Newtonian approximation always breaks down at large radii. This breakdown happens inside the matter cylinder whenever $M>1/8$ (in units with $c=G=1$, i.e., $M>1.68\ 10^{27}$gm/cm$=5.9\ 10^{4}M_{\odot}/R_{\odot}$). This feature is independent of the matter density.

A homogeneous cylinder of dust collapsing from rest will stay homogeneous and reach both infinite density and velocity within $t_c = 1.78/\sqrt{4\pi G\rho_0}$ sec, where ρ_0 is the initial density per unit length. As the collapse progresses the infall velocity first exceeds the speed of light at large radii. Despite this, for every $t<t_0$ one can find a small region for which the Newtonian approximation is valid. Addition of even the slightest amount of angular momentum will stop the collapse before the density becomes infinite. As in the spheroidal case (Lin, Mestel & Shu, 1975), a deviation from exact symmetry will grow during the dust collapse and it may prevent the dust from reaching infinite density.

C ENERGY

Before going on with the discussion of general relativistic systems it is useful to introduce, after Thorne (1965a,b), the concept of C energy.

$$U \equiv \frac{1}{8}\left[1 - \frac{\left(A,_{\mu}A,^{\mu}\right)}{4\pi^2|\xi|^2}\right] \quad , \qquad (1)$$

where A is the area of the cylindrical surface ($t,r,z+\gamma,\phi+\delta;0\leq\gamma\leq1, 0\leq\delta\leq2\pi$) and ξ,ζ are the Killing vectors along the z and ϕ directions respectively. (We use the definition of Thorne (1965a). Thorne (1965b) and following him Kuchar (1971) and others have used slightly different definitions: $\tilde{C} \equiv -1/8\ \ell n(1-8C)$). The C energy flux vector, P^{μ}, describes the energy density and energy flux:

$$P^{\mu} \equiv \left[\frac{1}{\sqrt{-g}} \epsilon^{\mu\alpha\beta\gamma} 2\pi A^{-2} U \xi_{\beta} \zeta_{\gamma} \right]_{;\alpha} \quad . \tag{2}$$

The C energy is a conserved quantity and in the Newtonian limit it equals the rest mass energy. This concept is extremely useful in both numerical and analytic work because it enables one to distinguish the presence of gravitational waves.

LEVI-CIVITA SOLUTION

The general static cylindrically symmetric vacuum solution for the Einstein field equations was discovered by Levi-Civita (1919) and was studied by Wilson (1920), Marder (1958a,b;1959) and Thorne (1965a). It is the cylindrical analog of the Schwarzschild solution with an infinite line singularity rather than a point singularity. (The singularity is physical but it is geodesically complete (Thorne, 1965a)). It is also the general static exterior gravitational field for a static infinite cylinder corresponding to the Newtonian potential. Unlike the spherical case, it is not the most general exterior solution for a cylinder. Gravitational waves can be generated by an imploding or exploding cylinder or can be superimposed on an exterior metric of a static cylinder. Such metrics are dynamic and are distinct from the Levi-Civita solution.

The Levi-Civita metric has the form:

$$ds^2 = (r/R)^{4c}\, dt^2 - (r/R)^{-4c(1-2c)}(dr^2+dz^2) - r^2(r/R)^{-4c} d\phi^2 \tag{3}$$

c and R are two independent constants. Other useful forms of this solution were given by Marder (1958) and Thorne (1965a). Levi-Civita (1919) compared this field with the Newtonian approximation and Wilson (1920) compared the geodesic equation and the equation for the Newtonian acceleration. They concluded that when the metric is the exterior metric of a cylinder with a finite radius, $c \simeq 2M$. Marder (1958a) pointed out that while R is a second constant it is not completely independent from c. R is also related to the distribution of matter in the cylinder. It has no Newtonian analog and in the Newtonian limit is equal to one.

The Levi-Civita metric is a static vacuum solution without gravitational waves, but as Thorne (1965a) pointed out it contains non zero C energy density.

GRAVITATIONAL WAVES

The most general metric for gravitational waves in vacuum with cylindrical symmetry was found by Jordon & Ehlers (1960) and independently by Kompaneets (1958) (see Stachel (1966) for extensive discussion of properties of these waves). Their line element is:

$$ds^2 = e^{2(\gamma-\psi)}(dt^2-dr^2)-e^{2\psi}(dz+\omega d\phi)^2-r^2e^{-2\psi}d\phi^2 \tag{4}$$

and the field equations are:

$$\begin{aligned}
\psi_{tt} - \psi_{rr} - \psi_r/r &= (e^{4\psi}/2r^2)(\psi_t^2 - \psi_r^2) \\
\omega_{tt} - \omega_{rr} + \omega_r/r &= 4(\omega_r\psi_r - \omega_t\psi_t) \\
\gamma_r &= r(\psi_t^2 + \psi_r^2) + (e^{4\psi}/4r)(\omega_t^2 + \omega_r^2) \\
\gamma_t &= 2r\psi_t\psi_r + (e^{4\psi}/4r)\,\omega_t\omega_r ,
\end{aligned} \tag{5}$$

where ψ_t, ψ_r are the t and r derivatives of ψ respectively. A special case of this metric, i.e. when ω vanishes, is the case of Einstein-Rosen waves. It was first found by Beck (1925) and rediscovered by Einstein & Rosen (1938). The more general Kompaneets solution describes gravitational waves with two polarization states (Stachel, 1966), while Einstein-Rosen waves have only one possible polarization.

Rosen (1954,1958) and Boardman & Bergman (1959) calculated the pseudo energy tensor for Einstein-Rosen waves in various coordinate systems. Weber & Wheeler (1957) and Marder (1958a,b;1959) studied the effect of these waves on particle motion and Thorne (1965a,b), and to some extent Marder (1958a) and Fierz (1957), calculated the C energy carried by these waves:

$$U = \frac{1}{8}(1-e^{-2\gamma}) \tag{6}$$

$$P_t = \frac{1}{8\pi r} e^{-2\gamma} \gamma_r , \tag{7}$$

$$P_r = -\frac{1}{8\pi r} e^{-2\gamma} \gamma_t . \tag{8}$$

They concluded that the Einstein-Rosen waves are real physical waves as they carry energy and can do work.

It is interesting to note that Boardman & Bergman (1959) have chosen a coordinate system in which the gravitational energy pseudo tensor gives the same energy flux as eq. (8). On the other hand the tttt component of the Bel Robinson tensor, which measures the energy density of gravitational waves in vacuum, does not give the same energy density as eq. (7);

$$T_{tttt} = \frac{\psi_r{}^2}{r^2} + \frac{3}{r}\,\psi_r(\psi_t^2 - \psi_r^2) + 6(\psi_t^2 + \psi_r^2) + 12\psi_r^2\,\psi_t^2$$

$$+\psi_{rr}{}^2 + \psi_{tt}{}^2 + 3(\psi_r{}^2 + \psi_t{}^2)(\psi_{rr} + \psi_{tt}) + 2\psi_{tr}{}^2 + 12\psi_t\psi_r\psi_{tr}$$

$$-2r\psi_r\left[3\psi_r{}^4 + 30\psi_r{}^2\psi_t{}^2 + 15\psi_t^4 + (\psi_{rr} + 2\psi_{tr} + \psi_{tt})(\psi_r{}^2 + 3\psi_t{}^3)\right]$$

$$+2r^2(\psi_t{}^2 + \psi_r{}^2)\left[(\psi_t{}^2 + \psi_r{}^2)^2 + 12\psi_t{}^2\psi_r{}^2\right] \quad . \tag{9}$$

STATIC MATTER CYLINDERS

The interior solution for a matter cylinder can be described by a line element analogous to Einstein-Rosen coordinates:

$$ds^2 = e^{2(\gamma-\psi)}(dt^2 - dr^2) - e^{2\psi}dz^2 - \alpha^2 e^{-2\psi}d\phi^2 \quad , \tag{10}$$

or alternatively by a diagonal line element in which $g_{rr} = g_{zz}$. Thorne (1965a) presented a static solution for matter with $p=\rho$. This solution resembles the isothermal Newtonian solution found by Ostriker (1965). In both solutions the mass per unit length is finite but the radius of the cylinder is infinite. Marder (1958a) found a family of static solutions for cylinders with finite radius. However, the equation of state in these solutions is not physically reasonable. As in the Newtonian case a dust cylinder can support itself against its gravitational attraction by centrifugal forces. van Stockum (1935) discovered a stationary rotating solution for a dust filled cylinder of finite radius and matched this interior solution to an exterior vacuum solution (Lewis, 1932). In a comoving coordinate system van Stockum's line element is:

$$ds^2 = dt^2 - e^{-\tilde{\omega}^2 r^2}(dr^2 + dz^2) - r^2(1 - \tilde{\omega}^2 r^2)d\phi^2 - 2\tilde{\omega}r^2 d\phi dt \quad , \tag{11}$$

where $\tilde{\omega}$ is a constant. This coordinate system rotates with angular velocity, $\tilde{\omega}$, relative to a frame which is inertial near the axis of rotation. Vishreshwara & Winicour (1977) found a larger family of differentially rotating dust solutions which contain the van Stockum solution as a special case. Thorne (1965a) described a numerical catalog of cylinders in equilibrium for the Harrison-Wheeler equation of state (Harrison, Wakano & Wheeler, 1958). His studies of the stability of these cylinders revealed instabilities when the central density is between 5 $10^{11}g/cm^3$ and 4 $10^{12}g/cm^3$. (This qualitatively compares with the instability of spherical configuration with similar central densities.) All spherical configuration with a physical equation of state and a central density greater than $\sim 10^{16}g/cm^3$ are unstable to collapse to a singularity (Harrison, Thorne, Wakano & Wheeler, 1965). It was conjectured by Thorne (1965a) that cylinders with a realistic equation of state are stable against such collapse. There is no proof of this conjecture.

CYLINDRICAL COLLAPSE

There have been several attempts to study the details of cylindrical collapse. Before describing this work we would like to discuss the coupling of matter motion and geometry in cylindrical systems. It was mentioned earlier that the Einstein-Rosen waves have only one possible mode of polarization. Such waves are generated by radial motion of matter. Combination of both radial and angular motion or radial and vertical (i.e., nontrivial $v_z = v_z(r)$, motion along the axis of the symmetry) motion is not sufficient for generating the other mode of polarization. This other mode (Kompaneets-Jordan-Ehlers waves) are generated only when the matter moves in angular, vertical and radial directions simultaneously.

The matter motion and the geometry are further decoupled when $p=\rho$. In this case the matter terms appear only in the equation for the metric component γ. Liang (1976) has shown that in this case matter oscillations will not induce the Einstein-Rosen waves, or vice versa.

Finally the linearity of the ψ equation in vacuum (eq. 5) allows superposition of solutions in this region. This feature enabled Marder (1958a) and others to study the propagation of Einstein-Rosen waves on a Levi-Civita background.

Marder (1958a) examined the effect of the passage of a weak wave packet through a matter cylinder by comparing the initial and final steady state solutions. He derived various relations between parameters of the wave packet and changes in the matter variables. Like his static solution this calculation was done for an unphysical equation of state.

Rao (1971) calculated the exterior field for pure radiation stress energy tensor: $T_{ab}=\rho k_a k_b$; $k_d k^d=0$. Cocke (1966) matched an interior dust solution (derived from Friedman universe) to an exterior vacuum solution. He found that the C energy flux, just outside the boundary of the dust, is in the same direction as the dust motion. He concluded, erroneously as we shall later see, that a collapsing cylinder absorbs gravitational energy.

Liang (1974) matched an interior collapsing dust solution to the exterior Levi-Civita metric. He concluded that the dust collapses in finite proper time to a naked line singularity. The assumptions used to obtain this result prohibit the generation of gravitational waves and thereby leave the result in question.

Thorne (1972) proved that a cylindrical system cannot have an event horizon, i.e. there is no cylindrical analog to the spherical black hole. Therefore a collapsing cylinder can either bounce and approach a steady state with an exterior Levi-Civita metric or collapse to a naked singularity.

NUMERICAL CYLINDRICAL COLLAPSE

We have recently developed (Piran, 1978a,b) a numerical code which traces the geometrohydrodynamic evolution of a general relativistic (rotating) perfect fluid cylinder with Γ law equation of state: $p=(\Gamma-1)ne$. p,n,e and Γ are the pressure, baryon number density, internal specific energy and adiabatic index respectively.

A brief description of the code is given in the Appendix. (See lecture by Wilson in this volume and Piran, (1978b) for details of the numerical methods).

Using this code we studied the collapse of cylinders. The initial configuration is a Newtonian cylinder in equilibrium whose internal energy is suddenly decreased. The matter collapses, reaching relativistic velocities. As expected, for values of Γ greater than 1, the matter bounces and does not collapse to a line singularity (Thorne, 1972). After the bounce a shock wave propagates outwards while the massive core oscillates, approaching a steady state. This general hydrodynamic behavior, even for low Γ values, resembles the features of spherical collapse with a hard equation of state (May & White, 1965). Typical variations in the central density are of order 10^4, but it ranges between $6\ 10^2$ and $6\ 10^6$ in extreme cases.

No gravitational waves are generated during the initial infall phase. However, large pulses of gravitational radiation are generated during the bounce period (see Fig.1a,b,c). The total energy of the emitted gravitational waves is highly dependent on the equation of state. While less than 1% of the initial rest mass energy is radiated when $\Gamma = 7/6$, about 65% of the initial rest mass energy is radiated as gravitational waves when $\Gamma = 2$. With high Γ value second and third pulses of gravitational waves carrying substantial amounts of energy are emitted during the oscillation of the core (see Fig. 1d).

A decrease in the mass per unit length makes the system more Newtonian. Keeping other parameters fixed, this results in a decrease in the efficiency of gravitational wave generation.

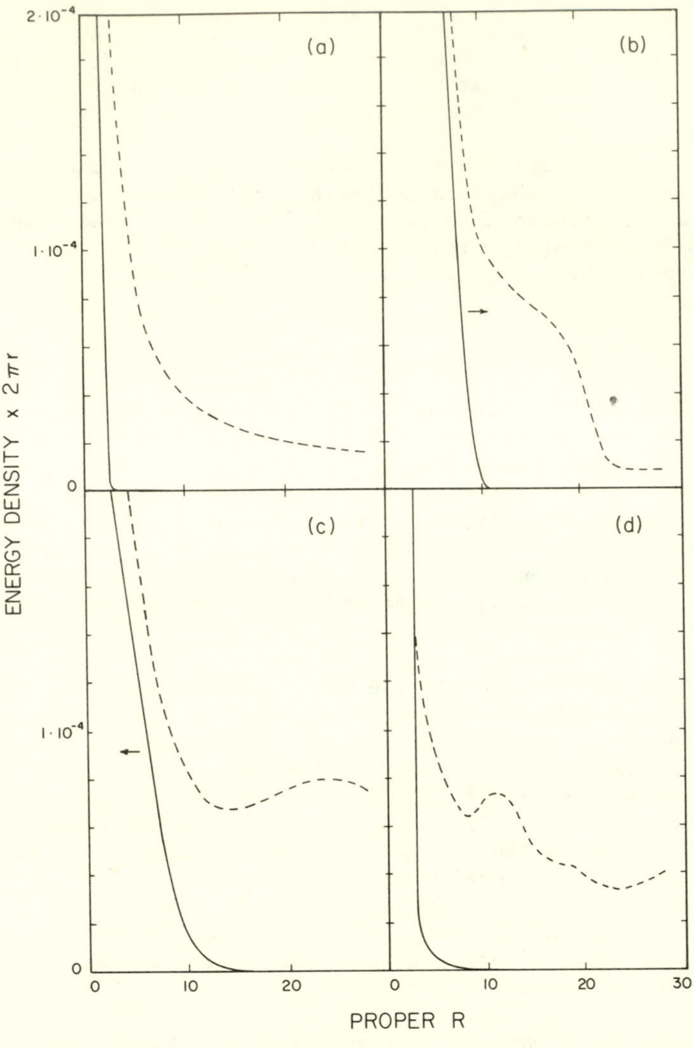

Figure 1

Matter energy density ρ (solid line) and geometric energy density (dashed line), multiplied by $2\pi r$, as a function of proper distance. The arrow describes the direction of motion of the matter. The adiabatic index is $\gamma = 5/3$ and the matter is nonrotating. The initial internal energy was reduced by a factor of 10. a: First bounce - moment of maximal contraction - no gravitational waves have been generated so far. b: The first gravitational wave pulse. c: Propagation of the first gravitational wave pulse. d: Propagation of the second gravitational wave pulse, during oscillations of the core.

CYLINDRICAL SYSTEMS

The basic features of the generation of gravitational waves presented here are in correspondence with the quadrupole moment approximation (Landau & Lifshitz, 1962). In this approximation the energy emitted as gravitational radiation is proportional to ΔT^{-6}, where ΔT is the time scale for changes in the quadrupole moment. Most of the radiation is emitted during the bounce and the oscillation when ΔT is shortest.

The initial exterior vacuum region has a Levi-Civita line element. Since the configuration is almost Newtonian $c \simeq 2M$, $R \simeq 1$. As long as there are no gravitational waves the exterior metric changes but keeps its Levi-Civita form. The main change is a decrease in R, which causes the non zero C energy distribution of the exterior region to simply redistribute itself. Therefore there is an inwards C energy flux. This led Cocke (1966) to the conclusion that the matter is absorbing energy while it collapses. Whenever a gravitational wave packet is emitted, there is a definite outgoing C energy flux (see Fig. 1). The region of this gravitational wave pulse is characterized by a sharp deviation from the initial situation of approximate conformal flatness (see Fig. 2).

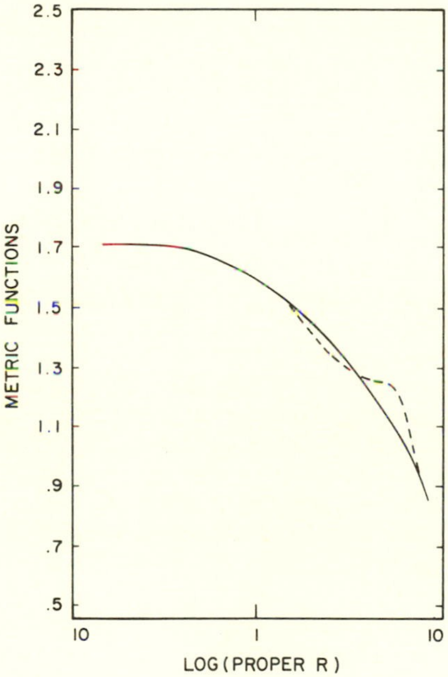

Figure 2

The metric functions $\sqrt{g_{rr}} = \sqrt{g_{zz}}$ (solid line) and $\sqrt{g_{\varphi\varphi}}/r$ (dashed line). A gravitational wave pulse is clearly observed in the region where $g_{rr} \neq g_{\phi\phi}/r^2$.

After the oscillations of the core die out and the gravitational waves propagate to "infinity" the exterior solution reaches a steady state with a Levi-Civita metric, $c \neq 2M$ and $R<1$ corresponding to the general relativistic nature of the final configuration.

Addition of rotation does not change the basic picture of collapse and bounce. However, the bounce is shorter and occurs at larger radii. We would expect this behavior from the nature of the centrifugal force, which stiffens the matter and can be compared with an equation of state with $\Gamma=2$. (For a cylinder with a constant density and $\partial p/\partial r \sim r^{-(2\Gamma+1)}$; $F_{grav} \sim M^2 r^{-3}$; $F_{cen} \sim r^{-5}$ where M and L are the mass and the radius and F_{grav} and F_{cen} are the Newtonian gravitation and centrifugal forces.)

The basic difference in the hydrodynamic picture is the ejection of both a large fraction (up to 40%) of the mass and an even larger fraction of the angular momentum. This results in smaller oscillations of the core.

The amount of gravitational waves emitted is larger than in the corresponding nonrotating configuration. A modest amount of angular momentum is most efficient in increasing the production of gravitational waves. High amounts of angular momentum halt the collapse very quickly before the matter manages to gather high enough infall velocity. Very small amounts of angular momentum do not have any significant effect and the collapse is halted by the increasing pressure, as in the nonrotating case.

CONCLUSIONS

A clear picture of cylindrical collapse now emerges. The initial static configuration has an exterior Levi-Civita metric. During the early infall phase the exterior metric changes but it keeps its Levi-Civita form. (As in the spherical case it seems that the constraint equations are sufficient to describe the system during this stage.) No gravitational waves are emitted during this period. The matter bounces, the faster and harder the bounce is the bigger the amount of gravitational energy radiated. After some oscillations the system settles down to a steady state with a different exterior Levi-Civita metric.

The fraction of rest mass energy which is emitted as gravitational wave energy is surprisingly large. This is very encouraging compared with previous estimates of the efficiency of gravitational wave generation. Clearly this high efficiency is partially due to the extreme asymmetry of the system and cannot be used as a quantitative estimate for a realistic astrophysical collapse. However, it is reasonable to assume that the qualitative picture will be the same. It is unlikely that any significant amount of gravitational wave energy will be emitted during the initial infall of any collapse, while one should expect generation of gravitational waves during a bounce, e.g. in the formation of a neutron star.

Finally, in none of the cases which we calculated did the interior become singular. This does not prove, but it suggests, that the pressure can halt a cylindrical collapse before a singularity is reached. This feature might be related to the Newtonian behavior of small enough regions near the axis. Clearly this gives further support to the cosmic censorship conjecture.

I would like to thank J. R. Wilson for communicating his results prior to publication and B. S. DeWitt, M. R. Brown and P. Gleichauf for fruitful discussions. This work was supported by NSF grant PHY77-22489.

APPENDIX: A NUMERICAL CYLINDRICAL CODE

The hydrodynamic variables (density, pressure and momentum densities) are determined by $T^{\mu}_{\nu;\mu}=0$ and by the equation of state. The geometry is described by, according to the ADM formalism (Arnowitt, Deser & Misner, 1962), the three metric of a time slice g_{ij} and by the extrinsic curvature tensor K_{ij} giving the imbedding of this slice in the four dimensional space time. Following Wilson (1977, and lecture in this volume), the coordinate freedom is used to specify the shift vector N^i in such a way that g_{ij} is diagonal with $g_{rr}=g_{zz}$. Apart from the numerical convenience this serves to eliminate an unphysical degree of freedom from the system. Maximal slicing (Lichnerowicz, 1944) is used to determine the lapse function N. Again this fixes $K^i_i=0$ so that one of the K^i_j's can be eliminated. The rest of the geometric variables are determined either from the constraint equations or from the ADM evolution equations. (With one possible state of polarization at least two geometric variables must be determined from evolution equations). Details of the numerical methods are discussed in Piran (1978b).

REFERENCES

Beck, G. (1925). Zür Theorie binärer Gravitationsfelder. Z. Physik, 33, pp. 713-721.

Boardman, J. & Bergman, P. G. (1959). Spherical Gravitational Waves. Phys. Rev., 115, pp. 1318-1324.

Chandresekhar, S. & Fermi, E. (1953). Problems of Gravitational Stability in the Presence of Magnetic Field. Ap. J., 118, pp. 116-141.

Cocke, W. J. (1966). Some Collapsing Cylinders and their Exterior Vacuum Metrics in General Relativity. J. Math. Phys., 7, pp. 1171-1178.

Einstein, E. & Rosen, N. (1937). On Gravitational Waves. J. Franklin Inst., 223, pp. 43-54.

Fierz, M. (1957). Quoted by Weber & Wheeler (1957).

Harrison, B. K., Thorne, K. S., Wakano, M. & Wheeler, J. A. (1965). Gravitation Theory and Gravitational Collapse. University of Chicago Press, Chicago, Illinois.

Harrison, B. K., Wakano, M. & Wheeler, J. A. (1958). In Onzieme Conseil de Physique Solvay, La Structure et l'evolution de l'universe. Stoop, Brussels, Belgium.

Jordan, P. Ehlers, J. & Kundt, W. (1960). Strenge Lösungen der Feldgleichugen der Allgemeinen Relativitatztheorie. Akad. Wiss. Mainz Abh. Math.-Nat. Kl. Jahrg., No. 2.

Kompaneets, A. S. (1958). Strong Gravitational Fields in Free Space. Zh. Eksp. l. Theort. Fiz., 34, pp. 953-955. (English translation: Soviet Physics J.E.T.P., 7, pp. 659-660.

Kuchar, K. (1971). Canonical Quantization of Gravitational Waves. Phys. Rev. D10, pp. 955-986.

Lamb, H. (1927). Hydrodynamics. Cambridge University Press, Cambridge, England.

Landau, L. D. & Lifshitz, E. M. (1962). The Classical Theory of Fields. Addison-Wesley Publishing Co., Reading, Mass.

Levi-Civita, T. (1919). Rend. Acc. Lincci, 28, pp. 3, 101.

Lewis, T. (1932). Some Special Solution of the Equations of Axially Symmetric Fields. Proc. Roy. Soc. Lond., A136, pp. 176-185.

Liang, E. P. T. (1974). Some Exact Models of Inhomogeneous Dust Collapse. Phys. Rev. D10, pp. 447-457.

Liang, E. P. T. (1976). Dynamics of Primordial Inhomogeneities. Ap. J., 204, pp. 235-250.

Lichnerowicz, A. (1944). L'integration des equations de la Gravitation Relativiste et le Probleme des n corps. J. Math. Pures et Appl., 23, pp. 37-63.

Lin, C. C., Mestel, L. & Shu, F. H. (1965). The Gravitational Collapse of a Uniform Spheroid. Ap. J., 142, pp. 1431-1446.

CYLINDRICAL SYSTEMS

Marder, L. (1958a). Gravitational Waves in General Relativity I. Proc. Roy. Soc. Lond., 244, pp. 527-537.

Marder, L. (1958b). Gravitational Waves in General Relativity II. Proc. Roy. Soc. Lond., 246, pp. 33-143.

Marder, L. (1959). Flat Space-Time with Gravitational Fields. Proc. Roy. Soc. Lond., 252, pp. 45-50.

May, M. M. & White, R. H. (1965). Hydrodynamic Calculations of General Relativistic Collapse. Phys. Rev., 141, pp. 1232-1241.

Ostriker, J. (1964). The Equilibrium of Polytropic and Isothermal Cylinders. Ap. J., 140, pp. 1056-1066.

Piran, T. (1978a). Cylindrical General Relativistic Collapse. To Phys. Rev. Lett. 41, pp. 1085-1088.

Piran, T. (1978b). Numerical Codes for Cylindrical General Relativistic Collapse. In preparation.

Rao, J. K. (1971). Radiating Levi-Civita Metrics. J. of Phys., A4, pp. 17-20.

Rosen, N. (1953). Bull. Research Council Israel, 3, pp. 328-340.

Rosen, N. (1958). Energy and Momentum of Cylindrical Gravitational Waves. Phys. Rev., 110, pp. 291-292.

Smarr, L. (1978). Lecture in this volume.

Stachel, J. J. (1966). Cylindrical Gravitational News. J. Math. Phys., 7, pp. 1321-1331.

Thorne, K. S. (1965a). Geometrodynamics of Cylindrical Systems. Ph.D. Thesis. Princeton University, Princeton, N. J. Unpublished.

Thorne, K. S. (1965b). Energy of Infinitly Long Cylindrically Symmetric Systems in General Relativity. Phys. Rev., 138, pp. B251-B256.

Thorne, K. S. (1972). Nonspherical Gravitational Collapse - A Short Review in Magic Without Magic, ed. J. R. Klauder. Freeman Company, San Francisco, CA.

van Stockum, W. J. (1937). The Gravitational Field of a Distribution of Particles Rotating about an Axis of Symmetry. Proc. Roy. Soc. Edin., 57, pp. 135-143.

Vishveshwara, C. V. & Winicour, J. (1977). Relativistically Rotating Dust Cylinders. J. Math. Phys., 18, pp. 1280-1284.

Weber, J. & Wheeler, J. (1957). Reality of the Gravitational Waves of Einstein and Rosen. Rev. Mod. Phys., 27, pp. 509-520.

Wilson, J. R. (1977). Lecture given at the GR8 Conference, Waterloo, Canada. Unpublished.

Wilson, J. R. (1978). Lecture in this volume.

Wilson, W. (1920). Space-Time Manifolds and Corresponding Gravitational Fields. Phil. Mag., 40, pp. 710-725.

A NUMERICAL METHOD FOR RELATIVISTIC HYDRODYNAMICS

James R. Wilson

Lawrence Livermore Laboratory

I. INTRODUCTION

The purpose of the project to be described is to develop a numerical scheme that will calculate the gravitational radiation emitted by stars that collapse and either bounce in a deep gravitational potential or form a black hole. The equations are naturally separated into two parts, the hydrodynamic equations and the gravitational field equations. The hydrodynamic equations have been developed in a series of works. In the papers (Wilson 1972, Wilson 1977) the axially symmetric magnetohydrodynamic equations were solved numerically for flow in a fixed metric background space. In particular some simple flows around black holes were investigated. Next the hydro numerical scheme was extended to calculating the hydrodynamics in a metric calculated as if the material flow were stationary. With this extension some calculations of supermassive rotating stars were performed (Wilson 1978). In the mean time methods for numerical treatment of the full dynamic material-free gravitational field evolution had been developed (Eppley, Smarr 1979) using the A.D.M. 3+1 system. The present work is a union of the relativistic hydrodynamics with the full treatment of the gravitational field equations. This work was done in collaboration with L. Smarr.

In order to calculate axial symmetric flows in which flow patterns become complicated it is felt necessary to use an Eulerian fluid description. Most previous work with strong field hydrodynamics has been done in a Lagrangian framework (May and White 1966, Pachner 1972). A Lagrangian fluid description is one in which the coordinates move with the material, in a Eulerian fluid description the material moves through a fixed coordinate system. Therefore the first point of the present project was to develop a spherical Eulerian calculational scheme that can be generalized to axial symmetry. In section II below we will describe this spherical program in full detail to show the numerical methods used. In section III the results of a few simple examples using the spherical program are given. Then in section IV the equations for the axially symmetric case are given along with a description of the numerical techniques that are peculiar to the two space dimensional case. Finally in section V the results of several axially symmetric calculations are presented. These results are very preliminary since the computer program development was completed just before the conference. In section V prospectus is given for the numerical program.

II. SPHERICALLY SYMMETRIC EULERIAN HYDRODYNAMICS

We will take as our basic hydrodynamics equations:

particle conservation $(\rho U^{\mu})_{;\mu} = 0$ (1)

Energy momentum conservation $T^{\mu\nu}{}_{;\mu} = 0$ (2)

Perfect fluid $T_{\mu\nu} = (\rho+\varepsilon+P)\, U_{\mu}\, U_{\nu} + P g_{\mu\nu}$ (3)

Velocity normalization $U^{\mu}\, U_{\mu} + 1 = 0$ (4)

Equation of state $P = P(\rho,\varepsilon)$ (5)

We then introduce as basic variables

$$D = \rho U^{t},\ E = \varepsilon U^{t},\ S_{\mu} = (\rho+\varepsilon+ P)\, U^{t}\, U_{\mu},\ V^{\mu} = U^{\mu}/U^{t}$$

in order to make the equations as similar as possible to the non-relativistic hydrodynamic equations. This allows the use of our experience with Newtonian hydrodynamics.

The equations of motion are now: (Wilson 1978).

$$\frac{\partial}{\partial t}\,(D\sqrt{-g}) + \frac{\partial}{\partial x^{i}}\,(DV^{i}\,\sqrt{-g}) = 0 \quad \text{(i is space index)} \tag{6}$$

$$\frac{1}{\sqrt{-g}}\left[\frac{\partial}{\partial t}\,(S_{i}\sqrt{-g}) + \frac{\partial}{\partial xj}\,(Si\ V^{j}\ \sqrt{-g})\right] + \frac{\partial\rho}{\partial x^{i}} +$$

$$\frac{1}{2}\ \frac{\partial g^{\alpha\beta}}{\partial x^{i}}\ \frac{S_{\alpha}S_{\beta}}{S^{t}} = 0 \tag{7}$$

$$\frac{\partial}{\partial t}\,(E\ \sqrt{-g}) + \frac{\partial}{\partial x^{i}}\,(EV^{i}\ \sqrt{-g}) = -P\left[\frac{\partial}{\partial t}\ (U^{t}\ \sqrt{-g}) +\right.$$

$$\left.\frac{\partial}{\partial x^{i}}\,(U^{t}V^{i}\ \sqrt{-g})\right] \tag{8}$$

These equations thus far imply no particular spacial symmetry. Note that in place of a total energy conservation equation we have chosen to evolve the thermal energy. With a thermal energy equation such as (8) it is easier to keep the entropy well behaved than with a total energy equation. Numerical errors will show up in the total energy non-conservation i.e. the mass of the star will drift in time.

For our example we take the spherical isotropic coordinate system defined by

$$ds^2 = -\alpha^2 dt^2 + A^2 (dr + \beta^r dt)^2 + A^2 r^2 d\Omega^2 \qquad (9)$$

$$\sqrt{-g} = \alpha r^2 A^3$$

The Einstein gravitational field equations determine (α, β^r, A) and will be discussed later.

We will endeavor to difference the hydrodynamics equations to second order accuracy in the time and space intervals. This is not always practical. In particular, as will be seen later, the convective terms will be differenced so as to approach first order for sharp spacial gradients. The momentum convective term (a convective term refers to a term such as $\partial_i (V^i D)$) is of mixed order because it is hard to have V^i at the correct time level. The artificial viscosity which will be introduced later in the difference equations is not properly time centered. In practice on test examples with Newtonian hydrodynamics, these deviations from second order have been found to introduce small errors. The reason is that over most of space and time for our problems the spacial gradients are moderate, the convective terms will be nearly second order, the artificial viscosity only acts over the region where there is a shock wave, and the velocity V^i is a smooth function except at shock fronts. That is, except at shock fronts the equations may be considered second order accurate and in practice first order equations handle shocks as well as or better than second order equations. See lecture of Smarr earlier in this volume for an introduction to the finite differencing.

We now proceed to write down the difference equations. In order to derive the difference equations we visualize two sets discrete radial coordinates

RA_J and RB_J where $RB_J = .5x(RA_J + RA_{J+1})$.

The velocity like quantities S_r, V^r, β^r, V_g^r are considered to be centered at RA_J and all the others $(D, E, U^t A, \alpha, K^i_j)$ are centered at RB_J. The quantity V_g represents the velocity of the grid point RA_J which is introduced to allow the numerical difference grid to follow the material. In practice V_g will be chosen as a smooth function of radius so that the zone intervals will be slowly varying. This is necessary in order that waves propagate well through the discrete mesh. The quantity K^i_j is the extrinsic curvature and will be discussed later in the field equation section.

Operator splitting will be used extensively. If we have time evolution equation

$$\dot{A} = B + C + D + \ldots.$$

in the difference equation, we replace this equation by the equations

$$\dot{A} = B,\ \dot{A} = C,\ \dot{A} = D,\ \dot{A} = \ldots.$$

This enables one to advance the variable A in several simple steps instead of in one complicated step.

First let us consider the number density equation (6)

$$\frac{\partial}{\partial t}(D\alpha r^2 A^3) + \frac{\partial}{\partial r}(D(V-V_g)\alpha r^2 A^3) = 0 \qquad (10)$$

First split off $\frac{\partial}{\partial t}(D\alpha) = 0$. When in the course of the calculation α is advanced in time D is advanced by $D^{new} = D^{old}(\alpha old/\alpha new)$. This procedure insures that particle number is conserved to numerical round-off. Similarly when r and A are changed a corresponding change in D is made.

In order to conserve particles the convective term in (10) is evaluated by finding a zone to zone particle flux and then taking out of zone the same number that goes into its neighbor. First define a first order upwind flux though the Jth zone boundary:

$$F_1 = RA_J^2(\alpha_J A_J^3 + \alpha_{J-1}A_{J-1}^3) \times (V_J - V_{gJ}) \times \begin{Bmatrix} D_J & \text{if } (V_J - V_{gJ}) < 0 \\ D_{J-1} & \text{if } (V_J - V_{gJ}) > 0 \end{Bmatrix} \times \frac{1}{2} \qquad (11)$$

Next define a second order flux

$$F_2 = RA_J^2(\alpha_J A_J^3 + \alpha_{J-1}A_{J-1}^3)(V_J - V_{gJ}) \cdot (D_J + D_{J-1} + 2(V_J - V_{gJ})\Delta t(D_J - D_{J-1})/(RB_J - RB_{J-1}))/4 \qquad (12)$$

where Δt is the time step.

Note that zone centered quantities (eg. α, A) are averaged to get a more accurate value on the zone boundary. The flux F_2 is second order accurate because the density is interpolated to a distance, $(V-V_g)\Delta t$, backwards from the zone boundary. This is the value of the density which would physically be carried by the fluid to the zone boundary in one time step. The upwind scheme uses D from the zone center upwind from the zone boundary, too large a distance back in general.

These fluxes are weighted in such a manner that the equations approach first order for large gradients.

$$\mathrm{WT} = \text{MAXIMUM OF} \begin{Bmatrix} |D_J - D_{J-1}| \;/(D_J + D_{J-1}) \\ |E_J - E_{J-1}| \;/(E_J + E_{J-1}) \end{Bmatrix}$$

The energy term comes in because the same weight should be used in the energy equation as in the density equation in order to keep the specific energy from diffusing with respect to the matter.

$$F_J = F_1 \; \mathrm{WT} + F_2 \; (1-\mathrm{WT}) \tag{13}$$

The weighted flux is then differenced between zone boundaries to update D:

$$D_J^{new} = D_J^{old} - (F_{J+1} - F_J) \; 3/(\alpha_J \; A_J^3 \cdot (RA_{J+1}^3 - RA_J^3)) \tag{14}$$

Note that in spherical coordinates we can express the denominator as a proper volume difference

$$\sqrt{-g} \;\; dR = \alpha \; A^3 \; d \; (r^3) \; /3$$

The energy equation is

$$\frac{\partial}{\partial t} (E \; \alpha \; r^2 \; A^3) + \frac{\partial}{\partial r} (E \; (V-V_g) \alpha \; r^2 \; A^3) + P \left[\frac{\partial}{\partial t} (\alpha \; r^2 \; A^3 \; U^t) + \frac{\partial}{\partial r} (r^2 \; A^3 \; U^t \; (V-V_g)) \right] = 0 \tag{14A}$$

The first two terms are treated the same as the corresponding terms in the density equation. We let $P = E(\Gamma-1) \; / \; U^t$ (Γ is assumed to be constant only over a time step). The third term can be expressed as

$$\frac{1}{E} \frac{\partial E}{\partial t} + \frac{\Gamma-1}{\alpha \; r^2 \; A^3 \; U^t} \; \frac{\partial}{\partial t} \; (\alpha \; r^2 \; A^3 \; U^t) = 0 \tag{15}$$

and so we let $E^{new} = E^{old} \; (\alpha^{old}/\alpha^{new})^{\Gamma-1}$ and so forth for A, r, U^t. The last part of the energy equation is

$$\frac{1}{E} \frac{\partial E}{\partial t} + \frac{\Gamma-1}{\alpha \; r^2 \; A^3 \; U^t} \frac{\partial}{\partial r} \left[\alpha \; r^2 \; A^3 \; U^t \; (V-V_g) \right] = 0$$

This could be viewed as convection of U^t and treated the way the convective terms in D are treated, but we have chosen instead the following procedure;

$$Q = \Big[(\alpha_{J+1} A^3_{J+1} U^t_{J+1} + \alpha_J A^3_J U^t_J)$$

$$RA^2_{J+1} (V_{J+1} - V_{gJ+1})$$

$$- (\alpha_J A^3_J U^t_J + \alpha_{J-1} A^3_{J-1} U^t_{J-1}) RA^2_J$$

$$(V_J - V_{gJ})\Big] /2$$

$$H = 3Q (\Gamma-1)\Delta t \,/\, \big[\alpha_J A^3_J U^t_J (RA^3_{J+1} - RA^3_J) \big]$$

$$E^{new} = E^{old} (1 - H/2)/(1 + H/2) \tag{16}$$

The momentum equation is:

$$\frac{1}{\alpha r^2 A^3} \left[\frac{\partial}{\partial t} (S \alpha r^2 A^3) + \frac{\partial}{\partial r} (S \alpha r^2 V^r A^3)\right] + \frac{\partial P}{\partial r} +$$

$$\frac{1}{2} \frac{\partial g^{\alpha\beta}}{\partial r} \frac{S_\alpha S_\beta}{S^t} = 0$$

The first two terms are treated the same as they were treated in the density equation, except that because of centering we use an average. We use S as short for Sr.

$$S^{new}_J = S^{old}_J (\alpha^{old}_J + \alpha^{old}_{J-1})/(\alpha^{new}_J + \alpha^{new}_{J-1})$$

and appropriate average velocities, etc. are used in the transport term. The pressure acceleration is simply

$$S^{new} = S^{old} - \Delta t \, (P_J - P_{J-1})/(RB_J - RB_{J-1}) \tag{17}$$

The gravitational acceleration is

$$\frac{\partial S}{\partial t} + \frac{1}{2} \left[- \frac{\partial}{\partial r} \left(\frac{1}{\alpha}\right) S^2_t + 2 \frac{\partial}{\partial r} \left(\frac{\bar{\beta}}{\alpha^2}\right) S\, S_t \right.$$

$$\left. + \frac{\partial}{\partial r} \left(\frac{1}{A^2} - \frac{\bar{\beta}^2}{\alpha^2}\right) S^2 \right] /S^t = 0 \tag{18}$$

The velocity normalization equation (4) is used to find S^t and S_t, and

$$\bar{\beta} = (\beta_{J+1} + \beta_J)/2$$

otherwise the metric gradient terms are evaluated the same as the pressure gradient was evaluated.

The velocity normalization equation is

$$g^{\alpha\beta} S_\alpha S_\beta + (D + E + PU^t)^2 = 0 \tag{19}$$

This equation is centered at RB_J where D,E,P are centered. We let

$$\sigma = D + E + PU^t$$

$$S_J = \left[S_J / (\sigma_J + \sigma_{J-1}) + S_{J+1}/(\sigma_J + \sigma_{J+1}) \right] \sigma_J$$

This average is used because in most situations the velocity is a much smoother function than momentum density. The above is somewhat like averaging velocities. Equation (19) becomes a quadratic expression:

$$-S_t^2 / \alpha^2 + 2 \bar{\beta} S_t \bar{S}/\alpha^2 + (1/A^2 - \bar{\beta}^2/ \alpha^2) \bar{S}^2 + \sigma^2 = 0$$

which is solved for S_t (20)

After S^t is found by raising the index. The new U^t is given by

$$U^t = \sigma/S^t \tag{21}$$

This completes the differencing of the differencial equations (1)-(4). We turn now to some important numerical considerations.

As stated earlier an artificial velocity is used in order to smear out shock waves so they can be represented in a finite mesh. A velocity pressure is determined by

$$Q_J = k\, D_J\, (V_{J+1} - V_J)^2 \tag{22}$$

where k is a constant of the order of unity. Since it is desirable to use artificial velocity only when material is compressing, Q_J is set to zero if either

$$\delta_J V \equiv (V_{J+1} - V_J) > 0 \qquad \text{or}$$

$$\delta_J \left[\alpha\, r^2 A^3 U^t (V - V_g) \right] > 0$$

Note, these are not a true test for expansion, but it is all right to have an occasional nonzero Q during expansion. Q is added to pressure in the acceleration equation for S (17) and the compression term for E (14A). It might be argued that Q should also be added to P in $(D + E + P\, U^t)$. However, the space integral of Q is always small and in the present program this correction is not made.

The grid velocity is in principle a completely arbitrary function. It could be set equal to the material velocity to obtain a Lagrangian fluid representation. However, we are interested in a system that will be appropriate to the two space dimension case where gravitational waves are present. A smooth grid is necessary to propagate waves well. As a way to generate a smooth grid and at the same time have the grid follow the material in some average sense, we let

$$V_{g2} = h\, RA_2(V_2/RA_2 + V_3/RA_3 + V_4/RA_4)/3 \qquad (V_1 = RA_1 = 0) \tag{23}$$

where h is an input constant that varies from 0 to 1 depending on how closely we desire the grid to follow the material. The subscript refers to zone numbers. The new position of the second zone is given by:

$$RA_2^{new} = RA_2^{old} + V_2 \tag{24}$$

All other RA are found by letting

$$RA_{J+1} - RA_J = g \ (RA_J - RA_{J-1}) \quad J>1 \tag{25}$$

where g is a constant selected so that RA_{JMAX} has the desired value. After new grid positions are found

$$V_{gJ} = (RA_J^{new} - RA_J^{old})/\Delta t. \tag{26}$$

After the new grid positions are found the densities (D,E,S) are changed according to the change in the RA such as to conserve mass, energy, and momentum.

The time step, Δt, must be limited both for accuracy and for stability. The Courant hydrodynamic stability condition requires

$$\Delta t < (RA_{J+1} - RA_J)/C \tag{27}$$

for all J, where C is the sound speed.

The momentum diffusion by artificial viscosity also puts a stability limit of

$$\Delta t < (RA_J - RA_{J-1})/(4k|V_J - V_{J-1}|) \tag{28}$$

where k is the constant that appeared in the artificial viscosity. For a discussion of the stability requirements see Richtmeyer (1957). A limit on Δt for accuracy in the transport terms is taken as

$$\Delta t < (RB_J - RB_{J-1})/(4|V_J - V_{gJ}|) \tag{29}$$

The stability requirements on Δt were determined by linearizing the equations. In practice it is found necessary to restrict Δt to about half the value found from equations (27) and (28) because of non-linearities.

We now discuss the gravitational equations. The metric has been set earlier in equation (9). Following the ADM 3 + 1 approach to gravity (see York 1979) we select α and β to get desirable coordinates. The lapse function is selected by the maximal slicing condition $K_1^1 = o$ (K_1^i is the extrinsic curvature tensor). This leads to the following equation for α:

$$\frac{1}{r^2}\frac{\partial}{\partial r}\left(r^2 A \frac{\partial\alpha}{\partial r}\right) = \alpha A^3 \Big[(\rho+\varepsilon)(\gamma^2 - .5) + P(\gamma^2 + .5)$$
$$+ K^i_{\cdot j} K^j_{\cdot i}\Big] \equiv \alpha A^3 \rho_\alpha/2$$
$$\gamma = U^t\alpha \quad G = 1/8\Pi \quad c = 1. \tag{30}$$

The term γ reduces to usual special relativistic γ in flat space. We select β to make the space part of metric be isotropic: ie,

$$g_{rr} = g\theta\theta/r^2 = g\phi\phi/r^2\sin^2 .$$

This leads to β satisfying the following condition

$$\beta = -\frac{3r}{2}\int_r^\infty \alpha K^r_r \, dr/r \tag{31}$$

The momentum constraint equation (see York 1979) is used to find the extrinsic curvature tensor

$$K^r_r = 1/(r^3 A^3)\int_o^r \alpha S\, r^3 A^3 dr. \tag{32}$$

The Hamiltonian constraint equation is used to find A. This equation can be put in the form:

$$\frac{1}{r^2}\frac{\partial}{\partial r}\left(r^2 \frac{\partial A^{1/2}}{\partial r}\right) = A^{5/2}\Big[(\rho+\varepsilon)\gamma^2 + P(\gamma^2 - 1) +$$
$$K^i_{\cdot j}K^j_{\cdot i}/2\Big] \equiv A^{5/2}\rho_H \tag{33}$$

These equations not being partial differential equations are easily solved and the differencing won't be discussed except for a peculiar point in equation (33). The quantity $A^3\rho_H$ is a quantity that should be preserved, equation (33) is solved for A in such a manner that the above quantity doesn't change. Also in equation (31) the integral is extended to infinity in r by assuming $K^r_r \sim 1/r^3$ outside the calculational grid. In the boundary condition for A we assume $K^i_{\cdot j} K^j_{\cdot i} \sim 1/r^6$

so that (33) can be interpreted for $r \to \infty$ and A is then identified with the mass by $A = 1 + m/2r$ as $r \to \infty$.

An alternative scheme that has also been applied is based on a coordinate system analogous to Schwarzschild coordinates. That is

$$ds^2 = -\alpha^2 dt^2 + A^2(dr + \beta\, dt)^2 + r^2\, d\Omega^2. \tag{34}$$

Then equation 30-33 are replaced by

$$\frac{A}{r^2}\frac{\partial}{\partial r}\left(r^2/A \frac{\partial\alpha}{\partial r}\right) = \alpha\rho_\alpha/2 \tag{35}$$

$$\beta = - \alpha \, r K^r_r / 2 \tag{36}$$

$$K^r_r = \frac{1}{r^3} \int_0^r S \alpha r^3 dr \tag{37}$$

$$A^2 = 1/ \left(1 - \frac{1}{r} \int r^2 \rho_H \, dr\right) \tag{38}$$

These coordinates schemes can be extended to axially symmetric systems (see Smarr 1979). Which method is best has not been resolved. Most work so far on the axially symmetric case has been done with coordinates isotropic in R,Z which works fine for non-black hole calculations, but hasn't worked well for black hole formation. (See section IV). Another alternative simple spherical coordinate scheme is

$$ds^2 = - \alpha^2 dt^2 + (dr + \beta dt)^2 + B^2 r^2 \, d\,\Omega^2 \tag{39}$$

This system has not yet been tried.

The way the metric is treated in the axially symmetric case corresponds in the case of spherical symmetry to evolving A by

$$\dot{A} = - \alpha \, A \, K^r_r + \frac{\partial}{\partial r} (\beta A) \tag{40}$$

Hence we also wrote the spherical programs finding A by equation (40). Actually this method works better for black hole formation than finding A by the Hamiltonian constraint equation. This result arises because A tends to large values when a black hole is formed, and sometimes, for example in the Schwarzschild coordinates, 2m/r can become larger than 1 and stop the calculation. P. Chrzanowski (private communication) has developed a Lagrangian spherical hydrodynamics computer code which uses maximal slicing. For strictly spherical calculations this latter approach is probably more accurate than the Eulerian method described above.

III. EXAMPLES OF SPHERICAL CALCULATIONS

In stellar evolution where relativistic effects are important, two limiting equations of state are appropriate. Adiabatic index $\Gamma = 4/3$ for supermassive stars and $\Gamma = 2$ for stiff neutron star cores. The first calculation we would like to present is for $\Gamma = 4/3$.

A test of the calculational method was done by setting the initial pressure very small compared to the pressure necessary for support. The proper time for the fall of central α_c to zero can be compared with the analytic results. For a polytropic density configuration with 1% of equilibrium pressure initially the α_c fell to zero within 2% of the correct time. This was for central density rise of a factor of 3×10^4. For a Lagrangian hydrodynamic code this would not be good, but for an Eulerian code with all the material flow in the grid we consider this good.

Next we looked at the question of how homologous is collapse for a supermassive star. Two calculations were performed, one with a N = 3 polytropic density distribution and another with a more peaked density distribution:

($\rho = \exp(-r/r_1)/(r + 2\,\Delta r)^2$ r is the central zone size and $r_1 \gg \Delta r$).

The results are shown in Figures 1 A, B. We see that for a polytropic distribution the collapse is almost homologous, and even for the very peaked distribution most of the matter is falling into the hole shortly after hole formation. In the computer program we don't allow α to fall below $\alpha = .001$. The boundary of the black hole is actually outside the point where $\alpha = .001$, but α is a very steep function of r for low α so it is a fair measure of the black hole extent. Note also the time scales are far from zero. The times of collapse are many times the free fall time associated with the initial central density.

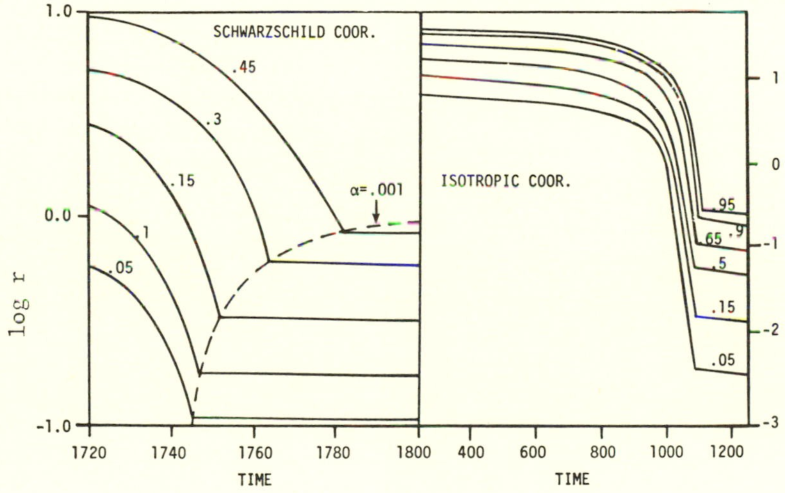

Figure 1 The r-t trajectories of several mass points are plotted for

a) $\rho = \exp(-r/r_1)/(r + 2\,\Delta r)^2$, G = 1.

b) ρ = (N = 3) polytrope, G = 1

The numbers near the curves are the mass inside the curves. The masses of the stars are one. Calculation a) was performed in Schwarzschild coordinates and b) was performed in isotropic coodinates. The dashed curve of Fig. a) is the locus of points where α goes to .001. Ordinate are $\log_{10}$ (r), r and t are in units of M.

We next turn to Γ = 2 equation of state to mock up stellar collapse and neutron star bounce. The calculations were done with an initial N = 1 polytropic density distribution. The lowest central α that a static star with Γ = 2 can sustain in equilibrium is about 0.5. It is of interest then to see in a bouncing situation how much deeper the star can fall and still recover. Since bounce is thought to be good for producing gravitational radiation, the maximum depth of bounce may be important. We started a collapse with an α = 0.85 and an inward velocity corresponding to free fall from infinity. The initial pressure was reduced from equilibrium by a constant factor over the star. By trial and error a calculation with a minimum α value of 0.20 was made in which the star bounced back. In Figure 2A, α versus time and radius versus time are shown. In Figure 2B a shock wave is seen to form at a time of 13, but it is not sufficient to stop the infall. Supposedly realistic collapses are more homologous and less supersonic and shocks form only in outer parts of star. See Wilson (1979).

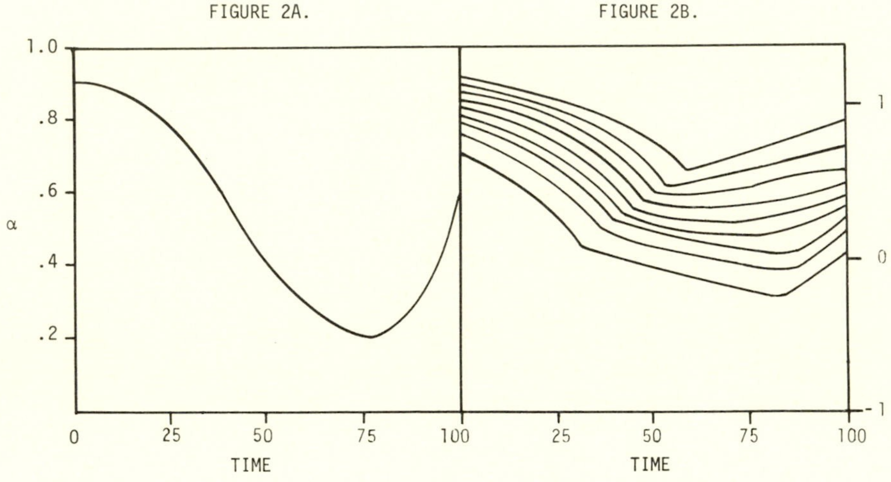

Figure 2 α and r versus time for a star bouncing in a strong gravitational potential. Ordinate of 2B is $\log_{10}$ (r).

Note how much shorter time scale is here than in Fig. 1.

We thus find stars can bounce back from a very deep potential, much deeper than might be inferred from the potential of static stars. At bounce the thermal energy is more than twice the particle rest energy. This means sound speed is about .80 the speed of light. The velocity of gravity waves is

$$\sqrt{1 - 4\pi G\rho/k^2}$$

for wave number k. It may be possible to have sound and gravity waves getting in step for deep bounces of asymmetric stars. (The shortest wave length for equal velocities is about the diameter of the star.)

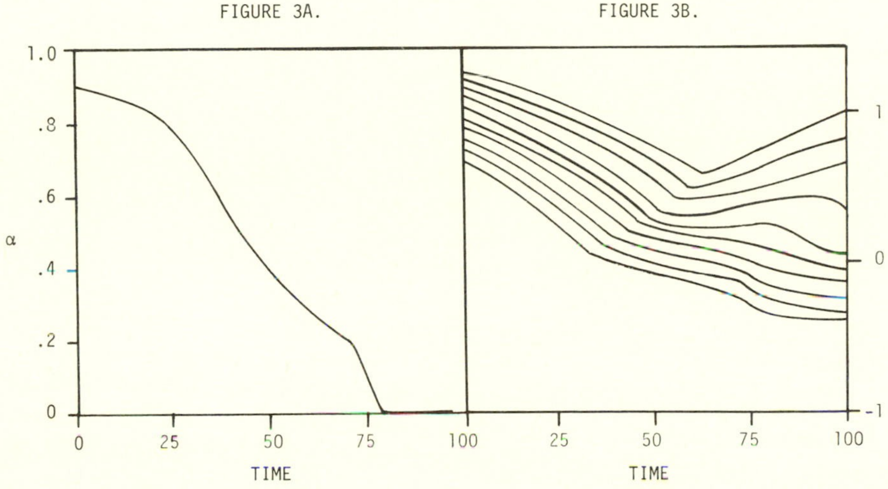

Figure 3 α and r versus time for a star collapsing to a black hole. Note the break in the α vs. t curve at the point where $\alpha \cong 0.2$.

In Figure 3 are shown α and r versus t for a $\Gamma = 2$ collapse that almost bounced, but then formed a black hole. At a time of about 18 enough slow down of the inner core occured so that a shock wave was produced in the outer layers strong enough to accelerate the outer 10% of the star's mass above escape velocity. This calculation suggests that it might be possible to make a supernova explosion and a black hole simultaneously.

IV. AXIALLY SYMMETRIC HYDRODYNAMICS

We chose the metric as

$$ds^2 = -\alpha^2 dt^2 + A^2 (\beta_R dt + dR)^2 + (\beta_Z dt + dZ)^2 + \tag{41}$$

$$B^2 R^2 d\phi^2$$

In order to put the metric in this form we use the coordinate conditions that $\dot{g}_{RZ} = 0$ and $\dot{g}_{RR} = \dot{g}_{ZZ}$ to give the equations for the shift vector,:

$$\nabla^2 \beta^R = \frac{\partial}{\partial R}\left[\alpha(K^R_R - K^Z_Z)\right] + 2\frac{\partial}{\partial Z}(\alpha K^R_Z) \tag{42}$$

$$\nabla^2 \beta^Z = -\frac{\partial}{\partial Z}\left[\alpha(K^R_R - K^Z_Z)\right] + 2\frac{\partial}{\partial R}(\alpha K^R_Z) \tag{43}$$

where

$$\nabla^2 = \frac{\partial^2}{\partial R^2} + \frac{\partial^2}{\partial Z^2}, \quad \nabla = \left(\frac{\partial}{\partial R}, \frac{\partial}{\partial Z}\right) \tag{44}$$

The lapse function, α , is chosen by the maximal slicing condition,

$$\dot{K}^i_i = 0. \tag{45}$$

$$\frac{1}{A^2 BR}\nabla . (BR\nabla\alpha) = \alpha\rho_\alpha/2 \tag{46}$$

See equations (30) and (33) in section II for definitions of ρ_α and ρ_H. The Hamiltonian constraint equation is

$$\frac{1}{A^2 BR}\nabla^2 (BR) + \frac{1}{A^2}\nabla . (\nabla \log A) = \rho H \tag{47}$$

we have taken $G = 1/8\Pi$ and $C = 1$. This equation is used to determine B. The momentum constraint equations can be put in the form;

$$\nabla^2 \left[A^2 BR (K^R_R - K^Z_Z)\right] = 2\left(\frac{\partial P}{\partial R} - \frac{\partial Q}{\partial Z}\right)$$

$$\nabla^2 (A^2 BRK^R_Z) = \frac{\partial P}{\partial Z} + \frac{\partial Q}{\partial R} \tag{48}$$

where

$$P = A^2 BR\left[\alpha S_R + \frac{1}{2B^3R^3}\frac{\partial}{\partial R}(B^3R^3K^\phi_\phi)\right]$$

$$Q = A^2 BR\left[\alpha S_Z + \frac{1}{2B^3}\frac{\partial}{\partial Z}(B^3K^\phi_\phi)\right] \tag{49}$$

The dynamic evolution equations are taken as

$$\frac{\partial A}{\partial t} = \beta . \nabla A + 1/2 (\alpha K^\phi_\phi + \nabla . \beta) \tag{50}$$

$$\frac{\partial K^{\phi}_{\phi}}{\partial t} = \beta \cdot \nabla K^{\phi}_{\phi} - \alpha \left(P + K^{i}_{j} K^{j}_{i}\right) + \frac{1}{A} \nabla^{2} \alpha + \frac{\alpha}{A^{2}} \nabla^{2} (\log A) \qquad (51)$$

The dynamic variables of the system are now D, E, S_R, S_Z for the hydrodynamics and A, K^{ϕ}_{ϕ} for the gravitational field. An equation of state completes the system of equations. We have put the equations in a form in which

$$\alpha \ , \ \beta^R \ , \ \beta^Z \ , \ B, \ K^R_Z \ , \ K^R_R - K^Z_Z$$

are found by solving elliptic differential equations and the other variables are advanced in time by hyperbolic equations (50)-(51).

To conceptualize how the difference equations are formed Figure 4 gives the centering of quantities in the computational grid.

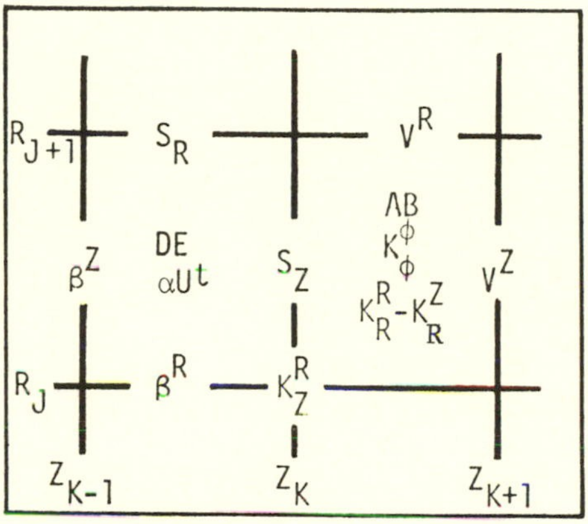

Figure 4 A portion of the grid into which R,S space is divided to indicate how quantities are centered. K,J are integers labeling the mesh. $R_1=0$, $Z_1=0$.

The elliptic equations are straight forward since ∇^2 is a simple operator. In a computational cycle first the hydrodynamic equations are solved to advance D,E,S_R,S_Z. The difference equations are a straight forward generalization of the spherical hydrodynamics outlined in section II. Next the six metric functions,

$$\alpha \ , \ \beta^R \ , \ \beta^Z \ , \ B, \ K^R_Z \ , \ K^R_R - K^Z_Z \ ,$$

are found by solving the six elliptic equations by the method of over-relaxation (see Smarr 1979). The old values are saved so that only a small number of iterations per cycle are needed. When α and B are changed the material densities are changed as indicated in section II. For the time evolution equations we start with the first part of the equation for A

$$\frac{\partial A}{\partial t} = \beta \cdot \nabla A \qquad (52)$$

In a convective term β acts the same as the grid velocity so we replace β by $V_c = \beta + Vg$. (Vg is the grid velocity). The difference equations are formed by taking A averaged a distance $V_c \Delta t$ up wind.

$$A_K^{new} = A_K + V_c^2 \Delta t \left[(1/2 + V_c^Z \Delta t/(Z_{k+1} - Z_k)) \right.$$
$$\cdot (A_{K+1} - A_K)/(Z_{K+3/2} - Z_{K+1/2})$$
$$+ (1/2 - V_c^Z \Delta t/(Z_{K+1} - Z_K)) \tag{53}$$
$$\left. \cdot (A_K - A_{K-1})/(Z_{K+1/2} - Z_{K-1/2}) \right]$$

The change due to $1/2\ (\alpha K_\phi^\phi + \nabla \cdot \beta)$ is straight-forward and is added in next. In the equation for K_ϕ^ϕ the convective terms are treated the same as the convective terms for A. The other terms in K_ϕ^ϕ are straight-forward. Log A is computed and then the difference form of ∇^2 log A is done in a straight-forward manner.

The boundary conditions for B, K_Z^R, $K_R^R - K_Z^Z$ are not straight-forward in difference form because of the R term in the Laplacian, i.e. $\nabla^2(RB)=S$ etc. The proper boundary condition for B on the axis is B = A. However, this leads to cusping on the axis so that what is done is set the first zone below the axis equal B in the first zone above the axis. Thus B and A drift apart on the axis with time. The proper boundary condition on $K_R^R - K_Z^Z$ comes from

$$K_R^R = 3K_\phi^\phi + \alpha R S_R \qquad R \to 0. \tag{54}$$

A straight-forward application of this boundary condition leads to instability so that in the first zone above the axis we let

$$K_R^R = 3/4\ (K_{\phi\ K-1}^\phi + 2K_{\phi K}^\phi + K_{\phi K+1}^\phi) + \alpha R S_R\ (J=2)/2A^2 B \tag{55}$$

where the indices are J = 1, K = K where not indicated. This averaging of K_ϕ^ϕ stabilizes the equations for K's. The boundary condition $K_Z^R = 0$ on axis doesn't work well in difference form because of the radius factor R inside the Laplacian for K_Z^R. We set K_Z^R at the first full zone above the axis by

$$K_Z^R = R \left[(\alpha_k + \alpha_{k-1})\ S_Z/2 + (B_k^3 + B_{k-1}^3)/(Z_k - Z_{k-1}) \right] \tag{56}$$

where K_Z^R, R are full zone above axis and α, S_Z, B are half zone above axis. We still have troubles on the axis. Bumps in the Ks grow slowly with time near the axis though they remain small enough that they probably don't hurt the calculation. If the zoning is made very fine on the axis the bumps disappear, but then the time step is uncomfortably small. This area of the numerics needs further research.

At the outside α is set by

$$\alpha = 1 - m/r + 1/2\ Q\ (2Z^2 - R^2)/r^5 \tag{57}$$

where

$$r = \sqrt{R^2 + Z^2}$$

and Q is the quadrupole moment

$$Q = \pi \int (2\ Z^2 - R^2)\ D\ \sqrt{-g}A^2\ dRdZ \tag{58}$$

Around the perimeter A is set equal to $1./\alpha$ at time zero and left unchanged with time. The K's and β's are currently set to zero on the outside. The code at present is set up so that it is symmetric about Z = 0.

Two methods have been used for the calculation of gravitation radiation. The first is based on the ADM mass flux formula

$$P^i = 1/4\ g_o^{ij}\ g_o^{ln}\ g_o^{mp}\ \dot{h}_{np}\ (2D_m\ h_{j\ell} - D_j\ h_{\ell m}) \tag{59}$$

where $g_{ij} = \overset{o}{g}_{ij} + h_{ij}$

This formula gives in our coordinate system $P = \dot{B}\ \nabla\ B$. This did not seem to work well since it involves taking a derivative of the function, B, which has a rather appreciable slope at the outside of the calculational grid. One is looking for a 1/r fall off in a function B that falls off as 1/r2 for a static system. Also B can be changing in time in a spherical calculation just due to β^i effects. We finally have chosen to look at the energy density term

$$K^i_j\ K^j_i \equiv K^2. \tag{60}$$

This has the advantage that the background fall off of K^2 (from spherical in and out motion) is $1/r^6$. In practice waves in K^2 stand out clearly. Since K^2 is an energy density and not a flux, it is necessary to assume that the flux of energy is

$$P = (K^2/2)\ (\nabla\ K^2/\ |K^2|) \tag{61}$$

The factor of half in this formula comes from comparing the energy flux in a plane wave with K^2 in the flat space limit. See (Eppley 1979, Smarr 1979)

V. AXIALLY SYMMETRIC CALCULATIONS

At present the code has been used primarily for the study of neutron star bounces. While the code will calculate the collapse of a star to form a black hole there are serious problems associated with estimating the emitted radiation. All bounce calculations were done with a $\Gamma = 2$ adiabatic index. The first test problem was run to compare a spherical bounce calculated on the axially symmetric (2D) code with the same problem calculated on the spherically symmetric code (1D). The adiabatic index $\Gamma = 2$ was chosen and the star was set up initially in equilibrium with constant entropy. Then the energy was reduced in all zones by a factor of 0.4 and the subsequent collapse followed on the two codes. The lapse function, α , for the two calculations is shown in Fig. 5.

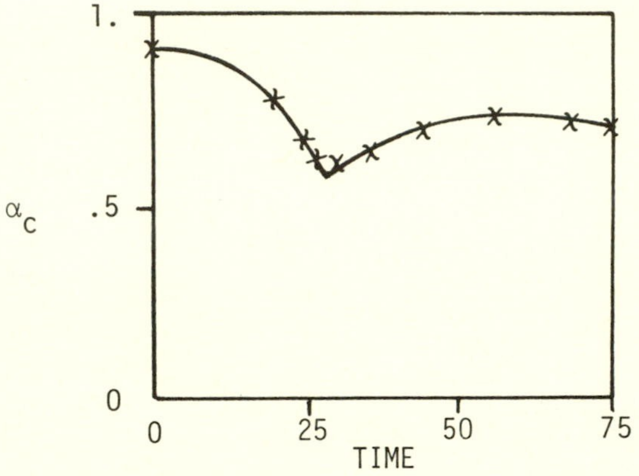

Figure 5 Central α as a function of time for spherical bounce. Solid line is the results of the 2D calculations and the X's are points from the 1D calculation. Time is in units of M.

The 2D calculation is not truly spherical because of numerical errors. The hydrodynamical quantities are very closely spherical. The extrinsic curvature components are much less spherical than the hydrodynamics variables in the combination of $K^i_{\ j}$'s that should be spherical. The emitted gravitational radiation energy in the 2D calculation is $1.x10^{-5}$M. This is to be compared to the later calculation which bounces at comparable densities and emits $1x10^{-2}$M in radiation. The next test was the collapse and bounce of an only slightly relativistic but deformed star. The objective of this test is to compare the radiation calculation of the code versus the usual quadrupole emission formula under conditions where the latter is almost valid. We define the quadrupole moment as

$$Q = \Pi \int (2Z^2 - R^2)\, D\alpha\, A^4\, BRdRdZ \tag{62}$$

The initial density distribution was chosen as

$$\rho = \rho_o \frac{\sin kx}{x} \quad x = \sqrt{R^2 + 2Z^2}\ ,$$

a distorted polytrope. The energy was taken as 10% of the energy that would be required to support the star in the Z direction. Several quantities are plotted versus time in Fig. 6.

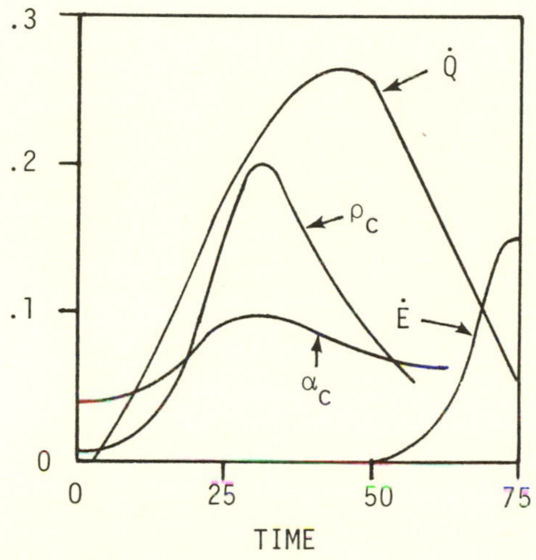

Figure 6 Central (1 - α), central density, ρc, time derivative of quadrupole moment Q, and rate of emission of gravitational emission E as functions of time. Scale is only for αc. This is the high α, 2:1 oblate collapse calculation.

The gravitational radiation emitted calculated by the quadrupole moment formula is 3.0×10^{-5}M. The radiation emitted as calculated from the flux of K^2 through the edit two-sphere near the outside of the computational grid was 2.5×10^{-5}M units. However, emission rate of radiation had just peaked and so it is estimated that the total emitted radiation would be close to 5×10^{-5}M. This additional 2.5×10^{-5} comes from an estimate of the space integral of K^2 in the wave that is inside the edit sphere. The same configuration was then rerun with a higher density so that the star would reach a minimum central α of 0.45 instead of a central minimum of 0.90 as in the previous calculation. See Fig.7

for α , Q, and E^{out} versus time.

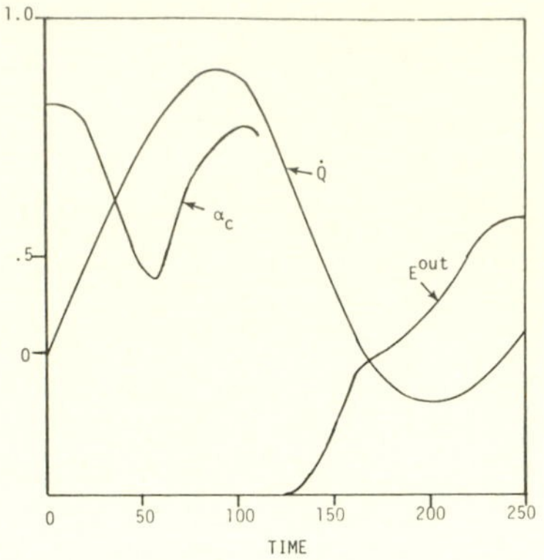

Figure 7 Central α, time rate of change of quadrupole moment and emitted gravitational energy for the case of a 2:1 oblate star bounce at small α. Left scale is for α_c, right for $\dot{Q}$.

Now the radiation emitted by the quadrupole formula is $5 \times 10^{-3}M$ and the flux out of K^2 is $1.2 \times 10^{-2}M$. The ratio of energy out as calculated by the quadrupole formula to the true energy out is seen to vary slowly as the system becomes more relativistic. It must be noted that we put the most factors of A and B in the quadrupole formula that seemed justifiable. The shape of the wave of K^2 is the shape in R,Z of a quadrupole radiation pattern. We also edited the flux of K^2 in the equator and assumed this was representative of a quadrupole radiation pattern. This gave the same total energy out to about 10 or 20%.

The next configuration run was a prolate star with a density profile defined by $\rho = \rho_o \sin kx/x$, $x = \sqrt{4R^2 + Z^2}$. This extreme example gave the largest emission of gravitational radiation, $1.2 \times 10^{-2}M$ by the quadrupole formula and $2.8 \times 10^{-2}M$ by the K^2 edit. See Fig. 8.

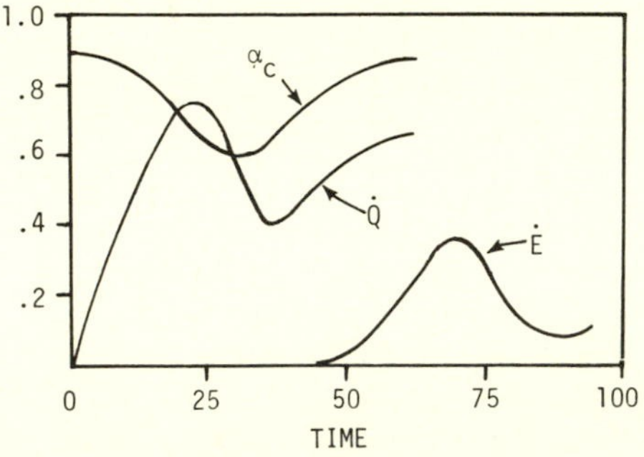

Figure 8 Central α, $\dot{Q}$, and $\dot{E}$ for prolate 4:1 collapse as functions of time.

The final calculation involved the collision of two neutron stars along the Z axis. Two spherical neutron stars were set up on the axis with radii 10M and separation of 40M. They were given a velocity as if they had fallen from infinity. It should be noted that our method of calculation automatically treats non-stationary initial data. With a non-stationary initial configuration unknown radiation may be lurking in the grid, but this radiation is presumed to be small.

As the neutron stars approach each other the minimum α slowly decreases (see Fig. 9).

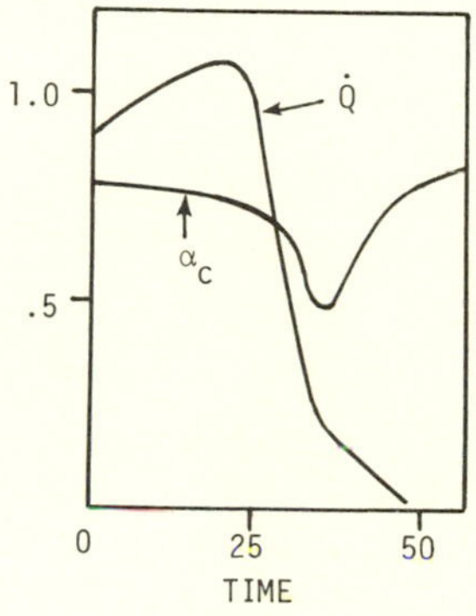

Figure 9 Central α and $\dot{Q}$ for the collision of two neutron stars as functions of time.

As the stars begin to coalesce α drops rapidly and then the newly formed neutron star rebounds. The calculation was not carried far enough to give a good estimate of the gravitational radiation but the quadrupole formula gave only 1.0×10^{-3}M emitted radiation.

In all these calculations we are only following the first bounce and of course a star might ring for several oscillations. We were looking to see if anything unusual happened for stellar bounces in strong fields. It appears that the quadrupole formula is good to a factor of 2 or 3 for neutron star bounces that are certainly quite extreme.

The oblate 2:1 density distribution was rerun with $\Gamma = 4/3$. Naturally a black hole was formed. Fig. 10 shows central α versus time for this case.

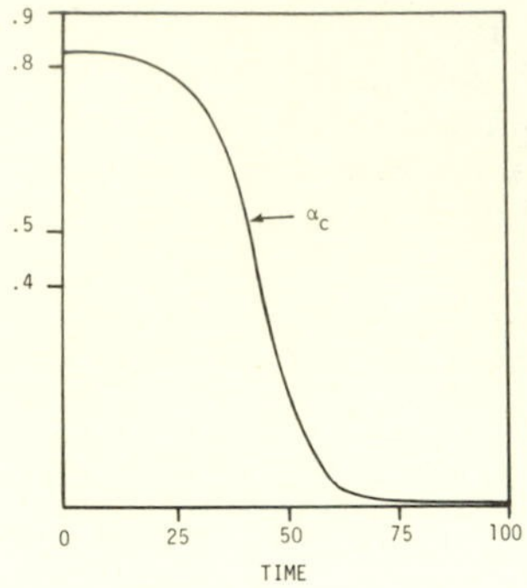

Figure 10 Central α as a function of time for the 2:1 oblate = 4/3 star collapse

However, after the central α went to near zero the region of low α kept on expanding (see Fig. 11).

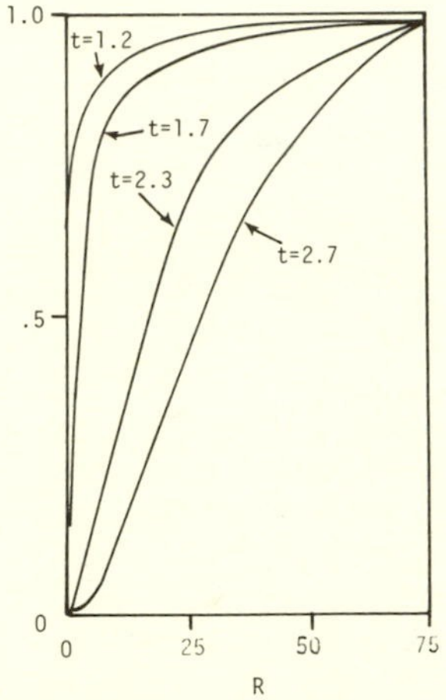

Figure 11 Lapse as a function of equatorial radius at several times for the Γ = 4/3 calculation

The coordinates fall into the hole too fast. An interesting effect occurs due to the large non radial momentum distribution. A region where $|\beta|>\alpha/A$ is formed where particles can no longer stand still as in an ergosphere. (For more detail, see Smarr and Wilson 1979.) For the radial motion this just corresponds to the particles being trapped. However, since the material momentum has a large non-radial component an apparent hyperbolic material motion is induced (a kind of "dragging on inertial frames") in the exterior region (r = a few M) shortly after central α goes to zero. Later the exterior material resumes radial infall. Much more work is needed on our computer program before we will be able to calculate radiation from black hole formation. Perhaps a different coordinate system will be necessary. We are working on a 2D code that has a metric more analogous to the Schwarzschild metric since in 1D the Schwarzschild-like metric worked better than the isotropic metric when a black hole is formed. In the spherical isotropic case β is a space integral (see equation 31) and doesn't go to zero in the region of small α until all matter has collapsed into the region of small α. The metric function A grows a lot even after a black hole is formed. While the Schwarzschild-like coordinates β goes to zero when α goes to zero and the inside freezes.

*WORK PERFORMED UNDER THE AUSPICES OF THE US DEPARTMENT OF ENERGY UNDER CONTRACT W-7405-ENG-48.

REFERENCES

Eppley, K., (1979) This volume.

May, M.M. and White, R.H., Stellar Dynamics and Gravitational Collapse, Meth. in Comp. Physics, 7, pp. 219-258.

Pachner, J. (1975). Numerical Integration of exact time-dependent Einstein equations with axial symmetry. General Relativity and Gravitation, ed. G. Shaviv and J. Rosen, John Wiley & Sons, New York. pp. 143-168.

Richtmyer, R.D., (1957). Difference Methods for Initial Value Problems, Interscience Publishers, New York.

Smarr, L., (1979). This volume.

Smarr, L., and Wilson, J. (1979), to be published.

Wilson, J.R., (1978). A numerical study of rotating relativistic stars. Proceedings of Int. Sch. of Phys. Fermi Cource LXV, North Holland, Amsterdam.

Wilson, J.R., (1977). Magnetohydrodynamics near a black hole, Marcel Grossman Lectures, North Holland Pub. Co., Amsterdam, pp. 393-414.

Wilson, J.R., (1972). Numerical study of fluid flow in a Kerr Space, Astrophys. J., 173, pp. 431-438.

York, J., (1979). This volume.

THE ROLE OF BINARIES IN GRAVITATIONAL WAVE PRODUCTION

John Paul Adrian Clark

Yale University Observatory

I. INTRODUCTION

Best estimates (Batten 1973) indicate that at least half the stars consist of binary or multiple systems. Since such systems have a time varying quadrupole moment they may be expected to generate gravitational waves (GW).

Peters and Mathews (1963) calculated the energy and angular momentum losses due to gravitational radiation in the weak field, slow motion limit of two point masses in a binary. The luminosity averaged over a period is

$$L_{GW} = 32\ G^4 m_1^2 m_2^2 (m_1 + m_2)\ f\ (e)\ /\ 5c^5 a^5 \tag{1a}$$

$$= (1.63\times10^{51} erg s^{-1})(m_1/m_\odot)^2 (m_2/m_\odot)^2 (m_1+m_2/m_\odot)(a/100km)^{-5} f(e) \tag{1b}$$

where a is the separation, m_1 and m_2 the masses of the two bodies, and f(e) is a function of the orbital eccentricity, and equals one for a circular orbit.

The waves are emitted at a frequency equal to twice the orbital frequency

$$\nu_{GW} = G^{1/2} (m_1+m_2)^{1/2} / \pi a^{3/2} \tag{2a}$$

$$= (164\ Hz)(m_1+m_2/2m_\odot)^{1/2} (a/100\ km)^{-3/2} \tag{2b}$$

The energy loss to gravitational waves causes the orbit to decay, and the components reach zero separation ("coalescence" or "collision") after a time (Peters 1964):

$$\tau_{GW} = 5c^5 a^4 / 256\ G^3 m_1 m_2 (m_1+m_2) \tag{3a}$$

$$=(2.0s)\ (m_1/m_\odot)^{-1} (m_2/m_\odot)^{-1} (m_1+m_2/m_\odot)^{-1} (a/100\ km)^4 \tag{3b}$$

Components of binaries, clearly have a real and very finite size, which restricts the separation of the components to be at least roughly as large as the Roche or tidal lobe of the more extended component, or a mass flow will ensue. Typically, for components of roughly equal mass, the Roche separation is approximately three times the radius of the extended component. A strong lower limit on the separation is that it should be larger than the sum of the radii of the components.

Table 1 shows the relative parameters for the closest possible binaries as a function of their most extended component, assuming two $\sim 1\ m_\odot$ objects.

Table 1: Parameters for Closest Possible Binaries

Most Extended Component	Typical radius (km)	L_{GW} (erg s^{-1})	ν_{GW} (Hz)	h @ 10 Kpc	Duration of binary
Main Sequence	10^6	$\sim 10^{29}$	3×10^{-5}	$\sim 3\times10^{-23}$	$\sim 10^6$ yr.
White dwarf	10^4	$\sim 10^{39}$	3×10^{-2}	$\sim 3\times10^{-21}$	$\sim 10^3$ yr.
Neutron Star	10	$\sim 10^{54}$	$\sim 1\times10^3$	$\sim 3\times10^{-18}$	~ 1 s.

The relative frequency of binaries in the galaxy, which will reach their Roche separation in less than 10^{10} years, however is approximately of the order of

$$N(m.s.) : N(w.d.) : N(n.s.) \simeq 1 : 10^{-2} : 10^{-5}$$

II. BINARIES AS PERIODIC GRAVITATIONAL WAVE SOURCES

In this paper we shall only discuss close compact object binaries. A thorough review of binaries as GW sources is given by Douglass and Braginsky (1979).

To date it has been common to portray periodic sources individually on a plot of dimensionless amplitude versus frequency. We recommend (Ron Drever, Ray Weiss, private communication) that the density of these sources $d^2N/d(\log h)d(\log \nu)$ should be plotted as a function of log h and log ν. The sum of $d^2N/d(\log h)d(\log \nu)$ over all known and inferred sources will provide us with complete information on periodic sources. Since h and ν normally cover many orders of magnitude, they are plotted logarithmically.

Consider white dwarf-white dwarf binaries close enough that they reach their Roche separation in less than a Hubble time ($\sim 10^{10}$yr.). Then, assuming the density of systems to scale as r^2 from 100 pc to 10 Kpc from the sun (i.e. disk population), the white dwarfs to have masses of 1 $m_{\odot}$ each, and denoting the mean time between formation of such systems by τ_{wd}, we get:

$$d^2N/d(\log h)d(\log \nu) \simeq 5\times10^3(\nu/10^{-3}\text{Hz})^{-4/3}(h/10^{-20})^{-2}(\tau_{wd}/1\text{yr})^{-1} \quad (4)$$

for:

$$9\times10^{-21}(\nu/1\text{ Hz})^{2/3} \lesssim h \lesssim 9\times10^{-19}(\nu/1\text{Hz})^{2/3} \quad [100\text{pc} \lesssim r \lesssim 10^4\text{pc}] \quad (5a)$$

$$5\times10^{-5}\text{ Hz} \leq \nu \leq 3\times10^{-2}\text{Hz} \quad [\tau_{GW} \lesssim 10^{10}\text{yr};\ a_{min} \gtrsim 3r_{wd}] \quad (5b)$$

The length of time between formation of such systems is probably of the order of decades to centuries. Figure 1 is a plot of equicontours of $d^2N/d(\log h)d(\log \nu)$, taking $\tau_{wd} = 10^2$yr.

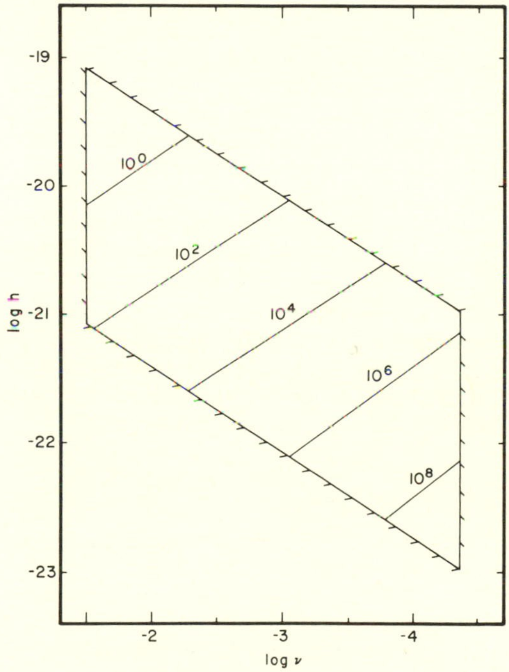

Fig. 1 - Equicontours of $d^2N/d(\log \nu)d(\log h)$ for close white dwarf-white dwarf binaries.

III. COMPACT OBJECT BINARIES

(i) Neutron Star - Neutron Star

Clark and Eardley (1977) studied the evolution of a binary system consisting of two neutron stars as their orbit decayed towards coalescence. In particular the loss of energy and angular momentum drove the components towards one another so that one of the neutron stars was within its tidal radius, whereupon tidal stripping ensued.

This scenario was envisioned as being the end product of either the orbital decay of a neutron star - neutron star binary whose initial period was less than about half a day, or the fissioning or fragmentation of a rapidly rotating core of a Type II supernova.

The calculation of the evolution of the components was carried out entirely using Newtonian dynamics, while allowing for gravitational wave energy and angular momentum losses. Mass loss from the less massive (more extended) neutron star as it spirals inside its tidal radius was found to stabilize the system against orbital decay. Depending on the masses of the components, mass and angular momentum losses from the system, and spin-angular momentum coupling, the system may be unstable to tidal breakup, otherwise prolonged mass transfer occurs. This substantially extends the duration of strong GW emission by the binary.

Extremely large neutrino and gravitational wave fluxes are generated for a few seconds until the components coalesce ($L_{\nu}^{peak} \sim 10^{56} erg.s^{-1}$; $L_{GW}^{peak} \sim 10^{54}$ erg.s^{-1}). Approximately 2% of the system rest mass is radiated as gravitational waves. Figures 2 and 3 show the time evolution of the neutrino and gravitational wave luminosities, and the GW frequency. Note that the GW signal for these events will be a "chirp" as the signal slides up the spectrum to ∿900 Hz and then back down again.

(ii) Black Hole - Neutron Star

Lattimer and Schramm (1976) considered the tidal disruption of a neutron star companion of a black hole as their orbit decayed towards coalescence. Unfortunately for our purposes they failed to estimate the resultant integrated GW flux during this event. However, to a first approximation, this problem is qualitatively the same as for n.s.-n.s. binaries, and thus probably has an efficiency of ∿2%, although some GW will be lost down the hole.

(iii) Black Hole - Black Hole

No detailed calculation of the final evolution of a binary

consisting of two black holes has yet been made. This is however a situation that will be rectified when researchers (cf. Eppley 1979 , Smarr 1979, Wilson 1979, these proceedings) in the field successfully complete a full 3-dimensional computer code.

Clark and Eardley (1977) crudely estimated the total efficency in this case to be ∿2% or ∿3%, for nonrotating or corotating black holes, respectively. Similarly, Zel'dovich and Novikov (1971) estimated the efficiency to be ∿3% for two approximately equal mass black holes. See also Detweiler, this volume.

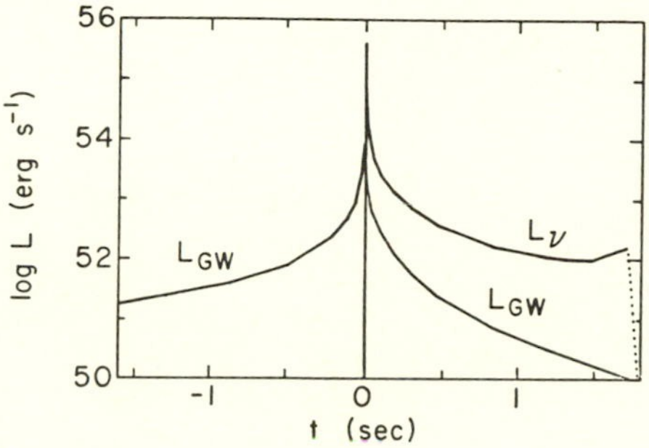

Fig.2

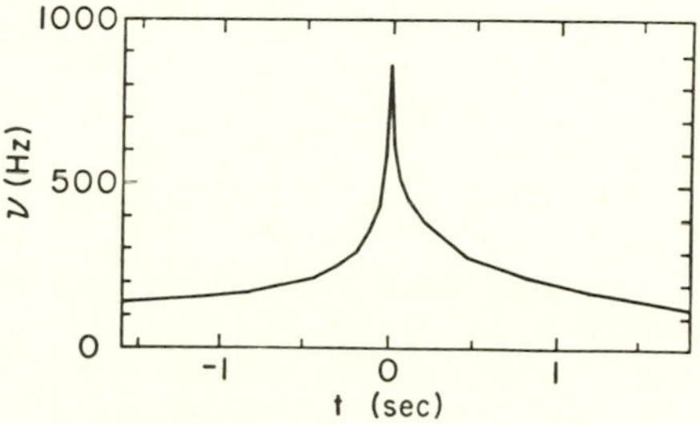

Fig.3

Fig.2 - Time evolution of a system with initial masses 0.8 and 1.3 $M_\odot$. Neutrino and gravitational wave luminosities. t=0 is the point of onset of mass stripping.

Fig.3 - Time evolution of a system with initial masses 0.8 and 1.3 $M_\odot$. Frequency of gravitational wave.

IV. FORMATION MECHANISMS FOR N.S. - N.S. PAIRS

(i) Binaries

The binary pulsar (PSR 1913+16) is widely assumed to be a pair of neutron stars in orbit about one another. Although the unseen companion is most likely a neutron star this has not yet been unequivocally confirmed, and it cannot be ruled out to be a helium star, white dwarf, or black hole (Smarr and Blandford 1976). The system is expected to evolve to the point of coalescence in $<10^9$ years due to GW losses, from its current separation of ∿1 $R_\odot$, and period of 7.75 hours (cf. Wagoner 1976).

Progenitors of n.s.-n.s. binaries (or other compact object binaries) are probably massive X-ray binaries (i.e. Cyg X-1, SMC X-1, Cen X-3), which originate from close binaries with primaries of mass $\gtrsim$ 15 $M_\odot$ (Flannery & van den Heuvel 1975, Smarr and Blandford 1976).

It is also possible that n.s.-n.s. binaries may form in dense globular clusters or galactic nuclei by exchange of a companion from a binary during the close passage of a third body.

(ii) Supernova Fission

O and B main sequence stars, which are the probable progenitors of Type II supernovae (SN) are observed to have $\sim 10^{51}$ erg.s of specific angular momentum. Since the collapsing core of such a SN will fission or fragment if its angular momentum is greater than $J_{fiss} \sim 10^{49}$erg.s, it is clear that a n.s. - n.s. pair (or group) will form unless a great deal of angular momentum is lost prior to collapse (cf. Wiita & Press 1976). The subsequent evolution of the system in this case is extremely similar to the neutron star binary case, except that the stars will probably still be hot, and collapsing, whereas in the former case (Clark and Eardley) they were assumed to be cold, and static.

A considerable body of observational evidence indicates however that neutron stars may be born slowly rotating and thus that SN fission events are rare (Kazanas and Schramm 1977, Lamb, Lamb, and Arnett 1975, Greenstein et al. 1977). This matter is still, however, very far from a closed book.

V. N.S. - N.S. COALESCENCE EVENT RATES

(i) SN Fission

Only those SN whose angular momentum exceeds J_{fiss} will form n.s. - n.s. pairs. Denoting the fraction of SN that doesn't fission by β_1, we see that a fraction $\beta_2 = 1-\beta_1$ will fission. Taking the SN event rate given in Clark, van den Heuvel & Sutantyo (1978) [see Arnett (1979, these proceedings) for discussion; also see Figure 4], we see that

$$r_{SN}(J \gtrsim J_{Fiss}) \simeq 0.1\ \beta_2 yr^{-1} \qquad d \lesssim 10\ \text{Kpc} \tag{6a}$$

$$\simeq \beta_2 \left(\frac{d}{10\text{Mpc}}\right)^3 \left(\frac{H_0}{100\text{kms}^{-1}\text{Mpc}^{-1}}\right)^3 yr^{-1} \qquad d \gtrsim 10\ \text{Mpc} \tag{6b}$$

(ii) Binaries

Clark et al. estimated the event rate of original binary coalescence events in a number of independent ways. First, the existence of the binary pulsar as the only known binary out of over 300 pulsars indicates that the most probable value of the ratio of the binary pulsar formation rate (r_{bp}) to the pulsar formation rate, R_{prob} is $3\text{x}10^{-3}$. The true value of R is poorly known due to small number statistics (one object!), and $-3.5 \leq \log R \leq -1.9$ at the 90% confidence level.

Clark et al. estimate that massive X-ray binaries have a 10% probability of remaining bound after the final SN explosion if $\log R_{prob} = -2.5$. This value is consistent with the survival probability derived by Sutantyo (1978) on the assumption that SN explosions are slightly asymmetric.

Approximating the pulsar birthrate by the SN rate (we shall thus be counting binaries containing black holes as well as neutron stars) Clark et al. estimate the event rate of binary coalescences to be

$$r_{bp} \simeq 2.9\times10^{-4}(315R)\ \mathrm{yr}^{-1} \qquad d \lesssim 10\ \mathrm{kpc} \tag{7a}$$

$$\simeq 3.2\times10^{-3}(315R)\left(\frac{H_o}{100\mathrm{kms}^{-1}\mathrm{Mpc}^{-1}}\right)^3\left(\frac{d}{10\mathrm{Mpc}}\right)^3\mathrm{yr}^{-1} \quad d \gtrsim 10\mathrm{Mpc} \tag{7b}$$

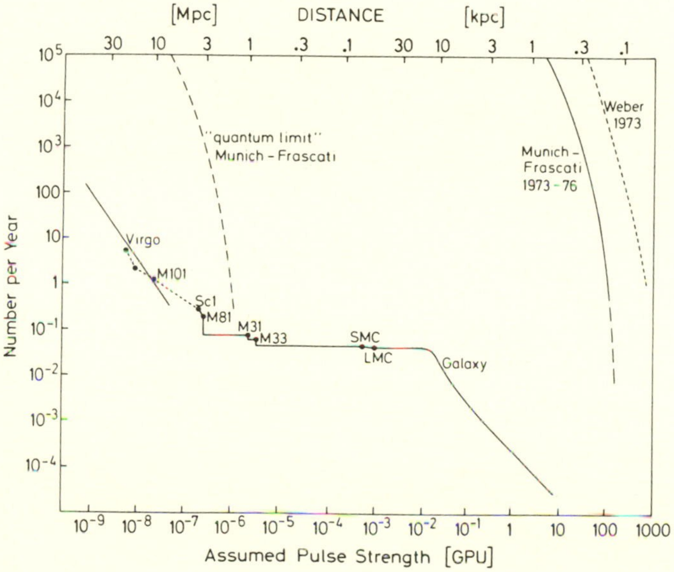

Fig.4 - Supernovae Event Rate (Kafka & Schnupp,1978)

VI. COMPARATIVE IMPORTANCE OF N.S. - N.S. COALESCENCES TO SN COLLAPSES

(i) Detectability

Recent calculations (Chia, Chau and Henriksen 1977, Shapiro 1977, Saenz and Shapiro 1978) indicate that peak gravitational wave efficiencies (η_{SN}) during core collapse probably do not exceed $\sim$0.1%.

Shapiro (1979, these proceedings) found recently that the effiency of homogeneous ellipsoidal SN cores was $\sim$1%, independent of core angular momentum J, if the cores bounced $\sim$5 times after the initial infall. The efficiency was high because the eccentricity grew on successive bounces. It is not yet possible to say if this important "eccentricity growth" effect occurs in real SN. If it does not, then the earlier calculations of Saenz and Shapiro (1978) for single-bounce collapse indicate that η_{SN} is a strong function of J. For collapse with a cold equation of state

$$\eta_{SN} \simeq 10^{-3}\ (J/J_{Fiss})^{3.5} \quad \text{for } J \lesssim J_{Fiss} \tag{8}$$

Consider two separate hypotheses. First, that the "eccentricity growth" effect occurs. And, second, that it does not. Then, taking γ_2 to be the fraction of SN with GW efficiencies of $\sim$1%, and $\gamma_1 = 1 - \gamma_2$, we see that $\gamma_1 = 0$, $\gamma_2 = 1$ under the first hypothesis, and that $\gamma_1 = \beta_1$, $\gamma_2 = \beta_2$ under the second hypothesis.

Neutron star - neutron star coalescences have efficiencies (η_{bp}) of the order of 2% , the same order as SN efficiencies if the "growth" effect operates, but otherwise, substantially larger.

Now let us compare the detection rate of GW events due to SN, and binary coalescences, by a detector of sensitivity S_o GPU_{-10} (10^{-10}GPU). In addition let us require that the detector be sufficiently sensitive to detect more than one event a year - so our coverage will extend out well past our galaxy, to where the number of sources varies directly with the volume observed. Then, the ratio of detected events due to binary coalescences to all detected events is (cf. Clark et al. for details)

$$\begin{aligned} Q &= \frac{dN_{bp}}{dt} \Big/ \frac{dN_{tot}}{dt} \\ &= (315R)/[(315R)+0.366\ \gamma_1\ (J_o/J_{Fiss})^{5.25} + \\ &\quad 129\ \gamma_2\ (\eta_{SN}^{Fiss}/10^{-2})^{1.50}] \end{aligned} \tag{9}$$

where J_o is a weighted mean specific angular momentum of all SN rotating slower than fission, and η_{SN}^{Fiss} is the efficiency of SN which fission and then suffer n.s. - n.s. coalescence.

The total detected event rate dN_{tot}/dt is

$$\begin{aligned} &(51\ \text{events.yr}^{-1})S_o^{-1.50}\ \left(\frac{H_o}{100\text{kms}^{-1}\text{Mpc}^{-1}}\right)\ [(315R) + \\ &0.366\ \gamma_1\ (J_o/J_{Fiss})^{5.25} + 129\ \gamma_2(\eta_{SN}^{Fiss}/10^{-2})^{1.50}] \end{aligned} \tag{10}$$

Consider two extreme cases: Case I, where all SN collapses occur near or above breakup angular momentum, i.e. $(J_o/J_{Fiss}) \sim 1$, and/or the "eccentricity growth" effect occurs. The $Q \sim 0.01$, i.e. only ~1% of all events will be due to binaries (R=1/315).

Case II, where no SN collapses occur near breakup, i.e. $(J_o/J_{Fiss}) << 1$, and the "eccentricity growth" effect is not present. Then $Q \sim 1.0$, i.e. 100% of all detected events are due to binaries.

Evidence pointing to the slow rotation of SN cores, thus indicates that the distinguishing criterion between the two cases above will be the existence or lack of "eccentricity growth".

Cases between these extremes may be anticipated if the "eccentricity growth" effect only occurs in some SN. A knowledge of the distribution of angular momenta of the other SN would be required in order to determine Q, whose dependence on γ_2 is shown in Figure 5.

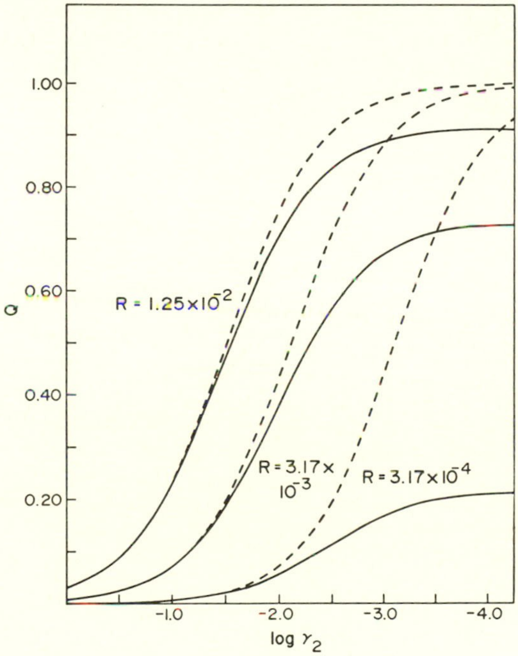

Fig. 5 - Q, the ratio of GW events due to binary pulsars to all GW events as a function of γ_2, the fraction of SN that have $\eta_{SN} \sim .01$, for (i) the most probable value of R (1/315), and (ii) 90% confidence limits for R (1/80; 1/3150). Solid lines represent log $(J_o/J_{Fiss})=0.0$, while dashed lines are for log $(J_o/J_{Fiss})=-1.0$.

A number of important conclusions can be drawn from this. In order to detect at least one event a year, a detector sensitivity of $\sim 10^{-7}$-10^{-8}GPU will be necessary, even if all SN have GW efficiencies of $\sim$1%. On the other hand, should SN be very inefficient generators of GW, one event a year will be detected at $\sim 10^{-9}$-10^{-10} GPU, but these bursts will be due to coalescence of binaries. The latter case corresponds to a dimensionless amplitude $h \sim 10^{-22}$ (cf. Figure 6).

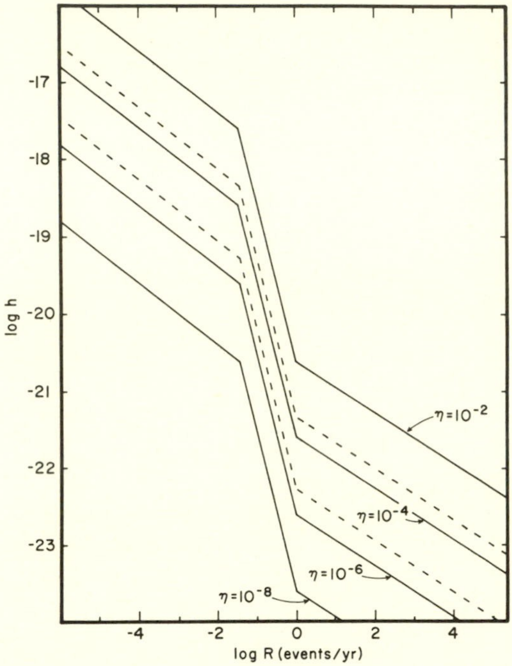

Fig.6 - Comparison of dimensionless amplitude h as a function of event rate. Solid lines are for SN collapses with efficiencies of 10^{-2}- 10^{-8}. Dashed lines are upper and lower estimates (99% confidence) for compact binary destruction events.

(ii) Frequency Spectrum

Clark and Eardley found that n.s. - n.s. coalescences generate waves with a peak frequency of $\lesssim$900 Hz.Clark et.al.showed that n.s.-n.s. pairs probably do not generate substantial amounts of gravitational waves above 1 kHz as mass transfer will tend to keep the bodies at least at the tidal limit of the body being stripped.

By contrast collapse of SN are expected to generate waves with a spectral peak of a few kHz (Thorne 1978). Saenz and Shapiro however find that for $J \sim J_{Fiss}$ the bulk of the energy is emitted below 1 kHz, although for more slowly rotating SN they find that the bulk of the energy is emitted at high frequencies. Since their model does not adequately describe conditions as $J \to J_{Fiss}$, this result should be regarded as tentative.

The expectation that SN will be high frequency sources has led observers to concentrate on developing detectors whose optimum performance is above 1 kHz, such as silicon and sapphire crystal detectors which operate best around $\sim$10 kHz.

If we assume that non-fissioning SN generate high frequency waves ($\gtrsim$1 kHz), whereas fissioning SN, and neutron star binaries generate low frequency waves ($\lesssim$1 kHz) we shall if anything understate the case in favor of low frequency waves (recalling that Saenz and Shapiro found non-fissioning SN generating low frequency waves). Then the fraction of detected events which are high frequency out of all events will be

$$dN_{SN}(J \lesssim J_{Fiss})/dt / \frac{dN_{tot}}{dt} = 0.366\ \beta_1 (J_o/J_{Fiss})^{5.25}$$

$$/[(315R) = 0.366\ \gamma_1 (J_o/J_{Fiss})^{5.25} + 129\ \gamma_2 (\eta_{SN}/10^{-2})^{1.50}] \quad (11)$$

Clearly a vast majority of detected events will be low frequency ($\lesssim$1 kHz) unless: (i) binary neutron star systems are substantially less frequent than estimated, (ii) very few SN have $J \gtrsim J_{Fiss}$, but (iii) all the remaining SN are rotating close to fission (i.e. $J_o \sim J_{Fiss}$). The simulataneous satisfaction of these three requirements seems rather unlikely particularly since one would expect quite a few fissioning SN if $J_o \sim J_{Fiss}$.

The above reasoning strongly challenges the thinking of the last decade, and indicates that most GW events detected will have their peak emission below 1000 Hz. This appears to indicate that the strategy of building detectors tuned to $\sim$10 kHz requires re-thinking unless the ease of construction at that frequency more than offsets the lower energy flux to be expected.

VII. CONCLUSIONS

Under the most optimistic assumptions a detector sensitive to $\sim 10^{-9}$ GPU or a dimensionless amplitude of $\sim 10^{-22}$ will be necessary to observe at least one compact binary event a year. This will, however, be detected in preference to SN events, if the SN efficiency is less than $\sim 10^{-5}$.

In contrast to our understanding of sources of gravitational radiation above $\sim$100 Hz as recently as two years ago (cf.Thorne

1978), current results indicate that compact binary objects may well be the most promising sources of gravitational waves, rather than SN collapses.

Furthermore, it appears that a majority of events will have peak frequencies below 1000 Hz, raising important questions about the design and implementation of the next generation of detectors. It is recommended that future discussions of periodic gravitational wave sources provide the density of such systems in dimensionless amplitude-frequency space, $d^2N/d(\log h)d(\log \nu)$.

REFERENCES

Arnett, W.D. (1979). This Volume.

Batten, A.H. (1973). Binary and Multiple Systems of Stars. Pergamon Press, Oxford.

Chia, T.T., Chau, W.Y. & Henriksen, R.N. (1977). Gravitational Radiation from a Rotating Collapsing Gaseous Ellipsoid. Astrophys.J., 214, pp. 576-583.

Clark, J.P.A. & Eardley, D.M. (1977). Evolution of Close Neutron Star Binaries. Astrophys.J., 215, pp.311-322.

Clark, J.P.A., van den Heuvel, E.P.J. & Sutantyo,W. (1978). Formation of Neutron Star Binaries and Their Importance for Gravitational Radiation. Astron.& Astrophys., in press.

Douglass, D.H. & Braginsky, V.B. (1979). Gravitational Radiation Experiments. In Einstein Centenary Volume, ed. S.W. Hawking & W. Israel.Cambridge University Press, Cambridge.

Eppley, K. (1979). This Volume.

Flannery, B.P. & van den Heuvel, E.P.J.(1975). On the Origin of the Binary Pulsar PSR 1913+16. Astron.& Astrophys.,39, pp.61-67.

Greenstein, J.L., Boksenberg,A., Carswell, R.& Shortridge, K. (1977). The Rotation and Gravitational Redshift of White Dwarfs. Astrophys.J., 212, pp.186-197.

Kafka, P. & Schnupp, L. (1978). Final Result on the Munich-Frascati Gravitational Radiation Experiment. Astron.& Astrophys., in press.

Kazanas, D. & Schramm, D.N. (1977). Neutrino Damping of Nonradial Pulsations in Gravitational Collapse. Astrophys.J.,214, pp. 819-825.

Lamb, D.Q., Lamb, F.K. & Arnett, W.D. (1975). Neutron Star Original Spin. Bull. Amer. Ast. Soc., 7, pp.545.

Lattimer, J.M.& Schramm, D.N. (1976). The Tidal Disruption of Neutron Stars by Black Holes in Close Binaries. Astrophys.J., 210, pp.549-567.

Peters, P.C. & Mathews,J. (1963). Gravitational Radiation from Point Masses in a Keplerian Orbit. Phys.Rev., 131,pp.435-440.

Peters, P.C. (1964). Gravitational Radiation and the Motion of Two Point Masses. Phys.Rev.,136, B1224-32.

Saenz, R.A. & Shapiro, S.L. (1978). Gravitational Radiation from Stellar Collapse: Ellipsoidal Models. Astrophys.J., 221, pp.286-303.

Shapiro, S.L. (1977). Gravitational Radiation from Stellar Collapse: The Initial Burst. Astrophys.J., 214, pp.566-575.

Shapiro, S.L. (1979). This Volume.

Smarr, L.L. & Blandford,R. (1976). The Binary Pulsar: Physical Processes, Possible Companions, and Evolutionary Histories. Astrophys.J., 207, pp. 574-588.

Smarr, L.L. (1979). This Volume.

Sutantyo, W. (1978). Asymmetric Supernova Explosions and the Origin of Binary Pulsars. Astrophys. & Sp. Sci., 54, pp.479-488.

Thorne, K.S. (1978). General-Relativistic Astrophysics. In Theoretical Principles in Astrophysics and Relativity, ed. N.R. Lebovitz, W.H.Reid & P.O.Vandervoort,pp.149-216. University of Chicago Press, Chicago.

Wagoner, R.V. (1976). A New Test of General Relativity. Gen. Rel. & Grav., 7, pp.333-337.

Wiita, P.J. & Press, W.H. (1976). Mass-Angular-Momentum Regimes for Certain Instabilities of a Compact Rotating Stellar Core. Astrophys.J., 208, pp.525-533.

Wilson,J. (1979). This Volume.

Zel'dovich, Ya.B. & Novikov,I.D. (1971). Relativistic Astrophysics Vol.I. University of Chicago Press, Chicago.

ASTROPHYSICAL SOURCES OF GRAVITATIONAL RADIATION

Jeremiah P. Ostriker

Princeton University Observatory

ABSTRACT

A general review of astrophysical mechanisms for producing gravitational waves is presented stressing three points: physical collisions between dense stars or other objects with dense stars is a promising source that has been somewhat neglected; the values of angular momentum and magnetic flux expected for stars prior to collapse may often be close to the ideal for maximal (but small) emission of gravitational waves (GW) which later is determined largely by geometrical factors and is insensitive to the initial entropy; line emitters of GW such as binaries or oblique rotators should have high priority, particularly those for which other measurements of period and phase are possible because it is only for these sources that independent tests are possible of the reality of the GW detection.

Following this, there is a speculative discussion of the evolution of a galactic nucleus having a few percent of its mass in neutron stars. It is argued that the neutron stars may collect at the center where, being unable to achieve equipartition, they will evolve very rapidly to a quite dense state. Until physical collisions amongst them dominate over other processes it is shown that the energy released in gravitational waves by free-free emission is a few percent of the cluster binding energy per relaxation time; the bound-bound emission in binaries produced by bound-free GW interactions is larger than the cluster binding energy! Ultimately a cluster of 10^8 neutron stars destroys itself in approximately $10^{6.5}$ years through physical collisions emitting $10^{-1}\ M_\odot c^2$ in thermal energy and $10^{-2}\ M_\odot c^2$ in gravitational radiation (bursts of 10^{52} ergs) in an optically violent quasar-like stage presumably ending as a massive but possibly uninteresting black hole.

1 INTRODUCTION

Since several relatively recent papers (e.g., Press and Thorne,

1977; Thorne, 1978; Clark, van den Heuvel & Sutantyo, 1978) have covered the physical processes and astrophysical sources that are likely to produce significant amounts of gravitational radiation, I will not attempt to present here an exhaustive review of the subject. Rather, I shall divide this paper into three parts. First, I shall describe two physical processes which, while probably quite uninteresting to the relativist, may be important in nature; they are not mentioned in the above reviews. One of them is directly related to the process of stellar collapse and indicates that the process may have been computed in an inappropriate manner in several recent papers, one that can substantially overestimate the radiation emitted in GW. Then I shall briefly survey the gamut of mechanisms and astrophysical environments available giving my own best estimates for the parameters necessary for observers wishing to detect the objects. Here I find myself a bit more optimistic than I had expected to be but readily admit that the estimated detectability could be incorrect by orders of magnitude. Finally, I shall present a totally speculative, but entertaining and I believe also plausible, scheme for the evolution of stellar systems containing large numbers of neutron stars. Gravitational radiation plays an important role in the evolution with perhaps 1% of the rest mass energy being emitted in pulses of 10^{52} ergs each, the system ending in a quasar-like state of violent collisions.

2 PHYSICAL PROCESSES

2.1 Physical Collisions

What would be emitted in gravitational waves if a rock of dimensions (ℓ_1, ℓ_2, ℓ_3) were thrown violently in the x direction against an immovable wall (or against its mirror image traveling in the -x direction)? Assuming that the velocity v is much larger than the sound speed in the missile (of density ρ) a standing shock will be set up of heated gaseous material between the still moving part of the missile and the wall. This obvious and apparently innocent remark is crucial in understanding gravitational collapse. The initial entropy per baryon is irrelevant in material which passes through a strong shock. Under the assumption of a strong shock and, taking for illustrative purposes $\gamma = 5/3$, we have the usual relations $P_2 = 3/4\ \rho u_1^2$, $\rho_2 = 4\rho_1$, $u_2 = 1/4\ u_1$, where (u_1, u_2) are measured in the shock frame. In the lab frame the shocked material is stationary so $u_1 = 4/3 v$, $u_2 = 1/3 v$ ($\dot{t} = -4/3 v$, $\dot{s} = 1/3 v$) = constant where t and s are the thicknesses of the unshocked and shocked portions of the missile. We shall calculate the gravitational radiation in the weak-field, slow-motion limit and thus require third-time derivatives of the moment of inertia tensors, $I_{ij} \equiv \int \rho x_i x_j dv$. Since the dimensions of the rock are given by ($\ell_3 \times \ell_2 \times s + t$) and (s, t) depend linearly on time, we need only calculate those parts of I_{ij} designated I'_{ij} which are cubic in (s, t). The only such element is

$I'_{11} = 1/3\rho\ell_2\ell_3[rs^3 + 3s^2t + 3st^2 + t^3]$. The only nonzero elements of the quadrupole moment tensor D_{ij} are $D_{11} = 2I_{11}$, $D_{22} = -I_{11}$, $D_{33} = -I_{11}$. Thus the rate of emission of gravitational radiation is

$$\frac{d\varepsilon_{GW}}{dt} = \frac{1}{45}\frac{G}{c^5}\left(\overset{\dots}{D}_{ij}\right)^2 = \frac{2^5 \times 7^2}{3^5 \times 5}\frac{Gm^2v^6}{\ell_1{}^2c^5} \tag{1}$$

which is emitted for the $\Delta t = \ell_1/v$ giving

$$\Delta\varepsilon = 1.29\,\frac{Gm^2}{\ell_1}\left(\frac{v}{c}\right)^5. \tag{2}$$

Consider the important special case of two particles each of mass m and escape velocity $v^2_{esc} \approx 2.28\ Gm/\ell_1$ which collide at velocity $v = \sqrt{2}v_{esc}$. Then

$$\frac{\Delta\varepsilon}{2mc^2} = 6.4\left(\frac{v_{esc}}{c}\right)^7 = 0.050\left(\frac{r_{sch}}{r}\right)\ ;\quad r_{sch} \equiv \frac{2G\,m}{c^2} \tag{3}$$

This is of course similar in form and slightly larger in value to the energy released by the collapse of a nonrotating spheroid of mass (2m) to a pancake where r = the final disc radius (cf. Thuan and Ostriker, 1974).

If we repeat the above calculations but assume instead that the missile was comprised of dust particles without extent and that no shock forms, then $\dot{t} = -v$, $(\dot{s}, s) = 0$ and formulae (1) through (3) are replaced by others with a numerical coefficient 2.54 smaller. It is at first surprising that the pressure retarded collision is so similar to one of a cloud of dust particles, but a little thought shows that this must be the case since the pressure is not set by the initial entropy in a strong shock but rather is $\sim \rho v^2$ due to entropy generated in the shock.

This simple set of calculations has important implications for estimations of GW released in real stellar collapse. Improved calculations which included pressure by Novikov (1975), Shapiro (1977) and Saenz and Shapiro (1978) maintained the approximation of homologous collapse used by Thuan and Ostriker (1974). Then, as the assumed initial entropy was lowered, it was found that the ratio of axes a/c at maximum compression increased. Since the bounce occurred in a shortened time scale of $(c/v_{free-fall})$, the GW radiation was increased over that found in the pressureless (dust) calculations by the large factor O(a/c). In the apparently similar model calculation of colliding missiles made in the present paper the pressureless and dust cases were quite similar. Why? The explanation lies in the treatment of entropy. Novikov (1975) and Shapiro (1977) maintained constant entropy; if the material bounces on collapse, the passage through a strong shock generates entropy. In any realistic computation of aspherical collapse (cf. Binney, 1977; Smarr and Wilson, this Conference) it is found that shocks do occur. Then the maximum compression is not large (four for $\gamma = 5/3$), the objects do not become very flat and we should expect that the pressureless dust calculation is accurate to within a factor of ~ 2. In sum, geometry is important but the local physics is much less important.

2.2 Gravitational Bremsstrahlung

Peters and Matthews (1963) treated eccentric binary motion, Peters (1970) small angle gravitational scattering, and several recent authors have considered relativistic small-angle scattering. However, large-angle scattering in the weak field slow motion limit is probably the most important process astrophysically (cf. Turner, 1977 for a more detailed discussion).

The result can be written (for equal masses*, m)

$$\Delta\varepsilon = \frac{G^3 m^4 v_{max}}{r_{min}^3 c^5} f(\eta) , \tag{4}$$

where

$$\eta \equiv \left(\frac{2pv_\infty^2}{Gm}\right) = \sqrt{e^2 - 1} \tag{5}$$

and

$$f(\eta) \rightarrow \begin{cases} \dfrac{37\pi}{15} & \text{for } \eta \gg 1 \\ \dfrac{85\pi}{24} & \text{for } \eta \ll 1 \end{cases} \tag{6}$$

$$r_{min} = \frac{\sqrt{1 + \eta^2} - 1}{\eta} p \ ; \qquad v_{max} = \frac{\eta}{\sqrt{1 + \eta^2} - 1} v_\infty \ . \tag{7}$$

An important special case is the grazing collision

---

*For unequal masses (m_1, m_2), $m^4 \rightarrow m_1^2 m_2^2$ in (4) and $m \rightarrow 2m_1m_2/(m_1 + m_2)$ in (5).

$$r_{min} = 2r \equiv \frac{4Gm}{v_{esc}^2}, \quad v_{max} = \frac{1}{2} v_{esc}$$

$$p_{min} = \frac{2Gm}{v_\infty^2} \left(\frac{v_\infty}{v_{esc}}\right) . \tag{8}$$

The energy radiated takes the familiar form

$$\Delta\varepsilon = \begin{pmatrix} 0.087 \\ 1.00 \end{pmatrix} mc^2 \left(\frac{v_{esc}}{c}\right)^7 \text{ for } r_{min} = \begin{pmatrix} 2 \\ 1 \end{pmatrix} r_* . \tag{9}$$

In general, for strong collisions

$$\Delta\varepsilon = 11.1 \frac{(Gm)^7}{c^5 v_\infty^7 p^7} \quad ; \quad r_{min} = \frac{p^2 v_\infty^2}{Gm} \tag{10}$$

The very steep dependence of $\Delta\varepsilon$ on impact parameter implies that the rare very close collisions are most important.

Both of the processes described in this section are important for nonrelativistic stellar systems.

3 GENERAL OVERVIEW OF ASTROPHYSICAL SOURCES OF GR

3.1 Continuum Sources

The total energy potentially available from gravitational collapse in the galaxy is quite large. The binding energy of a neutron star is $\Delta\varepsilon_{bind} \approx 10^{53}$ ergs. This energy must be released in or soon after the collapse, but only 10^{-2} of it appears in the supernova blast wave and still much less in optical emission. Most is emitted in some as yet unseen form and the total mean luminosity of the galaxy in this unknown form is quasar-like. The supernova (or pulsar formation) rate is of order $\nu_{SN} \sim 10^{-9}$ s^{-1} and thus the "collapse luminosity" is $L_{coll} = \Delta\varepsilon_{bind} \times \nu_{SN} \approx 10^{44}$ erg s^{-1}. On these grounds we might be optimistic concerning the detection of gravitational waves experimentally. However, most of the binding energy is probably emitted as neutrinos and the GW part is no doubt concentrated into brief intervals separated by the uncomfortably long period of $\sim$ 30 yrs. (See Arnett, this volume.)

On the basis of a straightforward lowest order calculation of galactic physical processes I have derived estimates of GW emission by various continuum processes. Results are summarized in Table 1. For each potential source I list the process and then below that the estimated energy released in solar masses, the duration of the event in seconds, and the estimated distance and time to the next event. Needless to say, all numerical estimates are highly conjectural, but they probably err on the side of over-optimism. The strength of the gravitational waves at the detector is given in terms of h_{-21}($\equiv h/10^{-21}$ the dimensionless wave amplitude h, cf. Press and Thorne, 1972) where

$$h_{-21} = 5 \times 10^3 \frac{(\Delta m/M_\Theta \Delta\tau)^{1/2}}{R_{10}}, \text{ and let } h_G \equiv h\ (R_{10} = 1). \qquad (11)$$

Here $(\Delta m/M_\Theta)$ is the mass radiated in solar masses, R_{10} is the distance to the source in units of 10 kpc, $\Delta\tau$ is the duration of the pulse (seconds) or for wave trains having period p and duration Δt, $\Delta\tau \equiv p^2/\Delta t$. Also listed are the expected event rates ν_G in units of yr^{-1}, for events in our galaxy. It is quite useful to note how the expected rate will depend on the sensitivity of the detector. The galactic visual luminosity is approximately $10^{10.3}$ L_Θ and the mean emissivity of extragalactic space is $\langle j\rangle = 10^{8.1}$ L_Θ mpc^{-3} (with $\langle j\rangle \approx 10^{8.4}$ within the sphere centered on the Virgo cluster and passing through our galaxy). Thus, there are $10^{-2.2}$ galaxies equivalent to our own in the average mpc^3 of the universe near us. Since we are quite a good representative of the luminosity weighted average field galaxy, the rate of GW events should be $10^{-2.2} \times \nu_G$ per mpc^{-3} giving an event rate within R_{mpc} of $\nu(R_{mpc}) = \nu_G R_{mpc}{}^3\ 10^{-1.8}$. The strength of the signal, h, falls off more slowly as $h(R_{mpc}) = h_G R_{mpc}^{-1} \times 10^{-2.0}$. So, for extragalactic sources

$$\nu(h) = \nu_G \left(\frac{h_G}{h}\right)^3 \times 10^{-7.8} \quad . \qquad (12)$$

This relation has the following simple interpretation. If a type of source can be barely detected within the galaxy it can obviously not be detected in external systems. If, however, the galactic source is stronger than the detection limit (in h) by a factor of 400 ($= (10^{-7.8})^{-1/3}$), then extragalactic sources will be detected at the same rate as galactic sources. As the detection limit is lowered further, the extragalactic sources dominate and the "log N-log S" relation is $\log \nu = \text{const} - 3 \log h_{lim}$.

3.1.1 Collisional sources. Several authors (e.g., Harwit and Salpeter, 1973) have noted the possibility that cometary size chunks of matter may, on occasion, impact upon otherwise quiescent neutron stars producing thereby staquakes, γ-ray or X-ray bursts or even neutrino bursts. While none of these explanations has met with great favor, it is not unreasonable that 10^{-12} $M_\odot$ chunks of matter with density ~ 1 and size 10^7 cm might impact occasionally. The energy release is miniscule but the nearest event could be as close to us as the nearest neutron star (5 pc?) and the frequency of events correspondingly high.

Degenerate dwarfs must exist in globular clusters and, since their masses are likely to be greater than those of the old main sequence stars remaining, they will tend to drift to the cluster center. The high mass-to-light stars required in the models of Illingworth and King (1977) are probably these objects. With a central density (for M15) of $\sim 10^7$ DD per pc^3 and an rms velocity of 30 km/s, the galactic collision rate estimated at 10^{-12} yr^{-1} is quite possibly unduly pessimistic.

3.1.2 Stellar collapse and/or explosion. As uncertain as is our general knowledge concerning stellar explosions, we are even less secure in predicting the gravitational radiation from these events since only nonspherical motions are important. Here asphericity produces two types of uncertainties; first, we do not know the asymmetry of the initial models even to order of magnitude, and second, the calculations become far more difficult although, with modern computational methods, the problem is certainly tractable. The available energy is large since the difference in binding energy between a degenerate dwarf and a neutron star ~ 0.1 $M_\odot$ c^2 and almost all of that energy is presumably radiated to infinity within a few seconds of the instant of collapse. However, much, probably most, is dissipated by various neutrino loss processes (cf. Schramm, this volume) and it is the problem of the theorist to estimate if the GR emission is closer to 10^{-2} $M_\odot$ c^2 or 10^{-10} $M_\odot$ c^2!

Two somewhat orthogonal approaches have been taken to the problem. In one the thermodynamics and hydrodynamics of collapse are calculated as accurately as possible (cf. papers by Arnett, Wilson and Nadozin in this volume) with the assumption of spherical symmetry and, depending on the nature of result, certain statements about probable GW losses are made. For example, if it is found that the "collapse" is almost hydrostatic and slow, that pressure support is always significant then, since GW losses depend on the inverse 5th power of the collapse time, it has been believed that they will be very small. I believe that this point of view is fundamentally in error due to a degeneracy, existing in the spherical case which is lifted by even a very slight and physically plausible degree of asymmetry. The reason is as follows. For spherical stars there exists a critical ratio of specific heats Γ_{crit} such that for $\Gamma < \Gamma_{crit}$ the fundamental radial mode (frequency σ) is unstable

$$\sigma^2 = 3 \langle\Gamma - \Gamma_{crit}\rangle |W|/I \tag{13}$$

(cf. Ledoux, 1945) where W is the gravitational energy of the star and I its spherical moment of inertia and brackets designate pressure weighted averages. For nonrotating nonrelativistic stars the value of $\Gamma_{crit} = 4/3$ which is, by coincidence, the same as Γ for any relativistic fluid. Since, during the late stages considered, the stars are very hot, radiation pressure is important and $\Gamma \approx \Gamma_{crit}$ implying that $|\sigma^2/4\pi G\rho| << 1$ and, usually, $\sigma^2 > 0$. General relativity increases Γ_{crit} and rotation (T_{rot}) decreases it so that

$$\Gamma_{crit} = \frac{4}{3} + k|W|/(Mc^2) - (2/3)T_{rot}/|W| \tag{14}$$

(cf. Thorne, 1978, for review) but for assumedly small values of $|W|/Mc^2$ and $T/|W|$ a quasi-static contraction to neutron stars may occur with little "bounce". The situation is drastically different if we alter the assumed symmetry since $\Gamma_{crit} = 1$ for cylindrical stars and $\Gamma_{crit} = 0$ for planar stars. In these, if the true Γ were $\approx 4/3$, then, however low the initial adiabat, $\Gamma > \Gamma_{crit}$ and (i) quasi-static contraction is unlikely and (ii) there will always be a bounce. Thus to the degree that the deviations from spherical symmetry are greater than $\Gamma - \Gamma_{crit}$ we must expect that aspherical collapse will be qualitatively different from spherical collapse.

In the opposite limit, Thuan and Ostriker (1974), Novikov (1975) and in an important recent paper by Saenz and Shapiro ("SS", 1978) have looked at aspherical collapse with increasing, but still very crude, attention to the thermodynamics (see earlier discussion) but some study of the geometrical effects expected in collapsing triaxial objects. Two limiting cases can be isolated. If the precollapse state is a rotating equilibrium body with no magnetic field, then knowledge of mass, density and initial angular momentum define the collapse. For values appropriate to a marginally stable degenerate dwarf SS find that the energy emitted (in solar masses) in gravitational waves is approximately

$$\Delta M_{GW} = 10^{-2.8}(J_{49})^4 M_\odot \tag{15}$$

for J_{49} (= angular momentum/10^{+49} erg sec) < 1. The steep dependence on J implies that, if the cores of massive stars are significantly coupled to their envelopes, then the initial values ($J > 10^{52}$) will be so reduced as to give quite uninteresting results. In one theoretical investigation Endal and Sofia (1977) found that the angular transfer mechanisms they studied were not efficient enough to reduce J_{49} to less than unity but, more significantly, we can relate J to the observed properties of degenerate dwarfs and estimate its value directly. From Ostriker and Tassoul (1969) we find that for an approximately 1 $M_\odot$ DD

$$J_{49} = (70/p(s)) \tag{16}$$

where p(s) is the rotation period in seconds. Most DD probably have periods much longer than this (Greenstein, Boksenberg, Carswell and Shortridge, 1977), but the observed sample is quite old ($t \gtrsim 10^9$ years) and there are probably efficient angular momentum loss mechanisms available so that the observed periods do not determine well the initial rotation rates. Young, hot DD's are not good candidates for spectroscopic determination of rotation period and other techniques will be required. It is thus of some significance that many of the young DD seen in binary-nova systems have periods which are plausibly attributed to rotation (cf. Herbst, Hesser and Ostriker, 1974; Katz, 1975) and the periods are typically of the order of 100 sec; the best studied case is DQ Her with a period of 71 seconds almost certainly due to rotation of the star as a whole (not just surface layers). It is possible but not likely that angular momentum transfer in these systems has significantly affected the periods. Further observational study (perhaps looking for variability associated with an oblique rotating magnetic field) of young DD's is quite important, but for the present it seems that rotation may be sufficient to allow significant radiation of gravitational waves in the first several bounces following collapse.

If $J_{49} >> 1$ then it is likely that, since $|T_{rot}/W| > 0.14$, the core will be subject to fission. This conjectured process has been studied recently by several authors (with an up-to-date review of the literature given in Clark and Eardley, 1977; Wiita and Press, 1976; and Clark, van den Heuvel and Sutantyo, 1978) who plausibly estimate that $\sim 10^{-2}\ M_\odot$ of GW will be emitted by the fragments as they spiral together; one is referred to the original papers for details.

Given our ignorance as to the value of J_{49}, we cannot separately discuss the relative frequencies of the two processes. The total rate is probably within a factor of two of the supernova rate, and that while not known accurately for the galaxy as a whole (1 per 30 years remains a good estimate), it is well determined locally from the historical supernovae. Given that four have been detected 1006, 1054, 1572 and 1604 at a mean distance of approximately 3 kpc in 1000 years, we should expect one within 10 kpc in $\sim$ 25 years.

In passing, we note that core flashes in red giants may produce some GW if the cores are somewhat flattened by rotation as seems plausible. The numbers in the table provide the roughest estimate of this process.

It has been noted for some time (cf. Ostriker, Spitzer and Chevalier, 1972) that many globular clusters have central relaxation times indicating that core collapse has occurred or will occur within a Hubble time. Lightman, Press and Odenwald (1978) have derived a globular cluster death rate from available statistics and the assumption that core collapse is somehow fatal (the process remains obscure) and estimate a current interval between deaths of $10^{8.3}$ years. Some fraction of the mass in the collapsed core ($\sim 10^3\ M_\odot$) may conceivably accumulate in a massive star which explodes leaving a relatively massive 10^2 to $10^3\ M_\odot$ black hole at the center. If all of this occurs, it is likely that there will have been sufficient angular momentum retained by the massive star to emit significant gravitational radiation. It might be interesting to search the galaxies within our supercluster for such stars (barely detectable) as possible candidates for GW events. If the analogy to the galaxy holds one would expect $\sim 10^{-18}$ events $yr^{-1}\ L_\odot^{-1}$ (sph. compt).

Very massive stars have been treated as possible models for QSO's and other phenomena since Hoyle and Fowler (1963). They are subject to a host of instabilities if found (cf. papers by K. Fricke) and might be significant sources of GW (cf. Fowler, 1964), but their existence is, to me, too unlikely to warrant further speculation. (See Blanford - this volume for discussion.)

3.1.3 Gravitational Bremsstrahlung. Since, as we showed earlier, the dominant contribution will almost always be made by large angle collisions, we shall only consider these, taking as cross section for the process $\sigma = 2\pi(GM_{tot})^2/v_{rel}^2c^2$. In the galaxy there may conceivably be a substantial mass ($\sim 10^{11}\ M_\odot$ within $R < 10$ kpc) of "ambient" pre-galactic holes (cf. Carr, 1978) with masses not in excess of $10^{4.5}\ M_\odot$ (Ipser and Price, 1977). At a relative velocity of ~ 300 km/s with respect to disc stars they would have a GW cross section of $10^{26}\ cm^2 = 10^{-11}\ pc^2$ for strong collisions releasing $\sim 10^{-2}\ M_\odot{}^2$. The collision time is unfortunately 10^9 years. If we examine collisions weak enough to give one event per 10 years, the energy emitted is unobservably small. Since the process is dependent on the square of the density, the rate is higher in more massive galaxies (e.g., in M87, it may be significant).

If, as conjectured earlier, some globular clusters contain black holes of mass $10^2 - 10^3\ M_\odot$, they will trap and consume stars at a significant rate (cf. Cohn and Kulsrud, 1978, and references therein). Normal stars will be tidally torn apart but degenerate dwarfs and neutron stars (comprising possibly 50% and 1% of the mass) will ultimately reach a region where gravitational radiation is the predominant energy loss mechanism. Since by this point the orbits will have become quite circular, this process belongs in the category of line rather than continuum emission (cf. Bahcall and Ostriker, 1976). Of lesser importance are stars on very elongated orbits happening to pass close enough to the hole to suffer significant GW loss.

The same processes may occur as stars are swallowed by massive black holes in galactic nuclei. If stars spiral in slowly

$\sim 0.1\ M_\odot\ c^2$ is available over many orbits ($\sim 10^8$), but if stars fall directly the total GW energy available is quite small.

3.2 Line Sources of Gravitational Waves

I have little to add here to the analysis presented in the reviews listed earlier except to note that Zimmermann (1978) finds the expected radiation from the Vela pulsar greater than that from the Crab and possibly detectable. Thorne (1978) has noted the various possible galactic astrophysical sources for pulsation, rotation and orbital energizing of GW. For several of these sources, particularly the close binaries and the rotating magnetic neutron stars, we have other direct means (optical, X-ray or radio) of noting precisely the period and phase expected of any gravitational waves. I would strongly urge that high Q detectors be developed if possible to study these sources since such a detection will have high credibility. If slight shifts of the receiver frequency to either side of the known frequency produce a null result, then the positive result at the expected frequency will be believed. Unfortunately if we do not have such confirmatory evidence from other independent observations, the gravitational wave "detections", however real, will remain for many scientists "a thing that went bump in the night" to be believed or not believed as a matter of taste and inclination. (See Weiss-this volume-on such detectors.)

For the continuum sources a coincidence in time or approximate direction of an event and one observed by other means (e.g., supernova in another galaxy) would have the same virtue. My impression is however that this route is more difficult, less precise (due to poor angular resolution) and less secure than that which seeks to observe relatively well understood periodic objects in our own galactic neighborhood. (See Weiss-this volume on such detectors).

4 FORMATION AND DESTRUCTION OF MILDLY RELATIVISTIC CLUSTERS

It has long been realized that, if clusters of relativistic objects could be made, they would have many interesting properties especially if the clusters themselves were very high density (cf. Zeldovich and Podurets, 1966; Zeldovich and Polnarev, 1974). Normal cluster evolution will in fact lead to high densities (cf. the famous "gravithermal collapse" of D. Lynden-Bell and Wood, 1978) on a timescale related to the central relaxation time.

$$T_{rel} = \frac{v_{rms}^3}{G^2 m_*^2 n_*} \qquad (17)$$

where numerical constants have been suppressed, v_{rms} is the cluster velocity dispersion, n_* and m_* the number density and mass of typical stars. Thus we may evaluate T_{rel} at the center of a stellar system and estimate that in a time $100 \times T_{rel}$ (r = 0) the core density will $\rightarrow \infty$ and its mass $\rightarrow 0$. Before this ensues, of course, physical collisions and other processes will occur to alter the evolution. It is very difficult however to concentrate a significant mass into a high density region because the process is essentially energy conservative; in the core $(GM_{core}/R_{core}) \approx$ const.

The process of equipartition or dynamical friction does provide a way to overcome this. If, in addition to the typical stars (n_*, m_*), there exists a group of heavier objects (n_N, m_N), then on an equipartition timescale, which is shorter than the reference relaxation time by $O(m_*/m_N)$, the heavier objects will tend towards the center giving up energy to the more numerous field stars. Then as was shown in an important paper by Spitzer (1969) there may be circumstances where it is impossible for equilibrium to be achieved. He showed that if $(N_N/N_*) > 0.16\ (m_*/m_N)^{5/2}$ the cluster of stars of type N will sink further and faster to the center until they are isolated from the rest of the stellar system and exist as a small and rapidly evolving independent system. It is easy to see that the criterion listed above is in fact satisfied for standard $1.4\ M_\odot = m_N$ neutron stars in a sea of normal old bulge stars $m_* = 0.6$. It is required that $N_N/N_* > 0.019$ which is satisfied for a normal initial mass function, although not by a large amount. In the center of M31 according to the parameters given in Tremaine, Ostriker and Spitzer (1975) the equipartition time for a $1.4\ M_\odot$ neutron star would be 2×10^9 years, less than the Hubble time, but it becomes too long to be of interest outside of a very small sphere ($\sim$ 10 pc), so equipartition in M31 will not lead to a significant accumulation of neutron stars in the core. Unfortunately, if we make the same calculation for M87 using the observed parameters for that system we find that the equipartition time is longer and the effect smaller. A preliminary method of concentrating neutron stars or their progenitors towards the center must be found. However, once a neutron star cluster is produced with relatively short relaxation time the above described dynamical runaway cannot be avoided. In such a cluster one can show from the formula derived in the earlier section on Bremsstrahlung that the energy output of the cluster is, remarkably enough,

$$\dot{E}_{GW} = \frac{E_{binding}}{T_{rel}} \left(\frac{v_{esc}}{c}\right)^5 \tag{18}$$

where the factor $(v_{esc}/c)^5 \approx 0.015$. Thus the emission of gravitational radiation proceeds more rapidly than the evolution of the cluster. The rate of such collisions is relatively small ($N \times (v_{rms}/v_{esc})^2$ per T_{rel}) and, since most of the radiated energy is stored in the binding energy of close pairs which then will merge through further emission of GR, the overall effect on the evolution of the cluster is small.

However, ultimately the physical collision rate will exceed all other processes. By the usual arguments one can show that this occurs when $(v_{rms}/v_{esc})^2 \approx 10$. At that point one can calculate that the cluster will have a radius of 7×10^{14} cm, a collision time of $10^{6.5}$ years, an energy emission rate of $10^{47.3}$ erg s^{-1} and a total reservoir of $10^{61.5}$ ergs — rather attractive parameters for a model quasar. We can imagine that collisions among the neutron stars (which are mildly relativistic) will copiously produce particles and fields as well as thermal emission from hot gas. After a violent period of $10^{6.5}$ years, during which all the neutron stars are destroyed, the system presumably settles into a relatively more quiescent state in which a central black hole slowly grows as it absorbs from a disc the remaining debris. Radiation emitted during this interval would be largely thermal, and the two stages might represent those of the violently variable quasars (and Lacertids) and the more common, relatively quiet, thermal quasar phase.

Many people have suggested that collisions among neutron stars in a dense cluster are attractive for modeling a quasar, most prominently Zeldovich and Novikov (1971). It appears however that normal, relatively well understood and checked stellar dynamical processes will lead quite naturally to this outcome in massive systems in which the neutron star equipartition times are relatively short.

REFERENCES

Bahcall, J. N. & Ostriker, J. P. (1976). X-ray pulses from globular clusters. Nature Phys. Sci., 262, 37-38.

Binney, J. J. (1977). Anisotropic gravitational collapse. Astrophys. J., 215, 492-496.

Carr, B. J. (1978). On the cosmological density of black holes. Comments on Astrophys., VII, 161-173.

Clark, J. P. A. & Eardley, D. M. (1977). Evolution of close neutron star binaries. Astrophys. J., 215, 311-322.

Clark, J. P. A., van den Heuvel, E. P. J. & Sutantyo, W. (1978). Formation of neutron star binaries and their importance for gravitational radiation. Astrophys. J., in press.

Cohn, H. A. & Kulsrud, R. M. (1978). The stellar distribution around a black hole: Numerical integration of the Fokker-Planck equation. Astrophys. J., in press.

Endal, A. S. & Sofia, S. (1977). Detectable gravitational radiation from stellar collapse. Phys. Rev. Lett., 39, 1429-1432.

Fowler, W. A. (1964). Massive stars, relativistic polytropes and gravitational radiation. Rev. Mod. Phys., 36, 545-555.

Greenstein, J. L., Boksenberg, A., Carswell, R. & Shortridge, K. (1977). The rotational and gravitational redshift of white dwarfs. Astrophys. J., 212, 186-197.

Harwit, M. & Salpeter, E. (1973). Radiation from comets near neutron stars. Astrophys. J. (Lett.), 186, L37-L46.

Herbst, W., Hesser, J. E. & Ostriker, J. P. (1974). The 71-second variation of DW Herculis. Astrophys. J., 193, 679-686.

Hoyle, F. & Fowler, W. A. (1963). On the nature of strong radio sources. Mon. Not. Roy. Ast. Soc., 125, 169-176.

Illingworth, G. D. & King, I. R. (1977). Dynamical models for M15 without a black hole. Astrophys. J. (Lett.), 218, L109-L112.

Ipser, J. & Price, R. H. (1977). Accretion onto pregalactic black holes. Astrophys. J., 216, 578-590.

Katz, J. (1975). The structure of DQ Herculis. Astrophys. J., 200, 298-305.

Ledoux, P. (1945). On the radial pulsation of gaseous stars. Astrophys. J., 102, 143-153.

Lightman, A. P., Press, W. H. & Odenwald, S. F. (1978). Present and past death rates for globular clusters. Astrophys. J., 219, 629-634.

Lynden-Bell, D. & Wood, R. (1978). The gravothermal catastrophe in isothermal spheres and the onset of red giant structure for stellar systems. Mon. Not. Roy. Ast. Soc., 138, 495-525.

Novikov, I. D. (1975). Gravitational radiation from a star collapsing to a disc. Astron. Zhur., 55, 657; English transl. Sov. Ast. A. J., 19, 398-399.

Ostriker, J. P., Spitzer, L. & Chevalier, R. A. (1972). On the evolution of globular clusters. Astrophys. J. (Lett.), 176, L51-L56.

Ostriker, J. P. & Tassoul, J. L. (1969). On the oscillations and stability of rotating stellar models. II. Rapidly rotating white dwarfs. Astrophys. J., 155, 987-997.

Peters, P. C. (1970). Relativistic gravitational bremsstrahlung. Phys. Rev. D., 1, 1559-1571.

Peters, P. C. & Matthews, J. (1963). Gravitational radiation from point masses in a Keplerian orbit. Phys. Rev., 131, 435-440.

Press, W. H. & Thorne, K. S. (1972). Gravitational wave astronomy. Ann. Rev. Astron. & Astrophys., 10, 335-376.

Saenz, R. A. & Shapiro, S. L. (1978). Gravitational radiation from stellar collapse: Ellipsoidal models. Astrophys. J., 221, 286-303.

Shapiro, S. L. (1977). Gravitational radiation from stellar collapse: The initial burst. Astrophys. J., 214,566-575.

Smarr, L. (1977). Space-times generated by computers: Black holes with gravitational radiation. Ann. N. Y. Acad. of Sci., 302, 569-604.

Spitzer, L. (1969). Equipartition and formation of compact nuclei in spherical stellar systems. Astrophys. J. (Lett.), 158, L139-L144.

Thorne, K. S. (1978). General Relativistic Astrophysics. In Theoretical Principles in Astrophysics and Relativity, ed. N. R. Lebovitz, W. H. Reid & P. O. Vandervoort, U of Chicago Press, Chicago, pp. 149-216.

Thuan, T. X. & Ostriker, J. P. (1974). Gravitational radiation from stellar collapse. Astrophys. J. (Lett.), 191, L105-L108.

Tremaine, S., Ostriker, J. P. & Spitzer, L. (1975). The formation of galactic nuclei. Astrophys. J., 196, 407-412.

Turner, M. (1977). Gravitational radiation from point masses in unbound orbits: Newtonian results. Astrophys. J., 216, 610-619.

Wiita, P. J. & Press, W. H. (1976). Mass-angular momentum regimes for certain instabilities of a compact rotating stellar core. Astrophys. J., 208, 525-533.

Zeldovich, Ya. B. & Novikov, I. D. (1971). Relativistic Astrophysics, U of Chicago Press, Chicago, p. 431.

Zeldovich, Ya. B. & Podurets, M. A. (1966). The evolution of a system of gravitationally interacting mass points. Sov. Ast. A. J., 9, 742-749.

Zeldovich, Ya. B. & Polnarev, A. G. (1974). Gravitational radiation from a superdense star cluster. Sov. Ast. A. J., 18, 17.

Zimmermann, M. (1978). Gravitational radiation from the Crab and

JEREMIAH P. OSTRIKER

Vela pulsars — A revised estimate. Orange Aid Preprint 508, California Institute of Technology.

	GALAXY	GLOBULAR CLUSTER	QSO, AGN
COLLISION	Comet × NS $10^{-25}\ M_\odot$, $10^{-3.5}$ s; $h_{-21} = 5 \times 10^{-6}$ 5 pc $10^{6}\ yr^{-1}$	DD × DD $10^{-11.5}\ M_\odot$, 10^{1} s; $h_{-21} = 3$ 10 kpc $10^{-12} yr$	NS × NS $10^{-2.4}\ M_\odot$, $10^{-3.0}$ s; $h_{-21} = 600$ $\nu_{lim} \gtrsim \nu_G$
COLLAPSE (±)	To NS or BH $10^{-3}\ M_0$ $h_{-21} = 200$ $\nu_{lim} \approx \nu_G$ Of DD $10^{-17}\ M_\odot$, $10^{0.7}$ s; $h_{-21} = 10^{-2}$ $\nu_G = 1\ yr^{-1}$	Massive Star Collapse $10^{-1}\ M_\odot$, 10^{-1} s; $h_{-21} = 10^{4.7}$ $\nu_{lim} = 2.5 \times 10^{6.4}\ \nu_G$ $= .012\ yr^{-1}$	Very Massive Star Formation and/or Collapse ?
G. BREMSSTRAHLUNG	Stars × Pregalactic BH $10^{-2}\ M_\odot$, 10^{0} s; $h_{-21} = 10^{4.7}$ 10 kpc $10^{-9.5}\ yr^{-1}$ $\nu_{lim} = 10^{-3.1}\ yr^{-1}$	DD × DD (As above) DD × BH Interesting	NS × NS (As above) Star × MBH

DD = Degenerate Dwarf
NS = Neutron Star
BH = Black Hole
h_{-21} for event at R = 10 kpc distance

DISCUSSION SESSION II: SOURCES OF GRAVITATIONAL RADIATION

Notes and summary: Reuben Epstein
Massachusetts Institute of Technology
J. Paul A. Clark
Yale University

(Chairman: J. Paul A. Clark)

[R. Epstein & J. P. A. Clark: The first part of this session consisted of four short prepared presentations by D. K. Nadyozhin and Z. F. Seidov, who have each provided their own summaries of their respective talks, and by M. Zimmerman and M. S. Turner. The remainder of this chapter is based on the informal discussion that followed. We have re-worked and substantially augmented this account in order to provide a more comprehensive review of gravitational radiation sources than was possible in a single conference session. In keeping with the discussion format, additional material prepared since the workshop is given as contributions by "Note", an imaginary participant, if it cannot be logically included in a contribution by an actual participant. We share responsibility for statements in square brackets.]

L. Smarr: This week we wish to get some consensus among theorists on possible sources of gravitational radiation.

PROPERTIES OF SUPERNOVA NEUTRINOS

L. Smarr: D. K. Nadyozhin is going to give us a few minutes on some properties of neutrino radiation during gravitational collapse.

D. K. Nadyozhin: The aim of this short communication is to report recent results concerning neutrinos radiated by collapsing stellar cores. The collapse calculations have been started from a $2.0M_\odot$ ($1.82M_\odot$ Fe core plus $0.18M_\odot$ oxygen envelope) initial stellar core with an initial radius, $R_0 \approx 5\times10^8$ cm. A complex set of hydrodynamic equations, taking into account neutrino heat conductivity, kinetics of oxygen burning, and neutrino energy and momentum deposition in external layers of the collapsing core, has been solved numerically (Nadyozhin, 1977a,b, 1978). Fig. 1 shows the resulting neutrino light curve.

The collapse can be conventionally subdivided into three stages; A, B, and C, as shown in Fig. 1. At the first stage, A, which lasts up to the establishment of a maximum $L_{\nu\bar{\nu}}$, only

$\sim 10^{52}$ ergs is radiated. The duration of stage A is equal to 0.04s. The bulk of the energy is radiated at the end of stage B and at stage C. This latter stage is characterized by the Kelvin cooling of the resulting hot hydrostatic neutron star. The time scale of this Kelvin cooling is about 15-20s.

The total energy, carried away by all the neutrinos (ν_e, $\bar{\nu}_e$, ν_μ, $\bar{\nu}_\mu$) amounts to 5×10^{53} erg nearly equally distributed among these neutrinos. The mean energy of emitted neutrinos turns out to be 10-12 MeV.

A recent more detailed investigation of the electron neutrino (ν_e) and antineutrino ($\bar{\nu}_e$) spectra (Nadyozhin & Otroshenko, 1978) has shown that these spectra can be approximated by the following expression

$$I_{\nu,\bar{\nu}} \sim \frac{E^3_{\nu,\bar{\nu}} \exp\,[-\alpha_{\nu,\bar{\nu}}(E_{\nu,\bar{\nu}}/kT_{\nu,\bar{\nu}})^2]}{1 + \exp\,(E_{\nu,\bar{\nu}}/kT_{\nu,\bar{\nu}})}\,, \tag{1}$$

where $I_{\nu,\bar{\nu}}\,dE_{\nu,\bar{\nu}}$ is the energy lost during gravitational collapse by neutrinos having energies between $E_{\nu,\bar{\nu}}$ and $E_{\nu,\bar{\nu}} + dE_{\nu,\bar{\nu}}$.

According to Nadyozhin and Otroshenko (1978), $\alpha_\nu \approx 0.005$, $T_\nu \approx$ 3.6 MeV for ν_e, and $\alpha_{\bar{\nu}} \approx 0.02$, $T_{\bar{\nu}} \approx 4.5$ MeV for $\bar{\nu}_e$, so the ν_e-spectrum is somewhat softer than the $\bar{\nu}_e$-spectrum. It is worthwhile to mention that the intensity of neutrino and antineutrino radiation is negligible at energies $E_{\nu,\bar{\nu}} \gtrsim 100$ MeV. For instance,

$$\frac{I_{\nu_e}(E_{\nu_e} = 100\,\text{MeV})}{I_{\nu_e}(E_{\nu_e} = 10\,\text{MeV})} \approx 3\times10^{-10}, \tag{2}$$

$$\frac{I_{\bar{\nu}_e}(E_{\bar{\nu}_e} = 100\,\text{MeV})}{I_{\bar{\nu}_e}(E_{\bar{\nu}_e} = 10\,\text{MeV})} \approx 10^{-10}\,. \tag{3}$$

Conclusions

1. The gravitational collapse generates a burst of neutrino and antineutrino radiation with total energy 5×10^{53} erg during 15-20s.
2. The mean energy of emitted neutrinos and antineutrinos amounts to 10-12 MeV.
3. The overall properties of neutrino radiation (i.e. -- the properties of neutrinos radiated during stages B and C of the collapse) are not sensitive to: (i) details of equation of state of stellar material, (ii) whether the pulsational

bounces of the collapsing core exist or not, (iii) whether an ejection of stellar envelope (i.e. -- supernova) occurs or not.

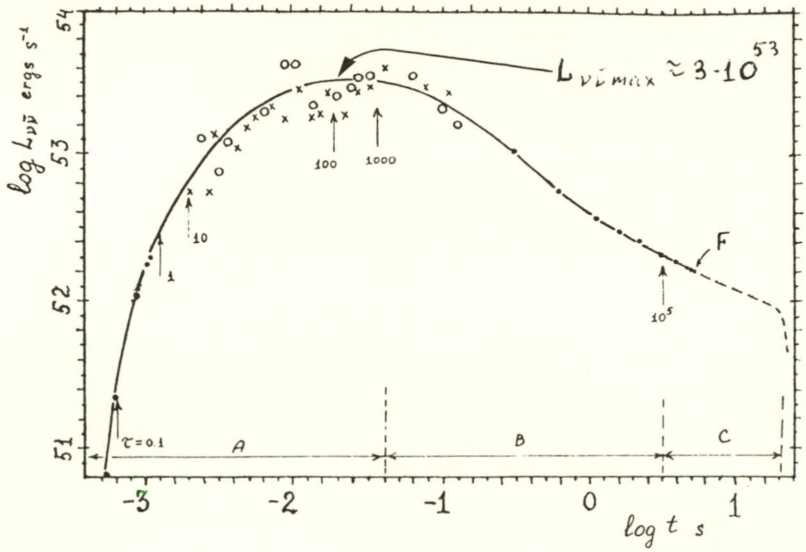

Fig. 1. The neutrino luminosity curve for the collapse of the given core. The numbers at the arrows indicate the values of the mean neutrino optical depth of the collapsing core. The horizontal axis gives the logarithm of time measured from the beginning of the collapse. The circles, crosses, and dots are results of calculations under different assumptions. The interpolating smooth curve is shown by a solid line, whereas the law of the Kelvin cooling of the resulting hydrostatic hot neutron star is extrapolated by the broken line.

D. Kazanas: A shock wave in the photosphere can make very high temperatures and hence very high energy neutrinos.

D. K. Nadyozhin: No. Neutrinos of no more than 10-15 MeV get through the outer layers. The outer layers are completely opaque to ~100 MeV neutrinos.

J. R. Wilson: The electron neutrinos and antineutrinos are still down around 10 MeV. There is a high-energy muon neutrino peak, but this is only a transient effect.

S. Shapiro: Does your calculated collapse slowly settle to a neutron star or does it bounce?

S. I. Blinnikov: The velocity profile clearly shows a strong shock, but the massive accreting outer envelope ultimately suppresses the shock. You have a bounce in the sense of gravitational radiation production, but not in sense of supernova ejection.

K. van Riper: If you have [a stellar core of] more than $2M_{\odot}$, your neutrino signal would certainly depend on whether or not a supernova occurs, because if one did not, then you would have all the neutrinos swallowed as the core made a black hole as the matter continued to accrete.

L. Smarr: Neutrinos will always come out, but they will not carry a lot of information about bounces, rotation, etc. On the other hand, gravitational radiation occurs only if the collapse is nonspherical, but it will carry a lot of information about the collapse.

K. S. Thorne: It doesn't tell you if your supernova model is right.

D. Kazanas: But the neutrino light curve can provide information on the physics of the bounce and shock. Landé has ~1 ms resolution, so he can observe the light curve.

D. K. Nadyozhin: Characteristic times of the reflected shock are several times 10^{-3}s, so the emerging neutrinos are not affected.

BOUNDARY CONDITIONS AND CORE COLLAPSE

L. Smarr: Dr. Seidov will now comment on boundary conditions of nonzero pressure in stellar core collapse.

Z. F. Seidov: At the advanced stages of evolution, the structure of a star with a degenerate core and nuclear-burning shells becomes very complex. Numerical calculations of the subsequent evolution become more and more difficult and computer-time consuming. However, the convergence of core evolution (its weak dependence on total mass of a star) has been found (Arnett, this volume). For this reason many authors have confined themselves to the consideration of the core only. The question arises; what boundary conditions should be used for the proper treatment of the role of the neglected envelope? (Notice that the core mass is always much less than the envelope mass.)

As a first approach to this problem, we investigated the equilibrium and stability of a star under constant external pressure, P_o.

Using the virial method, we have obtained the following expression for the second variation of total energy of the star:

$$\delta^2\varepsilon = \int_0^M \frac{P}{\rho}\,[(\gamma - \tfrac{4}{3})(2\tfrac{f}{r} + f')^2 + \tfrac{4}{3}(\tfrac{f}{r} - f')^2]\,dm + 16\pi P_o R f_o^2 \,, \qquad (4)$$

where $f \equiv \delta r(m)$ are small adiabatic variations of mass distribution, $f_o = f(R)$, $f' \equiv df/dr$, M and R are mass and radius of the star, and P, ρ and γ are pressure, density and adiabatic index of stellar matter, respectively. One can see from (4) that for $\gamma <$ 4/3, a star may be stable ($\delta^2\varepsilon > 0$) for a large enough external pressure P_o.

For γ = const. $\approx$ 4/3, one can assume $f \propto r$ and find for the critical value of γ, γ_c, the formula:

$$\gamma_c = \frac{4}{3} - \frac{4P_oV}{|W|} = \frac{4}{3} - 123.4\frac{P_o}{P_c} \,, \qquad (5)$$

where V, W and P_c are the volume, potential energy and central pressure of the star, respectively. For γ = const. $\approx$ 6/5, we have obtained the formula (Seidov, 1978):

$$\gamma_c = \frac{6}{5} - 0.7674(\frac{P_o}{P_c} - \frac{1}{64}) \,. \qquad (6)$$

The critical value of γ rapidly decreases from 4/3 to 6/5 when P_o/P_c increases from 0 to 1/64.

We conclude that the stabilizing effect of external pressure should be carefully taken into account in the case of collapsing stellar cores. The following properties should depend strongly on the choice of the boundary conditions:

(a) Critical density and temperature (but not critical mass) for stability loss,

(b) The number, amplitudes and periods of oscillations ("bounces") of the core before its ultimate collapse to black hole or neutron star,

(c) Peak values of neutrino and gravitational luminosities, the frequencies of their maximal spectral densities, etc.

In conclusion, we point out that the external pressure should also influence the stability and evolution of supermassive stars, because the mean adiabatic index is in this case close to 4/3 and because even the small general relativistic effects are important.

D. Eardley: How much difference can boundary conditions make in the collapse of a massive stellar core?

Z. F. Seidov: We cannot yet tell how much things are affected quantitatively.

SOURCE DIAGRAMS

[J. P. A. Clark & R. Epstein: It was agreed during the discussions that distinctions should be drawn between very short duration ("burst") gravitational radiation sources (referred to henceforth as impulsive sources, periodic sources (lasting some large number, e.g. $>10^3$, of cycles), and stochastic sources. The impulsive source diagram, Fig. 2, portrays, not only the dimensionless amplitude, h, vs. frequency, ν, but also event rate information. Rather than use 3-dimensional graphs, which would have been difficult to draw and read, it seems more appropriate to show h vs. ν for a given event rate. An event rate of one event per year seems to be a reasonable standard to adopt initially. The h values are for sources at a distance from the observer such that the spherical volume defined by this distance contains enough potential sources to give the desired event rate. For now, individual periodic sources are plotted in Fig. 3, although in the future, the frequency distributions of complete Galactic populations should be given. It was recommended that stochastic sources be plotted in Fig. 4 as $\nu^{1/2}S_h{}^{1/2}$, a dimensionless quantity, vs. ν, where S_h is the spectral density of h.

These diagrams are plotted to the same scale as the diagrams of observational limits in the lectures by R. Weiss (this volume), and it is left up to the reader to compare them and decide on the appropriate course that future gravitational radiation experiments should take.]

ESTIMATES OF THE MAXIMUM POSSIBLE RADIATION

M. Zimmerman: I will summarize a project that K. S. Thorne and I undertook at the instigation of R. Weiss and R. Drever. We tried to estimate the maximum possible gravitational radiation that could be present without violating a bare minimum of astrophysical knowledge and assumptions. We do not expect realistic sources to come within orders of magnitude of these "cherished belief" limits.

Our major assumptions are that sources cannot radiate more than their total mass; that we are not privileged observers in time or space; that the universe has roughly the closure density [$\rho_{CL} \sim 10^{-29}$ g cm^{-3}]; and that our Galaxy has a massive halo ($M_{Gal} \sim 10^{12} M_\odot$). For more details see Zimmerman and Thorne (1978).

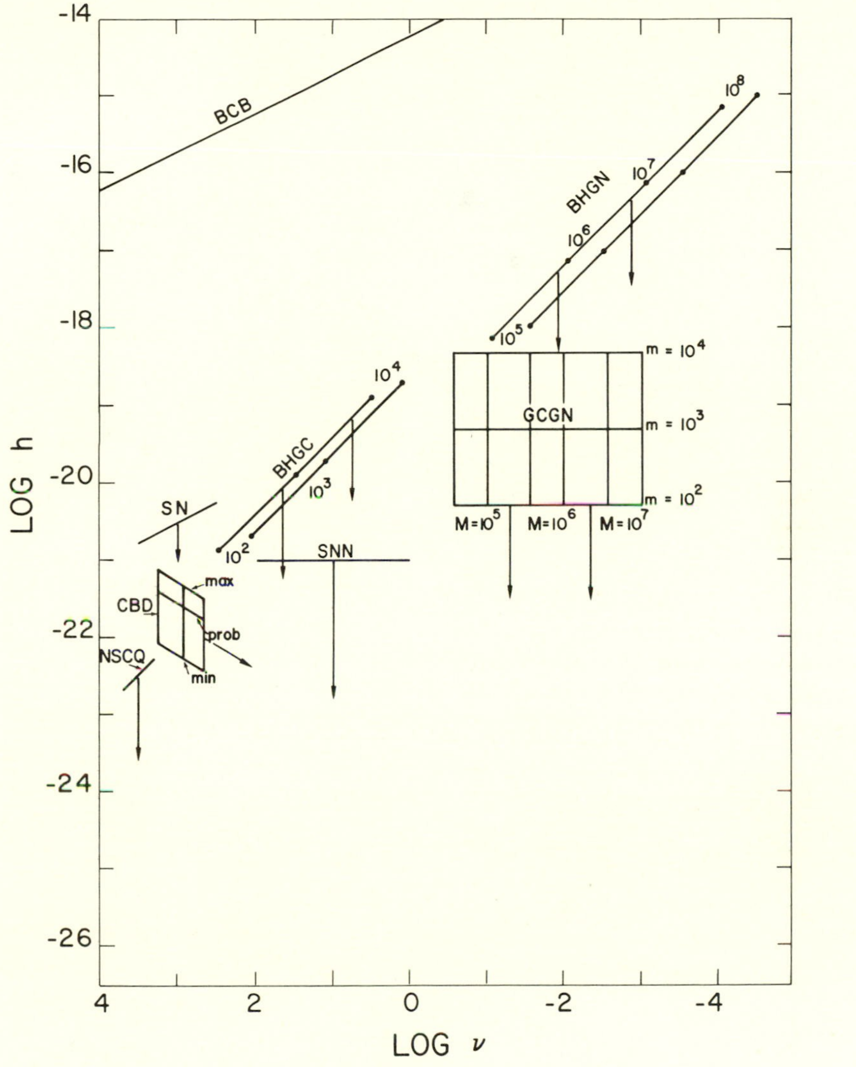

Fig. 2. Amplitude, h, versus frequency, ν, plotted for various impulsive sources, each denoted by an acronym defined in the text.

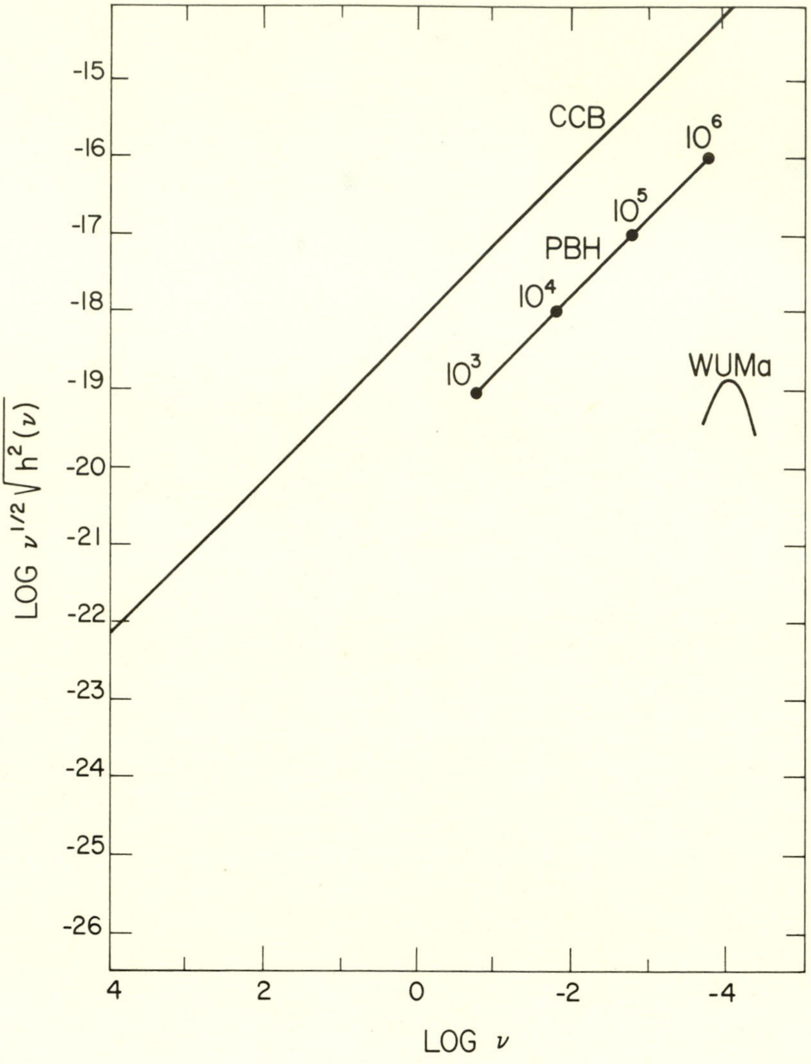

Fig. 3. Periodic sources, same as for Fig. 2.

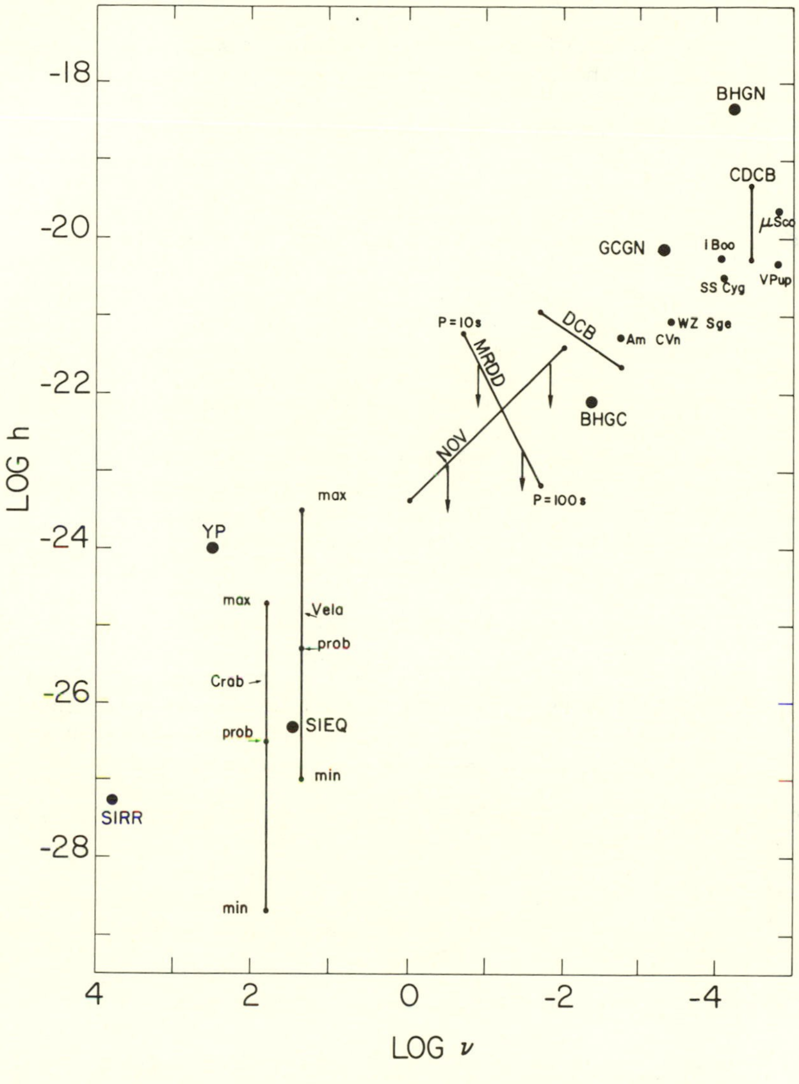

Fig. 4. Stochastic sources, same as for Fig. 2, except that the source strength is given as $\nu^{1/2}S_h^{1/2}$, where S_h is the spectral density of h. Here we use $S_h = h^2(\nu)$, consistent with the notation of Weiss (this volume).

Using these assumptions, we find that the S_h expected due to a closure density of stochastic background radiation is $\nu^{1/2}S_h^{1/2} \sim 6\times10^{-19}(\nu/1\ \mathrm{Hz})^{-1}$, where the energy spectral density is distributed uniformly over all frequencies, 0 through ν. (See Fig. 4, CCB.) Continuous sources in the Galaxy might give an S_h of 2 higher than for the cosmological closure density.

The limit for impulsive sources, where we assume $\Delta\nu \sim \nu$ as the spectral width of the source, for one burst per observation time of 1 year is $h \sim 6\times10^{-15}(\nu/1\ \mathrm{Hz})^{-1/2}$. (See Fig. 2, BCB.) The $\nu^{-1/2}$ dependence occurs because we assume that an abrupt burst lasts about a cycle.

We are assuming that all the impulsive sources in the Galaxy can radiate away at most one Galactic mass in a Hubble time. If we increase our observing time and still assume one burst per observing time, the bursts will be proportionately larger.

A plausible but unlikely source that in principle could give such amplitudes would be a massive Galactic halo of black-hole and neutron-star binaries, which are the remains of Population III stars that formed before the Galaxy was formed. With a reasonable distribution of orbital elements, 10^{10} yr would be the mean time for the stars of a binary to spiral together, releasing in the final moments 1-10% of their rest-mass energy into gravitational radiation. One could get these high h values in the 0.1 to 5 kHz range, depending on the masses of these stars.

L. Smarr: Do these burst events overlap in time?

M. Zimmerman: No. They are individual events that you could presumably resolve with a detector sensitive enough to detect them, and they are many orders of magnitude above what you would expect for a continuous background.

[Note: If we consider the limit for impulsive sources, assuming an event rate of ν, or one per cycle, so that the events overlap in time to form a stochastic signal, we estimate $\nu^{1/2}S_h^{1/2} \sim 2\times10^{-18}\ (\nu/1\ \mathrm{Hz})^{-1}$, which, as we would expect, roughly agrees with the S_h given above by M. Zimmerman for the stochastic signal from continuous Galactic sources.]

GRAVITATIONAL RADIATION FROM SUPERNOVA NEUTRINOS

M. S. Turner: I want to speak for R. Epstein and myself about gravitational radiation generated by supernova neutrinos. [Cf. Epstein (1978), Turner (1978).]

The key to gravitational radiation from neutrinos is the creation or scattering event from which the neutrino escapes into space. An electromagnetic analog to this process is the γ-

radiation from the escaping electron in β-decay. The emission of radiation by a creation or scattering event is discussed in terms of the zero-frequency limit (ZFL) of the relevant Feynman diagrams by S. Weinberg [For references and discussion, see Smarr (1977).] and also, for β decay, by J. D. Jackson (1975) and for gravitational radiation by L. Smarr (1977).

When we superimpose a large number of these escape events we find that the radiation contributions add coherently only at frequencies less than of the order of τ_{burst}^{-1}, where τ_{burst} is the characteristic time of the neutrino burst. This frequency is much lower than the frequencies characteristic of the creation and last-scatter events, so we are consistent in considering only the ZFL of each escaping neutrino. The result is just what one would obtain in a weak-field treatment of a classical energy density expanding in some high-velocity, nonspherical manner.

We will now use a simple model just to compute the expected dimensionless amplitude. This is simply an ellipsoidal neutrinosphere which gives an anisotropic neutrino emission distribution. Without this anisotropy, there is no net coherent gravitational radiation. The metric perturbation is of the order of

$$h_\nu \sim r^{-1} e_\nu^2 E_\nu \ , \tag{7}$$

where e_ν is the eccentricity of the neutrinosphere and where E_ν is the total energy emitted in neutrinos. The wave form is that of a rounded step function with a rise time of the order of τ_{burst}.

Now e_ν^2 may be anywhere from $\lesssim 1$, for the eccentricity amplification effect discussed by Shapiro, or less for more symmetric collapses. [The energy emitted in neutrinos, E_ν, is expected to be less than 0.1 Mc^2.] Then, $h_\nu \sim 10^{-21}(e_\nu/1)^2(d/10\ \text{Mpc})^{-1}(E_\nu/0.1\ Mc^2)$ at a frequency of 100 Hz or less (See Fig. 2, SNN.) The amplitude for the short burst of gravitational radiation from the hydrodynamic motion is

$$h_H \sim r^{-1} e_c^2 \ (\text{K.E.}) \ , \tag{8}$$

where e_c and (K.E.) are the eccentricity and maximum kinetic energy of the core during bounce, respectively. Thus, for example, the ratio h_ν/h_H is roughly $E_\nu/(\text{K.E.})$ for comparable degrees of asymmetry. The ratio of the gravitational radiation energy emitted by the neutrinos to that emitted by the bounce motion is

$$\frac{\Delta E_\nu}{\Delta E_H} \sim \left(\frac{E_\nu}{K.E.}\right)^2 \left(\frac{\tau_{bounce}}{\tau_{burst}}\right) , \tag{9}$$

where τ_{bounce} is the time scale of the bounce.

The biggest uncertainty here is not knowing what an asymmetrical neutrino burst really looks like. What we have heard this week suggests that there will be a short initial burst of $\sim 10^{-2} M_\odot$ with more neutrinos coming out on a cooling time scale of 10 to 20s.

K. S. Thorne: The wave amplitude of the radiation from neutrinos might be less than you suggest because the bulk of the neutrinos escape at late times. By then, the object may have settled into a relatively spherical state. The earlier times may be when you have the bulk of the anisotropy. D. K. Nadyozhin's calculation gives $E_\nu \sim 10^{52}$ erg or $0.006 M_\odot c^2$ in the first 0.04 s, which is still enough to be interesting.

L. Smarr: Given the importance of lower frequencies for some of the detection techniques discussed last week by R. Drever and R. Weiss (this volume), the low-frequency enhancement in your step-function wave form could be very important, especially at lower frequencies where the hydrodynamic radiation is negligible.

R. Drever: Is the change in the metric perturbation permanent?

R. Epstein & M. S. Turner: Yes, at least on observational time scales. On the scale of the light travel time from the source to the observer, one can no longer distinguish the radiation from changes in the background geometry due to the large-scale displacement of the neutrinos and other matter in space.

J. Wilson: With regard to determining what e_ν really looks like; if the neutrinos really come from a neutrinosphere, then I don't think core calculations have much to do with e_ν.

S. I. Blinnikov: One effect that might enhance the neutrino gravitational radiation is that at the poles of a flattened photosphere you have larger temperature gradients and hence a larger neutrino flux. This would increase the neutrino anisotropy.

R. Epstein: This is an adaptation of von Zeipel's theorem to supernova neutrinos in the diffusion approximation.

BLACK HOLE EVENTS

L. Smarr: One source that we should consider is that suggested by J. P. Ostriker (this volume) in which white dwarfs fall into the central black hole of a globular cluster.

K. Thorne: I can't believe that is very interesting because the total number of globular clusters involved out to the Virgo cluster is not more than 10^4 or 10^5, and each cluster will not eat more than about 10^2 white dwarfs in the lifetime of the cluster without building a large black hole. This gives, maybe 10^8 events per Hubble lifetime which is too few. Going beyond the Virgo cluster, h gets well below $h \sim 10^{-21}$.

[Note: Since the tidal radius of a white dwarf, $r_T = r_{wd}(m_{bh}/m_{wd})^{1/2}$, spiralling into a black hole of $\sim 10^3 M_\odot$ is substantially larger than the Schwarzschild radius of the black hole, the white dwarf will be disrupted prior to being "eaten". This will result in considerably lower gravitational wave production than one expects for an object swallowed whole, making this source less interesting.

Thorne and Braginsky (1976) calculated the gravitational radiation expected from black-hole formation or collisions involving one or more black holes of mass(es) $M \sim 10^6 M_\odot$ in galaxies with a redshift $z \approx 2.5$. The event rate is given by $\sim 17N(\phi_0/\mathrm{Mpc}^{-3})$ bursts per year, where N is the number of such events per galaxy, and where ϕ_0 is the density of galaxies which undergo such events. An upper limit to the event rate may be obtained by assuming all galaxies undergo one such event in a Hubble time. Taking $\phi_0 \sim 0.1\ \mathrm{Mpc}^{-3}$ (Blandford, this volume) and N = 1 gives, at most, roughly 1 event per year.

Noting that $.06 < \nu_{GW} M < .16$ for the range of Schwarzschild (nonrotating) through extreme Kerr (rapidly rotating) black holes (Detweiler, this volume), we see that

$$h \lesssim \begin{Bmatrix} 12 \\ 7 \end{Bmatrix} \times 10^{-18} (\eta/10^{-2})^{1/2} (M/10^6 M_\odot) \qquad (10a)$$

$$\nu \sim \begin{Bmatrix} 3 \\ 9 \end{Bmatrix} \times 10^{-3} (M/10^6 M_\odot)^{-1}\ \mathrm{Hz}\ . \qquad (10b)$$

The two numerical coefficients, above and below, refer to a Schwarzschild or extreme Kerr central black hole, respectively (See Fig. 2, BHGN.). The radiation efficiency is given by η.

Similarly, globular clusters may contain black holes of $M \sim 10^3 M_\odot$. If these form by collision or coalescence of two smaller holes, a substantial burst of gravitational waves may occur (Thorne, 1978). As an upper limit, we may assume this occurs once in all globular clusters (assume 300/galaxy), so we must look out as far as $d \sim 500$ Mpc to detect one event per year. This burst will have:

$$h \lesssim \begin{pmatrix}2\\1\end{pmatrix} \mathrm{x}10^{-20} (\eta/10^{-2})^{1/2} (M/10^3 M_\odot) \tag{11a}$$

$$\nu \sim \begin{pmatrix}12\\32\end{pmatrix} \mathrm{x}(M/10^3 M_\odot)^{-1} \text{ Hz} \tag{11b}$$

(See Fig. 2, BHGC.).]

L. Smarr: Another source like this is globular clusters with black hole cores spiralling into the nuclei of galaxies [See Blandford, this volume.].

K. S. Thorne: I would think those are interesting out to the Hubble distance.

[Note: The h and ν for events of this kind can be taken directly from Detweiler (this volume) for a particle spiralling into a large black hole:

$$h \lesssim 5\mathrm{x}10^{-20} (m/10^3 M_\odot)(d/10^9 \text{ pc})^{-1} \tag{12a}$$

$$\nu \sim \begin{pmatrix}1\\3\end{pmatrix} \mathrm{x}10^{-2} (M/10^6 M_\odot)^{-1} \text{ Hz} \tag{12b}$$

where m and M are, respectively, the globular cluster and galactic nuclei black hole masses (See Fig. 2, GCGN.).

It is well to remember that any of the three sources above involving massive black holes may be very much less frequent than the upper limits given, or even nonexistent. Although there exists reasonably suggestive evidence for black holes in galactic nuclei (Blandford, this volume), the case for black holes in globular clusters is even less certain (Ostriker, this volume). Furthermore, even if black holes are found in globular clusters and galactic nuclei, they could have formed with low efficiency ($\eta \ll 1$).

A very important point about these sources that has so far been overlooked is that they will be detectable as periodic sources prior to collision. If one event occurs per year, then the systems whose separation is smallest will have a lifetime of $\tau \sim 1$ year. Then, taking these events to occur at $d \sim 1$ Gpc, assuming that gravitational radiation is the dominant mechanism of orbital decay, and neglecting redshift effects, we have

$$h \sim 4\times 10^{-26}(m_1/M_\odot)^{3/4}(m_2/M_\odot)^{3/4}[(m_1+m_2)/M_\odot]^{-1/4}(\tau/1\ \mathrm{yr})^{-1/4} \tag{13a}$$

$$\nu \sim (0.2\ \mathrm{Hz})(m_1/M_\odot)^{-3/8}(m_2/M_\odot)^{-3/8}[(m_1+m_2)/M_\odot]^{1/8}(\tau/1\ \mathrm{yr})^{-3/8} \tag{13b}$$

(See Fig. 3, BHGN, BHGC, GCGN).]

NOVAE

D. H. Douglass: Ordinary novae should be given more attention as gravitational radiation sources. A nova explosion should leave the white dwarf vibrating and emitting gravitational radiation at $\nu \sim 10^{-2}$ to 1 Hz, which could persist for days or weeks.

There are a lot of cataclysmic variable stars in the galaxy, some of them quite close. SS Cygni, for example, fires off once every 30 or 40 days and is only 30 pc away. [For such a source, we have an estimate $h \sim 4\times 10^{-22}(\zeta/10^{-1})^{1/2}(\nu/10^{-2}\ \mathrm{Hz})^{-1}(E/10^{45}\ \mathrm{erg})^{1/2}(d/500\ \mathrm{pc})^{-1}$, where E is the energy of the explosion, and ζ the fraction of that energy which is deposited into vibration and dissipated as gravitational radiation (Douglass and Braginsky, 1979) (See Fig. 3, NOV.).]

I strongly urge more work on this source.

K. S. Thorne: Such work would be relatively difficult. The main uncertainties are how much energy is deposited into vibrations and how rapidly the various astrophysical damping mechanisms operate.

RADIATION INSTABILITY OF ROTATING STARS

K. S. Thorne: Bardeen, Friedman, Papaloizou, Pringle, Schutz, and others have pointed out that an instability can occur in rapidly rotating stars in which gravitational radiation emission causes a perturbation to grow (Papaloizou & Pringle, 1978). This scenario could possibly occur in a low-mass close binary system where a neutron star is accreting matter at a rate $\dot{M}$ from a companion and is thus being spun up.

If the neutron star has very little innate deviation from symmetry [and the neutron star has no residual magnetic field, otherwise the star will reach some equilibrium period $P_{eq} \sim$ 0.1-1 s (Savonije & van den Heuvel, 1977)], then the instability sets in at a period of $\sim 10^{-3}$s. An object of this kind at $d \sim$ 10 kpc gives $h \sim 5\times 10^{-28}$ at $\nu \sim 6\times 10^{3}$ Hz (see Fig. 3, SIRR).

I would think it likely that there would be enough nonsphericity that you would be braked by gravitational radiation. Assuming a small non-axial deformation of fractional size ε, and that all braking is due to gravitational radiation, then the neutron star reaches an equilibrium state with $h \sim 5\times10^{-27}(d/10\ \mathrm{kpc})^{-1}(\varepsilon/10^{-4})^{1/5}(\dot{M}/10^{-9}M_\odot\ \mathrm{yr}^{-1})^{2/5}$ and $\nu \sim 2.8\times10^{1}\ \mathrm{Hz}\ (\dot{M}/10^{-9}M_\odot\ \mathrm{yr}^{-1})^{1/5}(\varepsilon/10^{-4})^{-2/5}$ (See Fig. 3, SIEQ.).

Detecting these objects would be very rough. The prospects are much better to find a young neutron star that is in the process of spinning down.

SUPERNOVAE (SN)

[Note: See contributions by Shapiro, Arnett, Clark, and Wagoner in this volume for details of estimated SN efficiencies, event rates, and wave frequencies. The most serious unknown in determining the dimensionless wave amplitude is the efficiency, η. Denoting the bandwidth of the source by $\Delta\nu$, we have

$$h \sim 3\times10^{-21}(m_{SN}/1.4M_\odot)^{1/2}(\Delta\nu/1\ \mathrm{kHz})^{1/2}(\nu/1\ \mathrm{kHz})^{-1} \times (d/10\ \mathrm{Mpc})^{-1}(\eta/10^{-2})^{1/2}\ , \tag{14}$$

where $d \sim 10$ Mpc yields ~1 event yr^{-1} (See Fig. 2, SN.).

COMPACT BINARY DESTRUCTION

Note: Clark, van den Heuvel, and Sutantyo (1978) estimate the number of compact binary destruction events. One event per year will occur within ~40, 70 or 300 Mpc for the highest, probable, and lowest event rates (99% confidence), respectively. Such an event at $\nu \sim 1$ kHz will have

$$h \simeq 2\times10^{-22}(d/100\ \mathrm{Mpc})^{-1}(a_{min}/30\ \mathrm{km})^{-1} \tag{15}$$

for two $1.4M_\odot$ bodies whose minimum separation is a_{min} (See Fig. 2, CBD.).

CORE QUAKES OF NEUTRON STARS

Note: Solid neutron star cores frozen while still rapidly rotating may be expected to undergo core quakes every few years. An energy release of $\Delta E \sim 10^{45}$ ergs per quake might be expected, and one event per year as close as $d \sim 1$ kpc is indicated (Thorne, 1978). Such an event would generate an amplitude

$$h \sim 3\times10^{-23}(\Delta E/10^{45}\ \mathrm{erg})^{1/2}(\nu/3\ \mathrm{kHz})^{-1}(\hat{\tau}/1\ \mathrm{s})^{-1/2}$$

$$\times(d/1\ \mathrm{kpc})^{-1}(\zeta/1)^{1/2}\ , \qquad (16)$$

with $\nu \sim 3$ kHz where ζ is the fraction of ΔE deposited into torsional oscillations of the neutron star and subsequently radiated as gravitational radiation, and where $\hat{\tau}$ is the damping time of the oscillations (typically ~1 s) (Thorne, 1978). Our knowledge of ζ is close to nonexistent (See Fig. 2, NSCQ.).]

RAPIDLY ROTATING COMPACT OBJECTS

K. S. Thorne: Consider the youngest pulsar in the galaxy, say ten years old, put it at 10 kpc, and make the assumption that it is slowing down due to gravitational radiation at a rate which is given by its present angular momentum divided by its age. Then it will have $h \sim 10^{-24}$ [with a period of a few milliseconds for an initially rapidly rotating ($P \sim 1$ ms) pulsar] (see Fig. 3, YP).

D. Eardley: Let's remember that our newborn pulsars might be long-period.

J. P. A. Clark: In addition to the newborn rapidly rotating neutron stars, there are over 300 detected radio pulsars with periods ranging from $P \sim 0.1$ s to $P \sim 4$ s. Taylor and Manchester (1977) tell us that there are ~10^5 pulsars in our galaxy rotating more rapidly than the cutoff of $P \sim 4$ s. Thus, we may expect a growing density of pulsars at smaller h's than Crab and Vela. This could be illustrated by the $d^2N/d(\log h)d(\log \nu)$ technique (Clark, this volume).

K. S. Thorne: Let me give you the relevant numbers for the Crab and Vela. [Zimmerman (1978) gives $\nu = 22$ Hz, $h_{max} \sim 3\times10^{-24}$, $h_{prob} \sim 5\times10^{-26}$, $h_{min} \sim 1\times10^{-27}$ for Vela; and $\nu = 60$ Hz, $h_{max} \sim 2\times10^{-25}$, $h_{prob} \sim 3\times10^{-27}$, $h_{min} \sim 2\times10^{-29}$ for Crab.] (See Fig. 3.)

K. S. Thorne: J. P. Ostriker (this volume) has suggested that nearby magnetized rotating degenerate dwarfs with some nonaxial asymmetries may be promising gravitational wave sources [with $h \sim 6\times10^{-22}(\varepsilon/10^{-2})(I/10^{50}\ \mathrm{g\ cm^2})(P/10\ \mathrm{s})^{-2}(d/50\ \mathrm{pc})^{-1}$ and $\nu \sim 0.2\ \mathrm{Hz}(P/10\ \mathrm{s})^{-1}$, where I is the moment of inertia of the star.] (See Fig. 3, MRDD.)

BINARIES

D. H. Douglass: Doubly compact objects exist for a long time so we should look for more than just their final burst of radiation.

There is probably an unseen compact object in the galaxy at $\nu \sim 10^{-3}$ Hz and $h \sim 10^{-20}$ right now.

J. P. A. Clark: We can estimate the smallest separation of any doubly compact binary in the galaxy. If such binaries form at a rate r_{bp}, then there is currently such a binary with a decay time of r_{bp}^{-1}. Using the standard formula for decay by gravitational radiation emission we can infer its present separation, and thus its present h and ν. Because r_{bp} is so poorly known (3×10^{-6} yr^{-1} $\lesssim r_{bp} \lesssim 2 \times 10^{-3}$ yr^{-1}), the h and ν's corresponding to the minimum, probable, and maximum event rate (99% confidence) are shown on Fig. 3 (DCB). We assume that this binary is at least as distant as 10 kpc.

Alternatively, we may consider the h and ν of the compact binary closest to earth, which will decay in less than a Hubble time. This binary is selected statistically, assuming all doubly compact binaries decay in $\sim 10^{10}$ yr, so there are $10^{10} r_{bp}$ systems in the galaxy now (see Fig. 3, CDCB).

Unidentified: The binary with smallest separation is no better [in h] than classical binaries.

D. Eardley: But the periods [of classical binaries] aren't so short.

K. S. Thorne: Some of us at Caltech thought about contact binaries with masses ranging from $0.1 M_\odot$ on up. After estimating their space density, we tentatively decided that one is better off in terms of h with the more massive objects like i Bootis, but not by much.

J. P. A. Clark: However, these low-mass systems may still be easier to detect because they will have periods a factor of 10 shorter than the high-mass contact systems. (Separation ~ stellar radius, which increases with stellar mass.)

[Note: Some classical binaries (Douglass & Braginsky, 1979) have been plotted on Fig. 3.

STOCHASTIC SOURCES

Note: A source of gravitational radiation can be stochastic in any one of 3 ways: (1) the source is intrinsically stochastic or noisy, (2) the source is a sufficiently large population of sufficiently rapid burst sources that the bursts are no longer distinct in time, or (3) the source is a sufficiently large population of N periodic sources whose bandwidths, $\Delta\nu$, are large enough so that their line spectra overlap in frequency space, or $N\Delta\nu \gtrsim \nu$, where ν is the highest frequency of the population. This bandwidth is either the intrinsic bandwidth of each source or $\Delta\nu \sim 1/\tau_{int}$, the bandwidth determined by the observation time, τ_{int}, whichever is largest.

The cases (2) and (3) above are not true stochastic sources in the sense that a sufficiently large population of impulsive or periodic sources will always appear stochastic. The best example of an intrinsically stochastic source is the primordial turbulence of the universe. M. Zimmerman has provided the corresponding line, CCB, in Fig. 4 for a closure density of such radiation. Bertotti and Carr (1978), however, rule out the existence of any such large primordial background on nucleosynthesis grounds.

At this time, the only promising candidate for a stochastic burst source is the formation of massive ($M \sim 3x10^5 M_\odot$) pregalactic black holes (Thorne, 1978; Bertotti & Carr, 1978; and Blandford, this volume). If a fraction, ξ, of the mass of the universe is in black holes that condensed out of the pregalactic plasma at $z \sim 60$, then Thorne (1978) obtains $\nu^{1/2} S_h^{1/2} \sim 2.5x10^{-17}$ $x\ (\xi/0.1)^{1/2} (M/3x10^5 M_\odot)$. (See Fig. 4, PBH.) The mean frequency of this signal is $\nu \sim 3x10^{-3}$ Hz $(M/3x10^5 M_\odot)^{-1}$, and the bandwidth is $\Delta\nu \sim \nu$, assuming all the holes are of the same mass.

W Ursae Majoris stars are the most populous class of close binaries in the Galaxy. The $\nu^{1/2} S_h^{1/2}$ for these stars shown in Fig. 4 was taken from Douglass and Braginsky (1979). Given observation times of $\tau_{obs} \sim 1$ yr, the limiting line width, $\Delta\nu \sim 1/\tau_{obs}$, is much too large to resolve any but the very closest, and hence strongest, individual sources. At this line width, the more rare neutron star binaries at higher frequencies, $\nu \gtrsim 10^{-3.5}$Hz, will probably be individually resolvable. More information and/or work is needed to be certain. In any case, one should note that any reasonably uniform population of periodic sources will appear to be a stochastic source when observed with a large enough bandwidth.]

REFERENCES

Bertotti, B. & Carr, B.J. (1978). On the detectability of a cosmological background of gravitational radiation. Preprint.

Clark, J.P.A., van den Heuvel, E.P.J. & Sutantyo, W. (1978). Formation of neutron star binaries and their importance for gravitational radiation. Astron. & Astrophys., in press.

Douglass, D.H. & Braginsky, V.B. (1979). Gravitational radiation experiments. In Einstein Centenary Volume, ed. S.W. Hawking & W. Israel. Cambridge University Press. In press.

Epstein, R. (1978). The generation of gravitational radiation by escaping supernova neutrinos. Astrophys. J., 223, pp. 1037-1045.

Jackson, J.D. (1975). Classical Electrodynamics, 2nd ed. John Wiley & Sons, Inc., New York.

Nadyozhin, D.K. (1977a). The collapse of iron-oxygen stars: physical and mathematical formulation of the problem and computational method. Astrophys. & Space Sci., 49, pp. 339-425.

Nadyozhin, D.K. (1977b). Gravitational collapse of iron-oxygen stars with masses of $2M_\odot$ and $10M_\odot$. Astrophys. & Space Sci., 51, pp.283-301.

Nadyozhin, D.K. (1978). The neutrino radiation for a hot neutron star formation and the envelope outburst problem. Astrophys. & Space Sci., 53, pp. 131-153.

Nadyozhin, D.K. & Ostroshenko, I.V. (1978). Preprint of Institute of Mathematics, Moscow, U.S.S.R.

Papaloizou, J. & Pringle, J.E. (1978). Gravitational radiation and the stability of rotating stars. Mon. Not. Roy. Ast. Soc., 184, pp.501-508.

Savonije, G.J. & van den Heuvel, E.P.J. (1977). On the rotational history of the pulsars in massive X-ray binaries. Astrophys. J., 214, L19-L22.

Seidov, Z.F. (1978). In Russian, Pisma v Astronomichesky Zhurnal, 4, pp.316- . [Soviet Astron. Lett., in press.]

Smarr, L. (1977). Gravitational radiation from distant encounters and from head-on collisions of black holes: the zero-frequency limit. Phys. Rev. D, 15, pp.2069-2077.

Taylor, J.H. & Manchester, R.N. (1977). Galactic distribution and evolution of pulsars. Astrophys. J., 215, pp.885-896.

Thorne, K.S. & Braginsky, V.B. (1976). Gravitational-wave bursts from the nuclei of distant galaxies and quasars: proposal for detection using doppler tracking of interplanetary spacecraft. Astrophys. J., 204, L1-L6.

Thorne, K.S. (1978). General-relativistic astrophysics. In Theoretical Principles in Astrophysics and Relativity, ed. N.R. Lebovitz, W.H. Reid & P.O. Vandervoort, pp.149-216. University of Chicago Press, Chicago.

DISCUSSION SESSION II: SOURCES

Turner, M.S. (1978). Gravitational radiation from supernova neutrino bursts. Nature, 274, pp.565-566.

Zimmerman, M. (1978). Revised estimate of gravitational radiation from Crab and Vela pulsars. Nature, 271, pp.524-525.

Zimmerman, M. & Thorne, K.S. (1978). "Cherished beliefs" about the universe and the resulting upper limits on astrophysical gravitational radiation. Preprint.

INDEX